The Activation of Dioxygen and Homogeneous Catalytic Oxidation

ERRATUM

The figure on page 47 of this volume is oriented improperly. The figure should be turned 180 degrees. The labels to the figure and the figure caption are correct.

The Activation of Dioxygen and Homogeneous Catalytic Oxidation
Edited by Derek H.R. Barton *et al*.
Springer Science+Business Media, LLC

0-306-44591-3

The Activation of Dioxygen and Homogeneous Catalytic Oxidation

Edited by

Derek H. R. Barton
Arthur E. Martell
Donald T. Sawyer

Texas A&M University
College Station, Texas

Springer Science+Business Media, LLC

Library of Congress Cataloging-in-Publication Data

The Activation of dioxygen and homogeneous catalytic oxidation /
   edited by Derek H.R. Barton, Arthur E. Martell, Donald T. Sawyer.
      p.   cm.
   "Proceedings of the Fifth International Symposium on the
Activation of Dioxygen and Homogeneous Catalytic Oxidation, held
March 14-19, 1993, in College Station, Texas"--T.p. verso.
   Includes bibliographical references and index.
   ISBN 978-1-4613-6307-1   ISBN 978-1-4615-3000-8 (eBook)
   DOI 10.1007/978-1-4615-3000-8
   1. Oxidation--Congresses. 2. Catalysis--Congresses. 3. Oxygen-
-Congresses.   I. Barton, Derek, Sir, 1918-   . II. Martell, Arthur
Earl, 1916-   . III. Sawyer, Donald T. IV. International Symposium
on the Activation of Dioxygen and Homogeneous Catalytic Oxidation
(5th : 1993 : College Station, Tex.)
QD281.09A28 1993
541.3'93--dc20   93-34420
                                                            CIP

Proceedings of the Fifth International Symposium on The Activation of Dioxygen and Homogeneous
Catalytic Oxidation, held March 14-19, 1993, in College Station, Texas

ISBN 978-1-4613-6307-1

© 1993 Springer Science+Business Media New York
Originally published by Plenum Press, New York in 1993

SPONSORS

Air Products
Amoco
ARI Technologies, Inc,
Briston Meyers/Squibb
Dow
General Electric
Hoechst Celanese Corporation
ICI Americas
Johnson Matthey
Mallinckrodt
Merck Manufacturing Division
Molecular Structure Corporation
Monsanto Company
Solvay Interox America
Texas A&M University
Union Carbide
Unilever Research, U.S., Inc.

INTERNATIONAL ORGANIZING COMMITTEE

K. Dear, United Kingdom
C. Hill, U.S.A.
B. James, Canada
H. Ledon, France
J. R. Lindsay Smith, United Kingdom
B. Meunier, France
H. Mimoun, Switzerland
G. Modena, Italy
Y. Moro-oka, Japan
A. Nishinaga, Japan
D. Riley, U.S.A.
D. T. Sawyer, U.S.A.
R. Sheldon, The Netherlands - Chairman-Elect
L. Simandi, Hungary

LOCAL ORGANIZING COMMITTEE
D. H. R. Barton
A. E. Martell
D. T. Sawyer

HONORARY CHAIRMAN
Professor A. E. Shilov

PREFACE

This monograph consists of the proceedings of the *Fifth International Symposium on the Activation of Dioxygen and Homogeneous Catalytic Oxidation*, held in College Station, Texas, March 14-19, 1993. It contains an introductory chapter authored by Professors D. H. R. Barton and D. T. Sawyer, and twenty-nine chapters describing presentations by the plenary lecturers and invited speakers. One of the invited speakers, who could not submit a manuscript for reasons beyond his control, is represented by an abstract of his lecture. Also included are abstracts of forty-seven posters contributed by participants in the symposium. Readers who may wish to know more about the subjects presented in abstract form are invited to communicate directly with the authors of the abstracts.

This is the fifth international symposium that has been held on this subject. The first was hosted by the CNRS, May 21-29, 1979, in Bendor, France (on the Island of Bandol). The second meeting was organized as a NATO workshop in Padova, Italy, June 24-27, 1984. This was followed by a meeting in Tsukuba, Japan, July 12-16, 1987. The fourth symposium was held at Balatonfured, Hungary, September 10-14, 1990. The sixth meeting is scheduled to take place in Delft, The Netherlands (late Spring, 1996); the organizer and host will be Professor R. A. Sheldon.

This symposium, as was true of the previous symposium in Hungary, involved a large component of industrial applications of oxidation/oxygenation reactions, particularly with dioxygen and hydrogen peroxide as starting materials. These inexpensive, non-polluting reagents are prime candidates to replace halogen oxidants, and thereby reduce industrial wastes. Papers given at this symposium also reflect the considerable advances that have occurred in the oxygenation of alkanes with both heme and non-heme complexes as catalysts. Important advances have been made with non-heme iron and copper complexes to promote dehydrogenation/oxygenation reactions. Insights into the nature of such processes have developed from the study of oxidase, peroxidase, monooxygenase, and dioxygenase enzymes and their models. To summarize, the chapters and poster abstracts are characterized by both their large number and wide variety of topics, illustrating the rapid acceleration of the field of oxygen transport, activation, and catalysis.

Financial support for this symposium was provided by Texas A&M University and sixteen industrial sponsors. The organizers express their sincere thanks for this assistance. Without it the Symposium, and its publication, would not have been possible.

Our thanks go to Mary Martell for invaluable assistance with the organization, logistics, and registration for this symposium, and for handling the large amount of correspondence that a meeting like this requires.

D.H.R. Barton, A. E. Martell, D. T. Sawyer

College Station, Texas
May 14, 1993

CONTENTS

INTRODUCTION: THE DILEMMAS OF O_2 AND HOOH ACTIVATION

Derek H. R. Barton and Donald T. Sawyer

Department of Chemistry
Texas A&M University
College Station, Texas 77843

More than a billion years of life on earth under reducing conditions produced metal sulfides, iron(II), iron, and residual saturated hydrocarbons that could not be oxidized in the absence of dioxygen. This anaerobic form of life can still be found on earth under conditions such as the bottom of the Black Sea and in deep sea vents in the Oceans. It has not produced any organism with a complex central nervous system.[1]

The mutation that gave rise to the blue-green algae about three billion years ago permitted photosynthesis, which produced dioxygen as a by-product.[2,3] This changed the world forever. The iron, iron(II), and the hydrocarbons were oxidized as was the hydrogen sulfide. Life developed enzymes to activate dioxygen, and thereby achieve far more complicated compounds. Under reducing conditions, membranes used triterpenoids derived from squalene. With dioxygen squalene epoxide could be formed and cyclized to oxygenated triterpenoids, like lanosterol--from which by many specific oxidation processes cholesterol, the hormones of the adrenal cortex, the sex hormones and vitamin D could all be produced.

Dioxygen also permitted respiration (controlled combustion of carbohydrates), which produced more energy to be expended per unit of weight and of time. Evolution onto land and the development of warm-blooded animals soon produced much more complicated nervous systems until we eventually arrived at Mankind. Although modern man is a very late arrival on the scene, he has an extraordinary nervous system that has an absolute dependence on dioxygen for function and for survival. Dioxygen is perhaps the most important of all the reagents of life. How fortunate that it is a triplet molecule and not a singlet. How fortunate also that copper and especially iron complexes permit oxygen to be domesticated so that our bodies are oxidized--but only in just the right places.

The Activation of Dioxygen and Homogeneous Catalytic Oxidation,
Edited by D.H.R. Barton *et al.*, Plenum Press, New York, 1993

How fitting it is then that much work is done on the Chemistry and Biochemistry of dioxygen activation and reactivity. A vast interdisciplinary effort is being mounted to understand oxygenase enzymes of all kinds by isolating and studying the pure enzymes, and by developing model systems.

Industrial interest in oxygen chemistry has always been great. Here economics takes the dominant role. At least dioxygen (as air) is free and, in principle, could be used to convert a saturated hydrocarbon directly to ketone and water. Also industrial processes are not confined to copper and iron; and any element or combination of elements can be called into the service of man.

This book shows how all the elements of this complex multidisciplinary subject interact together to give a science well-suited to the end of the Century and the beginning of the next. The following chapter by Professor Roger Sheldon provides a concise history of these Symposia and the progress that has been mode during the past two decades to understand the metal-induced activation of dioxygen, and, in turn, to make use of this understanding in the design of catalysts and processes for the utilization of O_2. Several chapters follow that illustrate catalyst development for various industrial processes and focused chemical synthesis. Professors P. R. Ortiz de Montellano and Daniel Mansuy provide important summaries and a reminder that Nature is the supreme catalyst engineer for (a) the oxygenation of saturated hydrocarbons via *cytochrome P-450 monooxygenases* (cyt P-450)/O_2 and (b) the epoxidation of olefins via peroxidases (e.g., *horseradish peroxidase* (HRP)/HOOH ————▶ Compound I) or cyt P-450 Compound I. The latest insights into the structure and mechanism for the activation of O_2 by the *methane monooxygenase* (MMO) proteins are presented in the chapters by Professor S. J. Lippard and Professor L. Que, Jr. and co-workers.

Nature, in its ever present need for carbonaceous nutrients (food, fuel) for oxidative metabolism and respiration has developed these O_2-activation catalysts (cyt P-450 and MMO) for the controlled combustion of saturated hydrocarbons (initially to alcohols), e.g.,

$$CH_4 + O_2 + RedH_2 \xrightarrow{\text{MMO}} CH_3OH + H_2O + Red \qquad (1)$$
$$\text{(reductase)}$$

Likewise, HRP [an iron(III)-heme protein with an axial histidine] activates HOOH for the epoxidation of olefins(as does the cyt P-450/O_2/RedH_2$ system, but not MMO) via a porphyrin-cation-radical-ferryl intermediate [(por$^{+\cdot}$)Fe^{IV}=O, Compound I],[4]

$$c\text{-}C_6H_{10} + HOOH \xrightarrow{\text{HRP}} c\text{-}C_6H_{10}\text{-epoxide} + H_2O \qquad (2)$$

Much of the O_2-activation research during the past 20 years has focused on the viable reactive intermediates for the oxygenation of alkanes and the epoxidation of olefins.

Table I summarizes (a) several systems and their reactive intermediates for reaction with alkanes, (b) the initial product(s), and (c) the kinetic isotope effects [KIE, $k_{c-C_6H_{12}}/k_{c-C_6D_{12}}$].[5-10] Although many have suggested that hypervalent iron, ferryl ($L_xFe^{IV}=O$), or perferryl ($L_xFe^V=O$) are reactive with alkanes (and are reasonable reactive intermediates for MMO and cyt P-450), Compound I of HRP is unreactive (Professor Daniel Mansuy, this Symposium). The large dissociative bond energies (ΔH_{DBE}) for the C–H bonds of alkanes [e.g., H_3C–H (ΔH_{DBE}, 105 kcal mol^{-1}), Table II] require reactive intermediates (YO·) with sufficiently large bond-formation energies

Table I. Established and Proposed Reactive Intermediates for the Oxygenation of Alkanes (RH, c-C$_6$H$_{12}$).

System	Reactive Intermediate	Initial Product	KIE, $k_{c-C_6H_{12}}/k_{c-C_6D_{12}}$
pulse radiolysis[a]	HO·	R·	1.0
pulse radiolysis/O$_2$[b]	HO·	ROO·	1.0
Fe^{III}Cl$_3$/HOOH[c]	Cl$_3$Fe^V(OH)$_2$ (proposed)	ROH, RCl	2.9
Gif systems/O$_2$[d]	$L_xFe^V=O$, O$_2$ (pr)	ketone, ROH	2.5
Fenton reagent[e]	[$L_xFe^{II}OOH$, BH$^+$] (pr)	ROH [or (R)py]	1.1 → 1.7
Fenton reagent/O$_2$[f]	[$L_xFe^{III}OOH(O_2, BH^+)$] (pr)	ketone, ROH	2.1 → 10
HRP Compound I	Por^{+}·Fe$^{IV}=O$	NR	
cyt P-450	(R'S)Fe$^V=O$ (pr)	ROH	
MMO		ROH	

[a]Ref. 5. [b]Ref. 6. [c]Ref. 7. [d]Ref. 8. [e]Ref. 9. [f]Ref. 10.

(-ΔG_{BF}) for YO–H to cleave the C–H bond. On the basis of the data in Table II only HO· has sufficient radical strength to be reactive with CH$_4$. Although Fenton reagents[9] and Gif systems[8] react with 1°/2°/3° C–H bonds, they are unreactive with CH$_4$. Because methane is the primary substrate for MMO, the C–H bond breaking process must be other than an oxy-radical outersphere cleavage. Reactivity and product formation for Gif systems,[8] Fenton reagents,[9] and oxygenated Fenton systems[10] are facilitated via iron–carbon and iron–OO–carbon intermediates. Hence, the monooxygenation of alkanes by MMO and cyt P-450 must involve stabilization of the carbon radical (R·) via an iron–carbon bond (-ΔG_{BF}, 20–30 kcal mol^{-1}; A. Qiu and D. T. Sawyer, unpublished results). The electron-rich Nature of the iron centers of MMO and cyt P-450 and of the

Table II. Radical Strength of Oxygen Radicals (YO·) in Terms of Their YO–H Bond-Formation Free Energies ($-\Delta G_{BF}$), and Dissociative Bond Energies (ΔH_{DBE}) for H–R Molecules

Oxy radical (YO·)	Bond (YO–H)	$(-\Delta G_{BF})(aq)$,[a] (kcal mol^{-1})	Bond (H–R)	$(\Delta H_{DBE})(g)$,[b] (kcal mol^{-1})
HO·	HO–H	111	H–CH$_3$	105
$^-$O·$^-$	$^-$O–H	109	H–(n-C$_3$H$_7$)	100
·O$^-$	·O–H	98	H–(c-C$_6$H$_{11}$)	95
t-BuO·	t-BuO–H	97	H–(t-C$_4$H$_9$)	93
MeO·	MeO–H	96	H–CH$_2$Ph	88
PhO·	PhO–H	79	H–(c-C$_6$H$_7$)(CHD)	73
HOO·	HOO–H	82	H–C(O)Ph	87
O$_2$$^-$·	$^-$OO–H	72	H–Ph	111
·O$_2$·	·OO–H	51	H–SH	91
t-BuOO·	t-BuOO–H	83	H–SMe	89
MeOO·	MeOO–H	82	H–SPh	83
MeC(O)O·	MeC(O)O–H	98	H–OPh	86
(por$^+$·)Fe^{IV}=O	(por$^+$·)Fe^{III}O–H	85 (est)		
(por)Fe^{IV}=O	(por)Fe^{III}O–H	78 (est)		
Cl$_3$Fe^V=O	Cl$_3$Fe^{IV}–OH	88 (est)		
Cl$_3$Fe^V(OH)$_2$	Cl$_3$Fe^{IV}OH(HO–H)	96 (est)		

[a]Ref. 11. [b]Ref. 12.

·CH$_3$ radical apparently leads to a $-\Delta G_{BF}$ value of 35–45 kcal mol^{-1}, which would overcome the outersphere limits of Table II.

    The second question concerns the reactive intermediate for the epoxidation of olefins. Although there is general agreement on the formulation of Compound I of HRP[4] and its ability to epoxidize olefins stereospecifically[13] (see Arasasingham and Bruice, this volume), the possibility of alternative reactive intermediates remains (see Valentine, this volume). Table III summarizes (a) several HOOH/O$_2$ activating systems, (b) the proposed reactive intermediates for reaction with cyclohexene (c-C$_6$H$_{10}$) and cis-stilbene (cis-PhCH=CHPh), and (c) the observed products. Epoxide production appears to be limited to ferryl-like intermediates (L$_x$Fe=O), while Gif and Fenton systems yield alcohols and ketones. These two substrates appear to be effective diagnostic probes to characterize the reactive intermediate of epoxidation reagents.

A final concern in our quest for an understanding of oxygen activation and reactive intermediates is the tendency for many to formulate oxy anions (XO⁻) in aqueous solutions at pH 7.0. Table IV summarizes the aqueous acidities for various XO–H species,[16] and confirms that free superoxide ion (O_2^-·) and bicarbonate ion [HOC(O)O⁻] are the only dominant oxy anions at pH 7. Even hydrated electrons (e_{aq}^-) are as their conjugate acid (H·).

Table III. Cyclohexene (c-C_6H_{10}) and cis-Stilbene (cis-PhCH=CHPh) as Diagnostic Substrates for Reactive Intermediates for Epoxidation.

	A. c-C_6H_{10}	
System	Proposed Intermediate	Product
HRP–Compound I[a]	(por⁺·)Fe^{IV}=O	epoxide
Fe^{III}Cl$_3$/HOOH/MeCN (dry)[b]	Cl$_3$Fe^V=O	epoxide
(wet)	Cl$_3$Fe^V(OH)$_2$	c-C_6H_9OH, ketone
Gif systems[c]	L$_x$Fe^V=O/O_2	ketone, allylic c-C_6H_9OH
Oxygenated Fenton Systems[d]	L$_x$Fe^{II}OOH(BH⁺)(OH)	ketone, c-C_6H_9OH
	B. cis-PhCH=CHPh	
Fe^{III}Cl$_3$/HOOH/MeCN (dry)[b]	Cl$_3$Fe^{IV}=O	epoxide
Oxygenated Fenton Systems[d,e]	L$_x$Fe^{II}OOH(BH⁺)(O_2) L$_x$Fe^{IV}=O	PhCH(O) (75%) epoxide (25%)

[a]Ref 4, 13. [b]Ref 7. [c]Ref 8. [d]Ref 10. [e]Ref 14, 15.

Many of the contributions in this Symposium volume discuss important advances in metal-catalyzed activation of dioxygen and hydroperoxides. Several are concerned with proposed mechanisms and reactive intermediates for the oxygenation of alkanes, and alkenes. We hope that the preceding discussion will be helpful in the consideration of the latest results and mechanistic proposals.

Table IV. Acid-Base Character of XOH/XO⁻ Species in Aqueous Solutions.

XO–H (pH 7)	XO⁻	pK_a
H_2O	HO^-	45.7
ROH	RO^-	18
PhOH	PhO^-	10
FeOH	FeO^-	~14
HO·	·O⁻	11.9
HOOH	HOO^-	11.8
ROOH	ROO^-	12
FeOOH	$FeOO^-$	~11
HOO· (1%)	$·O_2^-$	4.9
HOC(O)OH (20%)	$HOC(O)O^-$	6.4
HOO⁻ (0.001%)	⁻OO⁻	>30
H·	e_{aq}^-	9.3

References

(1) Day, N. *Genesis on Planet Earth,* 2nd ed., New Haven: Yale University Press, 1984.

(2) Gilbert, D. L. (ed.) *Oxygen and Living Processes. An Interdisciplinary Approach,* New York: Springer Verlag, 1981, pp 1-43.

(3) Metzner, H. (ed.) *Photosynthetic Oxygen Evolution,* New York: Academic Press, 1978.

(4) Penner-Hahn, J. E.; Eble, K. E.; McMurry, T. J.; Renner, M.; Balch, A. L.; Groves, J. T.; Dawson, J. H.; Hodgson, K. O. *J. Am. Chem. Soc.* **1986,** *108,* 7819.

(5) Buxton, G. V.; Greenstock, C. L.; Helman, W. P.; Ross, A. B. *J. Phys. Chem. Ref. Data* **1988,** *17,* 513-886.

(6) Sheldon, R. A.; Kochi, J. K. *Metal-Catalyzed Oxidations of Organic Compounds,* New York: Academic Press, 1981, Chap. 2 and 3.

(7) Sugimoto, H.; Sawyer, D. T. *J. Org. Chem.* **1985,** *50,* 1785.

(8) Barton, D. H. R.; Doller, D. *Acc. Chem. Res.* **1992,** *25,* 504.

(9) Sawyer, D. T.; Kang, C.; Llobet, A.; Redman, C. *J. Am. Chem. Soc.* **1993,** *115,* 0000.

(10) Kang, C.; Redman, C.; Cepak, V.; Sawyer, D. T. *Bioorg. Med. Chem.* **1993,** *1,* 000.

(11) Sawyer, D. T. *J. Phys. Chem.* **1990,** *93,* 7977.

(12) Lide, D. R. (ed.) *CRC Handbook of Chemistry and Physics,* 71st ed., Boca Raton, FL: CRC, 1990, pp 9-86-98.

(13) Sugimoto, H.; Tung, H.-C.; Sawyer, D. T. *J. Am. Chem. Soc.* **1988,** *110,* 2465.

(14) Sheu, C.; Richert, S. A.; Cofré, P.; Ross, B., Jr.; Sobkowiak, A.; Sawyer, D. T.; Kanofsky, J. R. *J. Am. Chem. Soc.* **1990,** *112,* 1936.

(15) Tung, H.-C.; Kang, C.; Sawyer, D. T. *J. Am. Chem. Soc.* **1992**, *114*, 3446.

(16) Sawyer, D. T.; *Oxygen Chemistry*, New York: Oxford University Press, 1991, Chap. 2.

A HISTORY OF OXYGEN ACTIVATION: 1773-1993

R.A. Sheldon

Delft University of Technology
The Netherlands

INTRODUCTION

Controlled partial oxidation of hydrocarbons (alkanes, alkenes and aromatics) is the single most important technology for converting petrochemical feedstocks to industrial organic chemicals [1]. For economic reasons, these processes predominantly involve the use of molecular oxygen (dioxygen) as the primary oxidant. The success of these processes depends largely on the use of metal catalysts to promote both the rate of reaction and the selectivity to partial oxidation products. Both gas phase and liquid phase oxidations, employing heterogeneous and homogeneous catalysts, respectively, are practiced industrially. In biological systems a broad range of selective oxidations of hydrocarbon substrates by dioxygen are catalyzed by metalloenzymes, appropriately known as oxygenases [2,3]. A common feature of most of these processes is the involvement of a multivalent transition metal ion.

Structure and Reactivity of Dioxygen

The ground state of dioxygen is a triplet with two unpaired electrons with parallel spins. The first two electronically excited states are both singlets, formed by relocation and/or pairing of the unpaired electrons in the $2p\pi^*$ antibonding orbitals. The half-filled antibonding molecular orbitals of $^3O_2$ can accommodate two additional electrons. The addition of one electron affords the superoxide anion ($O_2^{\cdot-}$) and two-electron reduction gives the peroxide ion (O_2^{2-}).

The complete oxidation of organic materials by dioxygen, to give carbon dioxide and water, is thermodynamically very favorable. Fortunately, unfavorable kinetics preclude the spontaneous combustion of living matter into a puff of smoke. Thus, the direct reaction of $^3O_2$ with singlet organic molecules to give singlet products is a spin-forbidden process with a very low rate. One way of circumventing this energy barrier is via a free radical pathway. The reaction of a singlet molecule with $^3O_2$ (reaction 1) forming two doublets (free radicals) is a spin-allowed process. Reaction 1 is, however, highly endothermic (up to 50 kcal.mol^{-1}) and is observed at moderate temperatures only with very reactive substrates that form resonance stabilized radicals, e.g. reduced flavins (reaction 2). This is a key step in the activation of $^3O_2$ by flavin-dependent oxygenases.

The Activation of Dioxygen and Homogeneous Catalytic Oxidation,
Edited by D.H.R. Barton *et al.*, Plenum Press, New York, 1993

$$RH + {}^3O_2 \longrightarrow R\cdot + HO_2\cdot \qquad (1)$$

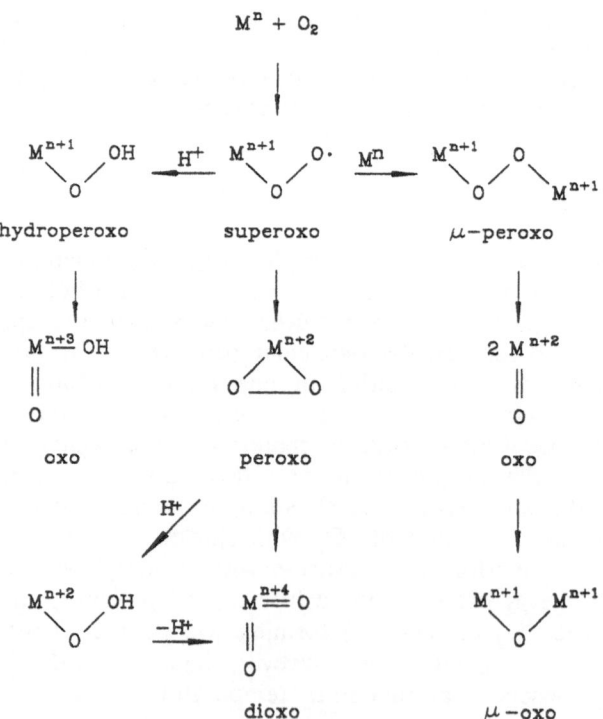

$$(2)$$

$$M^n + {}^3O_2 \longrightarrow \qquad (3)$$

A second way to overcome the spin conservation obstacle is for ${}^3O_2$ to combine with a paramagnetic transition metal ion (reaction 3). The expectation that the resulting metal-dioxygen complex may react selectively with organic molecules at moderate temperatures forms the basis for the extensive studies of oxygen activation by metal complexes during the last three decades [4-6]. The various oxygenated species that may play a role in metal-catalyzed oxidations with dioxygen are depicted in Figure 1.

$$M^n + O_2$$

hydroperoxo   superoxo μ-peroxo

oxo   peroxo   oxo

dioxo μ-oxo

Figure 1. Metal-oxygen species.

Why oxygen activation? The widespread interest in oxygen activation, as is evidenced by the more than 200 publications a year devoted to the subject, stems from the enormous commercial potential that exists for selective catalysts for the 'dream reactions' shown below (reactions 4-6).

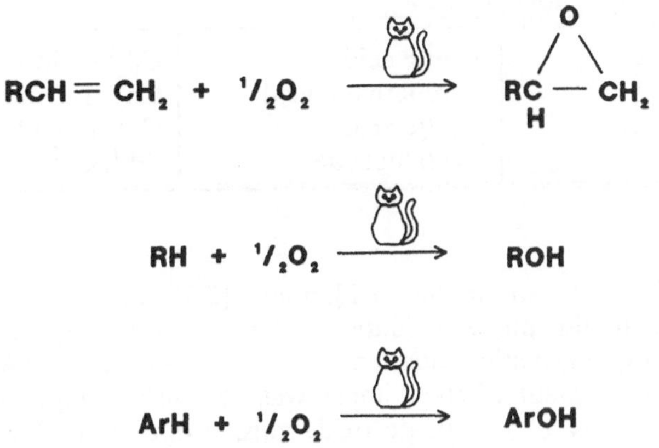

$$RCH = CH_2 + \tfrac{1}{2}O_2 \xrightarrow{} \underset{\underset{H}{\displaystyle |}}{RC} \overset{\displaystyle O}{\overbrace{}} CH_2 \qquad (4)$$

$$RH + \tfrac{1}{2}O_2 \xrightarrow{} ROH \qquad (5)$$

$$ArH + \tfrac{1}{2}O_2 \xrightarrow{} ArOH \qquad (6)$$

HISTORICAL DEVELOPMENT - THE FIRST 200 YEARS

Lavoisier's explanation of the phenomenon of combustion in 1774 signaled the demise of the phlogiston theory and the beginning of the modern era of chemistry. Following the rationalization of the phenomenon of catalysis by Berzelius in 1835, several gas-phase catalytic oxidation processes were developed, in the late 19th and early 20th century, for the production of inorganic bulk chemicals. Examples include the Winkler process (1875) for Pt-catalyzed oxidation of SO_2 to SO_3 and the Ostwald process (1902) for Pt-catalyzed oxidation of NH_3 to HNO_3. The Winkler process was later replaced by the BASF process (1915) which employed the cheaper V_2O_5 as the catalyst. One of the first industrial processes to involve controlled, catalytic oxidation of a hydrocarbon was the gas-phase oxidation of ethylene to ethylene oxide over a supported silver catalyst, discovered by Lefort in 1935.

Parallel to these developments, observations made in the 19th century linked the deterioration of many organic materials, such as natural oils and fats, to the absorption of dioxygen. Around the turn of the century it was recognized that these processes involved organic peroxide intermediates. Subsequently, detailed mechanistic studies with simple hydrocarbons led to the free radical chain theory of autoxidation [7]. Following close on the heels of these mechanistic developments several important catalytic oxidation processes, in both the gas and liquid phase, were developed in the period 1945-1960. Some examples are shown in Table 1.

Table 1. Catalytic oxidation processes with O_2.

Substrate	Product	Catalyst
Gas phase, heterogeneous catalyst		
Ethylene	Ethylene oxide	Ag/Al_2O_3
Propylene	Acrolein	Bi_2MoO_6
Propylene/NH_3	Acrylonitrile	Bi_2MoO_6
o-Xylene	Phthalic anhydride	V_2O_5/TiO_2
Liquid phase, homogeneous catalyst		
Acetaldehyde	Acetic acid	Co^{II} or Mn^{II}
p-Xylene	Terephthalic acid	Co^{II}/Br^- in HOAc
n-Butane	Acetic acid	Co^{II} in HOAc
Ethylene	Acetaldehyde	Pd^{II}/Cu^{II}

The evolution of enzymatic oxidations [2,3] proceeded parallel with developments in liquid phase oxidations. The intermediacy of iron-dioxygen complexes in many enzymatic oxidations was first proposed by Warburg around 1920. Further development of these ideas was seriously hampered by Wieland's dehydrogenation theory of enzymatic oxidations, proposed in 1932. This theory, widely accepted for more than two decades, held that the sole function of dioxygen in enzymatic oxidations is as an electron acceptor, forming water or hydrogen peroxide. This situation changed dramatically in 1955 when Hayaishi [8] and Mason [9] independently demonstrated the direct incorporation of dioxygen into the substrate in reactions (7) and (8).

$$(7)$$

$$(8)$$

An interesting historical quirk is the fact that the first oxygenase model system, the Udenfriend reagent (Figure 2), was reported in 1954 [10], one year prior to the discovery of the enzymes it emulates. Udenfriend and coworkers found that a mixture of Fe(II), EDTA, ascorbic acid and dioxygen is able to hydroxylate aromatic rings at neutral pH and under mild conditions. It was later found that the ascorbic acid can be replaced by a variety of hydrogen donors

$$\text{ArH} + \text{O}_2 \xrightarrow[\text{ascorbate}]{\text{Fe}^{II}/\text{EDTA}} \text{ArOH} + \text{H}_2\text{O}$$

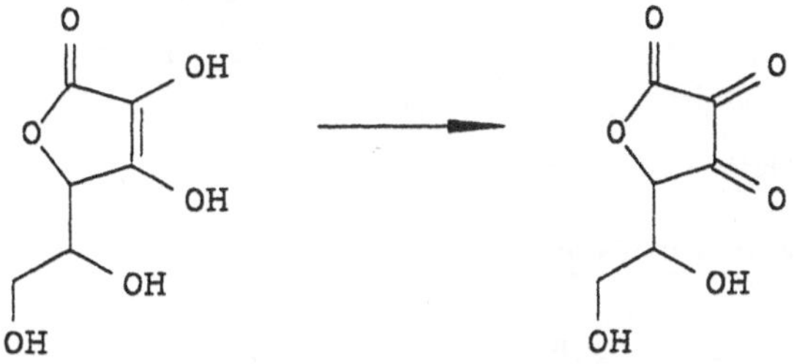

Figure 2. Biomimetic oxygenation - Udenfriend's reagent.

[1,11]. Indeed, one may conclude that in addition to Fe(II) and dioxygen all that is needed is a source of electrons and protons.

MECHANISMS OF OXIDATIONS WITH $^3\text{O}_2$

Abiological catalysis of oxidations with $^3\text{O}_2$ can be conveniently divided into three mechanistic types as illustrated in Figure 3. Liquid phase oxidations generally involve the free radical autoxidation mechanism with the exception of a minority of processes, e.g. the Wacker process for ethylene oxidation, that involve direct oxidation of the substrate followed by reoxidation of the reduced metal catalyst with dioxygen. Gas phase oxidations, in contrast, generally involve the so-called Mars-van Krevelen mechanism [12], i.e. direct oxidation of the hydrocarbon by an oxometal species ($\text{Mo}^{VI}=\text{O}$ or $\text{V}^V=\text{O}$) followed by regeneration with dioxygen. How can we account for this marked difference between the gas and liquid phase? In the liquid phase the facile free radical autoxidation is ubiquitous and difficult to compete with. In the gas phase, on the other hand, the concentrations of RH in the vicinity of the catalyst are much lower making radical chain processes less favorable. In practice, both types of process afford selective oxidations only with a limited number of relatively simple substrates. Thus, free radical autoxidation is a largely indiscriminate process and gives high selectivities only with molecules containing one reactive position, e.g. toluene. There is a great need, therefore, for catalytic methods that are able to compete with free radical autoxidation in liquid phase oxidations, i.e. to create gas phase conditions in the liquid phase. As we shall see later many biological oxygenations bear a marked resemblance to gas phase oxidations.

What do we mean by oxygen activation? A reasonable working definition is: catalysis of the oxidation of a hydrocarbon substrate by $^3\text{O}_2$ not involving a classical free radical autoxidation mechanism or direct oxidation by a metal salt. The latter stipulation is needed to exclude Wacker-type oxidation processes in which the oxygen in the product is, initially at least, derived from water. On the other hand, it should not matter whether dioxygen complex formation precedes or follows the oxidation of the substrate by an oxometal complex (see Figure 4). The former pertains to liquid and the latter to gas phase processes.

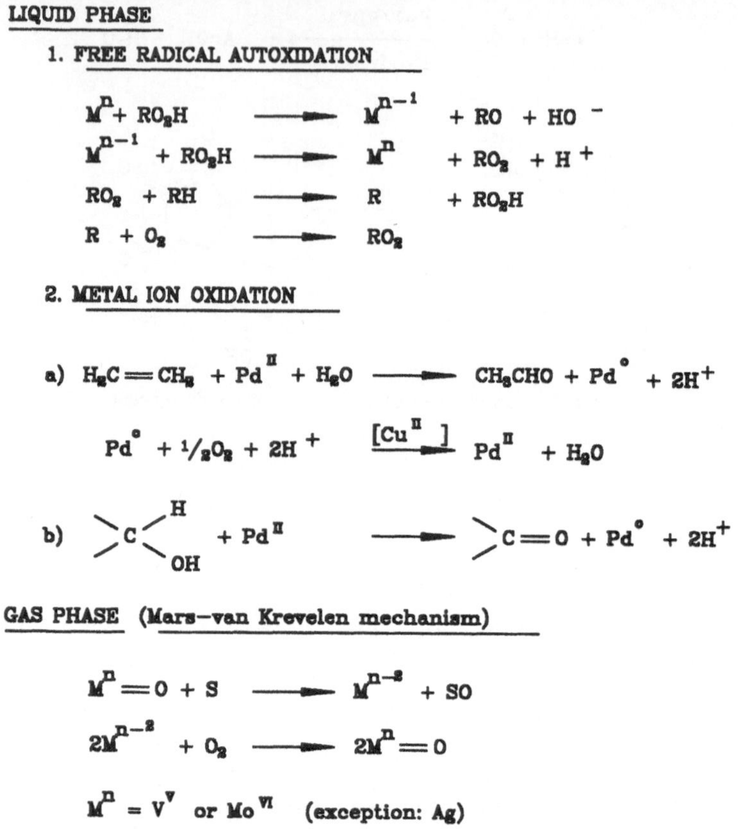

LIQUID PHASE

1. FREE RADICAL AUTOXIDATION

$$M^n + RO_2H \longrightarrow M^{n-1} + RO + HO^-$$

$$M^{n-1} + RO_2H \longrightarrow M^n + RO_2 + H^+$$

$$RO_2 + RH \longrightarrow R + RO_2H$$

$$R + O_2 \longrightarrow RO_2$$

2. METAL ION OXIDATION

a) $H_2C=CH_2 + Pd^{II} + H_2O \longrightarrow CH_3CHO + Pd^\circ + 2H^+$

$$Pd^\circ + 1/2 O_2 + 2H^+ \xrightarrow{[Cu^{II}]} Pd^{II} + H_2O$$

b) $\underset{}{>}C\underset{OH}{\overset{H}{<}} + Pd^{II} \longrightarrow \ >C=O + Pd^\circ + 2H^+$

GAS PHASE (Mars–van Krevelen mechanism)

$$M^n = O + S \longrightarrow M^{n-2} + SO$$

$$2M^{n-2} + O_2 \longrightarrow 2M^n = O$$

$$M^n = V^V \text{ or } Mo^{VI} \quad \text{(exception: Ag)}$$

Figure 3. Mechanisms of metal-catalyzed oxidations.

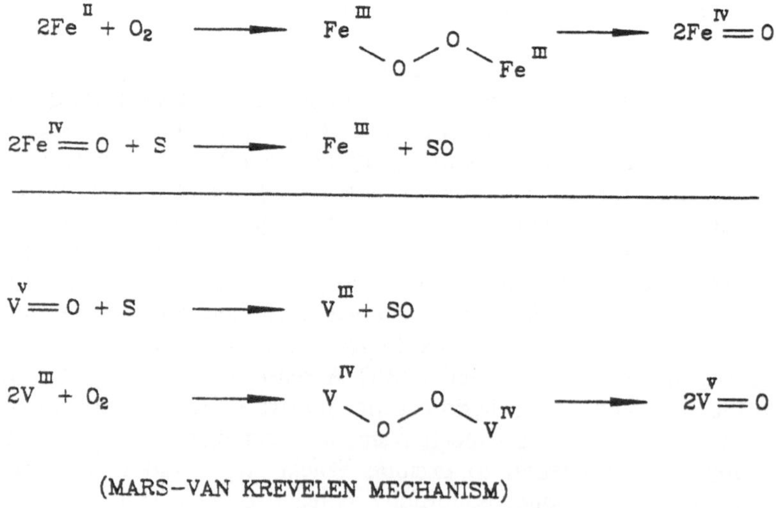

$$2Fe^{II} + O_2 \longrightarrow Fe^{III} \underset{O}{\overset{O}{\diagdown}}O - Fe^{III} \longrightarrow 2Fe^{IV} = O$$

$$2Fe^{IV} = O + S \longrightarrow Fe^{III} + SO$$

$$V^V = O + S \longrightarrow V^{III} + SO$$

$$2V^{III} + O_2 \longrightarrow V^{IV} \underset{O}{\overset{O}{\diagdown}}O - V^{IV} \longrightarrow 2V^V = O$$

(MARS–VAN KREVELEN MECHANISM)

Figure 4. What is oxygen activation?

REACTION OF DIOXYGEN WITH METAL COMPLEXES

In the late sixties a wide variety of low-valent transition metal complexes were shown to combine reversibly with dioxygen [13]. Some examples are shown in Figure 5.

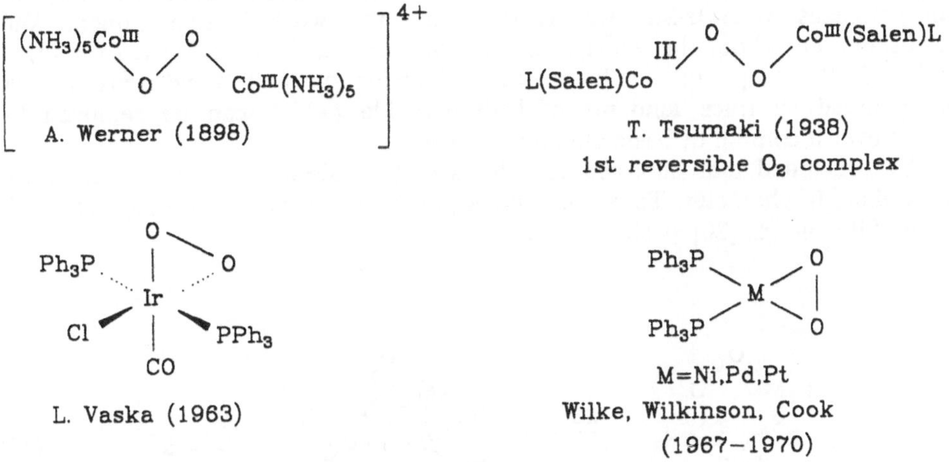

$$\left[(NH_3)_5Co^{III} \underset{O}{\overset{O}{\diagdown}} \diagup Co^{III}(NH_3)_5 \right]^{4+}$$

A. Werner (1898)

$$L(Salen)Co^{III} \underset{O}{\overset{O}{\diagdown}} \diagup Co^{III}(Salen)L$$

T. Tsumaki (1938)
1st reversible O_2 complex

L. Vaska (1963)

$$Ph_3P \diagdown M \diagup O$$
$$Ph_3P \diagup \overset{|}{O}$$

M=Ni,Pd,Pt
Wilke, Wilkinson, Cook
(1967–1970)

Figure 5. Metal-oxygen complexes.

Interest in the utilization of such complexes for the selective oxygenation of hydrocarbons was aroused by two publications. Collman and coworkers [14] reported in 1967 that the oxidation of cyclohexene in the presence of low-valent complexes of Ir, Rh and Pd afforded a mixture of cyclohexen-2-one and cyclohexene oxide (reaction 9). It was proposed that the reaction involved an 'oxygen activation' mechanism.

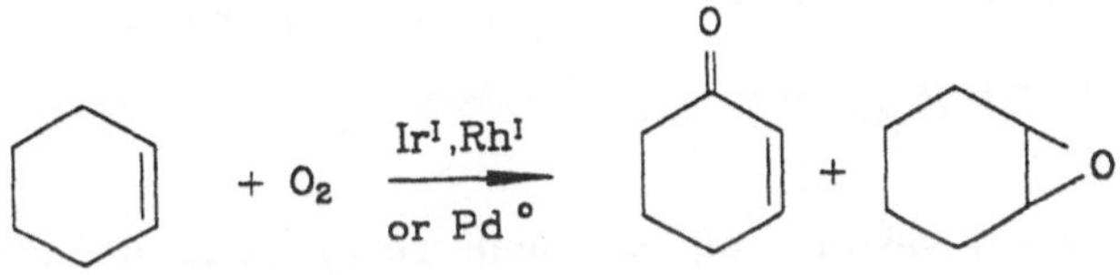

(9)

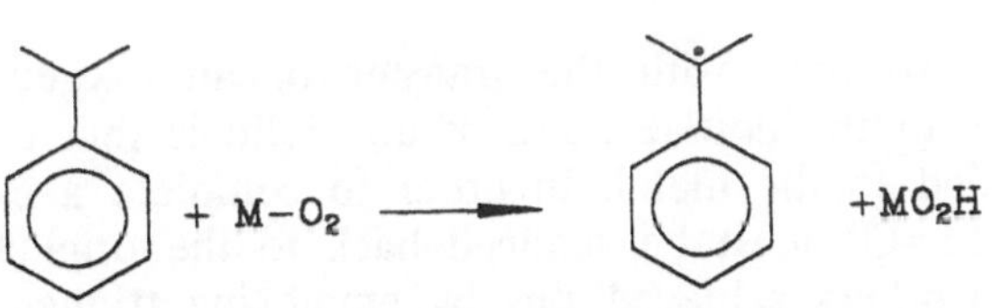

(10)

$$M\text{-}O_2 = (Ph_3P)_2PdO_2 \text{ or } CuPcO_2$$

In 1970 Stern [15] proposed that the key step in the $(Ph_3P)_4Pd$-catalyzed autoxidation of cumene at 35° involved hydrogen abstraction by the metal-dioxygen complex as shown in reaction 10. The same step had been proposed by Kropf [16] to account for catalysis of cumene autoxidation by metal phthalocyanines. However, careful kinetic studies [17-19] subsequently showed that these reactions all proceed via classical redox decomposition of trace amounts of hydroperoxides present in the cyclohexene or cumene. In the early seventies we carried out extensive studies of the oxidation of cyclohexene and 1-octene in the presence of low-valent complexes of Pd, Pt, Ir, Rh and Ru. In no instance did we observe results which were not consistent with a classical free radical chain autoxidation mechanism involving initiation via redox decomposition of hydroperoxides. One lesson we learned from this work is that cumene and cyclohexene are about the worst substrates that one could have chosen for these studies. Both compounds undergo facile autoxidation and, hence, are always contaminated by trace amounts of hydroperoxide (which can be removed by passing over a column of basic alumina prior to use).

We concluded that dioxygen complexes of low-valent group VIII metals are nucleophilic in character. Thus, they undergo a 3 + 2 cycloaddition reaction with electrophilic olefins [20] as shown below.

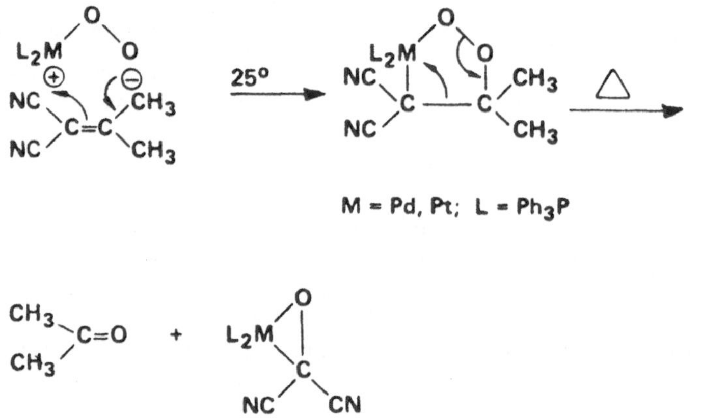

$$M = Pd, Pt; \quad L = Ph_3P \tag{11}$$

Similarly, these low valent complexes undergo cycloaddition reactions with a variety of inorganic molecules such as SO_2 and CO_2 [11]. They are obviously good models for the Winkler and Ostwald processes (see earlier) but not for hydrocarbon oxidation.

One problem associated with the transfer of an oxygen atom from the peroxometal species to the double bond of an olefin is that the second oxygen atom remains bonded to the metal. In order to complete a catalytic cycle this oxometal species (M=O) must be reduced back to the original oxidation state (M). Read and coworkers achieved this by employing triphenylphosphine as a coreductant [21]:

$$RCH = CH_2 \xrightarrow[\text{Ph}_3\text{P}]{O_2, \; RhClL_3} RCOCH_3 + Ph_3PO \tag{12}$$

Mimoun and coworkers [22] subsequently showed that no added reducing agent was necessary when Rh(III) perchlorate was used as the catalyst in alcoholic solution at room temperature. It was suggested that the reaction proceeded via a peroxymetallocycle intermediate which decomposes to give the methylketone and oxorhodium(III). The latter is converted to hydroxorhodium(III) which reacts with a second molecule of olefin in a Wacker type process to afford a second molecule of ketone (Figure 6).

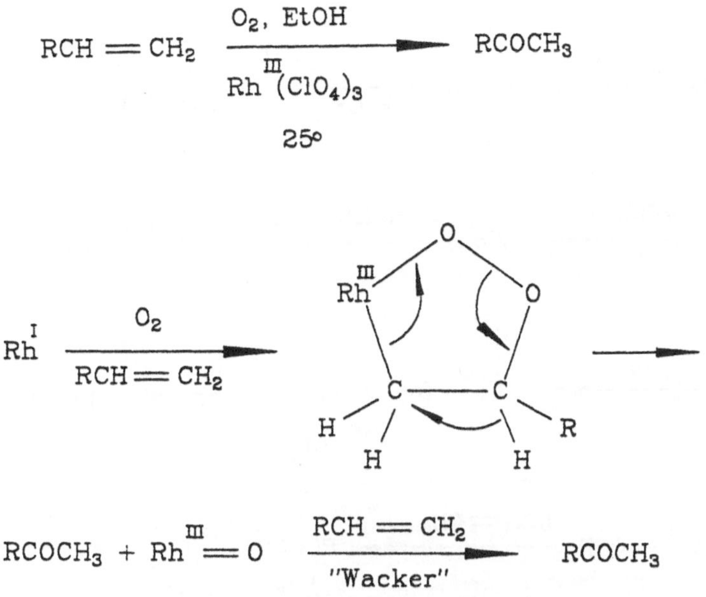

Figure 6. Rh-catalyzed oxygenation of an olefin.

More recently, Drago and coworkers [23] have shown that the alcohol solvent takes part in the reaction by reducing the Rh(III) to Rh(I). This is followed by reaction of Rh(I) with O_2 and a proton to give a Rh(III) hydroperoxide complex which oxidizes the olefin. Similarly, Drago [24] and Nishinaga [25] found that cobalt(II) Schiff base complexes catalyze the co-oxidation of olefins and primary alcohols (reaction 13).

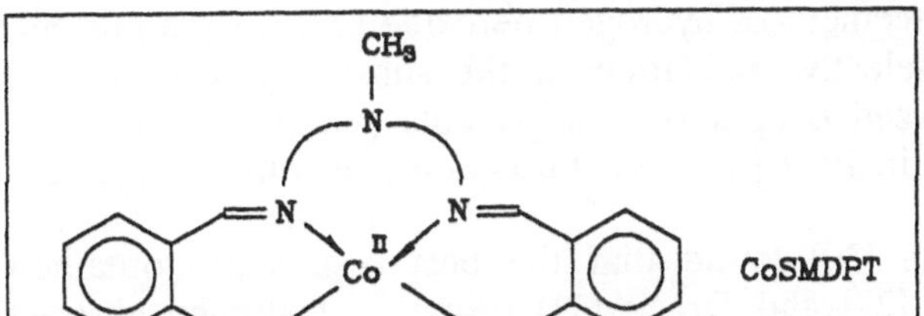

(13)

In addition to the methylketone the corresponding secondary alcohol, $RCH(OH)CH_3$, is also formed from the olefin. Nishinaga proposed [25] a cobalt hydride intermediate (see Figure 7) to explain the formation of $RCOCH_2D$ when $PhCD_2OH$ was used as the primary alcohol. Drago [24], in contrast, favors the addition of $LCo^{III}OOH$, formed by reaction of Co^I with O_2 and a proton, to the olefin. Whichever mechanism [27] is correct, one thing is clear: non-classical oxidation of olefins is observed only in the presence of a coreductant.

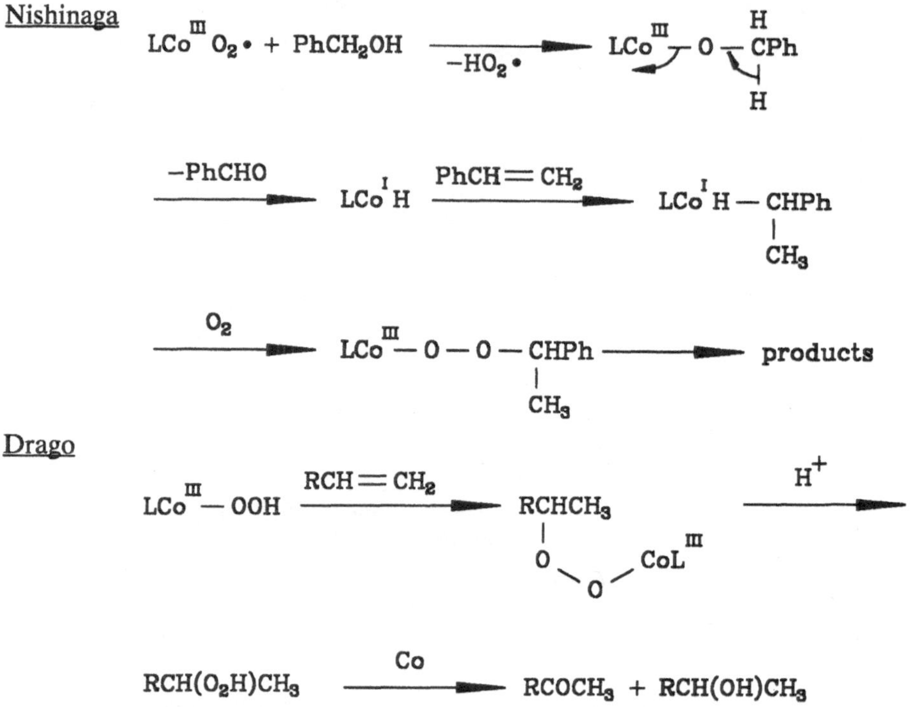

Nishinaga

$$LCo^{III}O_2 \bullet + PhCH_2OH \xrightarrow[-HO_2\bullet]{} LCo^{III} - O - \underset{\underset{H}{|}}{\overset{\overset{H}{|}}{C}}Ph$$

$$\xrightarrow{-PhCHO} LCo^{I}H \xrightarrow{PhCH=CH_2} LCo^{I}H - \underset{\underset{CH_3}{|}}{CHPh}$$

$$\xrightarrow{O_2} LCo^{III} - O - O - \underset{\underset{CH_3}{|}}{CHPh} \longrightarrow products$$

Drago

$$LCo^{III} - OOH \xrightarrow{RCH=CH_2} \underset{\underset{O \diagdown O \diagup}{\overset{|}{O}} CoL^{III}}{RCHCH_3} \xrightarrow{H^+}$$

$$RCH(O_2H)CH_3 \xrightarrow{Co} RCOCH_3 + RCH(OH)CH_3$$

Figure 7. Mechanism of cobalt Schiff base catalyzed co-oxidation of alcohols and terminal olefins.

CATALYTIC OXYGENATION OF PROTIC SUBSTRATES

Although metal dioxygen complexes do not react with hydrocarbons they do react readily with protic substrates, such as alcohols (see above), phenols and amines, with displacement of hydrogen peroxide [27]. In the presence of dioxygen this can lead to selective oxidation of the substrate. A typical example is the Co(II) Salen-catalyzed oxygenation of phenols (Figure 8) first described by van Dort and Geursen in 1967 [28] and subsequently extensively studied by Nishinaga [29].

We and others [30] found that the best results are obtained in DMF as solvent. Nishinaga [29] and Drago [31] proposed hydrogen abstraction from the phenol by the superoxocobalt(III) complex (reaction 14) as the initial step.

18

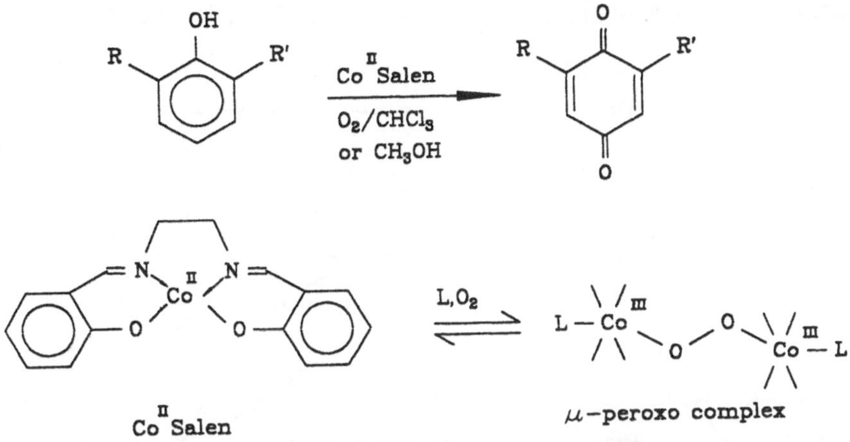

Figure 8. Co(II)Salen-catalyzed oxygenation of phenols.

$$LCo^{III}O_2^{\bullet} + ArOH \longrightarrow LCo^{III}O_2H + ArO\bullet \qquad (14)$$

We, on the other hand, proposed [1] a mechanism involving initial S_N2 displacement of peroxide by phenol on the μ-peroxo or superoxocobalt(III) complex, by analogy with other systems [27]. This is followed by oxygen insertion into the aryloxycobalt(III) intermediate and subsequent decomposition of the resulting alkylperoxycobalt(III) intermediate into the quinone (see Figure 9). Similarly, other protic substrates, that form ambident nucleophiles on deprotonation, also undergo catalytic oxygenation in the presence of cobalt and manganese Schiff base complexes. Some examples are shown below (reactions 15-17).

We suggest that a common mechanistic feature of all these reactions is S_N2 displacement of peroxide or superoxide at the metal center by the ambident nucleophile, followed by oxygen insertion into what is formally an organometal intermediate. It is interesting to note that reaction (17) proceeds only in the presence of the base triethylamine, i.e. under conditions in which the ambident nucleophile is generated.

ENZYMATIC OXYGENATIONS OF PROTIC SUBSTRATES

The transformations described above constitute models for several oxidations (see Figure 10) catalyzed by copper-dependent mono- and dioxygenases [2,3]. Reactions (18) and (19), for example, are models for trypophan dioxygenase (EC 1.13.11.11) and quercetin dioxygenase (EC 1.13.11.24), respectively. Other examples include dopamine monooxygenase (EC 1.14.17.1) and tyrosinase (EC 1.14.18.1).

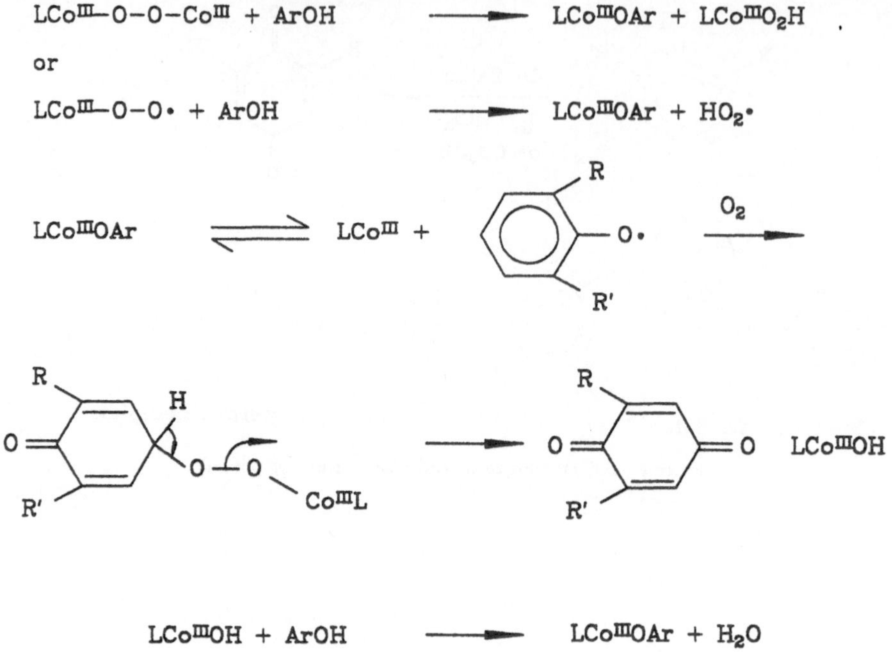
LCo^III–O–O–Co^III + ArOH → LCo^III OAr + LCo^III O2H
or
LCo^III–O–O• + ArOH → LCo^III OAr + HO2•
LCo^III OAr ⇌ LCo^III +
O•
O2
R
R'
H
O
Co^III L
LCo^III OH
LCo^III OH + ArOH → LCo^III OAr + H2O

Figure 9. Mechanism of Co(II) Salen-catalyzed oxygenation of phenols.

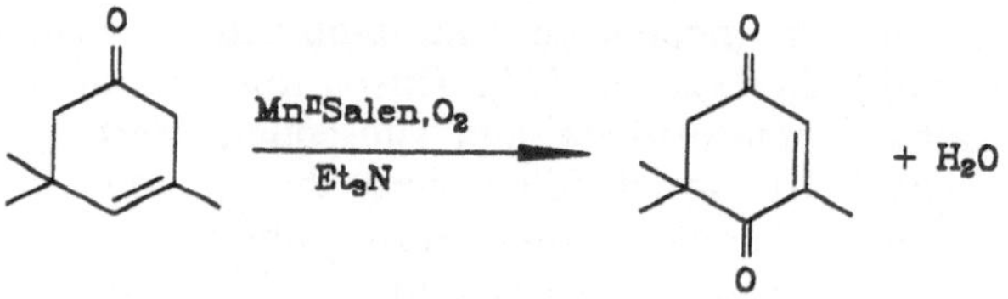
Mn^II Salen, O2
Et3N
+ H2O
(15)

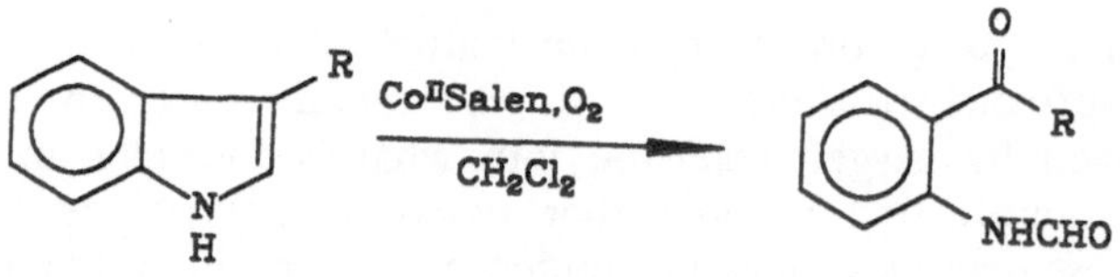
Co^II Salen, O2
CH2Cl2
NHCHO
(16)

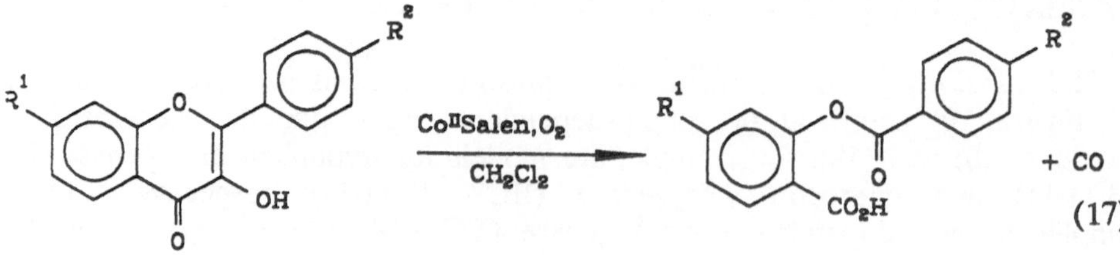
Co^II Salen, O2
CH2Cl2
OH
CO2H
+ CO
(17)

By analogy with the model systems we suggest that a common feature of all these enzymatic oxygenations is S_N2 displacement by the ambient nucleophile followed by oxygen insertion into a (formally) organocopper(II) species (reactions 18 and 19). In enzymatic systems the superoxometal complex ($MO_2\cdot$) is a more likely intermediate than the μ-peroxometal species (MOOM) due to steric constraints imposed by the protein ligand. Hence, we conclude that the distinction between a mono- and a dioxygenase is quite arbitrary and is determined by the fate of the organoperoxymetal (RO_2M) intermediate. Viewed from a mechanistic standpoint they may belong to the same class of reaction.

$$RH + MO_2\cdot \longrightarrow RM + HO_2\cdot \tag{18}$$

$$RM + O_2 \longrightarrow RO_2M \longrightarrow Products \tag{19}$$

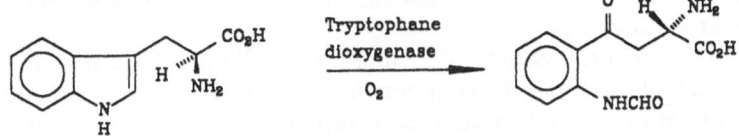

Figure 10. Oxygenation of protic substrates mediated by Cu-dependent oxygenases.

ENZYMATIC OXYGENATIONS OF HYDROCARBONS

In the case of unactivated, aprotic substrates (alkanes, alkenes and arenes) nucleophilic displacement at the metal is unfavorable and nature has had to find a different pathway for catalytic oxidation with dioxygen. A common feature of (almost) all of these systems is the intermediacy of a high-valent oxoiron species as the active oxidant. Both nonheme- and heme-dependent oxygenases are known. Examples of the former include methane monooxygenase [35], which mediates the selective oxygenation of methane to methanol and isopenicillin synthase which catalyzes a key step in the biosynthesis of penicillins [36]. The most well-known examples of the latter are the cytochrome P450-dependent monooxygenases [37-39] that mediate an amazing variety of oxidative in vivo transformations including olefin epoxidation and the hydroxylation of alkanes and arenes. Many of these processes are important steps in biosynthetic pathways, e.g. steroid hormone and prostaglandin biosynthesis, and the catabolism of foreign substances in the body.

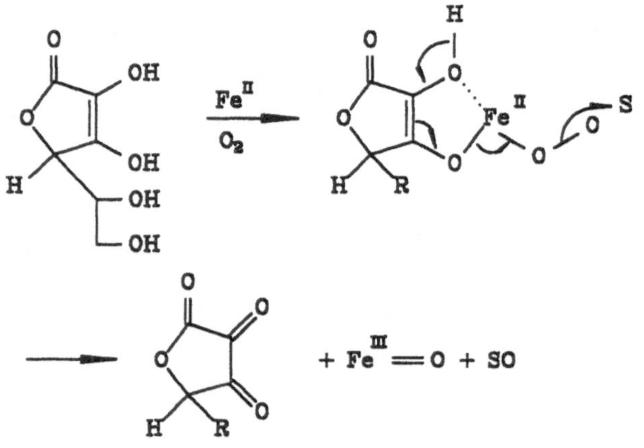

Figure 11. The 'oxenoid' mechanism.

In 1964 Hamilton [40] proposed an 'oxenoid' mechanism (Figure 11) for the hydroxylation of aromatics with Udenfriend's reagent (see earlier) which was considered a model for iron-dependent monooxygenases.   In hindsight it was a small step from the 'oxenoid' to the oxometal mechanism that is now widely accepted for cyt-P450 dependent monooxygenases. To our knowledge Ullrich and Staudinger [41] were the first to propose a formally oxoiron(V) heme (protoporphyrin IX) as the active oxidant. The widely accepted mechanism of oxygen transfer is illustrated in Figure 12.

In the 1970's several groups, e.g. those of Collman, Baldwin, Traylor and Momenteau [42] carried out elegant studies on model systems for the oxygen transport hemeproteins, hemoglobin and myoglobin. A primary aim of these studies was to prevent further reaction of the iron-dioxygen complex. As can be seen in Figure 12 a source of protons and electrons is needed in order to generate the active oxoiron(V) oxidant from the iron(II) dioxygen complex. In vivo they are provided by the cofactor NADPH.

METAL-CATALYZED OXYGEN TRANSFER

At about the same time that the model oxygen transport studies were being carried out other authors [43,44], notably Hrycay and coworkers [43], demonstrated that liver microsomal cytochrome P450 can catalyze the hydroxylation of hydrocarbon substrates using a variety of single oxygen donors, e.g. H_2O_2, RO_2H, chlorite, periodate and iodosylbenzene [44] as the primary oxidant.

This pathway, which later became known as the 'peroxide shunt', provided a means for circumventing the need for a coreductant (cofactor) in such systems. In

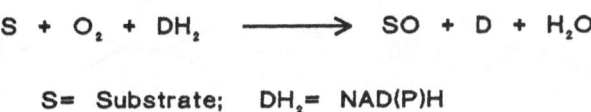

$$S + O_2 + DH_2 \longrightarrow SO + D + H_2O$$

$$S= \text{Substrate}; \quad DH_2= \text{NAD(P)H}$$

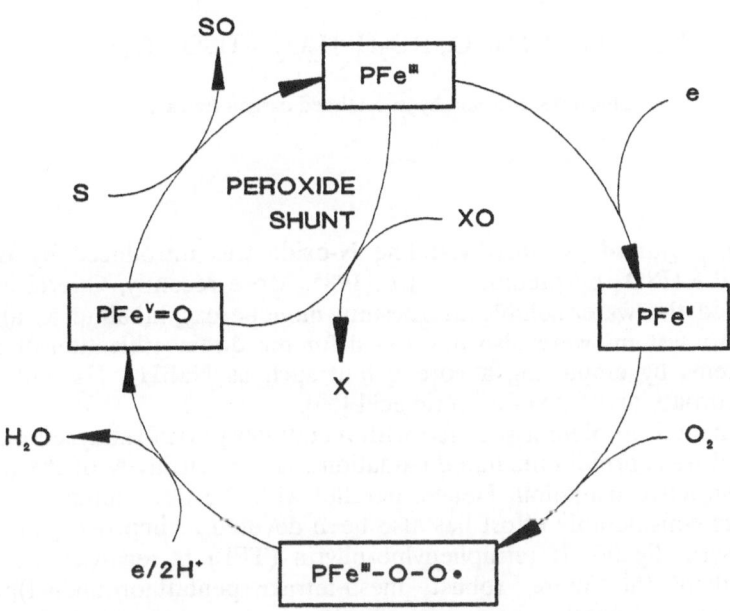

Figure 12. Mechanism of cyt-P450 catalyzed oxidations.

1979 Groves and coworkers [45] were the first to translate these results to a model system. They described the use of iron(III) meso-tetraphenylporphyrin (TPP) chloride in combination with iodosylbenzene for the epoxidation of olefins and hydroxylation of alkanes. Subsequently, chromium [46] and manganese [47] TPP complexes were shown to catalyze oxygen atom transfer from PhIO to an olefin or an alkane.

Following these seminal studies extensive investigations were carried out in the last decade on metalloporphyrin-catalyzed oxidations of olefins and alkanes (Figure 13) with a variety of single oxygen donors [48]. Manganese porphyrins which had been shown [47] to be superior to iron and chromium were generally the catalyst of choice. For example, several groups [49] developed the use of NaOCl in a biphasic (dichloromethane/water) system in the presence of a phase transfer catalyst. Cumyl hydroperoxide [50] and hydrogen peroxide [51] were introduced by Mansuy in 1984 and 1985, respectively. The presence of imidazole as an axial ligand was shown [50-52] to be essential for good performance. This was attributed [52] to its dual function as a stabilizing ligand and as a base in promoting the heterolysis of the M^{III}O-OH bond to form the putative M^V=O intermediate.

$$M = Fe, Cr, Mn \quad P = porphyrin$$

$$XO = PhIO, NaOCl, RO_2H, H_2O_2, KHSO_5, R_3NO$$

Figure 13. Mn porphyrin catalyzed oxygen transfer.

Similarly, p-cyano-N,N-dimethylaniline N-oxide was introduced by Bruice in 1982 [53] and $KHSO_5$ by Meunier [54] in 1985. More recently, Querci and Ricci [55] introduced the water-soluble magnesium monoperoxyphthalate as an oxygen donor. Various systems were also developed for the direct utilization of dioxygen in these systems by employing a coreductant such as $NaBH_4$, H_2 and colloidal platinum, ascorbate and zinc and acetic acid [56].

A fundamental problem associated with metalloporphyrin-catalyzed oxidations, inherent in all hemeprotein-mediated oxidations, is the sensitivity of the porphyrin ligand to destructive oxidation. Hence, parallel with the development of suitable oxygen donors considerable effort has also been devoted to improving the stability of the porphyrin ligand. If tetraphenylporphyrin (TPP) is regarded as the first generation then the more robust meso-tetrakis(pentafluorophenyl)porphyrin (TPFPP), meso-tetramesitylporphyrin (TMP) and meso-tetrakis(2,6-dichloro-phenyl)porphyrin (TDCPP) represent the second generation [48]. In the third generation ligands the stability is increased even further by replacing the hydrogens in the pyrrole rings by halogen, the ultimate example being the 'Teflon' ligand meso-tetrakis(pentafluorophenyl)-β-octafluoroporphyrin (F_8TPFPP) and related perhaloporphyrins [57].

Related epoxidations of olefins with PhIO in the presence of Salen and related complexes of chromium(III), manganese(III) and cobalt(III) have been reported by Kochi and coworkers [58]. The use of nickel(II) Salen in conjunction with NaOCl was also described [59]. More recently, these systems formed the basis for the development, by Jacobsen and coworkers [60], of chiral manganese(III) Salen complexes for the enantioselective epoxidation of prochiral olefins by ArIO or NaOCl. Similarly, asymmetric epoxidations with moderate to good enantioselectivities have also been described using chiral porphyrin ligands [61].

FROM GIF[I] TO GOAGG[III] AND BEYOND

Parallel with the frenetic activity in studies of model metalloporphyrins Barton and coworkers were busily developing model systems for nonheme iron-dependent monooxygenases [62]. The initial system reported in 1983 [63] consisted of iron(II) and dioxygen in pyridine solvent in combination with iron powder and acetic acid as a source of electrons and protons, respectively. This Gif[I] system was shown to selectively oxidize alkanes, showing an unexpected marked preference for oxidation of secondary C-H bonds to the corresponding ketones without the intermediacy of the corresponding alcohol. Further evolution culminated in the GoAGG[III] system comprising dipicolinic acid as a ligand for iron [62,64] and replacement of the dioxygen/coreductant with hydrogen peroxide as a single oxygen donor. The corresponding secondary alkyl hydroperoxide was shown to be a reaction intermediate. Quite surprisingly, it was shown that the oxygen in the hydroperoxide was derived from dioxygen formed by decomposition of H_2O_2 at some stage in the reaction. Barton [62] favors a mechanism involving insertion of an oxoiron(V) species into the secondary C-H bond followed by reaction of the alkyliron(V) intermediate with H_2O_2 to produce O_2 and an alkyliron(III) species. Insertion of O_2 into the latter affords an alkylperoxyiron(III) intermediate which decomposes to the ketone product (Figure 14). The preference for secondary vs tertiary C-H bonds was assumed to be due to steric control in the reaction of a bulky oxoiron(V) complex with the hydrocarbon substrate.

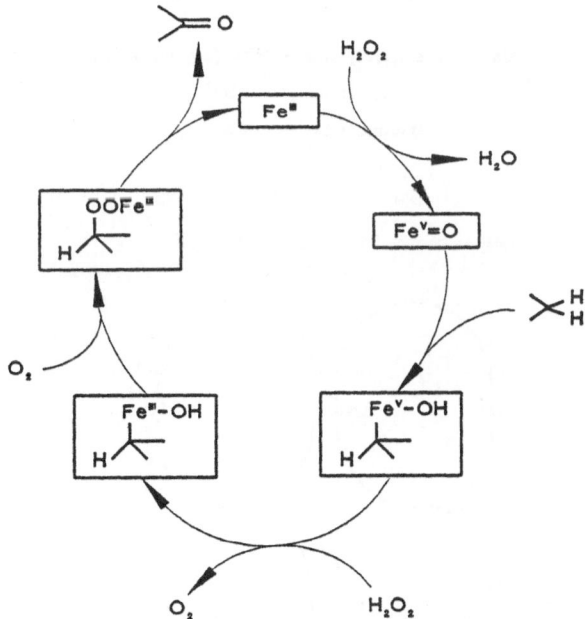

Figure 14. Mechanism of GoAGG[III].

DIRECT OXIDATION WITH DIOXYGEN IN MODEL SYSTEMS

The numerous model systems described in the preceding sections all involve, as do the in vivo systems, either dioxygen in conjunction with a coreductant as a source of electrons and protons, or a single oxygen donor. There still remains a definite need for direct oxygenation without the need for a coreductant. Two examples have been described in the literature which appear to fit this requirement. The first is the ruthenium tetramesitylporphyrin (Ru[II]TMP)-catalyzed epoxidation of olefins (Figure 15) reported by Groves and Quinn [65]. The second example is the selective hydroxylation of light alkanes (isobutane and propane) catalyzed by iron(III) perhalogenated porphyrins reported by Ellis and Lyons [66]. The suggested mechanism (Figure 16) is an example of a Mars-van Krevelen type mechanism operating in the liquid phase.

CONCLUDING REMARKS - FUTURE PROSPECTS

The quest for selective catalysts for the 'dream reactions' discussed at the beginning of this article continues unabated. There is still a great need for systems that create gas-phase conditions in the liquid phase. One approach is maybe to isolate redox metal ions, by isomorphous substitution, in the lattice of molecular sieves [67]. Such 'redox molecular sieves' may be viewed as 'inorganic enzymes' containing an active site in which there is no room for solvent molecules in addition to the substrate, i.e. gas phase conditions in the liquid phase.

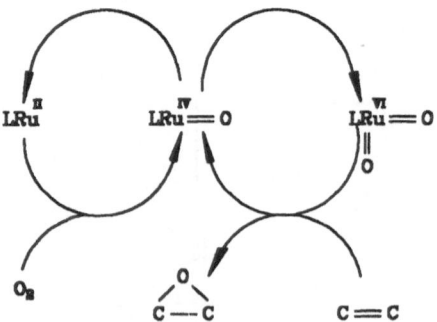

cis— —methylstyrene — 33% (cis epoxide)

norbornene — 43%

cyclooctene — 26%

TMP = meso—tetramesitylporphyrin

Figure 15. Ru(TMP)-catalyzed epoxidation with O_2.

$$(CH_3)_3CH + O_2 \xrightarrow[80^\circ, 3 \text{ h}]{[LFe^{III}X]} (CH_3)_3COH$$

L	X	CONV. (%)	SEL. %	TON
$TPPF_{20}$	OH	17	87	11300
$TPPF_{20}$-β-Br_8	Cl	28	83	17150

TON = Turnover number

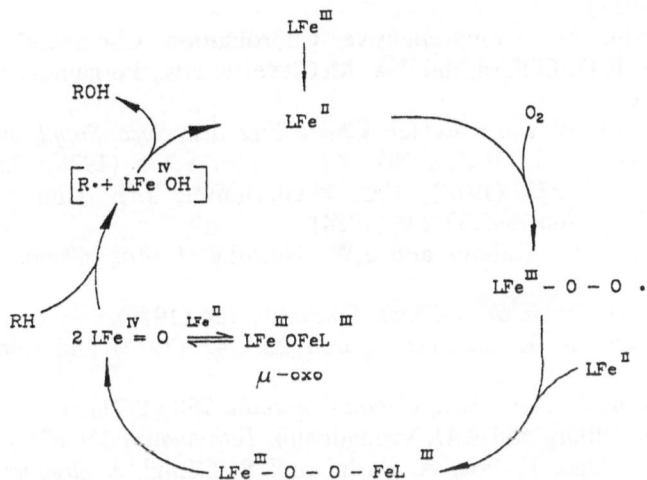

Figure 16. Fe($TPPF_{20}$)-catalyzed hydroxylation of alkanes by O_2.

Another approach is to improve the performance of heme-dependent oxidoreductases which could then provide practical methods for asymmetric oxidations. Thus, one should not forget that the protein component of a redox enzyme is a relatively inexpensive, complex chiral ligand.

REFERENCES

1. R.A. Sheldon and J.K. Kochi, "Metal-Catalyzed Oxidations of Organic Compounds", Academic Press, New York, 1981.
2. O. Hayaishi, ed., "Oxygenases", Academic Press, New York, 1962.
3. O. Hayaishi, ed., "Molecular Mechanisms of Oxygen Activation", Academic Press, New York, 1974.
4. L.I. Simandi, "Catalytic Activation of Dioxygen by Metal Complexes", Kluwer, Amsterdam, 1992.
5. D.T. Sawyer, "Oxygen Chemistry", Oxford University Press, Oxford, 1991.
6. A.E. Martell and D.T. Sawyer eds., "Oxygen Complexes and Oxygen Activation by Transition Metals", Plenum, New York, 1987; M. Chanon, M. Julliard, J. Santamaria and F. Chanon, New J. Chem. 16:171-201 (1992).
7. J.L. Bolland, *Q. Rev. Chem. Soc.* 3:1 (1949); L. Bateman, *ibid.* 8:147 (1954).
8. O. Hayaishi, M. Katagari and S. Rothberg, *J. Am. Chem. Soc.* 77:5450 (1950).
9. H.S. Mason, W.L. Fowlks and E. Peterson, *J. Am. Chem. Soc.* 77:2914 (1955).
10. S. Udenfriend, C.T. Clark, J. Axelrod and B.D. Brodie, *J. Biol. Chem.* 208:731 (1954).
11. H. Mimoun, *in*: "Comprehensive Coordination Chemistry", Vol. 6, G. Wilkinson, R.D. Gillard and J.A. McCleverty, eds., Pergamon, Oxford, 1987, pp. 317-410.
12. P. Mars and D.W. van Krevelen, *Chem. Eng. Sci., Spec. Suppl.* 3:41-57 (1954).
13. For reviews see: L. Vaska, *Acc. Chem. Res.* 9:175 (1976); J.S. Valentine, *Chem. Rev.* 73:235 (1973); E.C. Niederhoffer, J.H. Timmons and A.E. Martell, *Chem. Rev.* 84:137-203 (1984).
14. J.P. Collman, M. Kubota and J.W. Hosking, *J. Am. Chem. Soc.* 89:4809 (1967).
15. E.W. Stern, *J. Chem. Soc., Chem. Commun.* 736 (1970).
16. H. Kropf and K. Knaak, *Tetrahedron* 28:1143 (1972) and references cited therein.
17. R.A. Sheldon, *J. Chem. Soc., Chem. Commun.* 788 (1971).
18. W.J.T. van Tilborg and A.D. Vreugdenhil, *Tetrahedron* 31:2825 (1975).
19. A. Fusi, R. Ugo, F. Fox, A. Pasini and S. Cenini, *J. Organometal. Chem.* 22:219 (1970).
20. R.A. Sheldon and J.A. van Doorn, *J. Organometal. Chem.* 94:115 (1975).
21. C. Dudley, G. Read and P.J.C. Walker, *J. Chem. Soc., Dalton Trans.* 883 (1977).
22. H. Mimoun, *J. Mol. Catal.* 7:1 (1980); H. Mimoun, M.M. Perez Machirant and I. Seree de Roch, *J. Am. Chem. Soc.* 100:5437 (1978).
23. R.S. Drago, *J. Am. Chem. Soc.* 107:2898 (1985).
24. D.E. Hamilton, R.S. Drago and A. Zombek, *J. Am. Chem. Soc.* 109:374 (1987).
25. A. Nishinaga, *Tetrahedron Lett.* 29:6309 (1988).
26. For an excellent recent review see R. Drago, *Coord. Chem. Rev.* 117:185-213 (1992).
27. M. Pizzotti, S. Cenini and G. La Monica, *Inorg. Chim. Acta* 33:161 (1978).
28. H.M. van Dort and H.J. Geursen, *Recl. Trav. Chim. Pays-Bas* 86:520 (1967).

29. A. Nishinaga and H. Tomita, *J. Mol. Catal.* 7:179 (1980).

30. V.M. Kothari and J.J. Tazuma, *J. Catal.* 41:180 (1976).

31. C.L. Bailey and R.S. Drago, *Coord. Chem. Rev.* 79:321 (1987).

32. M. Constantini, A. Dromard, M. Jouffret, B. Brossard and J. Varagnat, *J. Mol. Catal.* 7:89 (1980).

33. A. Nishinaga, *Chem. Lett.* 273 (1975).

34. A. Nishinaga, T. Tojo and T. Matsura, *J. Chem. Soc., Chem. Commun.* 896 (1974).

35. J. Green and H. Dalton, *J. Biol. Chem.* 264:17698 (1989).

36. J.E. Baldwin, in "Recent Advances in Chemistry of Beta-Lactam Antibiotics", A.G. Brown and S.M. Roberts, eds., The Royal Society of Chemistry, 1985, pp. 62-85.

37. P.R. Ortiz de Montellano, ed., "Cytochrome P-450: Structure, Mechanism and Biochemistry", Plenum, New York, 1986.

38. P.N. White, *Bioorg. Chem.* 18:440-456 (1990).

39. M.J. Gunter and P. Turner, *Coord. Chem. Rev.* 108:115-161 (1991).

40. G.A. Hamilton, *J. Am. Chem. Soc.* 86:3391 (1964).

41. V. Ullrich and H.J. Staudinger, *in*: "Biological and Chemical Aspects of Monooxygenases", K. Block and O. Hayaishi, eds., Maruzen, Tokyo, 1966, pp. 235-249.

42. For key references see J.P. Collman, T.R. Halpert and K.S. Suslick, *in*: "Metal Ion Activation of Dioxygen", T.G. Spiro, ed., Wiley, New York, 1980, pp. 1-72; M. Momenteau, *Bull. Soc. Chim. Belg.* 100:731 (1991).

43. E.G. Hrycay, J.A. Gustafsson, M. Ingelman-Sundberg and L. Ernster, *Biochem. Biophys. Res. Commun.* 66:209 (1975) and references cited therein; see also G.D. Nordblum, R.E. White and M.J. Coon, *Arch. Biochem. Biophys.* 175:524 (1976).

44. F. Lichtenberger, W. Nastainczyk and V. Ullrich, *Biochem. Biophys. Res. Commun.* 70:939 (1976).

45. J.T. Groves, T.E. Nemo and R.S. Myers, *J. Am. Chem. Soc.* 101:1032-1033 (1979); see also J.T. Groves, W.J. Kruper, T.E. Nemo and R.S. Myers, *J. Mol. Catal.* 7:169-177 (1980); C.K. Chang and M.S. Kuo, *J. Am. Chem. Soc.* 101:3413 (1979).

46. J.T. Groves and W.J. Kruper, *J. Am. Chem. Soc.* 101:7613 (1979).

47. C.L. Hill and B.C. Schardt, *J. Am. Chem. Soc.* 102:6374 (1980); J.T. Groves, W.J. Kruper and R.C. Haushalter, *J. Am. Chem. Soc.* 102:6375 (1980).

48. For an excellent recent review see B. Meunier, *Chem. Rev.* 92:1411-1456 (1992); see also D. Mansuy, *Pure Appl. Chem.* 59:759 (1987); K.A. Jorgensen, *Chem. Rev.* 89:431 (1989); D. Ostovic and T.C. Bruice, *Acc. Chem. Res.* 25:314-320 (1992).

49. I.Tabushi and N. Koga, *Tetrahedron Lett.* 3681 (1979); E. Guilmet and B. Meunier, *Tetrahedron Lett.* 4449 (1980); J.P. Collman, T. Kodadek, S.A. Raybuck and B. Meunier, *Proc. Natl. Acad. Sci. USA* 80:7039 (1983); J.A.S.J. Razenberg, R.J.M. Nolte and W. Drenth, *J. Chem. Soc., Chem. Commun.* 277 (1986); S. Banfi, F. Montanari and S. Quici, *J. Org. Chem.* 54:1850 (1989).

50. D. Mansuy, P. Battioni and J.P. Renaud, *J. Chem. Soc., Chem. Commun.* 1255 (1984).

51. D. Mansuy, P. Battioni, J.P. Renaud and J.F. Bartoli, *J. Chem. Soc., Chem. Commun.* 888 (1985).

52. D. Mansuy, P. Battioni, J.P. Renaud, J.F. Bartoli and M. Reina-Artiles, *J. Am. Chem. Soc.* 110:8462 (1988); see also L.C. Yuan and T.C. Bruice, *J. Am. Chem. Soc.* 108:1643 (1986).

53. T.C. Bruice and M.W. Nee, *J. Am. Chem. Soc.* 104:6123 (1982).
54. B. Meunier, B. de Poorter and M. Ricci, *Tetrahedron Lett.* 4459 (1985).
55. C. Querci and M. Ricci, *J. Chem. Soc., Chem. Commun.* 889 (1989).
56. For a review see I. Tabushi, *Coord. Chem. Rev.* 86:1 (1988).
57. S. Tsuchiya and M. Seno, *Chem. Lett.* 263 (1989); see also P. Battioni, O. Brigaud, H. Desvraux, D. Mansuy and T.G. Traylor, *Tetrahedron Lett.* 32:2893 (1991).
58. T.L. Siddall, N. Miyaura, J.C. Huffman and J.K. Kochi, *J. Chem. Soc., Chem. Commun.* 1185 (1983); K. Srinivasan, P. Michaud and J.K. Kochi, *J. Am. Chem. Soc.* 108:2309 (1986); J.D. Koola and J.K. Kochi, *J. Org. Chem.* 52:4545 (1987); E.G. Samsel, K. Srinivasan and J.K. Kochi, *J. Am. Chem. Soc.* 107:7606 (1985).
59. H. Yoon and C.J. Burrows, *J. Am. Chem. Soc.* 110:4087 (1988).
60. W. Zhang, J.L. Loeback, S.R. Wilson and E.N. Jacobsen, *J. Am. Chem. Soc.* 112:2801 (1990).
61. J.T. Groves and R.S. Myers, *J. Am. Chem. Soc.* 105:5791 (1983); S. O'Malley and T. Kodadek, *ibid.* 111:9116 (1989); Y. Naruta, F. Tani, N. Ishihara and K. Maruyama, *ibid.* 113:6865-6872 (1991).
62. See. D.H.R. Barton and D. Doller, *Acc. Chem. Res.* 25:504-512 (1992) and references cited therein; see also C. Knight and M.J. Perkins, *J. Chem. Soc., Chem. Commun.* 925 (1991).
63. D.H.R. Barton, M.J. Gastiger and W.B. Motherwell, *J. Chem. Soc., Chem. Commun.* 41-43 (1983).
64. See also H.C. Tung, C. Kang and D.T. Sawyer, *J. Am. Chem. Soc.* 114:3445 (1992) and references cited therein.
65. J.T. Groves and R. Quinn, *J. Am. Chem. Soc.* 107:5790 (1985).
66. P.E. Ellis and J.E. Lyons, *Catal. Lett.* 3:389-398 (1989); J.E. Lyons and P.E. Ellis, *Catal. Lett.* 8:45-52 (1991).
67. R.A. Sheldon, *CHEMTECH* 566-675 (1991) and references cited therein.

INDUSTRIAL PERSPECTIVES ON THE USE OF DIOXYGEN:

NEW TECHNOLOGY TO SOLVE OLD PROBLEMS

Dennis Riley, Michael Stern, and Jerry Ebner

The Monsanto Company
800 N. Lindbergh Blvd.
St. Louis, MO 63167

INTRODUCTION

The selective catalytic oxidation of organic molecules continues as a very important reaction pathway for the synthesis of primary and specialty chemicals in the chemical industry worldwide. Catalytic utilization of molecular oxygen using both soluble metal compounds in liquid reaction media (homogeneous catalysis) and the surfaces of metals or metal oxide compounds in gas or liquid reaction media (heterogeneous catalysis) is very important today, and will become even more important in the future as worldwide environmental policies become more stringent. This will necessitate the development of new "no-waste" technologies which will provide economically viable syntheses of molecules of commercial importance. Clearly, selective catalytic oxidation with O_2 represents critical technology and will be an area in which continued research and technical breakthroughs will be required.

The economic driving forces for catalytic O_2 oxygenation chemistry have historically been raw material costs, reaction efficiency and reaction simplicity; i.e., the fewest number of synthesis steps. For example, the emergence of butane as a lower cost feedstock than benzene in the United States in the late 1970's resulted in the development and commercialization of a butane based route to maleic anhydride and not until the 1990's did similar events occur in Europe. For similar reasons, propene based chemistry replaced acetylene based chemistry in the 1960's for acrylonitrile, and o-xylene replaced naphthalene for phthalic anhydride synthesis. Such economic factors will continue to be primary forces, but an additional factor has emerged in the 1990's - the environmental compatibility of the process. The environmentally responsible chemical industry wants to reduce the pounds of by-products produced per pound of product. For example, the stoichiometric inorganic oxidants, such as dichromate, permanganate, chlorite, and chlorate, employed primarily in the production of fine chemicals, produce aqueous waste streams contaminated with high levels of high molecular weight inorganic salts. These salt containing by-product streams often contain low levels of organics and are not suitable for direct biotreatment; thus, they represent a significant cost to clean up.

The selective oxidation of organic compounds utilizing molecular oxygen will continue to be an area of great potential for the chemical process industry. Not only are there many commercial processes which utilize oxygen, but there is a continuing development of new and improved processes using oxygen which are driven both by its abundance and low cost and by its potential to be an environmentally friendlier oxidant than other oxidants such as chlorine. It should be noted that O_2 is an oxidant that, depending upon the mechanism of its action, can function as a one-electron, two-electron, or even as a four-electron oxidant. In contrast, hydrogen peroxide functions as a two-electron oxidant and chlorine as a one-electron oxidant. Currently pure oxygen

The Activation of Dioxygen and Homogeneous Catalytic Oxidation,
Edited by D.H.R. Barton *et al.*, Plenum Press, New York, 1993

is the least expensive source of oxidizing equivalents available: ~0.05¢/mole of oxidation equivalents as a four-electron oxidant. Oxidations involving oxygen will ultimately generate non-salt by-products (e. g., water). Traditionally, chlorine has been used extensively in the chemical process industry, and while it is also relatively inexpensive 0.90¢/mole of oxidation equivalents, its use generates chloride salt streams which are very difficult to clean up before releasing back into the environment. In this era of increasing environmental awareness such considerations have increased the cost of doing business with many old style chemical oxidations. The incentives to use oxygen to replace such chemistries is increasing and this has sparked a resurgence in research in catalytic oxygen oxidations in the chemical process industry worldwide.

It should be noted that the use of hydrogen peroxide (commercially generated from O_2), while more expensive than O_2 or Cl_2 (~4.5¢/mole of oxidation equivalents), is continuing tobecome a more competitive source of clean oxygen for oxidations, especially for high value-added chemicals. With the increased useage of hydrogen peroxide as a replacement for hypochlorite in the bleaching of pulp in the paper industry (for environmental reasons), hydrogen peroxide cost will continue to drop relative to other oxidants. As a consequence, we anticipate that research into selective uses of hydrogen peroxide, while not discussed here, will also be important.

At Monsanto, there have been developed several processes utilizing O_2 and there exists a continuing interest into new ways to utilize O_2 as a selective oxidant in a number of areas of critical interest to Monsanto. This report will summarize some chemical processes practiced at Monsanto which utilize O_2 (including recent developments in the conversion of butane to maleic anhydride, production of sulfuric acid, production of acrylonitrile, production of cyclohexanol, and the synthesis of N-phosphonomethylglycine-the active ingredient in the herbicide Roundup®). In addition, several recent areas of technical activity will be described which show new ways to utilize O_2 to solve an important process issue. The reactions to be described include new oxidative O_2-driven coupling technology in which substituted aromatic amines can be synthesized from nitrobenzene without the need for prior chlorination of nitrobenzene. This new technology is applicable to the synthesis of p-phenylendiamines without the generation of salt wastes. Similarly, new O_2-driven oxidative coupling technology will be described which makes it possible to directly form sulfur-nitrogen bonds without the need for chlorine. New homogeneous catalyst technology for the highly selective synthesis of 4,4'-dicarboxybiphenyl from the 4,4'-di-t-butylbiphenyl is also described. Finally, the development of new catalyst technology for the homogeneously catalyzed O_2 oxidation of N-substituted amino acids is described. This last reaction features the novel use of electron-transfer additives to control in a highly selective fashion the O_2-driven oxidative conversion of N-phosphonoiminodiacetic acid to N-phosphonomethylglycine.

TECHNOLOGIES PRACTICED AT MONSANTO UTILIZING O_2

Oxidative transformations driven by oxygen and catalyzed by transition metal complexes play a very important role in the control of selective oxidations. The role of the catalyst in oxidations is to interrupt the pathway leading to the most favorable thermodynamic products, water and carbon dioxide, and to provide a "low temperature" pathway for the controlled formation of the desired product. Monsanto has commercialized several O_2-driven oxidations which illustrate extremely well the principle of providing a selective pathway to desired products.

Butane to Maleic Anhydride

No process better illustrates the role of a catalyst in promoting selectivity than the Monsanto butane to maleic anhydride process.[1] In 1966 the remarkable catalytic oxidation of

$$C_4H_{10} + 3.5\ O_2 \longrightarrow \text{(maleic anhydride)} + 4\ H_2O \qquad (1)$$

butane to maleic anhydride using a vanadium phosphorus oxide catalyst was reported by Bergman and Frisch,[2] and this topic has been the subject of numerous scientific publications and several important reviews.[3-6] The butane reaction represents a 14 electron oxidation reaction with the removal of eight hydrogen atoms and the insertion of three oxygen atoms. The literature reveals reaction selectivities in the 70% range with reaction conversions in the 80-87% range when reaction feeds are 2+ mole % butane in air and reaction temperatures are 673 - 723°K, and the primary selectivity loss is to carbons monoxide and dioxide.

$$C_4H_{10} + 6.5 \ O_2 \ \text{-->} \ 4 \ CO_2 + 5 \ H_2O \quad\quad (2)$$

$$C_4H_{10} + 4.5 \ O_2 \ \text{-->} \ 4 \ CO + 5 \ H_2O \quad\quad (3)$$

Vanadium phosphorus oxide catalysts are the only catalyst systems successfully commercialized for this reaction chemistry, akin in this respect to the uniqueness of silver for epoxidation of ethylene. One common observation in a large number of published reports is good catalytic activity/selectivity is observed when the crystalline phase vanadyl pyrophosphate, $(VO)_2P_2O_7$, is present. The catalytic performance of the vanadyl pyrophosphate, however, is profoundly influenced by the method of preparation. Catalysts synthesized in organic media, which tend to possess higher BET surface areas, are superior to those formed in aqueous media in activity and selectivity to product. The catalysts are generally prepared from $[VOHPO_4]_2 \cdot H_2O$ precursors with approximately 5-20% excess phosphorus. There is general agreement that precursors prepared with excess phosphorus lead to higher selectivity catalysts. The precursor compound is transformed into vanadyl pyrophosphate by thermal treatment above 623°K, often in a butane containing atmosphere. These catalysts, after being run for several hundred hours, equilibrate to the reaction environment and typically have a vanadium oxidation state of $4.01\pm.01$, a bulk phosphorus to vanadium ratio of $1.00\pm.025$, XPS surface atomic P/V ratios > 1.0, and BET surface areas of 15-20 m^2/gm.

Although an exact molecular level description of the reactive oxygen presented by the active surfaces of VPO catalysts remains to be described, Ebner and Gleaves[7] have conducted extensive studies on various equilibrated VPO catalysts using $^{18}O_2$ and the Temporal Analysis of Products (TAP) microreactor and showed surface lattice oxygen is utilized in the dehydrogenation and oxygen insertion reactions of unsaturated C_4's to form the furan ring. In addition to this oxygen type, the TAP results suggested another form of activated oxygen is required for cleaving the sp^3 C-H bond of butane. This species was suggested to arise from the irreversible dissociative adsorption of oxygen producing a surface vanadium (+5) site. Pump/probe TAP experiments detected an additional short lived oxygen species, and it was found to enhance the rate of transformation of furan to maleic anhydride. This species could be an adsorbed, partially reduced superoxo or peroxo dioxygen that reacts with ring intermediates to form maleic anhydride. It is now generally accepted in the literature that the vanadium phosphorus oxide catalysts operate according to the Mars - van Krevelen mechanism, but the role of partially reduced dioxygen species remains to be shown experimentally.

Many authors in the field have suggested the active site region resides on the microcrystalline (1,0,0) surfaces of vanadyl pyrophosphate, and increasing the exposure of this surface correlates well with increases in activity; ie., greater active site density. Ebner and Thompson[8] have described the active site region as an ensemble of up to four isolated vanadium centers in a surface cleft formed by pendent surface pyrophosphate groups. The pendent pyrophosphate groups that define and overhang the ensemble of vanadium sites in the surface cleft present a total of twelve hydrogen atom binding sites as surface -P-O- in the unprotonated form. Models reveal the proximity of the adjacent surface pyrophosphate oxygen anions to each other, and illustrate such a configuration could provide hydrogen acceptor sites for transport of abstracted hydrogen atoms from the surface cleft region to sites of water formation and desorption.

Propene to Acrylonitrile

Idol first reported in 1959 that bismuth molybdenum oxides, compounds found in the Sohio patents for ammoxidation of propene to acrylonitrile, selectively oxidize propene to acrolein. Since that original discovery, Monsanto has developed its own proprietary catalyst composition for ammoxidation of propene to acrylonitrile.

$$2 \ CH_2=CHCH_3 + 3 \ O_2 + 2 \ NH_3 \ \text{---->} \ 2 \ CH_2=CHCN + 6 \ H_2O \quad\quad (4)$$

In the patent literature, numerous catalyst formulations for propene ammoxidation can be found, and the majority of these are formulated around molybdenum and/or antimony base oxides. The most effective catalysts for these reactions are complex metal oxide mixtures containing three to six metal components on a silica support, and result in propene based yields to acrylonitrile >80%. Air is the source of dioxygen for these reactions, and fluid bed reactors are most commonly used. The observation that selective ammoxidation of propene occurs in the presence or absence of molecular oxygen implicated the oxide of the catalyst structure as the source of selective oxygen. A redox mechanism accounting for this observation was first proposed by Mars-van Krevelen.[9] Many experiments conducted in laboratories around the globe using isotopically labeled dioxygen, $^{18}O_2$, have unambiguously proven that in the ammoxidation reaction surface lattice oxide is the primary player in removing hydrogens through C-H bond cleavage to produce water. The reduced surface sites are rapicly reoxidized by lattice oxygen, not by O_2. An electron-rich O vacancy passes through the structure to a separate surface site where replenishment occurs by activation and dissociation of O_2. This process is very fast compared to the reduction step, and rapid reconstitution of the subsurface and surface is believed critical to high selectivity catalysts. The multicomponent oxide catalyst systems have resulted from fine tuning this important redox property. In the ammoxidation reaction in which N insertion leads to formation of acrylonitrile, reactive M-NH groups are formed by ammonolysis of M-O groups, and these are believed to be the source of inserting N atoms.[10]

Sulfur Dioxide to Sulfur Trioxide (Sulfuric Acid)

Sulfuric acid, the largest volume commodity chemical produced, is synthesized by the catalytic oxidation of sulfur dioxide, derived from combustion of sulfur or hydrogen sulfide.[11] The sulfur trioxide product is hydrated to form sulfuric acid.

$$SO_2 + 0.5\ O_2 \longrightarrow SO_3 \tag{5}$$

$$SO_3 + H_2O \longrightarrow H_2SO_4 \tag{6}$$

Since the 1920's vanadium based catalysts have been used for this reaction chemistry, and after 70 years and over 1000 papers and patents, it seems reasonable to conclude that vanadium-based systems are uniquely well suited for this reaction chemistry. Monsanto, currently through Monsanto Enviro-Chem Systems, Inc., has been providing sulfuric acid catalysts and sulfuric acid plant engineering services for this industry for over 40 years.

Because the reaction of SO_2 to SO_3 is highly exothermic, the equilibrium becomes more unfavorable as temperature rises. In fact, since the catalysts must run at temperatures above 400°C, the equilibrium becomes problematic. Thus, multistaged adiabatic fixed bed reactor units are preferred to allow for interstage cooling, and oftentimes interstage SO_3 adsorption (double adsorption process). These engineering design features, which are incorporated to fight the reaction equilibrium problem, allow plants to operate at conversions of 99+%; a conversion level required to meet modern day worldwide air quality standards. Indeed, the need for plants to run at high conversion has created a niche for more expensive, high activity catalyst formulations, such as the Monsanto cesium promoted catalysts, which possess higher activities, and therefore reduce bed inlet temperatures leading to higher reaction conversions and reduced SO_2 emissions. For example, in a single adsorption sulfur burning plant, when beds four and five are replaced with a cesium containing formulation, and the reactors are operated at 410 versus 430°C, the conversion is raised from 98 to 99%, resulting in a 50% reduction in SO_2 emissions.

The typical catalyst for this reaction contains 6-9% V_2O_5, 6-12% M_2O (M=Na, K, Cs, Rb with M predominantly K), and 60-75% SiO_2. Most commercial catalysts have a K:V ratios of about 2 to 3.5. An important feature of this catalyst is that at reaction temperatures the actives operate in a molten-salt mixture supported on a porous silica pellet. The molten salt mixture is composed of vanadium oxides dissolved in alkali metal pyrosulfates. The physical parameters and thickness of the melt, which depend on temperature and gas compositions, are important to performance. It is important to maintain vanadium in the +5 oxidation state. The mechanism of the chemical reactions on the surface are not completely and unambiguously understood, largely because of the complexity of the liquid molten salt mixture under the wide range of conditions encountered in commercial reactors. Boreskov[12] proposes a binuclear complex of V(+5) binds two SO_2 molecules in the coordination sphere and dioxygen reacts with these SO_2 molecules to produce two SO_3 molecules. An alternate pathway, which involves the oxygen bound to the V (+5) oxidizing bound SO_2 to SO_3 with formation of V(+4), may predominate at low SO_2 conversion.

Cyclohexane to Cyclohexanol

The oxidation of cyclohexane to a cyclohexanone/cyclohexanol mixture (K/A oil) represents the first step in the two step process for the manufacture of adipic acid, a key monomer for the production of nylon. There are two major variants in this chemistry depending upon whether the autoxidation uses a metal catalyst such as Co or Mn or whether the direct autoxidation is carried out in the absence of a metal catalyst. The technology practiced at Monsanto, developed by Scientific Design (a division of Halcon International) utilizes boric acid as a stoichiometric reagent to trap either cyclohexanol or cyclohexyl hydroperoxide as a stable borate ester (Figure 1),[13] thereby preventing overoxidation yielding chain cleavage products. This chemistry is generally run to a few percent conversion (3-4%) in a continuous process (at 165-170°C and under 110-140 psig O_2 pressure) in order to achieve high selectivity to cyclohexanol (e. g., the cyclohexanol/cyclohexanone~12 to 1) with overall selectivity to $C_6H_{11}OH + C_6H_{10}O \sim 94\%$.

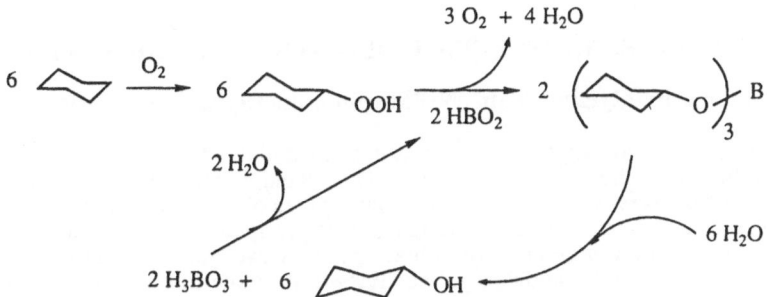

Figure 1. Boric acid promoted oxidation of cyclohexane to cyclohexanol.

The metaboric acid is fed to the oxidation train continuously and the mole ratio of boron added to O_2 utilized is kept in the 0.65 to 1 range. The primary role of the metaboric acid is to esterify the cyclohexanol, thereby preventing selectivity robbing overoxidation. The boric acid also serves to catalyze the de-peroxidation of the cyclohexylhydroperoxide to cyclohexanol in high yield (~95%) at the expense of other uncatalyzed decomposition products such as cyclohexanone. This effect arises from the ability of the boron compound to reduce the intermediate hydroperoxide to the corresponding cyclohexyl borate ester, dioxygen, and water (Scheme 1).[14]

Historically, Monsanto made cyclohexanol via reduction of phenol produced by the oxidation of cumene. The cheaper feedstock is cyclohexane and this has clearly driven the technology to the utilization of the low cost feed.

N-Phosphonomethylglycine from N-Phosphonomethyliminodiacetic Acid.

The utilization of oxygen as an oxidant for the synthesis of the commercially important amino acid N-phosphonomethylglycine (PMG or glyphosate), the active agent in the herbicide Roundup®, has been studied intensively at Monsanto. In the first few years of commercial production of glyphosate the process utilized an oxidative decarboxylation of N-phosphonomethyliminodiacetic acid which was driven by hydrogen peroxide.[15] As volumes increased and the need for a more cost efficient process became evident, research into the use of O_2 to drive the decarboxylation step was initiated. The catalytic route that has been successfully commercialized uses an activated carbon as a heterogeneous catalyst and this technology has been described in the patent literature.[16]

Water is the choice of solvent in this reaction, not only because it is oxidation resistant and inexpensive, but also because both PMIDA and PMG are virtually insoluble in all organic solvents. A key aspect of this technology is that the catalyst must give a high selectivity (>95%) at high substrate conversion (>99%) under commercially relevant conditions (<100°C and less than 100 psig O_2 pressure). This requirement arises because both PMG and PMIDA are so similar in their solubility properties; consequently, purification of the product would be very difficult without a very high conversion process. Another important consideration in the drive to reduce process costs and to maximize efficiency of this conversion is the need to minimize the volume of solvent; i. e., utilize as high a substrate payload as possible. Additionally, to minimize reagent costs and process operations, it is desireable to run with the free acid form of the substrate at its natural unbuffered pH (1-2). This presents the problem that both PMG and PMIDA are only sparingly soluble under such

$$H_2O_3PCH_2N(CH_2CO_2H)_2 \xrightarrow[\text{Carbon}]{O_2} H_2O_3PCH_2NH(CH_2CO_2H) + HCO_2H + H_2CO \qquad (7)$$

conditions, even at elevated temperatures. As a consequence, the heterogeneous catalyst scenario presents a problem; namely, the catalyst must be separated from the product by filtration of a dilute PMG solution. Also, to crystallize the product PMG requires subsequent ,removal of water-an energy intensive step.

RECENT ADVANCES IN OXYGEN UTILIZATION AT MONSANTO

N-Phosphonomethylglycine from N-Phosphonomethyliminodiacetic Acid

An obvious way to simplify the process described above for the conversion of PMIDA to PMG would be to use a homogeneous catalyst. This would, in principle, make it possible to carry out this catalytic conversion with the use of a high payload of the substrate. In fact, slurries would be doable in such a system since catalyst fouling due to precipitation or crystallization on and in the catalyst would not be a problem with a homogeneous catalyst. For theses reasons we have discovered and developed homogeneous catalysts to promote this selective oxidation in high conversion. The obvious process advantage of a homogeneous system lies in the ability to oxidize a high payload of substrate and simply filter off the solid product PMG. Recycle of the filtrate containing the catalyst back to the oxidation reactor offers great process saving over the heterogeneous catalyst process and would, in principle, be a much easier process to operate.

In our studies of homogeneous catalysts for the oxygen-driven conversion of PMIDA to PMG, we have discovered that the reaction is catalyzed by V(IV,V)[17] salts and by Co(II,III)[18] salts. The rate-determining steps in this catalytic chemistry is the oxidation of the reduced metal(PMIDA) complex with O_2 to produce hydrogen peroxide and either the V^V(PMIDA) or Co^{III}(PMIDA) complex. Both metals oxidize the bound carboxylate to yield an N-methylene radical which is then trapped by oxygen to yield the product via the formation of the N-formylPMG followed by its subsequent hydrolysis to PMG. The chemistry is only selective to the desired product if oxygen is present in high concentration (pressures >1500 psig) so as to trap the N-methylene radical (Figure 2). Competing H-atom abstraction yields the undesired N-MePMG.

Oxygen in this system functions in the dual role of not only driving the redox chemistry of the catalyst, but it also functions to intercept the N-methylene radical intermediate. This catalytic chemistry is not commercially viable at the pressures required for good selectivity, but the use of a co-catalyst which could be oxidized by O_2 rapidly and which could oxidize the radical offered the possibility of affording a catalytic process capable of operating at low pressure. Such a co-catalyst would eliminate the need for oxygen trapping of the intermediate N-methylene radical. Our study of co-catalysis has shown that redox active metal ions either have no effect on this chemistry or that they poison the reaction completely; e. g., iron or copper salts. We have discovered that derivatives of anthraquinone or methylviologen which function effectively as organic electron-transfer agents are very effective agents for increasing the selectivity to PMG in these systems when the reactions are performed under low oxygen pressure. Electron-transfer agents such as methylviologen and water soluble anthrquinones are able to oxidize the N-methylene radical to yield the iminium cation which hydrolyzes to PMG plus formaldehyde (Figure 3). The one-electron reduction product of the electron-transfer agent is then oxidized rapidly by O_2 to regenerate the oxidized form. This remarkable effect requires a catalytic amount of electron-transfer agent approximately equal to the amount of metal salt catalyst to achieve very high selectivities (>94%) to product PMG at modest pressures of O_2 (<200 psig). Since O_2 is a very efficient oxidant of of the

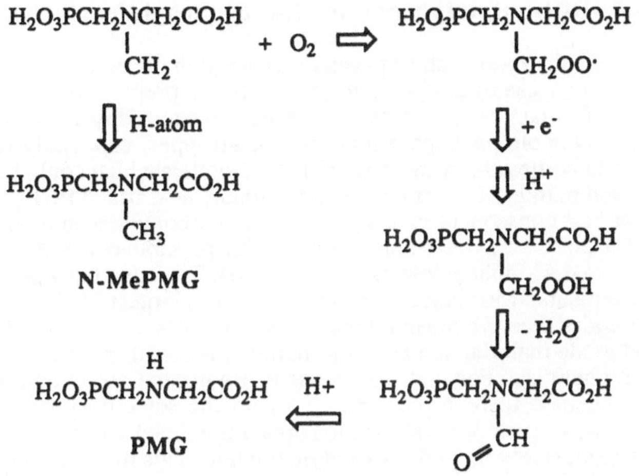

$$H_2O_3PCH_2NCH_2CO_2H$$
$$\overset{|}{CH_2\cdot}$$
$$+ \ O_2 \ \Rightarrow$$
$$H_2O_3PCH_2NCH_2CO_2H$$
$$\overset{|}{CH_2OO\cdot}$$

$$\Downarrow \ H\text{-atom}$$

$$\Downarrow \ + e^-$$

$$H_2O_3PCH_2NCH_2CO_2H$$
$$\overset{|}{CH_3}$$

N-MePMG

$$\Downarrow \ H^+$$

$$H_2O_3PCH_2NCH_2CO_2H$$
$$\overset{|}{CH_2OOH}$$

$$\Downarrow \ - H_2O$$

$$\overset{H}{\underset{}{H_2O_3PCH_2NCH_2CO_2H}}$$

$$\overset{H+}{\Longleftarrow}$$

$$H_2O_3PCH_2NCH_2CO_2H$$
$$\overset{|}{\underset{O=}{CH}}$$

PMG

Figure 2. Homogeneous catalyst mechanism for the oxidation PMIDA to PMG.

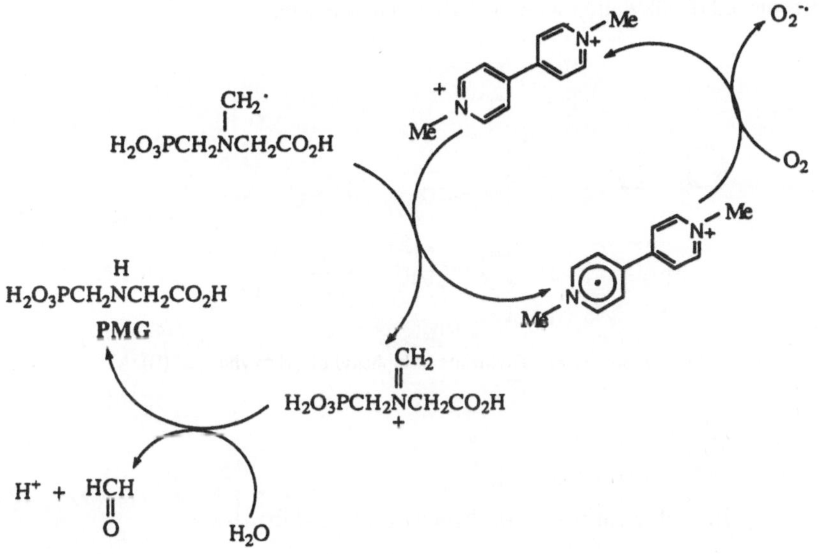

Figure 3. Electron-transfer agent promoted catalytic oxidation of PMIDA.

one-electron-reduction product of the methylviologen,[20,21] O_2 remains as the ultimate oxidant in these systems. In addition, such electron-transfer agents show excellent stability in these systems: undergoing repeated recycles with no loss in integrity.

The use of electron-transfer agents as co-catalysts for the interception of an intermediate in an oxygen-driven oxidation is an important concept and should have potential for lowering the pressures necessary for molecular oxygen oxidations.

Autoxidation of 4,4'-Di-t-butylbiphenyl to 4,4'-Biphenyldicarboxylic Acid

The autoxidation of substrates such as p-xylene to terephthalic acid finds commercial value due to the fact that the major use of terephthalic acid is in the preparation of polyester (poly-ethyleneterphthalate-PET) plastics for containers. There exists today a strong environmental driving force to have recyclable packaging materials of all types, especially recyclable plastic bottles. Since plastic soda bottles are made from PET, PET bottles which could be sterilizable (for recycle) could find a solid market. Unfortunately, terephthalic acid based PET plastic will deform at sterilization temperatures; consequently, new bifunctional carboxylic acid monomers which will give higher melting PET plastic are desireable. To meet this possible demand such diacids as 2,6-dicarboxynaphthalene and 4,4'-biphenyldicarboxylic acid (BDA) have been proposed as possible terephthalic acid replacements. Monsanto is one of the world's largest producers of biphenyl and an obvious high value-added use of biphenyl would be the conversion to BDA. Since the ultimate use would be in a food grade material, a major consideration is that there must not be any possible trace of halogenated biphenyls. For that reason use of traditional autoxidation catalysis using halide promoters was deemed to be a problem. To gain the selectivity for 4,4'-disubstituted biphenyls, a novel route was developed which relied upon the Friedel-Krafts alkylation of biphenyl with isobutene to give exclusively the 4,4'-di-t-butylbiphenyl. This material was then used as the substrate in a metal catalyzed autoxidation for the preparation of 4,4'-dicarboxybiphenyl.[22] The halide-free catalyst system was comprised of a mixture of Co(II) and Mn(II) acetate salts in the molar ratio ~20 to 1. When the reactions were run at 170°C for four hrs under 1000 psig air in acetic acid/propionic acid, a 65% conversion was achieved with about a 70% selectivity to 4,4'-dicarboxybiphenyl (Figure 4). This remarkably selective conversion of the t-butyl groups to carboxyl groups is novel, and represents an important extension of catalytic oxidation chemistry. Its utility stems from the need for a halide free product and the selective positional isomers which can only be achieved by alkylation with the bulky t-butyl groups.

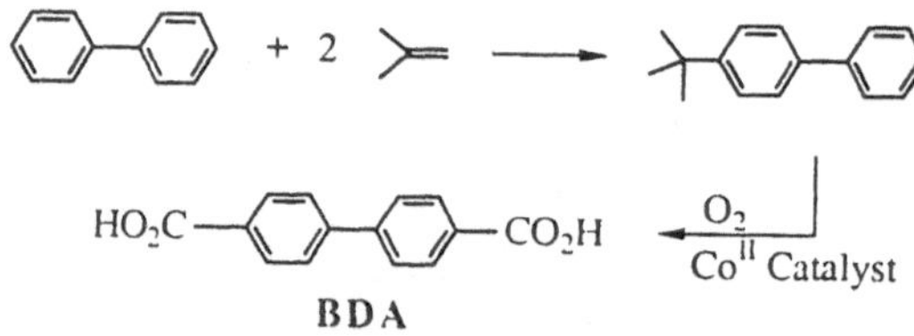

Figure 4. 4,4'-Di-isobutylbiphenyl route to biphenyl dicarboxylic acid (BDA).

Oxidative Coupling of Amines with Mercatobenzothiazoles

Currently, Monsanto's Rubber Chemicals Division manufactures and sells a family sulfenamide compounds (Santocures) which are used in the tire and rubber industry as anti-scorching compounds. The route to this materials relies on chlorine based coupling chemistry and is shown below in general terms in Figure 5. This chlorine based coupling affords the desired sulfenamide in the yield range of 85-90%, but it generates an aqueous waste salt stream containing trace levels of organics--a very difficult stream to cleanup.

We have developed a new patented catalytic oxygen based coupling route which not only eliminates the salt by-products (the only by-product is water), but allows us to directly couple mercaptobenzothiazole (MBT) directly with the desired primary or secondary amine in a quantitative yield under very mild conditions: room temperature to 70°C under ambient to 50 psig O_2, and the reaction times are short (< 1 hr). The activated carbon catalysts, described earlier for the production of the herbicide active glyphosate, are extremely efficient catalysts for this transformation. The only waste product in this reaction is water:

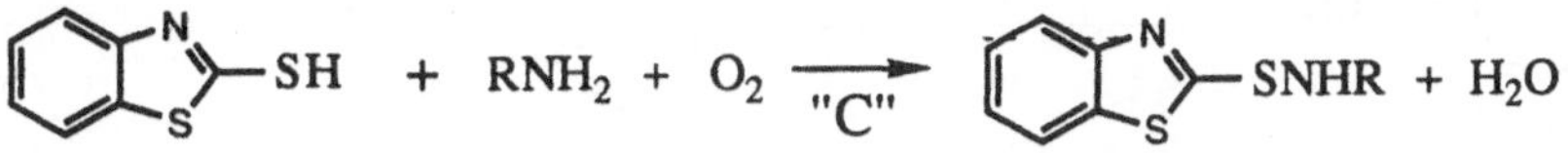

$$\text{MBT-SH} + RNH_2 + O_2 \xrightarrow{\text{"C"}} \text{MBT-SNHR} + H_2O \tag{8}$$

Nucleophilic Aromatic Substitution For Hydrogen: Oxidation of σ-Complex Intermediates

One of the oldest practiced industrial chemical reactions is the activation of aromatic C-H bonds by chlorine oxidation (Figure 6). The resulting chlorobenzenes can be further activated towards nucleophilic aromatic substitution by nitration producing a mixture of *ortho* and *para*-nitrochlorobenzene (PNCB). These intermediates are employed in a variety of commercial processes for production of substituted aromatic amines. Since neither chlorine atom ultimately resides in the final product, the ratio of pounds of by-products produced per pound of product generated in these processes are highly unfavorable. In addition, these processes typically generate aqueous waste stream which contain high levels of inorganic salts that are difficult and expensive to treat.

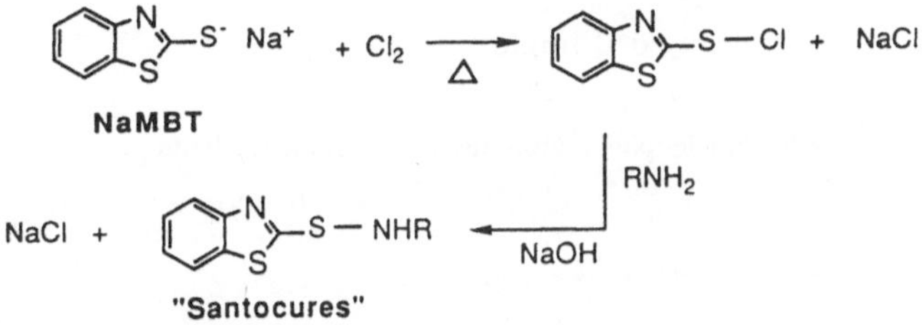

Figure 5. Current chlorine-based oxidative coupling route to sulfenamides.

By contrast, a more direct and atomically efficient route for the production of aromatic amines would be to eliminate the need for halogen mediated oxidation of benzene. This can be achieved by a class of reaction known as nucleophilic aromatic substitution for hydrogen (NASH-Figure 7). While this type of reaction has been known for over 100 years, this chemistry generally proceeds in low yields, give mixtures of ortho and para substitution products, and requires the use of environmentally unfavorable external oxidants.[23] This section will focus on two new examples of NASH chemistry applicable to the production of commercially relevant aromatic amines. The important step in these novel reactions is the facile oxidation of the σ-complex intermediate **1**.

The reaction of aniline and aniline derivatives with *p*-chloronitrobenzene is the critical coupling reaction practiced by Bayer and Monsanto for the manufacture of 4-nitrodiphenylamine, **2**. Hydrogenation of **2** produces 4-aminodiphenylamine (4-ADPA), **3**, which is a key intermediate in the *p*-phenylenediamine class of antioxidant used in rubber products (Figure 8).

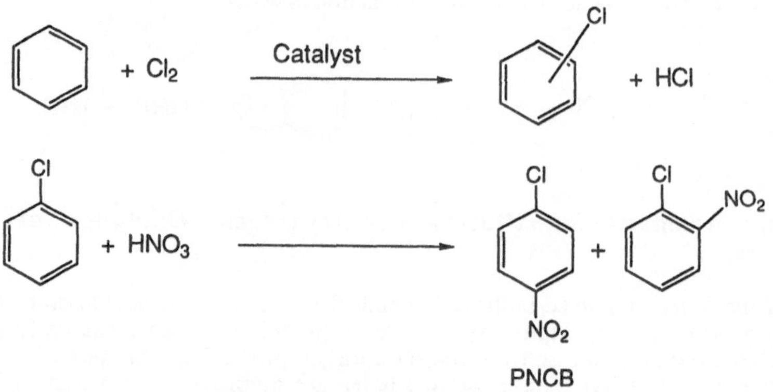

Figure 6. Commercial activation of benzene by chlorine oxidation.

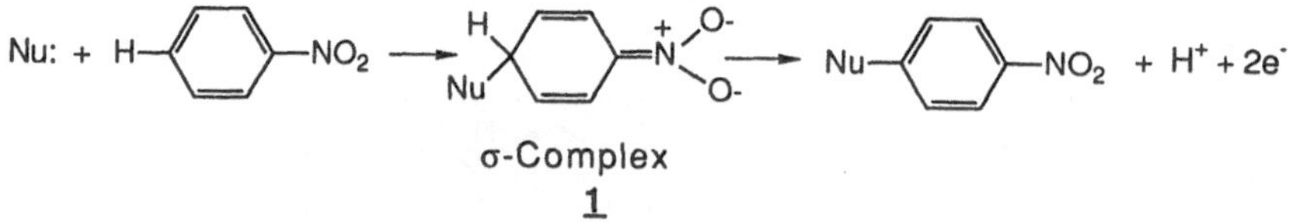

Figure 7. Nucleophilic aromatic substitution for hydrogen.

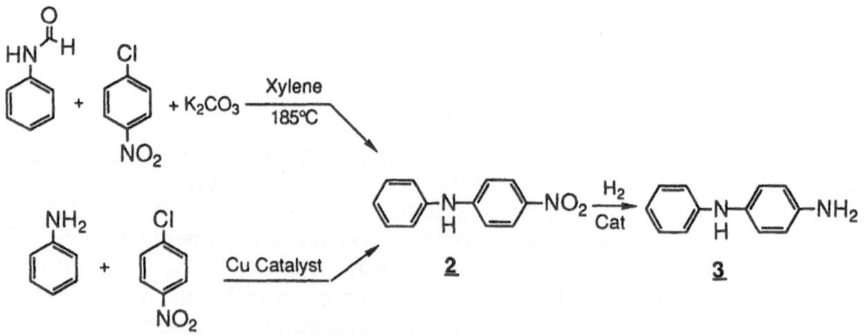

Figure 8. Commercial routes to 4-aminodiphenylamine.

This reaction suffers all the problems outlined above since these routes rely on chlorine to activate the aromatic ring towards nucleophilic attack. It has recently been discovered at Monsanto that the base catalyzed coupling of aniline and nitrobenzene via NASH chemistry is a superior route for the production of intermediates like 2.[24] Understanding the mechanism of the coupling reaction,[25] and in particular, the oxidation of σ-complex intermediate, 4, has allowed for the development of a commercial process based on this chemistry.

Our mechanistic studies revealed that oxidation of 4 can proceed by three separate pathways generating mixtures of 4-nitrosodiphenylamine 5 and 2 (Figure 9): 1) an intramolecular redox process with the nitro group of 4 functioning as the oxidizing agent generating 5, 2) an intermolecular pathway with free nitrobenzene functioning as the oxidant producing nitrosobenzene and 2, and 3) an oxygen driven pathway that also produces 2 and formally H_2O_2.

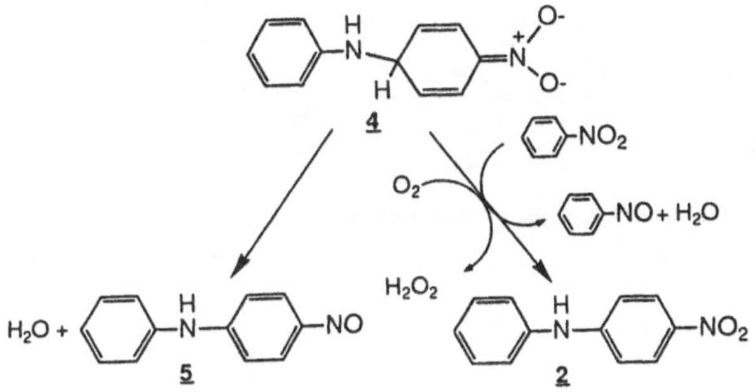

Figure 9. Production of 4-ADPA intermediates via nucleophilic aromatic substitution.

Attempts to drive this chemistry exclusively by the dioxygen pathway were unsuccessful since the aerobic oxidation of aniline to azobenzene by O_2 is extremely facile under the reaction conditions.[26] Thus, this chemistry was best suited to be run under anaerobic conditions utilizing the potential oxidizing capabilities of the nitro groups. Using these reactions conditions, selectivities to 5 and 2 in the 95% are routinely achieved. This reaction is unique in that it proceeds in high yield and selectivity under mild conditions (80 °C) without the need for an auxiliary leaving group. Accordingly, this process is halide free and extremely efficient with respect to raw materials consumed per pound of product generated making it a very attractive alternative for the commercial production of 3.

Another commercially important aromatic amine is p-nitroaniline (PNA) and its derivative p-phenylenediamine (PPD). PNA is currently produced at Monsanto by the reaction of ammonia with PNCB (Figure 10). Recently a new example of NASH chemistry directly applicable to the

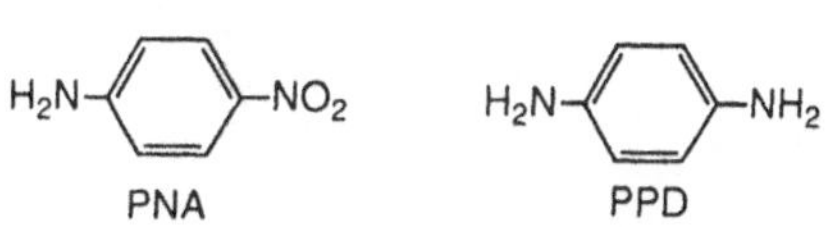

PNA   PPD

Figure 10. Commercially relevant aromatic amines.

41

production of PNA and PPD was discovered. We have found that the reaction of benzamide 6 and nitrobenzene in the presence of base under anaerobic conditions generated 4-nitrobenzanilide 7 in high yield under mild conditions. The only other observable product in this reaction was azoxybenzene 8 Simple treatment of 7 with methanolic ammonia results in the aminolysis of the amide bond generating PNA and benzamide. Thus, the overall stoichiometry for this series of reactions illustrate the formal amination of nitrobenzene with ammonia.

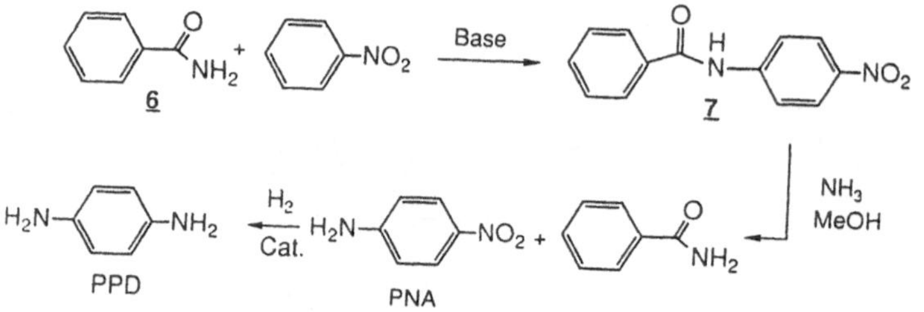

Figure 11. Amination of nitrobenzene via nucleophilic aromatic substitution for hydrogen.

That azoxybenzene is observed as a by-product of this reaction under anaerobic conditions indicates that nitrobenzene is functioning as the oxidant. A mechanism which explains the simultaneous formation of 7 and 8 is shown below (Figure 12). Intermolecular oxidation of the σ-complex 9 by nitrobenzene generates 7 and nitrosobenzene via disproportionation of the intermediate nitrobenzene radical anions. The ultimate formation of azoxybenzene is then governed by a cascade of electron transfer and nucleophilic reactions between the radical anions of nitrobenzene, nitrosobenzene and N-hydroxyaniline.

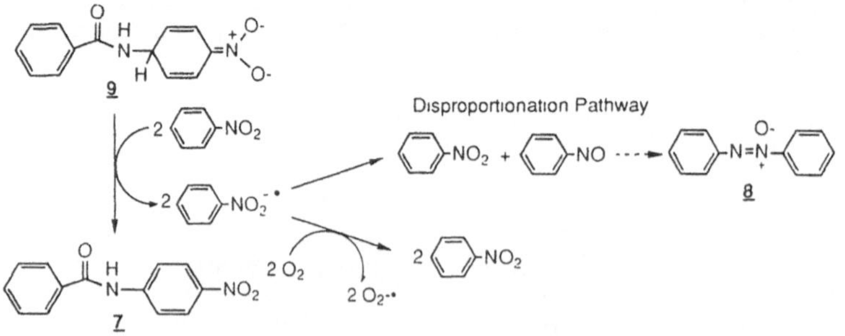

Figure 12. Aerobic oxidation mechanism for the production of 4-nitrobenzanilide

In contrast to the case where aniline is used as the nucleophile, the benzamide reaction can be improved by utilizing dioxygen in the reaction mixture since **6** is resistent to autoxidation. Under aerobic conditions the nitrobenzene radical anion is readily trapped by O_2 generating superoxide and nitrobenzene. This reaction pathway inhibits the formation of azoxybenzene by diverting the electron transfer cascade and ultimately utilizing dioxygen as the terminal oxidant. Thus, under aerobic reaction conditions **7** is the only observed reaction product. The formation of substituted benzanilides from the reaction of amides with nitrobenzene is the first example of the direct formation of aromatic amide bonds via nucleophilic aromatic substitution for hydrogen, and represents a new route for the amination of nitrobenzene. This reaction proceeds in high yield and regioselectivity, and does not require the use of halogenated intermediates, external oxidants or auxiliary leaving groups. Our mechanistic studies of NASH reactions have revealed that the controlled oxidation of σ-complex intermediates results in highly selective and environmentally favorable routes for the commercial production of aromatic amines.

CONCLUDING REMARKS

The more favorable environmental characteristics of catalytic oxidation reactions are providing a strong driving force for the chemical industry to expand the use of this reaction type. Minimizing non-selective reaction pathways will be important for reducing overall costs, which now encompass raw material usage, energy efficiency *and* environmental clean up. Although we have not discussed the production of hydrogen peroxide, commercially generated from oxygen, in this chapter, we also anticipate that hydrogen peroxide will continue to become a more competitive source of clean oxygen for catalysis, especially for higher value-added chemicals.

REFERENCES

1. J. C.Burnett, R. A. Keppel, and W.D. Robinson, "Commercial production of maleic anhydride by catalytic processes using fixed bed reactors", *Catalysis Today, 1*, 537 (1987).
2. R. L. Bergman and N. W. Frisch, US Patent 3293268 (1966), assigned to Princeton Chemical Research.
3. B.K.Hodnett, "Vanadium-phosphorus oxide catalysts for the selective oxidation of C4 hydrocarbons to maleic anhydride", *Catalysis Rev.-Science and Engineering, 27*, 373 (1985)
4. G.Centi, F. Trifiro', J. R. Ebner, and V. M. Franchetti, "Mechanistic aspects of maleic anhydride synthesis from C4 hydrocarbons over phosphorus vanadium oxides", *Chemical Reviews, 88*, 55-80 (1988).
5. G.J. Hutchins, "Effect of promoters and reactant concentration on the selective oxidation of n-butane to maleic anhydride using phosphorus oxide catalysts", *Applied Catalysis, 72*, 1 (1991) .
6. G. Centi, (ed), "Forum on vanadyl pyrophosphate catalyst", *Catalysis Today, 16*(1), (1993).
7. J. R. Ebner and J. T.Gleaves, "The activation of oxygen by metal phosphorus oxides-the vanadium phosphorus system", "Oxygen Complexes and Oxygen Activation by Transition Metals", (eds. A. E. Martell and D. T. Sawyer) Plenum Publishing Corp., New York, 273-292 (1988).
8. J. R. Ebner and M. J. Thompson, "An active site hypothesis for well-crystallized vanadium phosphorus oxide catalysts systems", *Catalysis Today, 16*, 51-60 (1993).
9. P. Mars and D.W. van Krevelen, "Oxidations carried out by means of vanadium oxide catalysts", *Chemical Engineering Science (Special Supplement), 3*, 41 (1954).
10. R. Grasselli and J. D. Burrington, "Selective oxidation and ammoxidation of propylene by heterogeneous catalysts", *Advances in Catalysis,* (eds. D.D. Eley, H. Pines, P. B. Weisz) Academic Press, New York, *30*, 133-163 (1981).
11. J. R. Donovan, R.D.Stolk and M.L. Unland, in *Applied Industrial Catalysis*, (ed.B.E. Leach) Academic Press, Inc., New York, 2 , 245-286 (1983) .
12. G.K Boreskov, "Catalytic activation of dioxygen", *Catalysis Science and Technology*, (eds J. R. Anderson and M. Boudart) Springer-Verlag, New York *3*, 39 (1982).
13. A. N. Bashkirov, V. V. kamzolkin, K. M Sokova, and T. P. Andreyeva, in "The Oxidation of Hydrocarbons in the Liquid Phase" (N. M. Emanuel, ed.), Pergamon, Oxford, p. 183 (1965).

14. H. Sakaguchi, Y. Kamiya, and N. Ohta, "Autoxidation of hydrocarbons in the presence of boric acids decomposition of aromatic hydroperoxides", *Bull. Jap. Pet. Inst.*, *14*, 71 (1972).

15. J. Franz, U.S. Patent 3954848, May, 1976.

16. A. Hershman and D. J. Bauer, U.S. Patent 4264776, Aug. 1976.

17. D. P. Riley, D. F. Fields, and W. Rivers, "Vanadium(IV,V) salts as homogeneous catalysts for the oxygen oxidation of N-phosphonomethyliminodiacetic acid to N-phosphonomethylglycine" *Inorg. Chem.*, *30*, 4191 (1991).

18. D. P. Riley, D. F. Fields, and W. Rivers, "Homogeneous catalysts for selective molecular oxygen-driven oxidative decarboxylations" *J. Amer. Chem. Soc.*, *113*, 3371 (1991).

19. D. P. Riley and D. F. Fields, "Electron-transfer agents in metal-catalyzed dioxygen oxidations: effective catalysts for the interception and oxidation of carbon radicals", *J. Amer. Chem. Soc.*, *114*, 1881 (1992).

20. The rate constant for the O_2 oxidation of the MV radical cation to MV is ~1.2 x 10^{+7} M^{-1} sec^{-1} at 12°C: P. Liu, Q. Zha, C. Xie, C. Li, and H. Wang, *Ciuhua Xuebao*, *4*(2) 131 (1983).

21. The rate constant for the O_2 oxidation of the 2,6-disulfo-9,10-anthraquinone radical anion in H_2O is 5 x 10^{+8} M^{-1} sec^{-1}: R. L. Wilson, *Trans. Faraday Soc.*, *67*, 3020 (1971).

22. R. A. Periana, and G. F. Schaefer, U. S. Patent 5068407 (1991).

23. L. F. Terrier, "Nucleophilic Aromatic Displacement," Feuer, H., Ed; VCH Publishers, Inc. New York (1991).

24. M. K. Stern and J. K. Bashkin, " Method of Preparing 4-Aminodiphenylamine" U.S. Patent 5,117,063, (1992).

25. M. K. Stern, F. D. Hileman and J. K. Bashkin, "Direct Coupling of Aniline and Nitrobenzene: A New Example of Nucleophilic Aomatic Substitution for Hydrogen" *J. Am. Chem. Soc.*, *114*, 9237 (1992).

26. J. Jeon and D. T. Sawyer, " Hydroxide-Induced Synthesis of the Superoxide Ion from Dioxygen and Aniline, Hydroxylamine, or Hydrazine," *Inorg. Chem.*, *46*, 12 (1990).

METAL PHOSPHATES: NEW VISTAS FOR CATALYSED OXIDATIONS WITH HYDROGEN PEROXIDE

A Johnstone, P J Middleton and R C Wasson[1]
R A W Johnstone, P J C Pires and G O Rocha[2]

[1]Solvay Interox R&D, Widnes, England
[2]Department of Chemistry, University of Liverpool
England

INTRODUCTION

Crystalline zirconium phosphates are potentially interesting as catalysts, since certain forms possess regular layer structures within which are strongly acidic sites which may undergo ion exchange with a range of metal cations. The ion exchange properties of zirconium phosphates were recognised in the 1950's[1,2,3]: however, the initial preparations were amorphous gels of variable composition and it was not until 1964[4] that zirconium phosphate was isolated in its crystalline form. Following on from this, other Group IV phosphates were then made in their crystalline forms. The phosphates have the general formula $M(HPO_4)_2 \cdot H_2O$ and possess a layer structure (Figure 1).

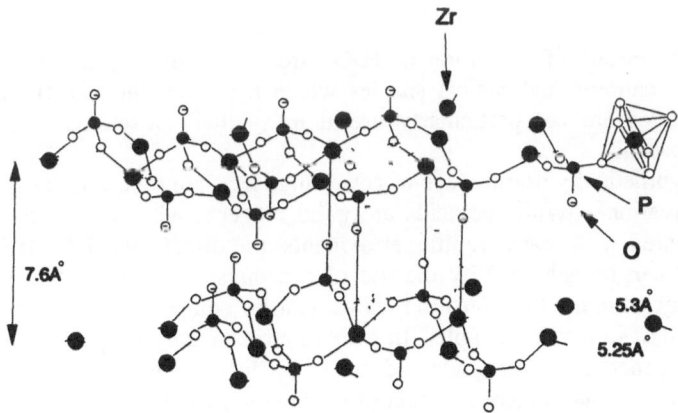

Figure 1 - Idealised Structure of α-Zirconium Phosphate

The Activation of Dioxygen and Homogeneous Catalytic Oxidation,
Edited by D.H.R. Barton *et al.*, Plenum Press, New York, 1993

These materials can be considered as strong inorganic solid acids and much of the catalytic activity so far observed in the literature has been attributed to their acidic nature. This acidity is attributed to the Brønsted acidity of the hydroxyl groups in the interlayers and to the Lewis acidity of the metal centre.

Hydrogen peroxide is particularly suitable to catalytic activation. Under neutral conditions at ambient temperature it is not very reactive but it may be converted 'in-situ' to a wide range of active species which will perform specific oxidations. It may also be used to re-oxidise other active oxidants within a process giving an overall catalytic effect. Other peroxygen reagents such as percarboxylic acids are generally more reactive than hydrogen peroxide. Some examples of how hydrogen peroxide is activated for use in chemical synthesis are given in Figure 2.

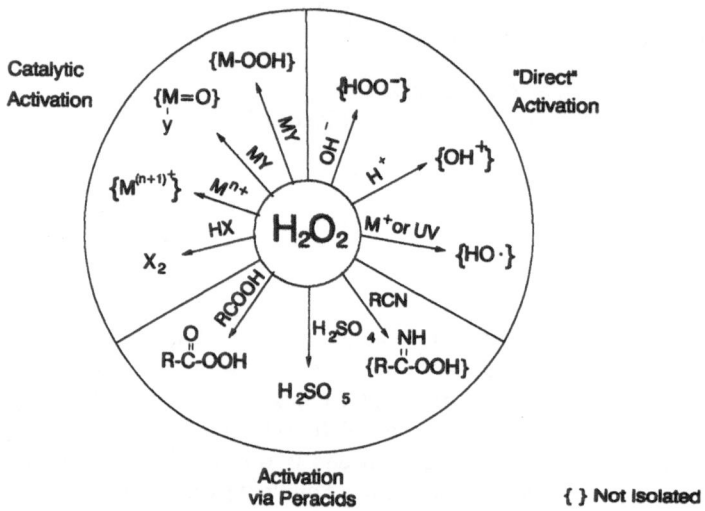

Figure 2 - Activation of H2O2

The simplest means of activation of H_2O_2 are by so-called direct methods. These produce anionic, cationic and radical species which carry out the oxidation. In general terms, these species are not particularly useful in synthetic reactions as they are often relatively non-specific.

The main synthetically useful ways of activating H_2O_2 are by conversion to peracids or by catalytic activation[5]. While peracids are good reagents with improved environmental properties compared with many traditional oxidants, the direct use of H_2O_2 is much more attractive. This can be achieved by the use of a catalyst to activate the oxidant. Many metals, especially amongst the transition series, can be converted to peroxo or oxo metal species, or take part in a redox couple. In general terms, catalytic systems have at least as wide a scope of reactivity as peracids and are being increasingly used in industry. Most systems are based on homogeneous catalysts, though recent research effort is focused towards finding a good heterogeneous catalyst. This would provide a 'zero-effluent' option, the only by-product of H_2O_2 oxidation being water. Tetravalent metal phosphates, being highly insoluble and having strongly acidic sites, were therefore candidates for study as H_2O_2 catalysts.

HYDROXYLATION

The hydroxylation of phenol by H_2O_2 is known to be catalysed by strong acids. This is a high capacity industrial process. Rhone Poulenc, Enichem and Ube produce catechol (CAT) and hydroquinone (HQ) from phenol by catalytic hydrogen peroxide processes (Figure 3).

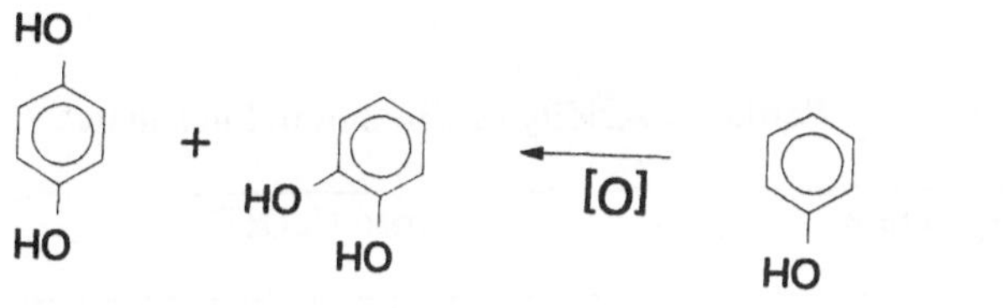

catechol   hydroquinone

Figure 3 - Phenol Hydroxylation

Acid catalysts[6], transition metal redox catalysts[7], and titanium zeolites[8] are all known to be effective for phenol hydroxylation. Acid catalysis proceeds by an ionic mechanism involving an intermediate hydroxonium ion ($H_3O_2^+$) whereas some transition metal ions promote the formation of hydroxyl radicals to effect substitution. However the introduction of a second hydroxyl substituent onto the aromatic nucleus tends to activate the molecule towards further reaction and this leads to the formation of unwanted, tarry by-products. The commercial solution is to use very low mole ratios of hydrogen peroxide to phenol and to recycle the unreacted phenol, ie. operate at low conversion. Some typical commercial methods are given in Table 1.

Table 1. Commercial Routes to Catechol and Hydroquinone

Catalyst	% Phenol Conversion	Selectivity dihydroxy	Ratio CAT:HQ
H_3PO_4/$HClO_4$ (Rhone Poulenc)	5	90	1.5:1
TS-1 (Enichem)	25	90	1:1
Ketone/acid (Ube)	<5	90	1.5:1

Early Experiments

Crystalline zirconium phosphate was prepared by the method described by Clearfield and Thakur[9]. Thermal activation was carried out at 100, 200, 300, 400°C. A portion of the amorphous material used to make the crystalline zirconium phosphates was retained for evaluation. Catalysts were characterised using titrimetric methods to measure acidity and x-ray diffraction to examine crystallinity. Thermogravimetric analysis (TGA) was used to determine phase changes on heat treatment.

The titrimetric results were as follows: The amorphous material would not give an end point, but the crystalline samples dried at 100°C were found to be the most acidic, acidity decreasing with heat treatment (Table 2).

Table 2 - Acidity of Zirconium Phosphates

Activation temperature		meq NaOH/g
Amorphous	100°C	no end point
Crystalline	100°C	8.65
	200°C	3.72
	300°C	2.93
	400°C	1.32

From our TGA experiments it appears that two moles of water, presumably of crystallisation, are displaced up to 200°C. Between 200 and 500°C a gradual change is seen with rapid loss of water seen between 500 and 600°C. There is a further gradual loss between 600 and 900°C.

Segaura et al[10] have postulated the following scheme:

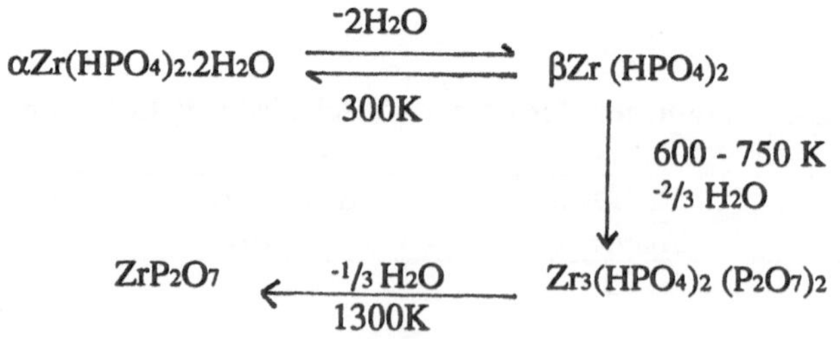

$$\alpha Zr(HPO_4)_2.2H_2O \underset{300K}{\overset{-2H_2O}{\rightleftharpoons}} \beta Zr(HPO_4)_2$$

$$\beta Zr(HPO_4)_2 \xrightarrow[-2/3 H_2O]{600 - 750 K} Zr_3(HPO_4)_2 (P_2O_7)_2$$

$$ZrP_2O_7 \xleftarrow[1300K]{-1/3 H_2O} Zr_3(HPO_4)_2 (P_2O_7)_2$$

X-ray diffraction analysis of the catalyst samples would tend to show that a chemical change takes place about 100°C. The crystalline sample has a very distinct diffraction pattern (Figure 4). However the sample heated to 400°C is pure zirconium pyrophosphate (Figure 5). At intermediate temperatures the samples are shown to be mixtures of zirconium phosphate and pyrophosphate. This contradicts Segaura's findings.

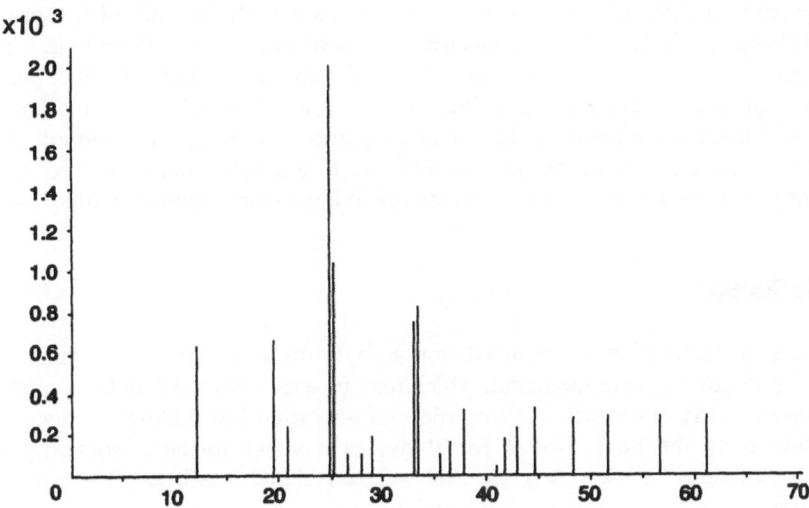

Figure 4 - XRD Analysis of Zirconium Phosphate Heated to 100°C

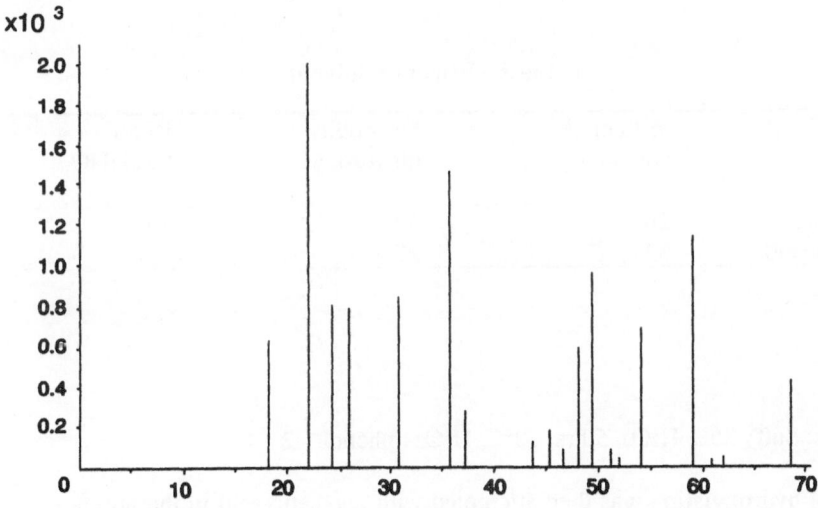

Figure 5 - XRD analysis of zirconium phosphate heated to 400°C

Five catalysts were assessed for activity in phenol hydroxylation. These were amorphous zirconium phosphate and the four crystalline samples heated at 100, 200, 300 and 400°C. The best catalyst in terms of conversion of phenol to catechol and hydroquinione was the crystalline sample heated at 100°C. This gave typically >90% selectivity to dihyroxybenzenes at 12% conversion of phenol and 2:1 mole ratio of phenol to H_2O_2[11]. Reactivity decreased with high activation temperatures.

Other phenols and phenol ethers were examined to assess the breadth of activity of this catalyst. Anisole was selected as an electron rich aromatic system though less so than phenol. A cleaner reaction at lower conversion was expected. Under similar conditions employed for phenol hydroxylation, a 20% conversion of anisole was measured with selectivity to 4-methoxy phenol of 15% and to guiacol of 42%. 1-naphthol was also assessed. No conversion was seen, presumably due to the bulky nature of the molecule. These findings are consistent with a mechanism involving an electrophilic oxidant species.

Mechanistic Studies

Zirconium phosphate was examined for activity in a range of solvents. Initial experiments were attempted in methanol, which may be used with the Enichem catalyst TS-1, but no reaction was observed. Acetonitrile was also tried but without success. Acetic acid was chosen as the next solvent for study, as a water miscible solvent, also the possibility of generating in-situ peracetic acid was considered feasible. This proved to be the best solvent.

Assuming a peracid to be the oxidising species, then propionic acid as solvent should show some activity. This was found to be the case. (Table 3).

Table 3 - Effect of Solvent

Solvent	% Phenol conversion	Selectivity dihydroxy	Ratio CAT:HQ
Acetic acid	26	59	1.4:1
Propionic acid	32	27	2.5:1

Conditions: SnⓅ, 35% H_2O_2, 5 hrs, 60°C, H_2O_2 : phenol 1:2

Phenol hydroxylation was then attempted with peracetic acid in the absence of a metal phosphate catalyst. Overoxidation of the substrate was observed, the major product being muconic acid. The phosphates were also found to be ineffective as catalysts for peracetic acid formations (from acetic acid and H_2O_2 alone). It was concluded that the mechanism was not a straight forward peracid oxidation.

A characteristic of the phenol hydroxylation reaction was that there was a lag between the disappearance of substrate and the appearance of products (Figure 6). No intermediates were observed. This led us to believe that the substrate was being absorbed into the interlayers and oxidation taking place there.

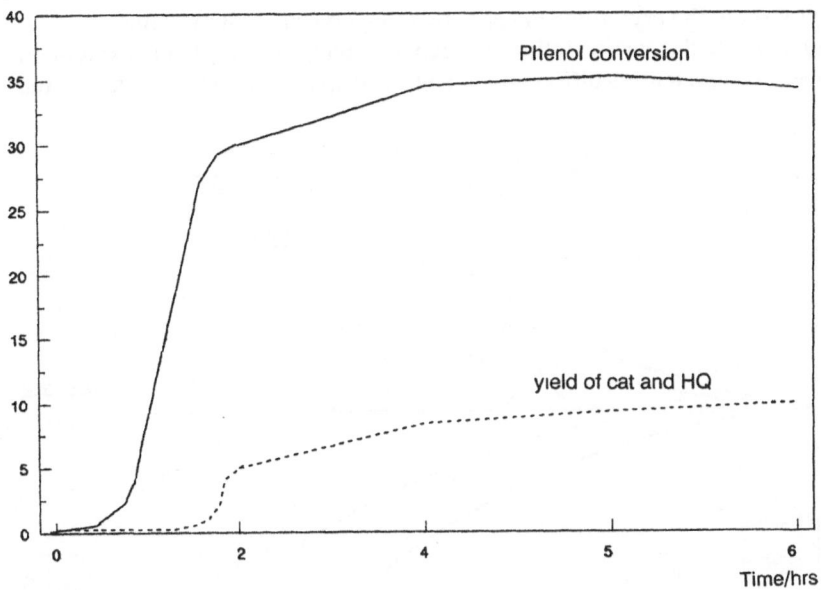

Figure 6 - Relationship of Phenol Conversion to Catecol and Hydroquinone Formulation

A further experiment was done to look at oxidant concentration in the solid and in solution (Table 4). It was found that, even after thorough washing, the peracetic acid formed was very much concentrated in the solid. This would explain why zirconium phosphate inhibits the formation of free peracetic acid, as it suggests that peracid is formed but held in the interlayers. The selectivity to dihydroxy products may then be attributed to adsorption selectivity. This is our current understanding of the mechanism.

Table 4 - Oxidant Concentration in the Solid and in Solution

Oxidant	g/kg oxidant in solid	g/kg oxidant in solution	Ratio
H_2O_2	3.5	340	~1:100
PAA	15.5	99	~1:6

Conditions: ZrⓅamorphous, 6 hrs, 70% H_2O_2, nt, AcOH

The existing method for preparing crystalline zirconium phosphate is firstly to precipitate the amorphous form from a mixture of phosphonic acid and zirconyl chloride and secondly to reflux the amorphous form in 12M phosphoric acid for several days after washing the amorphous form free of chloride. This gives the α-form. Not only is this a time consuming process, but the amorphous form is difficult to filter. We aimed to develop a one step process to the crystalline material by using crystal habit modifiers.

The rationale for this was as follows: Zirconium phosphate is able to exist in different forms. Since each crystallises differently, the initial disposition on the surface of the Zr^{4+}

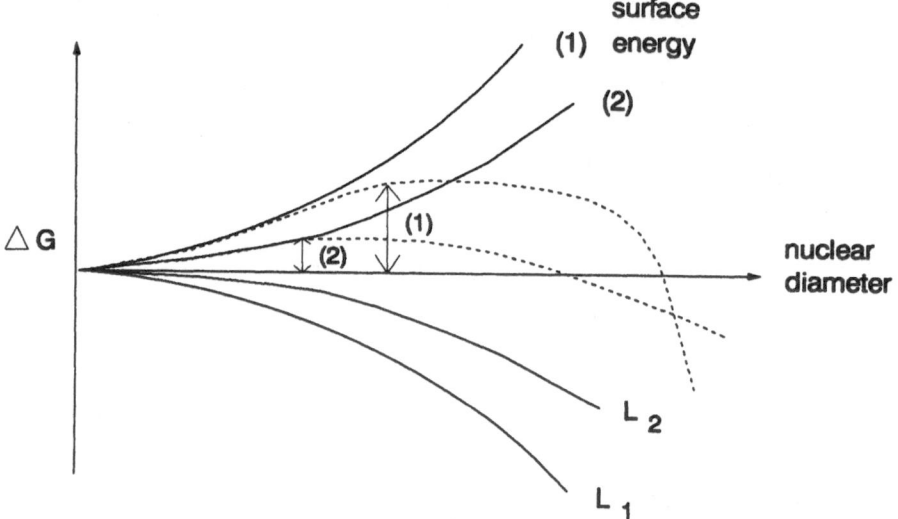

Figure 7 - Energy Diagram for Two Crystals

and HPO_4^{2-} ions is important. Kinetic control of the surface determines the form of the crystal obtained. Crystallisation almost invariably occurs on surfaces (of vessels, dust etc). If two crystals are considered in an energy diagram (Figure 7), the lattice L_1 is more stable than the lattice L_2, therefore L_1 is more insoluble than L_2. However since L_2 has a poorer lattice structure, water can still hydrate the surface and hydration leads to a more stable surface. Also (2) has a lower activation energy than (1) and a more stable nucleus, but as the crystals grow (1) becomes more stable than (2).

The above observations obey the Gay-Lussac Law which states that a compound that crystallises from a highly super-saturated solution first is the most soluble (from a saturated solution first it is the least soluble). In the above case, amorphous (2) is more soluble than (1). By choice of crystal habit modifier, it is possible to preferentially precipitate one form of the phosphate.

The addition of crystal habit modifiers to the original phosphoric acid/zirconyl chloride solution gave catalysts which were mainly crystallinein less than one hour but still had some amorphous character (Figure 8).

Nonetheless these materials were found to be effective phenol hydroxylation catalysts (Table 5) showing similar conversions and selectivities to a crystalline zirconium phosphate heated at 100°C.

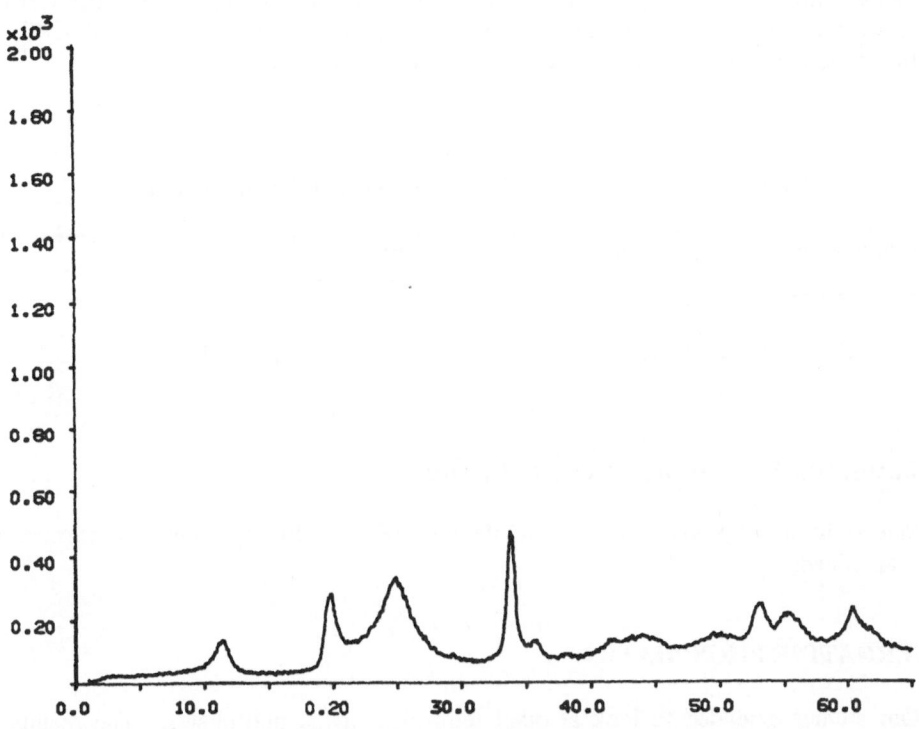

Figure 8 - X-ray Diffraction Pattern of Zirconium Phosphate Seeded with Cetyl Pyridinium Chloride (ZrⓅ CPC)

Table 5 - Phenol Hydroxylation Using Zirconium Phosphates Made with Crystal Habit Modifiers

Crystal Habit Modifier	% Phenol Conversion	Selectivity Dihydroxy	Ratio CAT:HQ
Cetyl pyridinium chloride	38	64	1.9:1
ALIQUAT 336	39	53	1.7:1
ETHYLAN CD919	43	49	1.6:1
None*	48	48	1.6:1

* 5 hrs

Conditions: ZrⓅ, AcOH solvent, 90°C, 6hrs, H_2O_2::phenol 1:1

Catalyst Recycle

An experiment was conducted to demonstrate that the zirconium phosphate catalyst was indeed working in a heterogeneous manner. After one recycle using Zr Ⓟ CPC, no significant reduction in catalytic activity was observed (Table 6).

Table 6 - Effect of Catalyst Recycle on Phenol Hydroxylation

Experiment No.	% Phenol conversion	Selectivity dihydroxy	Ratio CAT:HQ
1	32	53	1.5:1
2	35	47	1.5:1

Conditions: 50% H_2O_2, AcOH solvent, 80°C, 4 hrs

Selectivity to dihydroxy products and the ratio of catecol to hydroquinone formed was also maintained.

ALTERNATIVE PHOSPHATES

Our studies extended to look at other tetravalent metal phosphates. The results for those which were active phenol hydroxylation catalysts are given in Table 7[12].

Table 7 - Tetravalent Metal Phosphates as Phenol Hydroxylation Catalysts

Phosphate	% Phenol conversion	Selectivity dihydroxy	Ratio CAT:HQ
Zr	12	90	1.6:1
Sn	17	80	1.5:1
Ce	17	95	2.0:1
Ti	29	7	3.9:1

Conditions: H_2O_2 35-50%, AcOH solvent, 60-90°C, 6 hrs, H_2O_2:phenol 1:2

Other tetravalent metal species were examined and found to have no activity. These were vanadyl (VO_4^+), molybdenum, and mixed zirconium tungsten phosphates.

A comparison of the amorphous and crystalline forms of zirconium and tin phosphates was also made (Table 8). It is apparent that the crystalline forms show greater selectivity to the dihydroxy products. This is consistent with the oxidation largely taking place in the interlayers (which, of course, are not present to great extent in the amorphous form).

Table 8 - A Comparison of the Amorphous and Crystalline Forms of Zirconium and Tin Phosphates

Phosphate	% Phenol conversion	Selectivity dihydroxy	Ratio CAT:HQ
Zr amorphous[a]	34	56	0.9:1
Zr crystalline[a]	33	97	1.4:1
Sn crystalline[b]	26	59	1.4:1
Sn amorphous[c]	34	25	2.3:1

Conditions: 35% H_2O_2 , AcOH solvent, 5 hrs

a: 90°C, H_2O_2:phenol 1:1, b:100°C, H_2O_2:phenol 1:2, c: 60°C, H_2O_2:phenol 1:2

CONCLUSIONS

1. Both amorphous and crystalline zirconium, tin and cerium phosphates are active catalysts for phenol hydroxylation.
2. The activity shown is in line with the acidity of the metal centre.
3. Crystalline phosphates exhibit greater activity and selectivity than their amorphous counterparts in phenol hydroxylation.
4. Interlayer peracetic acid is implicated as the oxidising species.

ACKNOWLEDGMENTS

The authors would like to acknowledge the contributions of A Hackett, S Moores, M Service and S Wilson.

REFERENCES

1. K A Kraus and H O Phillips, *J Amer Chem Soc*, 78 (1956) 644.
2. C B Amphlett, L A McDonald and M J Redman, Chem Ind (London), (1956) 1314.
3. E R Russell, A W Adamson, J Shubert and A E Boyd, A.S.E.A.E.C. Rep. CN-508 (1943) (declassified 1957).
4. A Clearfield and J A Stynes, *J Inorg. Nucl. Chem.*, 26 (1964) 117.
5. 'Peroxygen Compounds in Organic Synthesis', a monograph available from Solvay Interox, ref AO12.
6. DE 2,658,545:Rhone Poulenc.
7. US 4,578,521:Mobil Oil.
8. DE 3,309,669:Anic.
9. A Clearfield and DS Thakur, *Appl. Catal* 26, (1986) 1.
10. K Segaura, S Nakata and S Asaoka, *Mater.Chem. and Phys*; 17 (1987) 101.

11. Solvay Interox Ltd, International Patent Application, publication No. WO 92/18449.
12. Solvay Interox Ltd, British Patent Application No. 9226494.4, filed 19.12.92.

THE DESTRUCTION OF TOXIC AQUEOUS ORGANIC COMPOUNDS BY $H_2O_2/Fe^{++}/UV/O_2$

C.A. Tolman, W. Tumas, S.Y.L. Lee, and D. Campos

DuPont Central Research and DuPont Chemicals
DuPont Company
Wilmington, DE 19880-0304

INTRODUCTION

Improved environmental performance continues to grow in importance to industry. We have been examining thermal and photochemical oxidation processes for the treatment of dilute aqueous organic wastes that are either generated in the course of chemical manufacturing or exist in contaminated groundwaters. Most of our work has focused on advanced oxidation processes[1] which employ oxidants and catalysts and depend to some degree on the in-situ generation of hydroxyl radicals ($\cdot$OH). Our program has involved: a) laboratory and treatability studies on model and actual wastewaters that compare chemical oxidation technologies, explore oxidation mechanisms, and utilize experimental designs aimed at delineating important parameters; b) pilot-scale field demonstrations on industrial streams and c) economic and engineering evaluations including comparisons with base technologies such as carbon adsorption and air or stream stripping/abatement. We have examined Fenton oxidation (Fe/H_2O_2), UV photochemical oxidation (UV/H_2O_2, $UV/Fe/H_2O_2$ and UV/ozone) and semiconductor (TiO_2) photocatalytic oxidation. Herein, we report on our investigations of the photochemically driven oxidation of a number of model aromatic wastes using iron catalysts, hydrogen peroxide, and oxygen. The goal of this work is to compare relative efficiencies of destruction, delineate the important factors that control destruction efficiencies and elucidate reaction mechanisms. We have chosen to focus on aromatic compounds because they react rapidly with hydroxyl radicals[2] and also have uv absorptions which allow one to follow the course of their disappearance. In addition, they tend to interfere with biodegradation at concentrations over 0.05% (500 ppm)[3] and thus are likely to require some alternative chemical treatment and/or pretreatment.

The Activation of Dioxygen and Homogeneous Catalytic Oxidation,
Edited by D.H.R. Barton *et al.*, Plenum Press, New York, 1993

The oxidation of most organic compounds by O_2 is highly favorable thermodynamically, as indicated in Table 1, where n is the number of gram-atoms of oxygen needed for complete combustion to carbon dioxide and water, shown for phenol in equation (1).

Table 1. Heats of Combustion of Organic Compounds

Compound	Formula	n[a]	ΔH[b]	$\Delta H/n$
Oxalic acid	$(CO_2H)_2$	1	60.2	60
Methanol	CH_3OH	3	170.9	57
Phenol	C_6H_5OH	14	732.2	52
n-Octane	C_8H_{18}	25	1302.7	52
Nitrobenzene	$C_6H_5NO_2$	14.5	739.2	51
Picric acid	$C_6H_2(OH)(NO_2)_3$	12.5	611.8	49
Methylene chloride	CH_2Cl_2	3	106.8	36
Carbon tetrachloride	CCl_4	2	37.3	19

[a] $n = 2*nC + 0.5*nH - nO$ (for O not in NO_2)
[b] In $kcal/mol^4$

$$C_6H_5OH + 7\,O_2 = 6\,CO_2 + 3\,H_2O \qquad (1)$$

The last column in Table 1 shows that the result of dividing the heat of combustion by n is a remarkably constant 50-60 kcal/mol of O atoms for all but the highly chlorinated compounds.

While oxidation is very favorable thermodynamically, it can be quite slow kinetically, requiring high temperatures or energy input, strong oxidants, or efficient catalysts to get useful rates. Table 2 gives a list of oxidation technologies for treating liquid wastes, divided according to the temperature of reaction. Incineration takes place at very high temperatures using air or oxygen as the oxidant and can be very effective at destroying organics, but can run into strong public opposition. Supercritical water oxidation (SCWO) takes place above the critical temperature of water (374°C), where water, oxygen, and organics are all miscible, avoiding mass-transfer limitations in the oxidation rate. SCWO has an advantage over incineration by not producing the nitrogen oxides (NO_x) which are inevitable in high temperature combustions using air. Wet air oxidation (WAO) takes place at lower temperatures and therefore lower pressures, but does not afford high degrees of destruction in short times like the higher temperature processes (times of the order of an hour are required).

Table 2. Oxidation Technologies for Liquid Phase Waste Treatment

Technology	Temperature (°C)	Oxidant	Rxn Time
Incineration	>1000	Air or O_2	secs
SCWO	374-600	"	"
WAO	160-320	"	hrs
Chemical Oxidation	25-100	Various	mins
Aerobic Biodegradation	~35	Air	days
Anoxic Biodegradation	~35	NO_3^-	"

Biodegradation which relies on oxidation is generally categorized as aerobic or anoxic depending on whether air or another oxidant, like nitrate, provides the electron sink. By using highly effective enzyme catalysts, these processes can operate near room temperature; however, they typically tend to be rather slow, requiring days. (Biological Oxygen Demand, or BOD, a common measure used to characterize waters containing biodegradable waste, is usually measured over a period of five days.)

Chemical oxidations operate anywhere from 25 to 100°C (or higher) and use powerful oxidants and/or some kind of stimulus, such as uv light, homogeneous or heterogeneous catalysts, sonication, a combination of uv with a semiconductor like TiO_2, or the ionizing radiation of an electron beam or γ-ray source. Some examples are shown in Table 3. Various combinations are possible, for example

Table 3. Components of Chemical Oxidation Systems

Oxidant	Stimulus
Cl_2, ClO_2	u v
NaClO	Fe^{++}/Fe^{+++}, Cu^+/Cu^{++}
$KMnO_4$	Pt^0
$K_2Cr_2O_7$[a]	Temp
O_3	Sonication
H_2O_2	TiO_2/uv
O_2	E-Beam
H_2O	γ-Radiation[b]

[a] Commonly used for COD (Chemical Oxygen Demand) analysis
[b] ^{60}Co is a popular source.[5]

Fe^{++}/H_2O_2 (Fenton's reagent), uv/H_2O_2, H_2O_2/O_3, $TiO_2/uv/H_2O_2$, etc. Water can be used as the oxidant with ionizing radiation in the form of high energy electrons or x-rays, or with ultraviolet light of sufficiently short wavelength (e.g. 185 nm) to dissociate water. We have looked at a number of possibilities, at both laboratory and pilot plant scales, but have concentrated on systems that can be described as $Fe^{++}/H_2O_2/uv/O_2$.

EXPERIMENTAL

Most runs were carried out in 1 liter ACE Glass photolysis reactor with a concentric quartz immersion well containing a 450 watt medium pressure Hg lamp. The reaction temperature was maintained by circulating water from a constant temperature bath. The reactor was stirred magnetically and sparged with gas at 100 mL/min through a fritted tube. The clock was started (t=0) when 100 mL of solution containing the H_2O_2 was added to give a total volume of 1 L; at that point the purge gas was switched from N_2 to air or O_2 if these gases were used. A minute was allowed for mixing, and the 450 watt uv lamp switched on. Samples were taken at various times for the measurement of pH, hydrogen peroxide concentration (determined by iodometric titration), uv spectrum, and Non-Purgable Organic Carbon analysis (NPOC), measured with a Rosemount Analytical (Dohrmann Division) DC-190 TOC Analyzer with an Automatic Sampling Module. Details were reported earlier.[6]

The concentrations of remaining aromatics were estimated from the uv spectra, correcting for the broad tailing absorption attributed to ferric oxalate complexes and other intermediates which were formed and subsequently degraded. In the case of phenolic compounds (e.g. p-nitrophenol), spectra of the pure compound were recorded over a range of pH spanning the pK_a, and the extinction coefficient at the isosbestic point between phenol and phenolate used to estimate the aromatic remaining in the degradation runs.

RESULTS

Phenol

In earlier detailed studies on phenol[6] we reported on a statistically designed experiments where we investigated eight major variables, whose ranges are indicated in Table 4, for the oxidation and eventual mineralization of 50 ppm phenol. Most of the variables had high, low, and middle values; in the case of the uv lamp power, the lamp was either on or off. Carbonate was either 0 or 10 mM, and was non-zero only when the pH was high; at the lower pH's of 2 and 5 carbonate would be converted to CO_2, and purged from the solution by the gas

Table 4. The Eight Major Variable Factors in the Oxidation of Phenol

Variable	Low	Med	High
UV Lamp Power (Watts)	0		450
Temp (°C)	25	52.5	80
pH	2	5	8
$[H_2O_2]$ (ppm)	20	110	200
Duration (Hrs)	0.5	1.5	2.5
$\%O_2$ in Purge	0	21	100
$[Fe^{++}]$ (ppm)	0	10	20
[Carbonate] (mM)	0		10

purge (N_2, air or O_2). When we use the term 'center point' (or C.P.) conditions, we mean that the variables allowed to assume three values were at their middle conditions; note that for $\%O_2$ this was air (21%) rather than 50%.

The measured responses included uv spectra (to measure the concentration of remaining phenol), the carbon concentration (measured as non-purgeable organic carbon, or NPOC), the peroxide (by titration), and the pH; in a few cases HPLC was used to measure the concentrations of oxidation intermediates.

Figure 1 shows the original design of thirty experiments used to explore the 8-dimensional space, where each point represents a set of experimental conditions. The no uv and uv conditions (each with a C.P.) are indicated by the large cubes on left and right. Within each large cube the major axes represent the variables: temperature, pH,

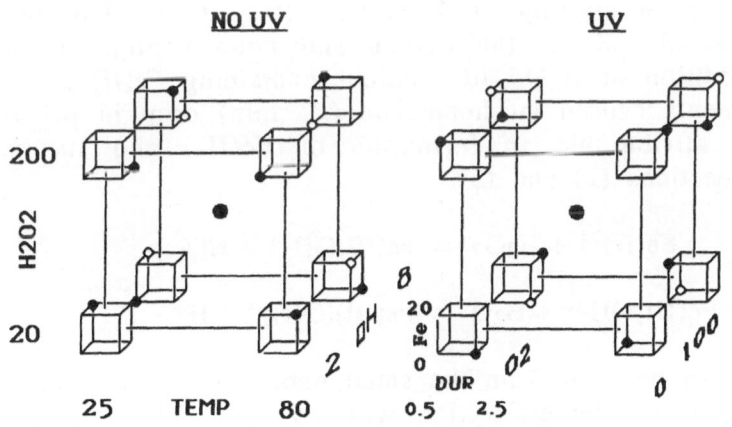

Figure 1. Initial experimental design showing center and corner points

and [H$_2$O$_2$]. For each high and low value of those variables, the small cubes have axes representing: duration of the experiment (from 0.5 to 2.5 hours), the %O$_2$ in the purge gas, and the [Fe^{++}] added. Each point represents a set of experimental conditions - the solid corner points for runs without added carbonate and the open ones those with 10 mM carbonate added (only at high pH). The center points are shown larger because they were run in triplicate to get an estimate of the experimental uncertainty.

The original design brought out a number of interesting facts, listed in Table 5. The high reactivity at the C.P.'s gives what is called

Table 5. Key Findings from the First Experimental Design

• In the no uv (Fenton chemistry) runs at least half of the phenol was left in all of the runs except the C.P., where it was gone when the solution was analyzed after 1.5 hours.
• There was little or no loss of NPOC with no uv except for the C.P., where about 30% mineralization occurred.
• The phenol was entirely gone in the uv runs, except for those at high pH.
• With uv, at least 30% mineralization occurred in all cases, with the greatest remaining carbon at high pH; the least remaining carbon, near 0%, was found at uv C.P. conditions.
• Added carbonate had a small effect - increasing remaining phenol and NPOC - but all of the high pH runs were quite slow in any case.
• Temperature had no measurable effect.
• %O$_2$ in the purge had little effect without uv, but significantly increased the extent of mineralization with uv.

---

curvature in the response. That is, a Taylor series expansion of an observed variable (like NPOC) in the experimental factors must contain one or more quadratic terms. By doing additional experiments at points in the space called star points, shown with the center points and measured NPOC concentrations in Figure 3, it was possible to show that the curvature comes primarily from the pH; runs starting at pH 5 are faster than those starting at 2 or 8. (Pignatello[7] has reported an optimum pH of 2.8 for the Fenton and photo-Fenton degradation of 2,4-D.) Addition of H$_2$O$_2$ to solutions containing Fe(II) and phenol in our experiments caused an immediate (<1 min) drop in pH from 5 to about 3.5, attributable to formation of Fe(III) and its hydrolysis, shown in reactions (2) and (3).

$$Fe(II)^{2+} + H_2O_2 = Fe(III)OH^{2+} + HO\cdot \qquad (2)$$

$$Fe(III)OH^{2+} + H_2O = Fe(III)(OH)_2^+ + H^+ \qquad (3)$$

It can also be seen from the small cubes in Figure 2 that duration is a factor in the mineralization with uv, but not without, and that with uv, the extent of mineralization is substantially less under N$_2$ than under either air or O$_2$ in 1.5 hours. The reason that duration is not a factor in the absence of uv could be seen by following the reactions with time. Fenton chemistry is fast under C.P. conditions, with loss of all phenol by the time the uv spectrum of the first sample, taken one minute after H_2O_2 addition, could be recorded. The H_2O_2 was all gone by the t=10 minute sample. (Recall that duration in the original design had values of 0.5, 1.5, and 2.5 hours.) With uv C.P. conditions (air purge) 90% mineralization could be achieved in 45 minutes. This time could be decreased by a factor of five, to 9 minutes, by going to an O_2 purge. (Air and O_2 appear to be equally effective (see the lower right hand cube in Figure 2) because the carbon results shown along the y-axis (%O_2) are for 1.5 hours; 2-3 ppm carbon is a typical background number with our instrument with pure water, and indicates essentially complete mineralization.

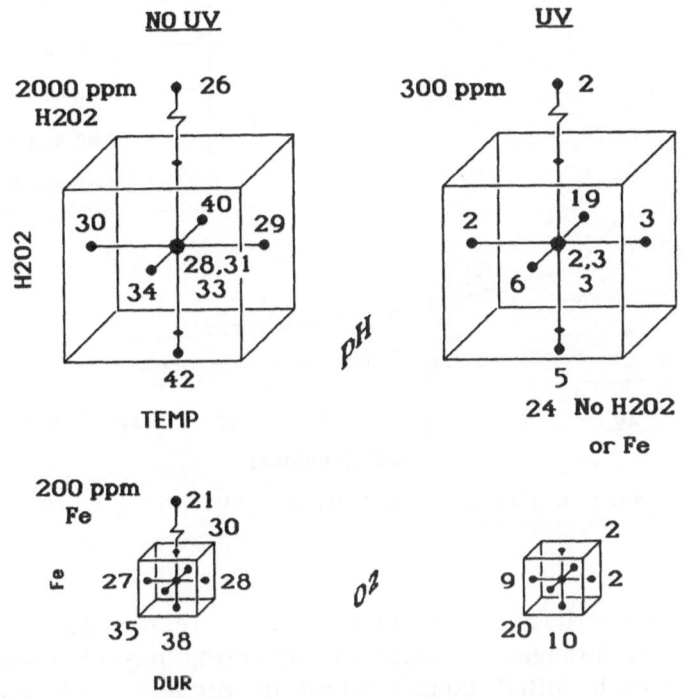

Figure 2. Center and star points: measured NPOC (ppm).

Other Organic Compounds

The oxygen effect was originally a surprise, because we thought that organic radicals would be captured so rapidly by O_2 that it would not matter how much oxygen was present in solution, as long as the solution was not oxygen-starved. We wanted to see how general the effect was, so we have studied the mineralization of a variety of other organic compounds at uv C.P. conditions and at star points using N_2 or O_2 purges. Figure 3 shows the NPOC as a function of time for a series of such experiments with p-nitrophenol. For runs with an air purge

(Runs p-NP #1, #3, and #4) note that mineralization is faster at an initial pH of 5 than at 2, which is faster than 8. For runs at pH, 5 in which the gas purge was varied (Runs #1, #2 and #5), the rate decreased in the order O_2 > air > N_2. Note that the curves (except for Run #4 at pH 8) are nearly superposed for the first five minutes, and then diverge. Titrations showed that the peroxide was gone by the t=5 sample for all of the runs except #4; at the high pH the H_2O_2 lasted for 30 minutes.

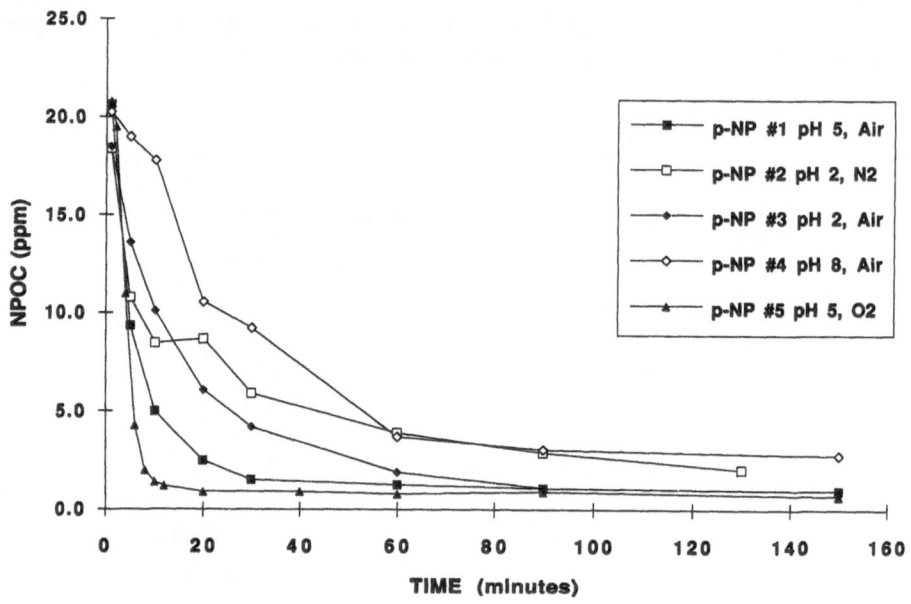

Figure 3. Mineralization of p-nitrophenol. Star and C.P. conditions.

Table 6 summarizes the results of the organic degradation runs starting at pH 5, arranged in order of increasing oxygen content in the purge gas for each initial concentration or organic, with the organics arranged in order of decreasing ease of mineralization, aromatic before aliphatic. Runs with 50 ppm organic, 10 ppm Fe^{++}, ca. 110 ppm H_2O_2, and an O_2 purge are indicated in bold-faced type. Though most runs started with 50 ppm of organic, some started with 100. Doubling the concentration of nitrobenzene (compare Runs NB #1-3 with #4-6) without increasing the iron or peroxide increased the times to 90% mineralization by factors of three or more. The same sort of effect can be seen in comparing Runs p-Cr #1-2 with #3-4 and BA #1-2 with #3-4. In the case of aniline the slowing effect was much more marked; the very dark solutions formed on adding peroxide to the 100 ppm solutions may have effectively cut off uv. Doubling the concentrations of iron and peroxide, along with benzoic acid, gave mineralization rates which were actually faster (compare Runs BA #5 and 6 with #3 and 4).

Table 6. Summary of Organic Degradation Runs[a]

Organic Compound	Run No.	Conc. (ppm)	[Fe^{++}] (ppm)	[H$_2$O$_2$] (ppm)	O$_2$ (%)	$t_{0.9}$[b] (min)	$t_{0.9}$/$t_{0.9}$(O$_2$)
p-Nitrophenol[c]	p-NP #2	50	10	101	0	90	15.0
	p-NP #1	50	10	101	21	17	2.8
	p-NP #5	50	10	101	100	6	1.0
Nitrobenzene	NB #1	100	10	110	0	>150	>5
	NB #2	100	10	110	21	60	2.1
	NB #3	100	10	110	100	28	1.0
	NB #4	50	10	110	0	150	16.7
	NB #5	50	10	110	21	17	1.9
	NB #6	50	10	110	100	9	1.0
Phenol[d]	Phen #54	50	10	110	0	>150	>15
	Phen #31	50	10	110	21	45	5.0
	Phen #52	50	10	110	100	9	1.0
p-Cresol	p-Cr #1	100	10	110	21	140	ca. 5
	p-Cr #2	100	10	110	100	ca. 30	1.0
	p-Cr #3	50	10	110	21	32	2.1
	p-Cr #4	50	10	110	100	15	1.0
Aniline	An #1	100	10	110	21	>150	
	An #2	100	10	110	100	>150	
	An #3	50	10	110	21	40	2.2
	An #4	50	10	110	100	18	1.0
Benzoic Acid	BA #1	100	10	110	21	>150	
	BA #2	100	10	110	100	>150	
	BA #3	50	10	110	21	80	3.0
	BA #4	50	10	110	100	27	1.0
	BA #5	100	20	220	21	71	7.1
	BA #6	100	20	220	100	10	1.0
1,4-Dioxane	Diox #1	50	10	110	0	>150	>9
	Diox #3	50	10	110	21	25	1.5
	Diox #2	50	10	110	100	17	1.0
Propionic Acid	PA #1	50	10	110	0	140	7
	PA #2	50	10	110	100	20	1.0
2-Butanone	MEK #1	50	10	110	0	150	3.1
	MEK #2	50	10	110	100	48	1.0

[a] At 52°C, initial pH 5, with 450 watt lamp on at t=1 min
[b] $t_{0.9}$ is the time after addition of H$_2$O$_2$ at t=0 for 90% mineralization.
[c] Runs p-NP #3 and 4 at pH 2 and 8, are not included here. See Figure 3.
[d] Results of earlier work.[6]

The last column in Table 6 shows $t_{0.9}$ relative to the time for 90% mineralization with O_2. Using O_2 decreases the time required by factors of 2-5 relative to air in most cases, and by factors of at least 5 and sometimes greater than 15 relative to N_2. A bar chart comparing values of $t_{0.9}$ for selected 50 ppm solutions is shown in Figure 4, arranged in order of increasing $t_{0.9}(O_2)$. It can be seen that though the aromatics are usually more rapidly mineralized under these conditions, the aliphatic and aromatic compounds overlap. We were initially surprised by the slowness of mineralization of methylethyl ketone (MEK), because of the possibility of $\pi \rightarrow \pi^*$ electronic excitation of the carbonyl to give Norrish Type I and II reactions,[8] but at these low concentrations the absorbance was too small.

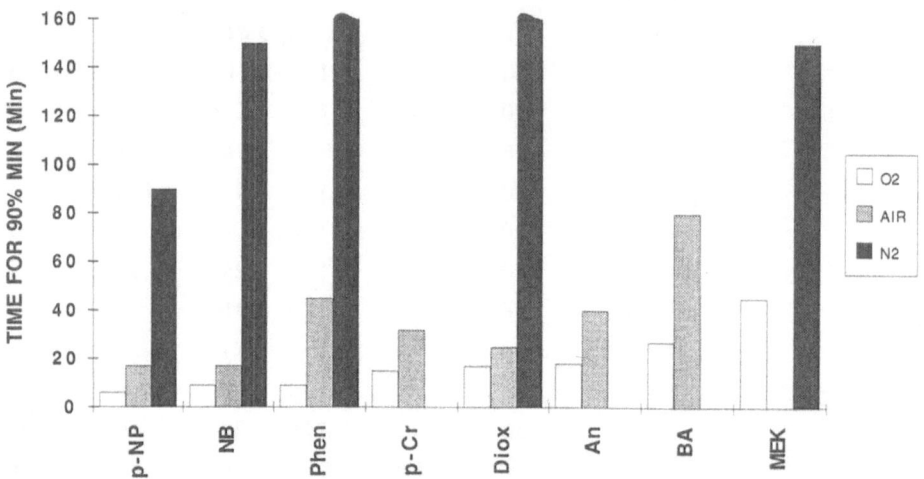

Figure 4. Effect of gas purge on the mineralization of various organics with initial concentrations of 50 ppm.

Vis/uv spectra of samples taken from the reactor after the starting compound was gone, but before complete mineralization, usually showed broad tailing absorbtion going out to about 500 nm which we attributed largely to ferric oxalate complexes; oxalate was identified by HPLC as the penultimate oxidation product in our phenol studies.[6] With phenol we found that Fenton chemistry (no uv) alone rapidly converted about 30% of the carbon to CO_2, and most of the rest to oxalic acid, which resisted further reaction in the dark. Even additions of high concentrations of H_2O_2 and Fe^{++} did not significantly increase the degree of mineralization, as can be seen from the NPOC numbers on the left side of Figure 2 above both large and small cubes, where for these star points the concentrations of peroxide and iron were increased to 10X the original design. We proposed that the oxygen effect seen with uv was a consequence of the effect of O_2 on oxalate mineralization.

Oxalate Degradation

We investigated the degradation of oxalic acid at 52°C and pH 5, with 10 ppm iron added as ferric perchlorate. We chose 100 ppm oxalate because that gives 27 ppm C, about the same concentration obtained in the dark using Fenton chemistry on initially 50 ppm phenol (38 ppm C). Figure 5 shows uv spectra obtained with 100 ppm oxalic acid (OA) or 10 ppm Fe^{+++} at pH 5, by themselves and together.

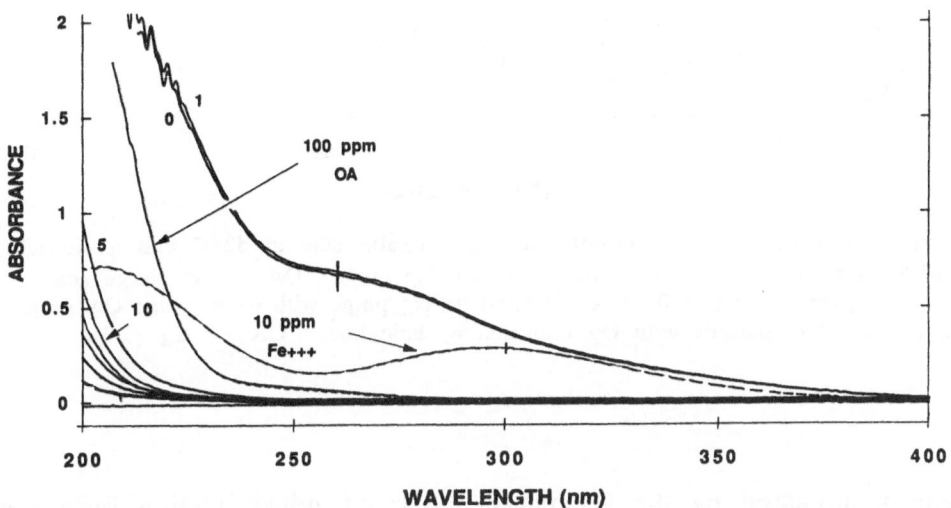

Figure 5. Uv spectra of 100 ppm oxalic acid and 10 ppm Fe^{+++} (added as the perchlorate) in pH 5 solutions in a 1 cm cell. Top spectra and those at the lower left are of samples at various times from Run OA #4: 0, 1, 5, 10, 15, 20, 30, 60, and 90 min. after starting an O_2 purge at t=0 and uv at t=1.

The broad band of the dashed Fe^{+++} spectrum at 300 nm is attributed to $Fe(III)OH^{2+}$,[9] while the top spectrum with a shoulder at 260 nm is assigned to $Fe(III)(C_2O_4)_3^{3-}$. The spectra in the lower left hand corner are of samples taken at various times from the O_2 purged solution during uv radiation. The t=5 minute sample (4 minutes after turning on the lamp) already shows the absorbance at 210 reduced by more than 80% relative to what it was in the 100 ppm OA solution; by t=10 it is reduced about 90%. Rapid mineralization is confirmed by the NPOC measurements seen in curve OA #4 in Figure 6.

A much slower rate of mineralization can be seen in Run OA #1 under N_2, where $t_{0.9}$ is about 65 minutes. The curve for OA #2 which is under N_2 but had 110 ppm H_2O_2 added, falls right on top of the curve with O_2 (OA #4) up till t=10, when titration shows that the H_2O_2 was all consumed. Run OA #3 started off like #1, but the lamp was turned off and the purge switched to O_2 between t=5 and t=10; then the uv and N_2 were switched back on. Dark/O_2 and uv/N_2 periods were alternated during the course of the run. It can be seen that though no mineralization takes place during the dark/O_2 periods, the

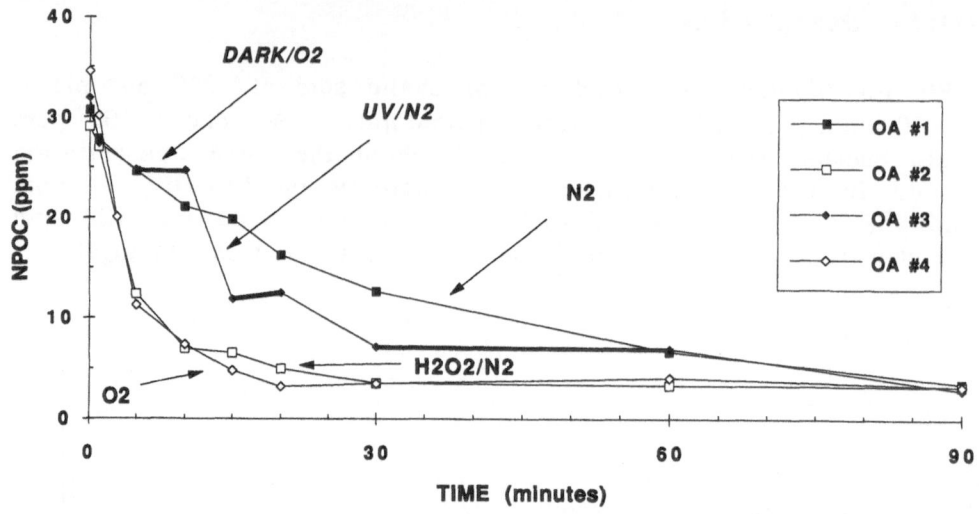

Figure 6. Mineralization of initially 100 ppm oxalic acid at 52°C and an initial pH of 5; uv lamp on at t=1 in all runs. OA #1: N_2 purge. OA #2: N_2 purge and 110 ppm H_2O_2 added at t=0. OA #3: alternate N_2 purge with uv on , and O_2 purge with uv off (dark periods with O_2 indicated by bold line). OA #4: O_2 purge.

system is prepared by the O_2 purge for rapid mineralization once the lamp is turned back on.

DISCUSSION

We proposed previously[6] that the oxygen effect is caused by a competition between O_2 and ferric oxalate complexes for an oxalate radical anion $(C_2O_4^{-\cdot})$ produced by electron transfer from coordinated oxalate to Fe(III) in a photoexcited ferrioxalate complex, Fe(III)Ox*, as shown in Figure 7. The proposal was based on the work of Zepp and coworkers,[10] who showed that Fe(II) photogenerated from Fe(III) oxalate is rapidly oxidized back to Fe(III) by H_2O_2, and that of Zuo and Hoigne,[11] who showed both that H_2O_2 is produced by reactions of O_2 with intermediates formed in the photoreactions of Fe(III) oxalate complexes, and that the dark reaction of O_2 with Fe(II) in the presence of oxalate is very slow. Very slow oxidation of ferrous iron by O_2 in the absence of organic ligands was reported earlier by Mathews and Robbins.[12] For this scheme to explain the results shown in Figure 6 it would be necessary to have O_2 in the solution at the same time that the lamp was on. Though we did switch to N_2 when we turned on the lamp in Run OA #3, it could take a few minutes with the gentle purge used to sparge all of the O_2 from solution.

Whatever the origin of the oxygen effect, it has important practical consequences for anyone wishing to mineralize toxic organic compounds using $H_2O_2/Fe^{++}/uv$. $D_{0.9}$, the electrical energy dose (in kWH/1000 gallons) required to mineralize 90% of the carbon, is given by equation (4). The dose increases in our system by about 28

$$D_{0.9} = 3875(L/kgal) \times 0.45(kW/L) \times (t_{0.9}-1)(min)/60(min/hr) \quad (4)$$

kWH/kgal for each minute of radiation, or about $1.40/kgal-min for electricity costing $0.05/kWH. Irradiation of a waste stream for more than 5-10 minutes becomes prohibitively expensive. (For comparison, 110 ppm of H_2O_2 is about 1 lb/kgal - costing about $1/kgal for $1/lb peroxide.)

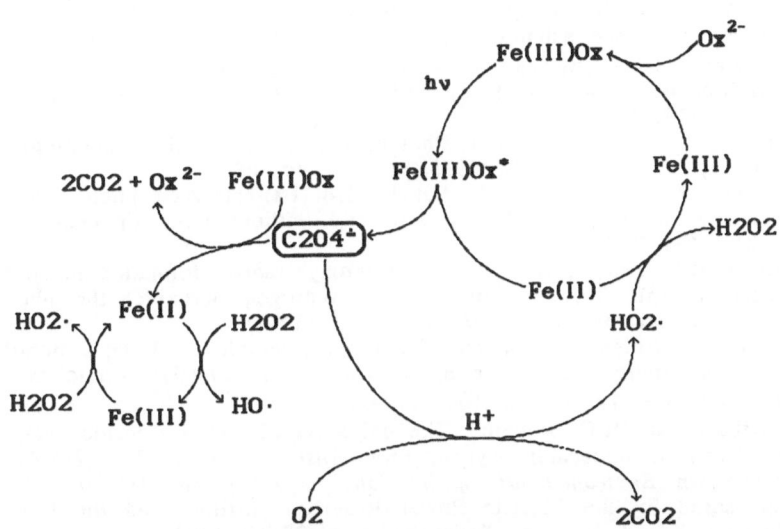

Figure 7. The proposed photo-Fenton mechanism with competition of O_2 and Fe(III)Ox for $C_2O_4^{-\cdot}$. Ox is oxalate.

We learned at a recent EPRI/NSF conference[13] that there are now about 100 commercial uv/H_2O_2 systems in operation for water treatment, with about half of them used for groundwater remediation. Because of the relatively high cost of mineralizing organic compounds using uv, we see the greatest opportunities for uv-based AOP technologies in the treatment of relatively dilute (<50 ppm), low flow (<100 GPM) streams which are not amenable to biodegradation because of toxicity. Combined technologies, such as partial chemical oxidation followed by biodegradation of the detoxified material, are likely to be the most economical.

ACKNOWLEDGEMENTS

We thank Dave Rothfuss, Beth Fenner, John Marcone, and William Lloyd for expert technical assistance. We also thank Fran Robertaccio and Jim Dyer for helpful discussions.

REFERENCES

1. a) ACS Symposium Series 422, "Emerging Technologies in Hazardous Waste Management," D.W. Tedder, ed., ACS Publications. 1989.   b) P. Jackman and R.L. Powell, "Hazardous Waste Treatment Technologies," Noyes Publications, Park Ridge, NJ, 1991.
2. W.R. Haag and C.C.D. Yao, Rate constants for reaction of hydroxyl radicals with several drinking water contaminants, *Environ. Sci. Technol.* 26:1005 (1992)
3. Reference 1b, p. 79.
4. "Handbook of Chemistry and Physics," 41st Ed., C.D. Hodgman, ed., Chemical Rubber Pub. Co., Cleveland, 1959-60, pp. 1913-1921.
5. J.W.T. Spinks and R.J.J. Woods, "An Introduction to Radiation Chemistry," John Wiley & Sons, New York, 1964.
6. C.A. Tolman, W. Tumas, S.Y.L. Lee, and D. Campos, The destruction of phenol in dilute aqueous solution by $H_2O_2/Fe++/uv/O_2$, presented at the *Symposium on Emerging Technologies for Hazardous Waste Management*, Atlanta, Sept. 21-23, 1992, and submitted for publication in the "A.C.S. Symposium Series."
7. J.J. Pignatello, Dark and photoassisted Fe^{3+}-catalyzed degradation of chlorophenoxy herbicides by hydrogen peroxide, *Environ. Sci. Technol.* 26:944 (1992).
8. N.J. Turro, "Modern Molecular Photchemistry," Benjamin Cummings Publishing Co., Inc., Menlo Park, CA, 1978, pp. 528 and 368.
9. K. Ohkubo, Y. Arikawa and S. Sakaki, Iron(III)-catalyzed photo-hydroxylation of benzene in aqueous solution with and without ionic or neutral surfactants, *J. Mol. Cat.* 26:139 (1984).
10. R.G. Zepp, B.C. Faust and J. Hoigne, Hydroxyl radical formation in aqueous reactions (pH 3-8) of iron(II) with hydrogen peroxide: the photo-Fenton reaction, *Environ. Sci. Technol.* 26:313 (1992).
11. Y. Zuo and J. Hoigne, Formation of hydrogen peroxide and depletion of oxalic acid in atmospheric water by photolysis of Iron(III)-oxalato complexes, *Environ. Sci. Technol.* 26:1014 (1992).
12. C.T. Mathews and R.G. Robbins, The oxidation of aqueous ferrous sulphate solutions by molecular oxygen, *Proc. Aust. Inst. Min. Met.*, 242:47 (1972).
13. *Symposium on Environmental Applications of Advanced Oxidation Technologies*, sponsored by the Electric Power Research Institute and the National Science Foundation, San Francisco, Feb. 22-24, 1993.

HOMOGENEOUS CATALYZED SELECTIVE OXIDATIONS
BASED ON O₂ OR H₂O₂. NEW SYSTEMS AND
FUNDAMENTAL STUDIES

Craig L. Hill,[1] Dean C. Duncan,[1] Eric A. Hecht,[1] and Ira
Weinstock[2]

[1]Department of Chemistry
Emory University
Atlanta, GA 30322
[2]Forest Products Laboratory
Madison, Wisconsin 53705-2398

INTRODUCTION

We present here results that impact our fundamental knowledge regarding
metal catalyzed O_2 based oxidations and systems that address the mechanism of
the Ishii/Venturello H_2O_2 based catalytic epoxidation chemistry. Our efforts
continue to focus on fundamental energetic and mechanistic issues that define
and indeed limit further development in the burgeoning area of selective
catalytic oxidation. One particular application of a selective O_2 based oxidation
applied to an issue of practical and growing importance, the selective bleaching
of wood pulp for paper manufacture, is a collaborative venture between our
laboratories at Emory University and those of Dr. Weinstock and collaborators
at the U.S. Department of Agriculture Forest Products Laboratory.

The general goal is to develop catalytic systems that facilitate the selective,
rapid, and sustained oxidation of organic substrates, RH, with optimal (inexpensive and environmentally benign) oxidants and reaction conditions.[1-4] In
addition, we seek systems that are versatile; that is, they have rationally

The Activation of Dioxygen and Homogeneous Catalytic Oxidation,
Edited by D.H.R. Barton *et al.*, Plenum Press, New York, 1993

71

tunable reactivities and are flexible in formulation. We have recently focused our attention on some of the most difficult aspects of homogeneous catalytic oxidation, namely, the oxidation of highly inert or sensitive and complex organic or biological molecules using the two most desirable oxidants, O_2 and H_2O_2. Unfortunately, O_2 and H_2O_2 are among the most difficult of terminal oxidants to use. These oxidants have more modes of reaction, more background (uncatalyzed) reactions with substrate, and have a tendency to oxidize by mechanisms that are not readily or rationally controlled.[1,2] Both H_2O_2 and O_2 can function as oxygen donors and in other oxidative capacities.

It is in context with the use of H_2O_2 and O_2 that we now enumerate six of the most significant limitations or difficulties with respect to the development of optimal, selective, metal-complex-based catalytic oxidation systems. Some of the limitations are illustrated in generic equation form below (L_xM = transition metal center with x ligands L; OX and RED are the oxidizing agent and its complementary reduced form).[1-7] First, the catalyst must be very stable under turnover conditions. Of particular concern is irreversible oxidative degradation of organic ligands or matrices (equation 1). All organic materials including many fluorocarbons are thermodynamically unstable with respect to further oxidation.[8] Second, the formation of dead-end intermediate oxidation states can be a problem. In catalytic aerobic oxygenation (no reducing agent), intermediate oxidation state complexes, i.e. $L_xM^{(n+1)+}$, can be a problem as such complexes often do not react readily with O_2 and are often reduced or oxidized slowly if at all. Such species can be generated by direct reaction with O_2 (equation 2) or peroxy species (equation 3), by reduction of metal oxo species (equation 4) or by direct outer sphere oxidation (equation 5). Third, the catalyst can inactivate by μ-oxo dimer formation, for example, equation 6. Many such dimers are inactive, less active, or still active but less selective in oxidant activation and/or substrate oxidation than the starting monomer. Fourth, there may be diffusional problems. This is particularly an issue when the catalyst is an extended rigid matrix such as a zeolite. One subset of this problem is that products can often be substantially larger than substrates and can be too large to diffuse out of pores in such catalysts. Fifth, inhibition by one or more species generated during catalysis can be a real problem (e.g. equation 7) where the adduct complex, $L_x(RO)M^{n+}$, is either unreactive or reactive but less selective. The sixth and final limitation, and the most significant one for metal-facilitated oxidation by O_2 in general, is the initiation of autoxidation if, in fact, this process is not desired.

Many of the catalytic approaches and systems under investigation in our laboratories are based on soluble early transition metal oxygen anion clusters.

$$L_xM + OX \longrightarrow (L)_{x-1}(LO)M + RED \qquad (1)$$

$$[L_xMO_2] \longrightarrow L_xM^{(n+1)+} + O_2{}^{-} \qquad (2)$$

$$L_xMOOY \longrightarrow L_xM^{(n+1)+} + YOO^-, \; Y = M \text{ or } H \qquad (3)$$

$$L_xM^{(n+2)+}O + RED + Y^+ \longrightarrow L_xM^{(n+1)+} + YO^- + OX \qquad (4)$$

$$L_xM^{n+} + OX \longrightarrow L_xM^{(n+1)+} + RED \qquad (5)$$

$$[L_xM^{(n+2)+}O] + L_xM^{n+} \longrightarrow [L_xM^{(n+1)}\text{-}O\text{-}M^{(n+1)}L_x] \qquad (6)$$

$$L_xM^{n+} + RO \rightleftharpoons L_x(RO)M^{n+} \qquad (7)$$

These complexes, which we will refer to as polyoxometalates for convenience, are unusually attractive for a number of reasons:[9,10] (1) Many of them are readily prepared in quantity from inexpensive starting materials; (2) Most are very low in toxicity; (3) They have a range of properties directly relevant to catalytic oxidation that can be altered synthetically including ground state redox potentials, shapes and molecular charges (of potential value in the selective recognition of substrates, etc.), solubility, and other properties; (4) They can be derivatized with organic groups and readily immobilized by adsorption or covalent attachment.

Additional features of polyoxometalates impact some of the six limitations. First, polyoxometalates are oxidatively resistant as a great majority of them are composed of d^0 transition metal ions (most commonly W(VI), Mo(V), and/or V(V)) and oxide ions. Second, the unusually high stability of polyoxometalates coupled with the tunability of their catalytically relevant properties defines a considerable ability to avoid problems associated with intermediate oxidation states, μ-oxo dimers, and product inhibition. Illustrations of two representative polyoxometalates, $W_{10}O_{32}{}^{4-}$, an isopolyoxometalate (or isopolyanion) and $(TM)XW_{11}O_{39}{}^{n-}$, a transition metal (TM) substituted heteropolyoxometalate (or heteropolyanion) in polyhedral notation are given in Figure 1.

NON-RADICAL-CHAIN SELECTIVE OXIDATIONS BY O_2 CATALYZED BY POLYOXOMETALATES

Many highly oxidizing polyoxometalates, particularly those containing vanadium(V), have long been known to directly oxidize a range of organic substrates, R, (equation 8) and that reduced forms could be reoxidized by a variety of oxidants, OX, and under some conditions, by dioxygen (OX = O_2), equation 9. The net reaction, equation 10, (the sum of equations 8 and 9) is

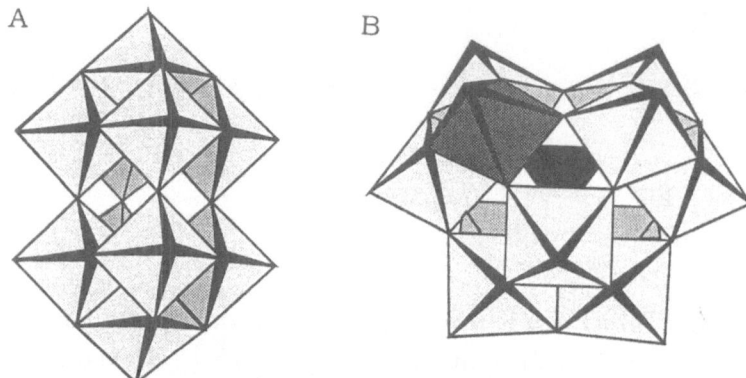

Figure 1. Representative polyoxometalates in polyhedral notation. **A.** the isopolyanion decatungstate ($W_{10}O_{32}^{4-}$) (D_{4h} point group symmetry) and **B.** the heteropolyanion family, $(TM)PW_{11}O_{39}^{5-}$, where TM is a first row divalent transition metal ion, P is the heteroatom (C_s point group symmetry). In the latter class of complexes, which constitute functional oxidatively resistant inorganic metalloporphyrin analogs,[11] P is one of many elements that can function as the heteroatom. The darker octahedron on the surface and the very dark internal tetrahedron of **B** represent the TM ion and the heteroatom, respectively. In polyhedral notation a complementary notation to ball-and-stick or bond representations, the vertices of the polyhedra, principally WO_6 octahedra, are the nuclei of the oxygen atoms. The metal atoms lie inside each polyhedron.

highly attractive as it could form the basis of selective aerobic oxidations (*vide infra*). Several years ago, the Novosibirsk group reported direct oxidations of sulfur compounds and bromide by polyoxometalates (i.e. equation 8).[12] In 1989 three groups reported the homogeneous oxidation of a variety of organic substrates catalyzed by the representative and readily available heteropolyanion, $H_5PV_2Mo_{10}O_{40}$ (equation 10 via equations 8 and 9). Our group investigated the mechanism of oxidation of thioethers, RSR', with

$$R + P_{ox} \longrightarrow R_{ox} + P_{red} \qquad (8)$$

$$P_{red} + OX \longrightarrow P_{ox} + RED \qquad (9)$$

$$R + OX \xrightarrow{\;\;P_{ox}\text{ (catalyst)}\;\;} R_{ox} + RED \qquad (10)$$

extraordinarily high (>99%) selectivity, by the widely used commercial oxidant t-butylhydroperoxide (OX = TBHP).[13] Brégeault and co-workers reported the highly selective aerobic oxidative cleavage of some ketones with little or no mechanistic investigation[14,15] and Ronny Neumann and co-workers reported the aerobic oxidation of amines, some alcohols, some alkylaromatic hydrocarbons and bromide by equations 8-10.[16-19]

In a recent collaborative venture between Emory and the U.S.D.A. Forest Products Laboratory in Madison, Wisconsin, the chemistry in equations 8-10 was applied to a problem of considerable magnitude and environmental importance, the bleaching of wood pulp in conjunction with the manufacture of paper.[20] Wood is a highly organized composite of carbohydrate polymers (cellulose and hemicelluloses) and lignin. In the production of high quality printing and writing paper, most of the lignin is removed via chemical pulping. During this process, a variety of highly colored conjugated aromatic structures are generated, principally from lignin. The purpose of bleaching is to remove these colored materials along with any residual lignin not removed during pulping. At the present time, wood pulp is bleached by chlorine or chlorine-based oxidants and although these oxidants achieve the desired chemistry - the selective oxidative bleaching of the lignin with minimal damage to the cellulose - they have a serious problem. Varying quantities of chlorinated organic by-products are produced and while some of these are relatively benign, others, such as the dioxins, are among the most deleterious of substances. Nearly all chlorinated materials are toxic or carcinogenic to a measurable degree. When one considers the amount of wood pulp bleaching that takes place in the developed countries, it is not hard to see that the environmental impact of chlorine based bleaching is rapidly becoming socially, politically, and economically unacceptable.

In preliminary work, we have demonstrated that polyoxometalates can catalyze equations 8-10 where the substrate, R, is wood pulp and the oxidant, OX, is either H_2O_2 or O_2. This approach is effective as the lignin is selectively oxidized and removed; the cellulose is minimally affected. Figure 2 gives molecular representations of lignin and cellulose as well as the general selective oxidation process (equation 8) for bleaching wood pulp by polyoxometalates. Currently, vanadium containing polyoxometalates ("V-polyoxo" in Figure 2) have proven to be the most effective at combining both high selectivity and reasonable rates for equation 8 with the ability to be reoxidized with O_2 (slow at ambient temperature) or H_2O_2 (rapid at ambient temperature). One of the most pleasing aspects of this chemistry is that the number and diversity of polyoxometalates that could, in principle, facilitate

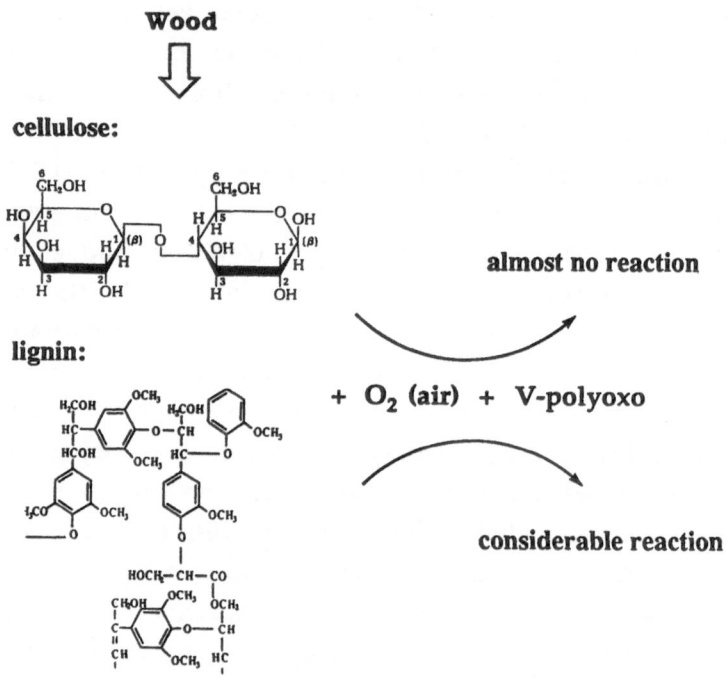

Figure 2. Basic scheme for environmentally friendly bleaching of wood - a new and catalytic bleaching technology for paper manufacture. No chlorinating agents are used. Wood is composed of two principal biopolymers cellulose and lignin whose structures (representative repeating units) are illustrated.

equations 8-10 is so substantial that the oxidative reactivity and selectivity (equation 8) can be fine tuned for a particular application (e.g. lignin versus cellulose). A telling point here is that 30 or so new polyoxovanadate structures have been synthesized and characterized in the last 4 years alone. During this brief interval, the number of isopolyvanadates that contain only V(V) has increased from 2 to 6.[21-23] The molecular diversity and range of redox characteristics among the vanadium-containing polyoxometalates alone appears to be vast and it is still largely unexplored and undeveloped.

A key in the use of dioxygen as a terminal oxidant in catalyzed oxidations lies in equations 8-10, namely the separation of catalyst (polyoxometalate) reduction - substrate oxidation, equation 8, from the reduced catalyst reoxidation by O_2, equation 9. In part, as the reduced forms of the polyoxometalates are usually low in reactivity and very stable under turnover conditions, equations 8 and 9 can be separated from one another in time and/or in space. As radicals and other reactive species that can initiate radical chain oxidation by O_2 (autoxidation), the dominant mode of organic oxidation by this oxidant, are generated in equation 8, autoxidation can be avoided by separating equations 8 and 9. This fact has been appreciated by other groups working in this area. We turn now to another aspect of the chemistry in equations 8-10 that is subtle but has considerable potential consequences for the metal-catalyzed or facilitated O_2-based oxidations and that is the nature of the O_2 reoxidation step, equation 9.

The reoxidation of reduced polyoxometalates by O_2 was first examined by Matveev and co-workers who concluded that reduction proceeded by a multi-electron transfer in a complex involving O_2 and the reduced polyoxometalate (a "heteropoly blue").[24,25] Within the last two years two other groups, those of Papaconstantinou and Neumann, also addressed equation 9 where OX = O_2 for soluble reduced polyoxotungstates. Hiskia and Papaconstantinou investigated the reoxidation of three one-electron-reduced heteropolytungstates in water and reported interesting rate-pH profiles. Little hard mechanistic information was inferred, however, including whether or not oxygen atoms from O_2 were incorporated into the polyoxometalate or not.[26] Neumann and co-workers inferred a mechanism involving binding and then incorporation of oxygen atoms into the polyoxometalate skeleton versus an "outer sphere" mechanism not involving incorporation of oxygen atoms into the polyoxometalate skeleton based on infrared ^{18}O-labeling experiments.[19]

Two very curious but heretofore unexplained phenomena relate to the interaction of reduced polyoxometalates with O_2 and are addressed by our experiments described right below. The first involves the lack of oxygenated products seen in the photochemical functionalization of heptacyclotetradecane (HCTD) in the presence of O_2 by Christina Prosser-McCartha.[27,28] The second involves two observations by Neumann and Levin,[19] namely, the suppression of tetralin autoxidation and the clean conversion of α-terpinene to p-cymene in the presence of O_2 without formation of oxygenated products.

Shown in Fig. 3 is one of several preliminary results obtained recently in our laboratory addressing the general mechanistic features of reduced polyoxometalate reoxidations. Here, two distinct species[29] of the two electron reduced decatungstate $[H_xW_{10}O_{32}]^{(6-x)-}$, where x = 0 or 2, were oxidized with

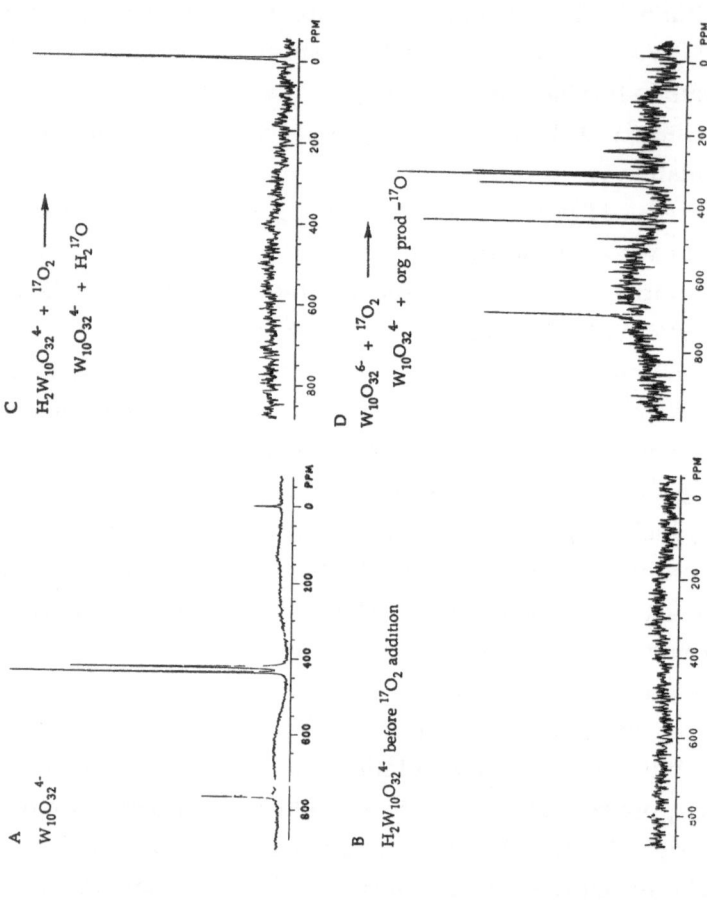

Figure 3. A. Natural Abundance ^{17}O NMR spectrum of $Q_4[W_{10}O_{32}]$ in acetonitrile where $Q = (n\text{-}C_4H_9)_4N^+$. Spectrometer frequency = 67.8234288 MHz; pulsewidth = 20 μs (θ = 37°); 2K data points recorded in double precision; relaxation delay = 20 Hz; line broadening = +/- 33,333 Hz; line broadening = 20 Hz; receiver gate = 1 ms; receiver gate = 29 μs; dwell time = 15 μs; acquisition time = 15.36 ms; spectral width = +/- 33,333 Hz; line broadening = 20 Hz; temperature = 65 °C; 1,048,576 acquisitions (4.5 hrs); 10-mm tube; concentration = 0.13 M. B. Two mL of 0.005 M $Q_6[W_{10}O_{32}]$ in 0.5 M QPF_6/CH_3CN containing two equivalents of triflic acid (added as a 0.1 M CH_3CN solution). Same parameters as in A except the following: relaxation delay = 21 ms; receiver gate = 13 μs; spectral width = 35,714 Hz; dwell time = 14 μs; acquisition time = 14.34 ms; 50,000 acquisitions (29.5 min); temperature = 24 °C; line broadening = 60 Hz; unlocked (spectrometer drift < 0.2 ppm over 24 hours). C. Sample B completely oxidized with a 10-fold excess (50 μmol) of 20 atom % $^{17}O_2$. The H_2O peak increased with time and no other signals were observed during the oxidation. Same parameters as in B. D. Identical sample to C without triflic acid. Same parameters as in B and C except: receiver gate = 27 μs.

78

^{17}O-labeled O_2 in acetonitrile and the transformations studied by ^{17}O NMR. Fig. 3A shows the natural abundance ^{17}O NMR spectrum of the oxidized isopolyanion $[W_{10}O_{32}]^{4-}$ in CD_3CN. Using peak integration[30] to aid in assignment, the peaks at 765 and 732 ppm correspond to the belt and cap terminal oxygens respectively; the resonances at 434 and 420 ppm are assigned to the three types of μ_2 bridging oxygens where the latter peak is unresolved; and the signal at -1.6 ppm is due to the two interior μ_5 bridging oxygens which lie on the C_4 axis. Further details concerning this spectrum will be published elsewhere. Spectra B, C, and D show the oxidation of $[H_xW_{10}O_{32}]^{(6-x)-}$ with $^{17}O_2$. Specifically, Fig. 3B exhibits no signal for an acetonitrile solution of $[H_2W_{10}O_{32}]^{4-}$ after acquiring 50,000 transients. Following the addition of a 10-fold excess of 20 atom % $^{17}O_2$ (Fig. 3C), the only signal observed both during and after the oxidation was at -8.5 ppm due to H_2O formation.[31] In contrast, the oxidation of $[W_{10}O_{32}]^{6-}$ with $^{17}O_2$ (Fig. 3D) resulted in the formation of several oxygenated organic products (the main peaks are 697, 443, 425, 337, 313, and 306 ppm) at the expense of water. No resonances attributable to oxygens of the isopolyanion framework were detected. This evidence suggests the dominant mechanism for O_2 reoxidation of $[H_xW_{10}O_{32}]^{(6-x)-}$ does not involve direct incorporation of oxygen from O_2 into the polyoxometalate. More interestingly, the presence of H^+ leads to suppression of oxygenated organic products giving only H_2O as the product of O_2 reduction. This result suggests that equations 8 and 9 may not need to be separated; that is, in the presence of both H^+ and O_2, P_{ox} may selectively oxidize a substrate followed by O_2 reacting exclusively with P_{red} to generate H_2O thereby completing the catalytic cycle without concomitant oxygenation of the substrate.

Similar results were obtained for $H_5[PV_2Mo_{10}O_{40}]$ and $H_3[PMo_{12}O_{40}]$ in the presence of the substrate 1,3-cyclohexadiene. In particular, during the reoxidation of reduced $H_5[PV_2Mo_{10}O_{40}]$ (0.1817 g/2 mL degassed CD_3CN + 1 eq $^{17}O_2$ + 5 eq 1,3-cyclohexadiene), no increase in the polyoxometalate peak intensities was observed within the first seven spectra taken (total reoxidation time at this point is 245.6 min) whereas the water resonance had increased by a factor of nine. Furthermore, the initial rates of three polyoxometalate peaks (these could be clearly observed in the first spectrum, $i.\ e.$ without $^{17}O_2$) were measured giving relative rates ($V_{H_2^{17}O}$ formation/$V_{P_{ox}-^{17}O}$ incorporation) of 5.4, 6.8, and 7.7 for 930, 925, and 544 ppm respectively. In a control experiment, similar initial rates for ^{17}O incorporation were obtained for $H_5[PV_2Mo_{10}O_{40}]$ + $H_2^{17}O$ with respect to the reoxidation reaction. Consequently, both the presence of a substantial induction period and the relative rate data strongly suggest that the source of ^{17}O in $[PV_2Mo_{10}O_{40}]^{5-}$ is from $H_2^{17}O$ and not $^{17}O_2$;

therefore, an outer sphere mechanism is operable for O_2 reoxidation in this system contrary to the recent report by Neumann and Levin.[19]

INVESTIGATION OF THE MECHANISM OF THE VENTURELLO-ISHII EPOXIDATION SYSTEM

The selective epoxidation of alkenes including terminal alkenes by hydrogen peroxide catalyzed by mixtures of phosphate and tungstate under phase transfer conditions was first reported by Venturello et al. in 1983[32] and subsequently developed by this group.[32-34] Five years later, the group of Ishii in Japan began to develop catalysis for the same process but using $PW_{12}O_{40}{}^{3-}$ as a catalyst precursor rather than a mixture of $PO_4{}^{3-}$ and $WO_4{}^{2-}$ used by Venturello and co-workers. The Ishii group subsequently developed this system extensively, a research effort that continues to this day.[35-37] Both the Venturello and Ishii systems involved two liquid phases, water and chlorocarbon, with phase transfer cations.[32-37] The key point is that both these systems gave remarkably selective epoxidation and this chemistry is now used commercially. As promising as this selective homogeneous catalytic oxidation chemistry appears at present, very little is known about what transpires at the molecular level in this chemistry.

In 1985 the group of Venturello et al. reported the isolation from their system and the X-ray crystal structure of $\{PO_4[W(O)(O_2)_2]_4\}^{3-}$.[33] They proposed that it was the active species based on some suggestive but not convincing experimental evidence. Ishii later came to believe this was also the active species in his system. Again, he offered some suggestive but not definitive evidence on this point.[36] Brégeault and co-workers then prepared and spectroscopically characterized some peroxotungstates and concluded that the $\{PO_4[W(O)(O_2)_2]_4\}^{3-}$ was indeed the active species.[38] Unfortunately, no kinetics or time resolved studies were done by any of these groups. As a consequence, nothing was known with certainty regarding the active epoxidizing specie(s) in the Venturello and Ishii systems. A point that can not be made too strongly is that isolated and characterized metal complexes are rarely the true active catalysts. They usually represent kinetic cul-de-sacs. The true catalytic species are often present only in small concentrations and not amenable to characterization as they are by definition, highly reactive. It was against this backdrop that we decided to investigate the kinetics not only of formation of $\{PO_4[W(O)(O_2)_2]_4\}^{3-}$ under the Venturello-Ishii epoxidation reaction conditions but also of reaction of this and other peroxotungstates with the alkene substrates. The details of this chemistry, including the experimental protocols, results and discussion will be published elsewhere. Research to date has clarified the following points, however, which we feel confident in reporting here.

(1) The complex $\{PO_4[W(O)(O_2)_2]_4\}^{3-}$, can be isolated from a reaction of $H_3PW_{12}O_{40}$ and H_2O_2 after going through several higher nuclearity species derived from sequential breakdown of the parent Keggin polyanion.

(2) $\{PO_4[W(O)(O_2)_2]_4\}^{3-}$ is a catalytically important species and perhaps the most reactive epoxidizing agent in the system.

(3) $\{PO_4[W(O)(O_2)_2]_4\}^{3-}$ is unstable to degradation in the presence of H_2O but is more stable in the presence of H_2O_2 (50 equivalents).

(4) Species formed after the reaction of $\{PO_4[W(O)(O_2)_2]_4\}^{3-}$ with alkene are also active epoxidizing agents under the reaction conditions although collectively less so than $\{PO_4[W(O)(O_2)_2]_4\}^{3-}$ itself.

(5) These subsequent epoxidizing species can reform $\{PO_4[W(O)(O_2)_2]_4\}^{3-}$ upon reaction with as little as one equivalent of H_2O_2.

(6) The Venturello-Ishii chemistry is extraordinarily complex. There may be a multitude of weaker epoxidizing agents that are peroxotungstates of higher nuclearity.

EXPERIMENTAL

Q$_4$[W$_{10}$O$_{32}$] was prepared by the method of Chemseddine et. al.[39] All chemicals and solvents purchased were of the finest quality available and no further purification was required. A solution of 0.005 M [W$_{10}$O$_{32}$]$^{6-}$ was prepared by controlled potential electrolysis at -2.10 V vs. 0.01 M Ag/AgNO$_3$ in acetonitrile containing 0.5 M QPF$_6$. Triflic acid was added as a 0.1 M acetonitrile solution. ^{17}O NMR parameters may be found in the Figure 3 caption. Increasing the delay time in the natural abundance ^{17}O NMR spectrum of Q$_4$[W$_{10}$O$_{32}$] (and several other polyoxometalates) led to no increase in relative peak integrals; consequently, none of the resonances are saturated under the experimental conditons as suggested from linewidth measurements and assuming $T_2=T_1$.[40] In the $^{17}O_2$ reoxidation experiments, the acquisition parameters were chosen to allow for complete relaxation of H_2O (and therefore also [W$_{10}$O$_{32}$]$^{4-}$) while still maximizing the signal-to-noise for observing any ^{17}O incorporation into [W$_{10}$O$_{32}$]$^{4-}$ (i. e. long delay time, short acquisition time). Several of the oxygenated organic peaks in Fig. 3D, however, are much narrower than H_2O under these conditions and some saturation of these resonances is evident. Further experimental details including the electrochemical synthesis and characterization of $[W_{10}O_{32}]^{6-}$ and its protonated derivatives will be published elsewhere.

ACKNOWLEDGMENTS

CLH wishes to thank the National Science Foundation (grant CHE-9022317), the Army Research Office (DAAL03-87-K-0131), and Interox Corporation for support of the different research programs addressed in this article. We thank Mark Weeks in our group for the synthesis of several vanadium containing polyoxometalates. R. Carlisle Chambers in our group initiated the work on the kinetics of the Venturello-Ishii chemistry.

REFERENCES

(1)   Hill, C. L.; Khenkin, A. M.; Weeks, M. S.; Hou, Y. In *ACS Symposium Series on Catalytic Selective Oxidation*; S. T. Oyama and J. W. Hightower, Eds.; American Chemical Society: 1993, in press.

(2)   Sheldon, R. A.; Kochi, J. K. *Metal-Catalyzed Oxidations of Organic Compounds*; Academic Press: New York, 1981, Chapter 3.

(3)   Sheldon, R. A. *ChemTech* **1991**, 566-576.

(4)   Drago, R. S. *Coord. Chem. Rev.* **1992**, *117*, 185-213.

(5)   References 2-4 and 6-7 are general reviews of oxidations catalyzed by metal complexes.

(6)   Jørgensen, K. A.; Schiøtt, B. *Chem. Rev.* **1990**, *90*, 1483-1506.

(7)   Meunier, B. *Chem. Rev.* **1992**, *92*, 1411-1456.

(8)   For example, see Christe, K. O. *Chem. & Eng. News* **1991**, *[October 7th Issue]*, 2.

(9)   Pope, M. T. *Heteropoly and Isopoly Oxometalates*; Springer-Verlag: Berlin, 1983.

(10) Pope, M. T.; Müller, A. *Angew. Chem., Int. Ed. Engl.* **1991**, *30*, 34-48.

(11) Hill, C. L.; Kim, G.-S.; Prosser-McCartha, C. M.; Judd, D. In *Polyoxometalates: From Platonic Solids to Anti-retroviral Activity*; M. T. Pope and A. Müller, Eds.; Kluwer Academic Publishers: Dordrecht, Netherlands, 1993; submitted for publication.

(12) Kozhevnikov, I. V.; Matveev, K. I. *Applied Catalysis* **1983**, *5*, 135 and references cited.

(13) Hill, C. L.; Faraj, M. In *Proceedings of the 1989 U.S. Army Chemical Research, Development and Engineering Center Scientific Conference on Chemical Defense Research*; US Army Armament, Munitions & Chemical Command, 1990; pp 131-134.

(14) Ali, B. E.; Brégeault, J.-M.; Mercier, J.; Martin, J.; Martin, C.; Convert, O. *J. Chem. Soc., Chem. Commun.* **1989**, 825-826.

(15) Ali, B. E.; Brégeault, J.-M.; Martin, J.; Martin, C. *New J. Chem.* **1989**, *13*, 173-175.

(16) Neumann, R.; Lissel, M. *J. Org. Chem.* **1989**, *54*, 4607.

(17) Neumann, R.; Levin, M. *J. Org. Chem.* **1991**, *56*, 5707.

(18) Neumann, R.; Assael, I. *J. Chem. Soc., Chem. Commun.* **1988**, 1285-1287.

(19) Neumann, R.; Levin, M. *J. Am. Chem. Soc.* **1992**, *114*, 7278-7286.

(20) Weinstock, I. A.; Hill, C. L.; Minor, J. L. In *Second European Workshop on Lignocellulosics and Pulp*; Grenoble, France, September 14, 1992; General Abstracts.

(21) Day, V. W.; Klemperer, W. G.; O. M. Yaghi, O. M. *J. Am. Chem. Soc.* **1989**, *111*, 5959-5961.

(22) Hou, D.; Hagen, K. S.; Hill, C. L. *J. Am. Chem. Soc.* **1992**, *114*, 5864-5866.

(23) Hou, D.; Hagen, K.; Hill, C. L. *J. Chem. Soc., Chem. Commun.* **1993**, 426-428.

(24) Berdnikov, V. M.; Kuznetsova, L. I.; Matveev, K. I.; Kirik, N. P.; Yurchenko, E. N. *Koord. Khim.* **1979**, *5*, 78.

(25) Kozhevnikov, I. V.; Burov, Y. V.; Matveev, K. I. *Izv. Akad. Nauk. SSSR, Ser. Khim.* **1981**, 2428.

(26) Hiskia, A.; Papaconstantinou, E. *Inorg. Chem.* **1992**, *31*, 163-167.

(27) Prosser-McCartha, C. M.; Hill, C. L. *J. Am. Chem. Soc.* **1990**, *112*, 3671-3673.

(28) Prosser-McCartha, C. M. Ph.D. Thesis, Emory University, 1992.

(29) Complete characterization will be reported elsewhere.

(30) Peak integrations were normalized for uneven power distribution over the spectral window using the following equation: $A_i = A_0 \left\{ \dfrac{\sin x_i}{x_i} \right\}$, where A_i is the power for a given value of x_i and A_0 is the power for $x_i = 0$ ($x_i = \pi \Delta \nu_i \tau$; $\Delta \nu_i$ is the carrier offset frequency in kHz and τ is the pulsewidth in μs).

(31) The H_2O chemical shift in CH_3CN was found to vary with concentration and temperature (- 6 to - 8.8 ppm).

(32) Venturello, C.; Alneri, E.; Ricci, M. *J. Org. Chem.* **1983**, *48*, 3831-3833.

(33) Venturello, C.; D'Aloiso, R.; Bart, J. C.; Ricci, M. *J. Mol. Catal.* **1985**, *32*, 107.

(34) Venturello, C.; Gambaro, M. *Synthesis* **1989**, *4*, 295.

(35) Ishii, Y.; Yamawaki, K.; Ura, T.; Yamada, H.; Yoshida, T.; Ogawa, M. *J. Org. Chem.* **1988**, *53*, 3587-3593.

(36) *Hydrogen Peroxide Oxidation Catalyzed by Heteropoly Acids Combined with Cetylpyridium Chloride*; Ishii, Y.; Ogawa, M., Eds.; MYU: Tokyo, 1990; Vol. 3, pp 121-145.

(37) Sakaue, S.; Sakata, Y.; Nishiyama, Y.; Ishii, Y. *Chem. Lett.* **1992**, 289-292.

(38) Aubry, C.; Chottard, G.; Platzer, N.; Brégeault, J.-M.; Thouvenot, R.; Chauveau, F.; Huet, C.; Ledon, H. *Inorg. Chem.* **1991**, *30*, 4409-4415.

(39) Chemseddine, A.; Sanchez, C.; Livage, J.; Launay, J. P.; Fournier, M. *Inorg. Chem.* **1984**, *23*, 2609-2613.

(40) The dominant mechanism for nuclear relaxation in ^{17}O NMR is via quadrupolar relaxation; thus, for rapid molecular tumbling, the condition $T_1 = T_2$ holds. See (a) Butler, L. G. in ^{17}O *NMR Spectroscopy in Organic Chemistry*, Boykin, D. W., Ed., CRC Press: Boca Raton, 1991, Ch.1; (b) Abragam, A. *Principles of Nuclear Magnetism*, 2nd ed. Clarendon Press: Oxford, 1983, p. 314.

ELECTROPOX: BP'S NOVEL OXIDATION TECHNOLOGY

T.J. Mazanec

BP Research and Environmental Science Center
4440 Warrensville Center Road
Cleveland, Ohio 44128
USA

ABSTRACT

Solid oxide electrochemical cells and oxygen transporting membranes have been used to conduct oxygen separation and partial oxidations simultaneously in a single reactor. As the substrate hydrocarbon never comes into contact with free oxygen, enhanced selectivity and unusual reactivity are to be expected. Important mechanistic insights and the development of Electropox, a new process for converting methane to syngas, have resulted from this work.

Studies of methane oxidation using solid oxide fuel cells show that high selectivities to C_2+ can be obtained by appropriate choice of electrocatalyst. In all cases selectivity decreases with increasing conversion, just as it does in the heterogeneously catalyzed reaction. A consecutive reaction scheme is proposed to account for the limit of the methane coupling yield, and co-feed experiments using CH_4/C_2H_4 mixtures support the mechanism. Operation of the cells at higher temperature results in a process integrating oxygen separation and syngas production into a single step, which we call Electropox.

Improvements in cell design and engineering have paralleled the Electropox process development. Advanced cell concepts have evolved that incorporate both electronic and ionic conducting functionalities into a single membrane. Oxygen transport rates have been demonstrated with these materials that may permit their use as oxygen separating membranes in chemical reactors.

INTRODUCTION

A great variety of chemicals and fuels are prepared by oxidation processes. Engineering of the admixing of dioxygen and hydrocarbon in these processes has generated a number of different reactors and concepts that allow selective reactions to be conducted that would otherwise remain unselective. A design that has only recently received

The Activation of Dioxygen and Homogeneous Catalytic Oxidation,
Edited by D.H.R. Barton *et al.*, Plenum Press, New York, 1993

85

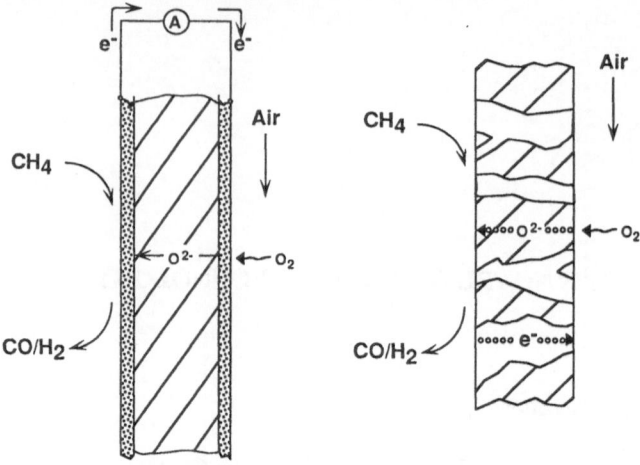

Figure 1 Externally Short Circuited Cells and Dual Phase Membranes

significant attention in this regard is the solid membrane reactor concept in which oxygen separation from air and catalytic oxidation are conducted simultaneously.

In a ceramic electrochemical reactor oxygen activation is physically separate from substrate activation. Activation of dioxygen occurs at a cathodic surface, oxygen traverses the electrolyte as oxide ions, the ions react with a substrate on the anodic face, and the electrons return to complete the circuit. Two types of electrochemical membrane reactors are under study which differ in the route by which the electrons travel from anode to cathode; these are illustrated schematically in Figure 1. In the shorted fuel cell reactor the electrons return through an external circuit, from which electrical power can be extracted. Electroceramic membranes allow the electrons to make their way from anode to cathode via electrically conductive phases within the membrane proper.

Activation of dioxygen in either of these electroceramic reactors offers the opportunity to introduce oxygen exclusively as oxide ions. This is in sharp contrast to conventional homogeneous and heterogeneous processes wherein dioxygen is activated in the presence of the substrate. In conventional dioxygen activation species such as superoxide, peroxide, or oxygen radicals can be formed by succesive reduction of oxygen, as shown in Figure 2. By directly introducing oxide ions ceramic electrochemical reactors access the 'back door' of the manifold of oxygen species compared to conventional processes. It is hoped that the ability to introduce oxygen as oxide ions will lead to unusual reactivity or selectivity in catalytic oxidations.

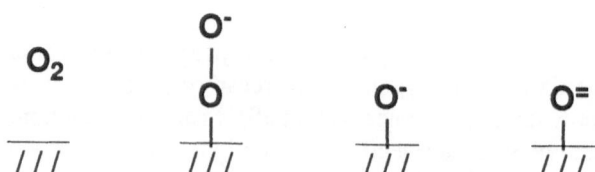

Figure 2 Oxygen Activation Manifold

A particularly challenging oxidation that we have studied in solid electrochemical membrane reactors is the partial oxidation of methane. Methane, the chief constituent of natural gas, is a very robust hydrocarbon. Products of methane reactions are generally more reactive than methane, limiting schemes for direct methane conversion to useful materials.[1-2] With this in mind, attempts to upgrade methane have begun to focus on the relative rates of oxidation of methane compared to its products[3-4] and attention has been directed toward the fundamental kinetic limitations of conventional processes. Gas phase molecular oxygen and partially reduced oxygen species often have been cited as responsible for the destructive oxidation of the desired products. If this is the case a membrane reactor has the potential to overcome the kinetic limitation of the methane oxidation process by keeping free molecular oxygen separated from the products.

A number of studies of the methane oxidation reaction have been reported using solid oxide fuel cells.[5-18] Results of these studies demonstrate that fuel cells can act as oxygen membranes to produce useful products from methane. The present contribution considers the important question of whether the promise of enhanced selectivity can be achieved in methane oxidation using oxygen membrane reactors. Results are presented from studies on shorted fuel cells as the oxygen separating device and from studies using advanced membrane concepts to separate oxygen from air.[19-22] An exciting new application of these devices has been discovered to be the partial oxidation of methane to synthesis gas in a process referred to as Electropox.[23]

RESULTS AND DISCUSSION

Tubular fuel cell reactors can be prepared easily by applying anode and cathode layers to the opposing faces of dense yttria stabilized zirconia (YSZ) tubes. The cathode and

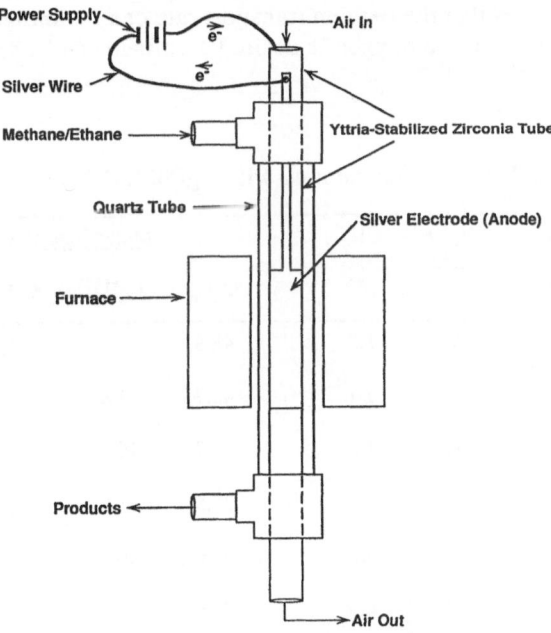

Figure 3 Tubular Electrocatalytic Reactor

anode surfaces must be connected via conductive leads to complete the circuit, as shown in Figure 3. In order to enhance the reactivity of the anode surface toward the substrate hydrocarbon it can be modified by inclusion of catalytic metals. The cathode can likewise be modified to enhance its ability to activate dioxygen.

The tubular fuel cell is fitted into a sleeve of quartz or other inert material to form the reactor. Methane can be fed to one side of the cell while air is fed to the other side. It matters little which side is chosen as cathode or anode, except that it is much easier to modify and analyze the external tube surface, so for practical reasons the external surface was used as the anode.

Methane oxidation reactions are strongly temperature dependent. Table 1 summarizes the effect of temperature on the conversion of methane in a short circuited tubular electrochemical reactor. The anode in this cell consisted of silver ink to which PbO and Mn_3O_4 had been added. As expected, the current, which is a measure of oxygen transport, increases with temperature due to the increased conductivity of YSZ at higher temperatures. This means that the methane/oxygen ratio is decreasing with temperature in these experiments. (Note that all of the oxygen fed in all experiments described herein is fed through the electrolyte and not mixed with the feed. For ease of understanding and comparison to conventional results the oxygen feed rate is converted to cc of oxygen as if it were fed as a gas.)

As the temperature is raised up to 750 C the selectivity to C2's increases and then appears to reach a plateau. The fraction of C2 product that appears as ethylene likewise increases with temperature. Oxygen consumption in all of these experiments is 100%. None of these observations is inconsistent with other studies of heterogeneously catalyzed oxidative methane coupling[24-26]. The increased conversion of methane follows from the increased fraction of oxygen entering the system. Increasing amounts of ethylene can be attributed to thermal cracking of ethane taking place in the gas phase.

Thermal cracking should increase with increased residence time at constant temperature. To investigate the role of thermal cracking in the electrocatalytic methane oxidation, the residence time was varied for reactions conducted under two conditions, one at constant oxygen transport rate and one at constant CH_4/O_2 ratio. A convenient feature of the external circuit cells is that the oxygen transport rate can be limited by adjusting the resistance in the circuit. If more oxygen is desired a voltage can be applied to drive the oxygen transport.

Table 1 Effect of Temperature with AgPbMn/YSZ/Ag Cell.

Temp.	Oxygen Rate		CH_4/O_2	CH_4	Molar Selectivity, (%)			
	Current	$[O_2]$	Ratio	Conv				
(°C)	(mA)	[cc/min]		(%)	C_2H_6	C_2H_4	C_3H_6	CO_2
700	58	0.22	92	1.0	48.9	—	—	51.1
725	79	0.30	67	1.4	42.7	11.8	—	45.5
750	99	0.37	54	1.8	41.7	16.6	—	41.7
775	129	0.48	41	2.5	37.5	21.8	—	40.7
800	177	0.67	30	3.3	32.0	26.4	—	41.6
825	226	0.85	23	4.3	24.9	28.9	4.1	42.1

CH_4 Feed Rate = 20 cc/min

Table 2 Effect of Residence Time at Constant Current.

Res. Time (sec.)	Oxygen Rate Current (mA)	Oxygen Rate [O_2] [cc/min]	CH_4/O_2 Ratio	CH_4 Conv (%)	Molar Selectivity, (%) C_2H_6	Molar Selectivity, (%) C_2H_4	Molar Selectivity, (%) C_3H_6	Molar Selectivity, (%) CO_2
2.8	125	0.47	47.3	3.2	29.7	30.8	5.5	34.0
7.1	126	0.47	18.4	6.4	15.6	31.2	5.7	47.5
12.7	130	0.49	10.2	10.6	9.2	25.2	6.5	59.1

800 C, Ag/YSZ/Ag cell.

Table 2 summarizes the results obtained when the oxygen transport rate was held constant by adjusting cell resistance and the methane feed rate varied. As the residence time at 800 C increases the ethylene/ethane ratio increases, although the selectivity to C2's decreases. This is consistent with the hypothesis of thermal cracking of ethane following the oxidative coupling of methane. Similar results have been observed with heterogeneous catalysts.

Variation of residence time at constant CH_4/O_2 ratio was achieved by applying a voltage and controlling the methane feed rate with mass flow controllers. Results of these experiments, summarized in Table 3, demonstrate that the methane conversion and selectivity to C2's remain nearly constant as the residence time is changed under constant CH_4/O_2 ratio conditions. The ratio of ethylene to ethane increases with residence time consistent with increased thermal pyrolysis at the longer residence time.

A feature of the electrochemical cells that make them very attractive as selective catalytic reactors is the ability to control the nature of the catalytic site by adjusting the potential as well as the composition of the anode. In heterogeneous catalysis the work function of the active site is detrmined by the electronics of neighboring atoms and the atmosphere to which the catalyst is exposed, but the surface of the anode of an

Table 3 Effect of Residence Time with Constant CH_4/O_2 Ratio.

Res. Time (sec.)	Voltage Volts	Oxygen Rate Current (mA)	Oxygen Rate [O_2] [cc/min]	CH_4/O_2 Ratio	CH_4 Conv (%)	Molar Selectivity, (%) C_2H_6	Molar Selectivity, (%) C_2H_4	Molar Selectivity, (%) C_3H_6	Molar Selectivity, (%) CO_2
2.4	0.6	385	1.45	17.4	6.7	25.7	28.0	4.2	42.1
2.9	0.5	343	1.29	16.7	7.3	23.8	29.5	4.2	42.1
3.3	0.4	301	1.13	16.7	7.2	23.0	31.4	3.4	42.2
3.8	0.3	258	0.97	16.7	7.1	22.0	32.5	5.7	39.8
4.9	0.2	206	0.77	16.4	7.2	19.6	33.9	5.5	41.0
6.1	0.1	154	0.58	17.4	6.7	18.5	33.2	5.4	42.9
7.9	0.0	128	0.48	16.7	6.9	17.6	32.2	5.9	44.3

800 C, AgBi/YSZ/Ag Cell

electrochemical cell is responsive also to its potential relative to the cathode which can be adjusted by simply adjusting the voltage of the circuit. In some cases rate enhancements beyond that expected from added oxygen have been noted. An elaborate theory, the NEMCA effect (Non-Faradaic Electrochemical Modification of Catalytic Activity), has been proposed to describe these electrocatalytic rate enhancements.[27] According to the theory materials can exhibit either positive or negative enhancements depending on their electrophilicity.

Data collected in Table 3 can be used to evaluate the proposed NEMCA effect in the oxidation of methane since the CH_4/O_2 ratio was held constant and the only variables that are changing are the potential and residence time. According to the theory a significant change in the methane conversion should be observed as the voltage on the cell is changed. However the methane conversion values are all clustered around 7.0 ± 0.3 % and can be considered identical within the error of the measurement. Thus the cell under consideration provides no evidence of the NEMCA behavior.

Table 4 Effect of Anode Composition for AgM/YSZ/Ag Cells.

| Additive | Oxygen Rate | | CH_4 | Molar Selectivity, (%) | | |
| | Current | $[O_2]$ | Conv | | | |
M	(mA)	cc/min	(%)	C_2+	CO	CO_2
-	196	0.74	4.3	58.1	-	41.9
Sm	262	0.98	3.5	9.3	22.2	68.5
Ho	263	0.99	2.9	14.8	-	85.2
Pb	278	1.05	4.6	42.3	-	57.7
Li/Mg	231	0.87	4.0	49.5	-	50.5
Bi	251	0.94	3.8	70.3	-	29.7

800 C, CH4 flow rate = 20 cc/min

The effect of the electrocatalyst composition was studied as well. A series of Ag electrode cells was modified by incorporating a second metal or metal oxide. The results for inclusion of Sm, Ho, Li/Mg, and Bi, along with the naked Ag anode cell are summarized in Table 4. These cells all had about the same current and methane flow rate, so that the effect of the residence time and CH_4/O_2 ratio was minimized. The methane conversions were clustered between 2.9% and 4.3%, but the selectivity to C2+ product ranged from 9.3% for Sm to 70.3% for Bi. Thus, there is a dramatic effect of catalytic metal on the selectivity to higher hydrocarbons.

In the search for a guiding principle to correlate all of the experimental observations, a graph of the dependence of C2 selectivity on CH_4 conversion for a variety of tubular cells was constructed as shown in Figure 4. It appears that all of the data points are bounded by a single, limiting curve. This implies that there is a fundamental mechanistic limitation to the conversion of methane to higher hydrocarbons that cannot be exceeded.[1,2,28]

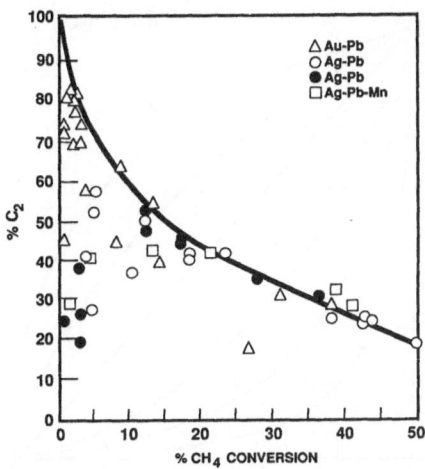

Figure 4 Methane Coupling Selectivity and Conversion

Our interpretation of these results is summarized in the simplified mechanistic scheme shown in Figure 5. In this scheme the conversion of methane to ethane/ethylene and then to CO_2 occurs by a series of consecutive reactions. The relative rates of methane and ethylene activation, k_1 and k_2 in Figure 5, determine the upper limit of C2 yield. (Since ethane can convert to ethylene readily by non-oxidative pyrolysis and ethylene is more stable than ethane the rate of ethylene activation is taken to be the limiting rate.) The direct conversion of CH_4 to CO_2 has been ignored. Assuming this rate is negligible is the most optimistic assumption possible as far as catalysis is concerned since conversion of CH_4 to CO_2 cannot contribute to C2 yield.

Lunsford[29,30] and others have determined that homolytic C-H bond rupture is the rate limiting step in heterogeneously catalyzed CH_4 activation. It is hard to imagine a mechanism by which a catalytic site could discriminate between methane C-H bonds and the C-H bonds in other compounds under the conditions of methane oxidation. Therefore, it can be assumed that carbon-hydrogen bonds of gas phase ethylene can be activated in a similar manner. (Any enhanced association of ethylene with the catalytic surface due to π interactions will increase the effective concentration of ethylene and thus increase its destruction, so the most favorable situation will be one in which ethylene does not selectively interact with the surface.) The consecutive reaction scheme reduces to two rate limiting steps, one for methane activation and one for ethylene destruction, which occur by the same mechanism. Since the active site cannot distinguish among C-H bonds, in the limiting rate equations for CH_4 and C_2H_4 activation the same concentration of oxidation sites, [Ox], can be used as shown in Equations 1 and 2.

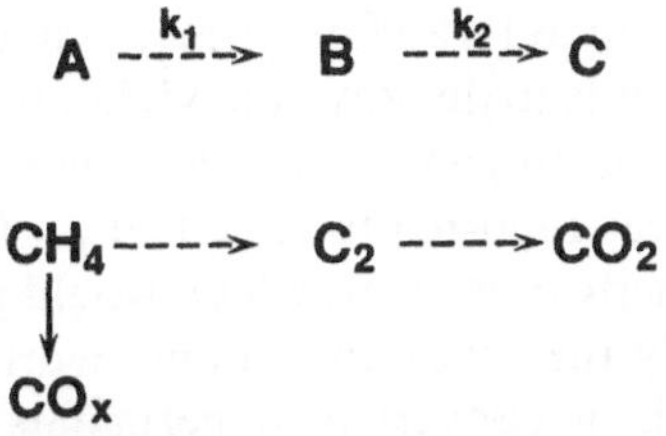

$$A \xrightarrow{k_1} B \xrightarrow{k_2} C$$

$$CH_4 \dashrightarrow C_2 \dashrightarrow CO_2$$
$$\downarrow$$
$$CO_x$$

Figure 5 Consecutive Reaction Mechanistic Scheme

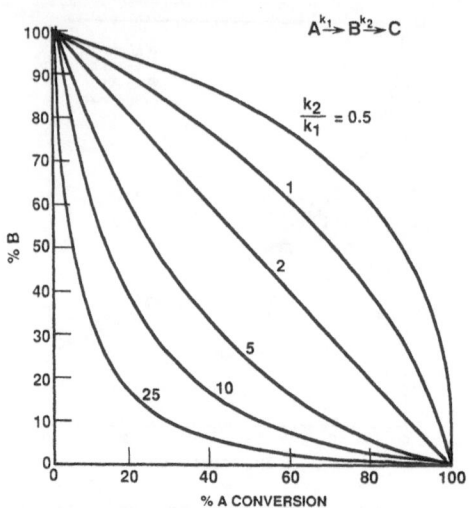

Figure 6 Selectivity/Conversion Relation as a Function of k_2/k_1

$$\text{Rate } CH_4 = -k_1 [CH_4] [Ox] \qquad (1)$$

$$\text{Rate } C_2H_4 = -k_2 [C_2H_4] [Ox] \qquad (2)$$

This greatly simplifies analysis of the system since the two rates differ only in the fundamental rate constant and the concentration of the reactants. Computer simulation of the consecutive reaction mechanism allows one to determine the k_2/k_1 ratio needed to achieve the desired yield of C_2H_4. Figure 6 shows how the selectivity to B varies as a function of conversion of A for various k_2/k_1 ratios. (Since oxidative activation of ethane to give an ethyl radical results in an ethylene product, it only serves to consume oxidant and can be ignored. If direct ethane conversion to CO_2 occurs this will reduce C2 yield.) The results in Figure 4 are fit well by a k_2/k_1 ratio of about 6.

Experimental test of this mechanism was conducted by performing a competition study with ethylene/methane mixtures in the tubular reactor. The results, summarized in Table 5, demonstrate that ethylene oxidation competes readily with methane oxidation under the experimental conditions of the electrocatalytic cell. The ratios of k_2/k_1 calculated for these experiments are 4.0 and 4.6. This is in reasonable agreement with the ratio derived from the methane coupling experiments. Thus, the consecutive reaction mechanism can be applied successfully to systems of this type. The inescapable conclusion is that methane dimerization is limited by the relative rates of methane and ethylene activation.

The vast literature on the methane coupling reaction support the notion of a limit to the coupling yield of about 30% (or $k_2/k_1 = 2$) based on consecutive reaction schemes, but little experimental data on comparative rates of methane and ethylene conversion are available. One of the few attempts to disprove the yield limit by directly measuring CH_4 and C_2H_4 activation rates over heterogeneous catalysts is by Burch and Tsang.[3] In their study a LiCl/MnOx catalyst was reported to have $k_2/k_1 = 0.42$ at 700 C, by separately measuring the CH_4 and C_2H_4 activation rates, which would provide a maximum C2 yield of about 57%. Unfortunately the two experiments were conducted at different O_2 consumption rates, actual methane conversion experiments reached a maximum of only 12% yield, and no examples of the predicted high yield have been reported.

Table 5 Ethylene Cofeed Experiment.

| Feed (Mole %) | | Oxygen Rate | | Effluent (mole %) | | K_2/K_1 |
CH$_4$	C$_2$=	Current (mA)	[O$_2$] cc/min	CH$_4$	Ethylene	
100	–	885	3.33	82.4	3.9	-
91.4	8.7	620	2.33	86.4	6.8	4.0
82.7	17.3	490	1.84	79.5	14.2	4.6

800 C, AgPb/YSZ/Ag cell, CH$_4$ feed rate = 20 cc/min, applied voltage = 2.0 V

How low does the k_2/k_1 rate need to be to achieve economic attractiveness? Since stoichiometry dictates that four times as much oxygen is consumed by combustion as is used in partial oxidation, the selectivity needs to be about 80% to avoid wasting methane and oxygen.

$$2 O_2 + CH_4 = CO_2 + 2 H_2O \tag{3}$$

$$\frac{1}{2} O_2 + CH_4 = \frac{1}{2} C_2H_4 + H_2O \tag{4}$$

And the per pass conversion of methane needs to be about 80% to minimize the cost of separation. At the 80% conversion/80% selectivity point the yield is 64% and k_2/k_1 is about 0.2. Since it seems unreasonable to propose a mechanism by which C-H bonds of methane are selectively attacked over those of ethylene at the temperatures required for reaction (700-900 C), the yields of C2 products appear limited to modest, economically unattractive quantities.

Our studies turned from the methane coupling reaction to the investigation of higher temperature methane reactions. Using tubular electrocatalytic cells with Pt anodes, the conversion of methane becomes more selective with CO as the major product at high temperatures. As summarized in Table 6 methane conversions can approach 100% with CO selectivities up to 97% at 1100°C. This suggested that the electrochemical cells could be used for the partial oxidation of methane to synthesis gas. We call this process Electropox, for electrocatalytic partial oxidation.

Table 6 Effect of Temperature on Methane Conversion.

Temp °C	CH$_4$ Rate cc/min	Applied Potential Volts	Oxygen Rate Current mA	[O$_2$] [cc/min]	CH$_4$ Conv %	Product Selectivity C2's	CO	CO$_2$
800	20.6	0.0	107	0.40	3.1	22.6	56.6	20.8
800	8.0	0.0	227	0.85	23.5	-	95.9	4.1
1000	8.0	0.0	340	1.28	40.4	-	99.9	0.1
1100	8.0	0.0	330	1.24	39.2	-	99.9	0.1
1100	8.0	2.0	994	3.74	99.9	-	96.8	3.2

CH4//Pt-YSZ/YSZ/Pt//Air cell

Table 7 Current Density, Syngas Yields in Short Circuit Cells.

Cell	Feed	Oxygen Rate		Resistance Ω-cm^2	CO Yield %
		Current mA/cm^2	[O_2] [cc/min]		
Pt/YSZ-4/Pt (tube)	CH_4	15	0.056	50	99
Pt/YSZ-4/LSM	CH_4	347	1.30	2.65	-
PtBi/YSZ-4/LSM	CH_4	525	1.97	1.34	56
PtBi/YSZ-8/LSM-LSM	CH_4	552	2.08	1.25	61
PtBi-cermet/YSZ-8/LSM	CH_4	705	3.14	0.8	-
Ni-cermet/YSZ-8/LSM-Pt	CH_4	930	2.85	-	83
	H_2	1245	4.68	0.45	-

For the production of chemicals and fuels from methane the Electropox process represents a chance to reap a tremendous benefit. By combining the oxygen separation with methane oxidation to syngas, an entire process segment, the oxygen separation unit, can be eliminated from a conventional plant. The savings could be as much as 25% of the capital investment in a natural gas to fuels plant. In order to achieve this savings oxygen separation would need to be done in a relatively small oxidation reactor, requiring high oxygen transport rates. Efforts to prove the technical feasibility of the process began by focusing on the oxygen transport rate.

The practical limitation of the tubular YSZ electrolyte reactor is that the current densities (oxygen fluxes) are very low. This is due in part to the thick tube wall (1-1.5 mm) and in part to the poor adherence of the electrodes to the tube, resulting in very large contact resistances. By utilizing an externally shorted disc reactor the interfacial resistance could be greatly reduced as shown in Table 7. Current densities of greater than 1 A/cm^2 were obtained. Increasing the oxygen flux greatly reduces the size of the reactor required to convert a given quantity of natural gas, and makes the process economically attractive.

Another way of improving the oxygen flux is to introduce an internal short circuit. Since for most natural gas upgrading schemes electricity is not a valuable product due to the remote location of the plant, the external circuit is superfluous. A short circuit is obtained by introducing into the ionically conducting electrolyte a second phase to conduct electrons. This is demonstrated schematically in Figure 1. Oxide ions can migrate from air to natural gas via the ionically conducting phase and electrons can return via the electronically conducting phase.

Dual phase membranes have been tested that exceed the preliminary oxygen flux targets for a commercial Electropox process. Table 8 summarizes some of the oxygen flux results obtained with dual phase membranes separating hydrogen and air at 1100°C. Membranes with oxidic as well as metallic electronic conductor phases have been tested. Ease of fabrication and high oxygen flux make these ideal candidates to form into membranes of all shapes for conducting the Electropox process.

Table 8 Oxygen Flux Results for Dual Phase Membrane Discs at 1100 C

Membrane Electronic/Ionic Phase	Feed	Oxygen Rate		Thickness
		Current mA/cm^2	$[O_2]$ [cc/min-cm^2]	mm
$(Pd)_{.5} / (YSZ)_{.5}$	H_2	555	2.1	0.8
	CH_4	530	2.0	0.8
$(Pt)_{.5} / (YSZ)_{.5}$	H_2	467	1.8	0.8
$(B\text{-}MgLaCrO_x)_{.5} / (YSZ)_{.5}$	H_2	114	0.4	0.8
$(In_{90}Pr_{10})_{.4} / (YSZ)_{.6}$	H_2	275	1.1	0.8
$(In_{90}Pr_{10})_{.5} / (YSZ)_{.5}$	H_2	601	2.3	0.8
	H_2	1458	5.5	0.3
	H_2	1611	6.1	0.25
$(In_{95}Pr_{2.5}Zr_{2.5})_{.5} / (YSZ)_{.5}$	H_2	2083	7.8	0.3

ACKNOWLEDGEMENT

I thank the British Petroleum Company, PLC for permission to publish this manuscript, and Elsevier Science Publishers for permission to reproduce copyrighted Tables and Figures.

REFERENCES

1. J.A.Labinger and K.C.Ott, J. Phys Chem., 1987, 91, 2682.

2. J.A.Labinger, Cat. Lett., 1988, 1, 371.

3. R.Burch and S.C.Tsang, Appl. Cat., 1990, 65, 259.

4. P.R.Pereira, V.De Gouveia, F.Rosa, Preprints Petr. Div. ACS, 1992, 37, 200.

5. R.A.Goffe, D.M.Mason, J. Appl. Electrochem., 1981, 11, 447.

6. B.G.Ong, C.C.Chiang, D.M.Mason, Sol. State Ionics, 1981, 3/4, 447.

7. B.C.Nguyen, T.A.Lin, D.M.Mason, J. Electrochem. Soc., 1986, 133, 1807.

8. K.Otsuka, S.Yokoyama, A.Morikawa, Bull. Chem. Soc. Jpn., 1984, 57, 3286.

9. K.Otsuka, S.Yokoyama, A.Morikawa, Chem. Lett. (Japan), 1985, 319.

10. K.Otsuka, A.Morikawa, Japan Pat. 61-30688, 12 Feb 1986.

11. K.Otsuka, K.Suga, I.Yamanaka, Catal. Today, 1990, 6, 587.

12. M.Stoukides, C.G.Vayenas, J. Catal., 1980, 64, 18.

13. S.Seimanides, M.Stoukides, J. Electrochem. Soc., 1986, 133, 1535.

14. D.Eng, M.Stoukides, Catal. Lett., 1991, 9, 47.

15. N.Kiratzis, M.Stoukides, J. Electrochem. Soc., 1987, 134, 1925.

16. H.Nagamoto, K.Hayashi, H.Inoue, J. Catal., 1990, 126, 671.

17. D.J.Kuchynko, R.L.Cook, A.F.Sammells, J. Electrochem. Soc., 1991, 138, 1284.

18. N.U.Pujare, R.L.Cook, A.F.Sammells, US 4,997,725, 5 Mar 1991, assigned to Gas Research Institute.

19. T.J.Mazanec, T.L.Cable, US 4,802,958, 7 Feb 1989, assigned to the Standard Oil Co.

20. T.J.Mazanec, T.L.Cable, J.G.Frye, Jr., US 4,793,904, 27 Dec 1988, assigned to the Standard Oil Co.

21. T.J.Mazanec, T.L.Cable, US 4,933,054, 12 Jun 1990, assigned to the Standard Oil Co.

22. T.L.Cable, T.J.Mazanec, J.G.Frye, Jr., Eur. Pat. Appl. 0399833, 28 Nov 1990, assigned to The Standard Oil Co.

23. T.J.Mazanec, T.L.Cable, J.G.Frye, Jr., Solid State Ionics, 1992, 53-56, 111.

24. K. Otsuka, K. Jinno, A. Morikawa, J. Catal., 1986, 100, 353.

25. V. D. Solokovskii, Catalysis Today, 1992, 14, 415; and references therein.

26. Y. Amenomiya, V. I. Birss, M. Goledzinowski, J. Galuzska, A. R. Sanger, Catal. Rev. Sci. Eng., 1990, 32, 163.

27. C. G. Vayenas, S. Bebelis, I. V. Yentekakis, H-G. Lintz, Catalysis Today, 1992, 11, 303; and references therein.

28. J. S. Lee, S. T. Oyama, Catal. Rev. Sci. Eng., 1988, 30, 249.

29. D.J.Driscoll, J.H.Lunsford, J. Phys. Chem., 1985, 89, 4415.

30. J.H.Lunsford, Catal. Today, 1990, 6, 235; and references therein.

KINETICS OF REACTION OF DIOXYGEN WITH LITHIUM NICKEL OXIDE, AND THE ROLE OF SURFACE OXYGEN IN OXIDATIVE COUPLING OF METHANE

Y. -K. Sun,[1] J. T. Lewandowski,[1] G. R. Myers,[1] A. J. Jacobson,[2] and R. B. Hall[1]

[1] Corporate Research Laboratories
Exxon Research and Engineering Company
Annandale, New Jersey 08801

[2] Chemistry Department, University of Houston
Houston, Texas 77204

1. INTRODUCTION

Selective, catalytic processes to convert methane into higher hydrocarbons or oxygenates are of great industrial importance because they would enable the conversion of abundant, remote natural gas reserves to useful petrochemicals and fuels.[1-6,7a] One of the most studied conversion processes is oxidative coupling, the formation of ethane, ethylene, and higher hydrocarbons (abbreviated as C_{2+}) from the reaction of methane with oxygen.[2-14] Alkali-promoted metal oxides are relatively selective catalysts for this conversion. A wide variety of oxides have been studied, including: basic oxides such as MgO,[7] rare-earth oxides such as Sm_2O_3,[9,10] and transition metal oxides such as Mn oxides[2,3,11] and NiO.[12-14] With all of these materials, deep oxidation of methane to CO_2 and H_2O competes with oxidative coupling. The highest yields (methane conversion times selectivity to C_{2+}) achieved to date are below 30%. It is estimated that methane conversions in excess of 60% and C_{2+} selectivities greater than 80% are necessary for an economic conversion process.[1] One of the challenges in achieving better yields is to design or discover catalytic materials that have a high intrinsic selectivity for oxidative coupling, and a minimum activity for deep oxidation.

The Activation of Dioxygen and Homogeneous Catalytic Oxidation,
Edited by D.H.R. Barton *et al.*, Plenum Press, New York, 1993

In order to design selective catalysts, it is necessary to understand the influence of catalyst structure on reaction pathway and reaction rate. Some of the fundamental aspects of oxidative coupling that need to be understood include: what is the nature of the surface site responsible for methane activation, how is this site produced by reaction of the material with gas-phase oxygen, what is the nature of the surface site for deep oxidation to produce CO_2, how is it produced, is it the same as the site for methane activation, how do these sites depend on catalyst structure? It is important to determine whether activation and oxidation occur on the same site. If they do, production of CO_2 is inescapable.

The variety and complexity of the structures of most catalysts make it difficult to determine the nature of the active sites and whether activation and deep oxidation occur at the same site. For alkali-doped MgO and CaO, Lunsford has shown a correlation between methane activation and the existence of O$^-$ species, identified by ESR.[7b,7d] Under oxidative coupling conditions, this O$^-$ site is produced to a significant extent only in the presence of gas-phase oxygen. No significant conversion occurs in the absence of gas-phase oxygen. Similar behavior is observed for other non reducible oxides. In contrast, a number of reducible metal oxides can for a time selectively convert methane to C_{2+} hydrocarbons in the absence of gas-phase oxygen. This suggests that for these materials a lattice oxygen site at the surface is responsible for selective activation. (Lattice oxygen is subsequently replenished by reacting the reduced material with gas-phase oxygen.) Examples include supported and promoted manganese oxides, with C_{2+} selectivities in excess of 80% at low methane conversion[3,11] and $LiNiO_2$, for which C_{2+} selectivities near 100% have been reported.[12d] The nature of the site responsible for deep oxidation is less well understood. It is possible that the active sites for reducible materials are different from those for non-reducible materials. In both cases however, it is clear that the reaction of dioxygen with these materials is a key step in producing active sites.

Of the wide range of catalyst materials studied, lithium-substituted nickel oxides are unusual in that they form sufficiently simple structures that a relatively complete characterization is possible.[15-17] In addition, the origin of the selectivity imparted in nickel oxide by lithium substitution is particularly interesting. NiO readily reacts with CH_4 at temperatures above 873 K, but only CO_2 and H_2O are formed.[12,22] This is quite different than the case of MgO, which has little or no reactivity toward methane without the addition of alkali metal. Somehow, substitution of Li cations for Ni cations converts nickel oxide from a material which is highly active and selective for complete combustion to one that can be highly selective for hydrocarbon production.

Recently, we have shown that the selectivity of lithium nickel oxides depends on the structure, especially on the long-range order of the cations.[17] Here we report results on studies of the kinetics of reaction of gas-phase oxygen and of methane with an ordered structure that has high C_{2+} selectivity. The catalyst material used in this study has the composition $Li_{0.44}Ni_{0.56}O$, close to the theoretical limit for complete substitution, $Li_{0.5}Ni_{0.5}O$. We confirm for this composition that lattice oxygen can be highly selective for methane activation and production of C_{2+} hydrocarbons. However, a proper accounting of CO_2 that remains on the catalyst as a carbonate, reveals that the selectivity is only 75%, in contrast to an earlier study in which a nearly 100% C_{2+} selectivity was reported.[12d] In addition, we identify a surface oxygen state in equilibrium with gas-phase oxygen that participates in the production of CO_2. We characterize the reaction kinetics of these two sites using transient isotope-switching techniques and steady state rate measurements.

2. EXPERIMENTAL

The transient isotope-switching apparatus has been described previously.[18] The reactor was a plug-flow, fixed-bed quartz microreactor, 8 mm in diameter, with a total volume of 0.8 cc. Approximately 0.4 g of catalyst was supported in the middle of the reactor on a fused quartz frit. Temperature in the catalyst bed was measured with a chromel-alumel thermocouple shielded in a quartz jacket. The reactant gas stream was controlled by a mass flow control switching system capable of switching one or more of the reactants to its isotopically labelled counterpart in approximately 1 s. The partial pressures and flow rates of the reactants are not altered by this switch so that the steady-state concentrations of reactants and products in the gas phase, *and on the catalyst surface*, are not disturbed.

The lithium nickel oxide catalyst used in this study was synthesized by reacting lithium carbonate and nickel oxide in air at 1073 K. The catalyst was characterized by x-ray diffraction to determine the purity of phase, and by microanalysis to determine the elemental composition. The nominal composition of the catalytic material used here was $Li_{0.44}Ni_{0.56}O$. The short and long range order of the material was characterized by small angle neutron scattering, EXAFS, and x-ray scattering.[15,17] The catalyst is black in color and has a surface area of approximately 2 m^2 g^{-1}, as determined by gas adsorption. To remove surface carbon contamination and any carbonate phase remaining from the synthesis or formed during methane conversion, the catalyst was heated to 923 K in a flow stream of 20% oxygen in either argon or helium.

Rates of consumption of reactants and formation of products were monitored continuously with a quadrupole mass spectrometer (Extranuclear) through a capillary sampling tube located at the outlet of the reactor, and periodically by a fast gas chromatography system (MTI M200). This GC system was capable of determining the entire product distribution through C_4 hydrocarbons every 60 s. Oxygen $^{16}O_2$ (99.997%) was obtained from Airco, Ar (99.999%) from Airco, and 20% $^{18}O_2$ (97%) in Ar (99.998%) from ICON.

3. RESULTS AND DISCUSSION

The technique of transient isotope-switching can provide fundamental information about the kinetics of catalytic reactions that is difficult or impossible to obtain with other methods. Following a gas-phase isotope switch at steady state, transients of products labeled with the original isotope result from reactions of intermediates that have been left behind on/in the catalyst after the gas-phase switch. The area under the transients provides a direct measure of the steady-state concentrations of intermediates on/in the catalyst at the time of the switch. The temporal shape of the transients provides information on the reaction kinetics of the intermediates. In the present study, we compliment the transient techniques with steady-state measurements to obtain information on the rates of the various reactions over a temperature range wider than that accessible by the isotope-switching method.

We first present, in Sec. 3.1., results concerning the exchange reactions of gas-phase dioxygen with the surface of the catalyst. In Sec. 3.2., we present results on the kinetics of exchange with lattice oxygen. Exchange reactions with lattice oxygen are relatively slow and require higher temperatures than that required for exchange with the surface state. The rapid exchange between O_2(gas) and surface oxygen, and the slower exchange between surface oxygen and lattice oxygen, allows us to preferentially populate the surface state with labelled oxygen, and thereby investigate the contribution of each species to catalytic oxidation reactions. Experimental results on the reactions of these species with methane are presented in Sec. 3.3..

3.1. SURFACE OXYGEN STATE

3.1.1. Evidence for a Surface-Bound Oxygen State

In Fig. 1. we show the time dependence of the mass spectrometric signals for gas-phase $^{16}O_2$, $^{16}O^{18}O$, and $^{18}O_2$ following a switch from 20% $^{16}O_2$ in Ar to 20% $^{18}O_2$ in Ar at 678 K and at a total flow rate of 40 cc min^{-1}. The mass spectrometric intensities are converted to the partial pressures of each species in the reactor. The calibration against absolute number density in the reactor is readily accomplished because the feed gas provides an accurate, internal reference point. In the upper panel of Fig. 1, it is shown that upon switching, the partial pressure of $^{16}O_2$ decreases rapidly with a complementary increase in the partial pressure of $^{18}O_2$, the result being a constant oxygen pressure. In the lower panel, it is shown that $^{16}O^{18}O$ evolves from the reactor, with a relatively rapid rise in rate followed by a slower decay. To check whether gas-phase reactions or reactions at the reactor wall contribute to the isotope exchange, a blank experiment was conducted with the catalyst replaced by quartz chips. No $^{16}O^{18}O$ was observed at temperatures up to 1123 K. The full width at half maximum (FWHM) of the $^{16}O^{18}O$ evolution peak is ~ 29 s at 678 K, much longer than the ~ 1 s of the isotope switching transient. There is also a decay in the $^{16}O_2$ intensity, and a rise in the $^{18}O_2$ that occurs over a similar time. This is less obvious in Fig.1 because the amount of O_2 desorbing from the surface is a small fraction of the partial pressure of O_2 coming into the reactor. Subtracting the signals due to rapidly changing feed gas components following the switch, which decay/rise exponentially with a time constant of ~ 1 s, we find that the area under the $^{16}O_2$ decay is approximately equal to the area "missing" in the $^{18}O_2$ rise. Furthermore, the areas are roughly 1/2 that of the area under the $^{16}O^{18}O$ transient. Thus, we find that the distribution of isotopes in the products is consistent with that expected for statistical exchange, i.e. $^{16}O_2$: $^{16}O^{18}O$: $^{18}O_2$ (taken up by catalyst) = 1 : 2 : 1.

These results show that there is a bound state of oxygen in or on the catalyst that can exchange with gas-phase oxygen. Prior to the switch of the gas-phase label, all of these are ^{16}O. After the switch to $^{18}O_2$ in the gas-phase, ^{16}O is gradually lost from the catalyst, and ^{18}O gradually builds up. ^{16}O can be lost in two ways, either as $^{16}O_2$, or as $^{16}O^{18}O$. Eventually, the exchangeable state loses all of its ^{16}O and becomes fully saturated with ^{18}O, at which point no further desorption of $^{16}O^{18}O$ is detected.

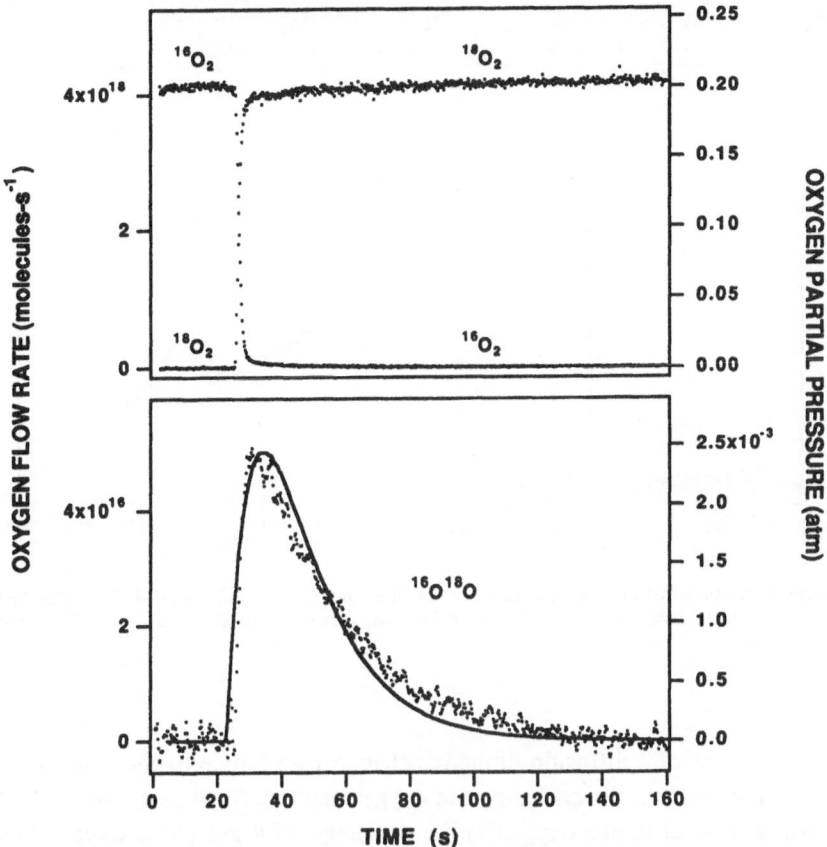

Fig. 1. Transient responses of mass spectrometric intensities of oxygen at 678 K upon switching from 20% $^{16}O_2$ in Ar to 20% $^{18}O_2$ in Ar. A background in the $^{16}O^{18}O$ signal due to $^{16}O^{18}O$ in the oxygen-18 feed has been appropriately subtracted. The flow rate was 50 cc-min^{-1}. The catalyst was $Li_{0.44}Ni_{0.56}O$ (weight = 0.36 g). The solid curve is a result from a model, see Sec. 3.1.2. for detail.

The number of ^{16}O atoms contained in the $^{16}O^{18}O$ transient is 1.8×10^{18}. This number is obtained from the integral of the $^{16}O^{18}O$ signal over time. It is shown below that for a statistical exchange of isotopes an equal number of ^{16}O atoms will desorb as $^{16}O_2$. Hence, at 678 K, the total number of exchangeable oxygen atoms is 3.6×10^{18}. Given a catalyst weight in this experiment of 0.36 g, a surface area of ~ 2 m^2 g^{-1}, and a surface oxygen atom site density of 1.6×10^{15} cm^{-2} [for the (100) surface[15]], this corresponds to a coverage of ~ 0.3 monolayers (~ 5×10^{14} cm^{-2}) of ^{16}O.

An important characteristic of the exchange process at temperatures below about 850 K, is that if the gas-phase label is switched from $^{16}O_2$ to $^{18}O_2$ and then back again to $^{16}O_2$, the shapes and amplitudes of the transient signals do not change, irrespective of the time the catalyst is exposed to $^{18}O_2$. This shows that the reservoir of exchangeable oxygen is fixed. Extended exposure to $^{18}O_2$, up to 400 s, does not lead to an increase in the amount of oxygen exchanged. An example of this is shown in the left panel in Fig. 2. This argues

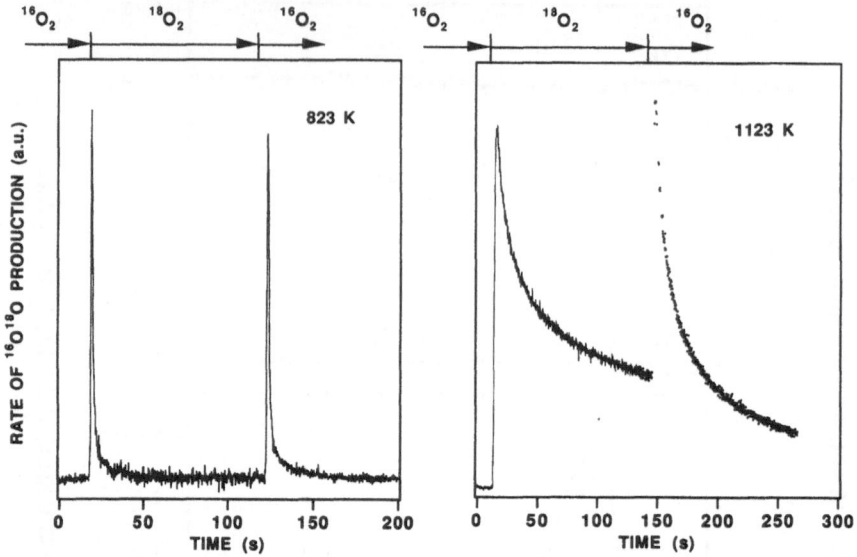

Fig. 2. Transient responses of mass spectrometric intensities of $^{16}O^{18}O$ at 823 and 1125 K upon switching from 20% $^{16}O_2$ in Ar to 20% $^{18}O_2$ in Ar, followed by switching back to 20% $^{16}O_2$ in Ar. The total flow rate was 40 cc-min^{-1}.

against the existence of a diffusion limited exchange with lattice oxygen atoms near the surface, and indicates that the exchangeable oxygens are confined to the surface. Above 850 K, participation of lattice oxygen in the exchange with gas-phase oxygen becomes detectable and the shape of the transient does depend on the exposure time. This is shown in Fig. 2b, and will be discussed in Sec. 3.2..

Another important characteristic below 850 K (this will be discussed in more detail in Sec. 3.1.3.), is that the catalyst is saturated with exchangeable oxygen for oxygen partial pressures ranging from 0.2 to less than 0.002 atmospheres. Consequently, the kinetics of $^{16}O^{18}O$ production in Fig. 1 is determined by the desorption kinetics of the surface oxygen, rather than by the the adsorption kinetics of the gas-phase oxygen. This is consistent with the observation by Lambert and coworkers[18a] that oxygen chemisorbs on activated lithium nickel oxide even at room temperature, which indicates that the activation barrier to adsorption is rather low.

To recapitulate the characteristics of the isotope exchange at temperatures below about 850 K, we find that

1. There is a fixed reservoir of exchangeable oxygen. Extended exposure to labeled gas-phase O_2 does not increase the amount of exchange. This indicates that sub-surface lattice oxygen does not participate in the exchange. We conclude that the exchangeable oxygen is confined to the surface.

2. The number of exchangeable oxygen atoms is approximately 5×10^{14} cm^{-2}, equivalent to about one-third of a monolayer Thus, the adsorption site is not a minority, defect site, as has been proposed.[19a] The population of these sites under reaction conditions can be quite high in the presence of gas-phase O_2, and this chemisorbed oxygen can contribute significantly to the catalytic reactivity of these materials.

3. The statistical distribution of isotopes in the products is consistent with an exchange mechanism consisting of dissociative adsorption of dioxygen to produce surface-bound oxygen atoms, together with recombinative desorption of chemisorbed atoms to give O_2(gas). This is not unequivocal however. Based on the statistics of the exchange alone, it is not possible to rule out a 4-center exchange reaction involving a chemisorbed, dinuclear oxygen species, such as O_2^-.[20] Exchange of isotopes between dinuclear species would result in the same labelling statistics as atom recombination. A four-center reaction mechanism, however, is considered to be highly improbable based on bonding, orbital overlap/symmetry and entropy considerations. We therefore assume that the exchange proceeds through a surface-bound atomic oxygen. We designate this as O(s), and distinguish it from lattice oxygen, designated O(l), which may exist at the surface but which does not participate in the exchange below roughly 850 K.

4. The O(s) state is saturated over the range of oxygen partial pressures investigated. The adsorption rate is sufficiently high over the entire range that when desorption occurs the sites are rapidly repopulated. Thus, for these temperature and pressure conditions, desorption is the rate-limiting step in the isotope exchange.

3.1.2. Kinetic Analysis of Oxygen Isotope Switching Results

The following sections provide a kinetic analysis of the transient responses based on an atomic state for the chemisorbed oxygen, O(s). We show that this approach allows us to account for the qualitative features of the results described above, the temperature dependence of the rate of isotope scrambling under steady-state conditions, and results from temperature programmed desorption (TPD) experiments performed at very low pressure. The steady-state exchange and TPD experiments are described in Sec. 3.1.3.. The kinetics of isotope exchange of O_2 (gas) with oxide materials have been reviewed by Gellings and Bouwmeester.[20] Readers are referred to this work and references therein for a more comprehensive discussions of the mechanisms and kinetics involved in more complex systems.

The steps in the oxygen exchange reaction when restricted to the catalyst's surface can be written as:

$$^{18}O_2(g) \rightleftharpoons 2\,^{18}O(s) \tag{1}$$
$$^{18}O(s) + {}^{16}O(s) \rightleftharpoons {}^{16}O^{18}O(g) \tag{2}$$
$$^{16}O(s) + {}^{16}O(s) \rightleftharpoons {}^{16}O_2(g). \tag{3}$$

The desorption rates of $^{16}O^{18}O$ and $^{16}O_2$ are given by

$$R_{^{16}O^{18}O} = 2k_d n^2 \theta \theta^* \tag{4}$$

and

$$R_{^{16}O_2} = k_d n^2 \theta^2, \tag{5}$$

where θ and θ^* are the relative surface coverages (relative to the coverage at saturation) of atomic ^{16}O and ^{18}O, respectively; k_d is the rate constant for the second-order recombinative desorption of surface oxygen atoms; and n is the density of surface sites where oxygen atoms adsorb. Material balances for the two oxygen isotopes on the surface are given by

$$\frac{d\theta}{dt} = 2S\frac{F(t)}{n} - 2k_d n\theta^2 - 2k_d n\theta\theta^*$$

(6)

and

$$\theta + \theta^* = \theta_0,$$

(7)

where θ_0 is the total relative coverage of the surface oxygen at steady state; F(t) is the incident flux of $^{16}O_2(g)$ upon the surface; S is the probability of dissociative chemisorption per collision. At the time of the switch, t = 0, F(t) declines exponentially with a time constant of roughly 1 second. (The flux of $^{18}O_2(g)$ increases in a complementary fashion.) Note that terms associated with the reverse reaction in Eq. 2 are not included in the above material balance. This is expected to be valid under conditions where readsorption of $^{16}O^{18}O$ can be neglected, which is certainly the case in the current experiments where the pressure of desorbed $^{16}O^{18}O$ is less than 1% of the steady-state pressure of oxygen in the gas phase.

Upon switching, although the gas-phase isotopic label is changed from 16 to 18, the overall steady state between dissociative adsorption and recombinative desorption is maintained — the total pressure of O_2 and the total coverage of surface oxygen remain constant. Using Eqs. 6 and 7, and taking F(t) as a step function rather than an exponential decay, we have:

$$\theta = \theta_0 e^{-2k_d n\theta_0 t}$$

(8)

$$\theta^* = \theta_0 (1 - e^{-2k_d n\theta_0 t}).$$

(9)

It is interesting to note that despite the second-order kinetics associated with the recombinative desorption of the surface oxygen atoms, surface coverage of ^{16}O deceases exponentially upon switching from $^{16}O_2$ to $^{18}O_2$. The rate expressions for production of $^{16}O_2$ and $^{16}O^{18}O$, Eqs. 4 and 5, become

$$R_{^{16}O^{18}O} = 2k_d n^2 \theta_0^2 e^{-2k_d n\theta_0 t}(1 - e^{-2k_d n\theta_0 t})$$

(10)

$$R_{^{16}O_2} = k_d n^2 \theta_0^2 e^{-4k_d n\theta_0 t}.$$

(11)

Eq. 10 shows that immediately following the isotope switch, the rate of production $^{18}O^{16}O$ increases rapidly, reaches a maximum when the surface concentrations of ^{18}O and ^{16}O are equal at $t = 0.35(k_d n\theta_0)^{-1}$, and then declines more slowly as the concentration of ^{16}O at the surface falls off exponentially. The FWHM of the peak is equal to $0.88(k_d n\theta_0)^{-1}$. We can obtain a value for k_d at 678 K from the FWHM of the $^{18}O^{16}O$ desorption transient shown in Fig. 1, or by fitting the entire curve using Eq. 10. The value for k_d obtained from fitting the curve is 6×10^{-17} cm^2 s^{-1}. The calculated $^{16}O^{18}O$ desorption transient is shown in Fig. 1 (solid curve), and is in reasonable agreement with the data.

Using the value for k_d obtained by fitting the $^{18}O^{16}O$ (g) transient in Fig. 1, the dependence of the coverages of ^{16}O and ^{18}O, and the desorption rates of $^{16}O_2$ and $^{16}O^{18}O$ as a function of time can be calculated using Eqs. 8 to 11. The results are shown in Fig. 3. We see that upon switching from $^{16}O_2$ to $^{18}O_2$, the coverage of ^{16}O decreases exponentially as the coverage of ^{18}O increases. When they become equal, the production rate of $^{16}O^{18}O$ reaches a maximum. At longer times the production rate of $^{16}O^{18}O$ declines exponentially, as does the production rate of $^{16}O_2$.

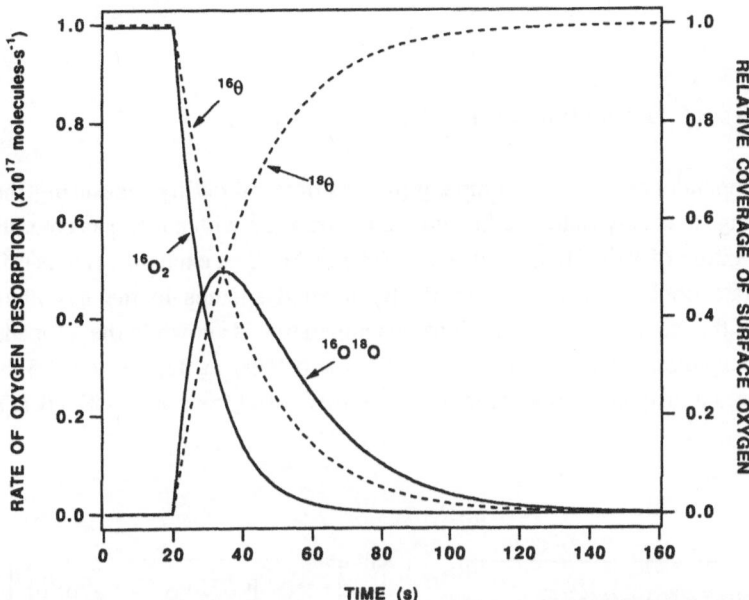

Fig. 3. Results from a model that describes the kinetics involved in oxygen isotope switching from 20% $^{16}O_2$ to 20% $^{18}O_2$. The surface coverages of ^{16}O and ^{18}O, and the desorption rates of $^{16}O_2$ and $^{16}O^{18}O$ are plotted as a function of time.

The number of molecules in the $^{16}O^{18}O$ transient can be determined accurately because the signal is not obscured by a high pressure of feed gas with the same isotopic composition. To determine the total oxygen surface coverage at steady state we also need to know the amount of ^{16}O that desorbs as $^{16}O_2$. The percentage of ^{16}O atoms that desorb as $^{16}O^{18}O$ ($N_{^{16}O^{18}O}$) and that desorb as $^{16}O_2$ ($N_{^{16}O_2}$) can be determined from integration of Eqs. 10 and 11. The amounts are given by

$$N_{^{16}O^{18}O} = \int_0^\infty R_{^{16}O^{18}O}\, dt = \frac{1}{2} n\theta_0 \tag{12}$$

and

$$N_{^{16}O_2} = 2\int_0^\infty R_{^{16}O_2}\, dt = \frac{1}{2} n\theta_0 \tag{13}$$

105

These show that 50% of the ^{16}O atoms on the surface desorb as $^{16}O_2$, and the other 50% desorbs as $^{18}O^{16}O$. Experimentally, we find that the area under the $^{16}O_2$ curve in Fig. 1, corrected for the contribution from the exponentially declining $^{16}O_2$ (g) feed, is 0.5 ± 0.2 of that under the $^{16}O^{18}O$ transient, in good agreement with the calculation.

In principle, the dependence of k_d on temperature can be determined by conducting the same isotope switching experiment at different temperatures. In practice, this approach is limited by the narrow temperature range that can be used. As the temperature increases, the FWHM of the $^{16}O^{18}O$ desorption transient rapidly becomes comparable to or shorter than the isotope switching time. The variation of k_d with temperature can be determined over a wider range of temperatures using the steady-state method described in the next section.

3.1.3 Kinetics of Desorption of the Surface Oxygen

The dependence of k_d on temperature was determined by measuring the rate of isotope exchange at steady state as a function of temperature with a feed gas consisting of an equi-molar mixture of $^{16}O_2$ (10%) and $^{18}O_2$ (10%) in Ar. The exchange rate is reflected in the steady-state concentration of isotopically labeled species in the gas phase. The dependence of the gas-phase concentrations on temperature is shown in the left panel of Fig. 4. Mass spectrometric signals of $^{16}O_2$, $^{18}O^{16}O$ and $^{18}O_2$ at 32, 34, and 36 amu were monitored continuously as the reactor temperature was raised from 500 to 873 K at a rate of

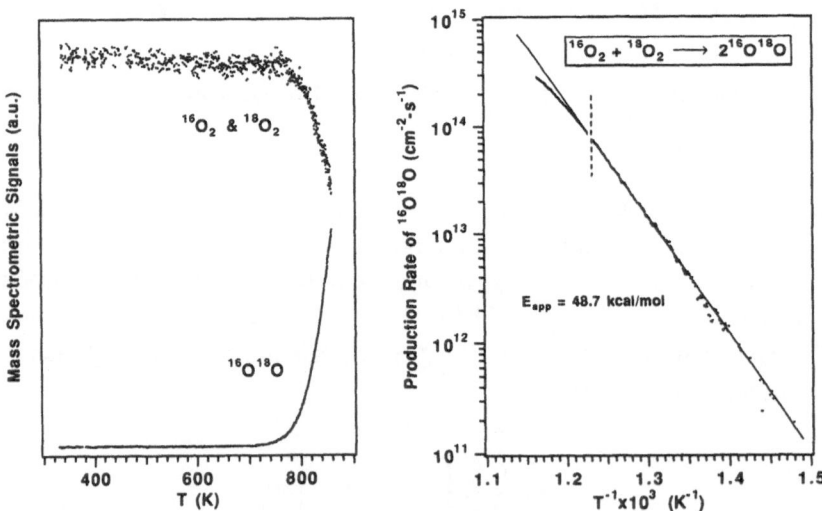

Fig. 4. Isotope exchange reaction of $^{16}O_2 + {}^{18}O_2 \longrightarrow 2^{16}O^{18}O$ over $Li_{0.44}Ni_{0.56}O$ (weight = 0.35 g). Reactants: 10% $^{16}O_2$ and 10% $^{18}O_2$ in Ar. Flow rate: 40 cc min^{-1}. Left: mass spectrometric signals of amu 32, 34 and 36 as a function of temperature. Heating rate: ~ 0.3 K s^{-1}. Right: Arrhenius plot of the steady-state isotope exchange rate.

~ 0.3 K s^{-1}. The reactor was then gradually cooled. Signals obtained during cooling (not shown) follow exactly those obtained during heating, indicating that the rates measured were indeed steady-state rates. As shown in the left panel of Fig. 4, signals of $^{16}O_2$ and $^{18}O_2$ decrease equally as temperature increases. This supports that the exchange reaction below 873 K involves only the surface oxygen, not the lattice oxygen, since otherwise, the mass 36 signal would decrease faster than that of mass 32. The decrease in the $^{16}O_2$ and $^{18}O_2$ signals is accompanied by a corresponding increase in the $^{16}O^{18}O$ intensity.

An Arrhenius plot of the rate of $^{18}O^{16}O$ formation as a function of temperature from 673 to 873 K is presented in the right panel of Fig. 4. The vertical dash line is a marker for 10% conversion. Rates to the right of the marker (from 670 to 813 K) can be described very well by the solid straight line, which has a slope that corresponds to an apparent activation energy of 48 kcal mol^{-1}, and an intercept of 1×10^{27} molecules-cm^{-2}-s^{-1}. Rates to the left of the marker fall below the straight line due to loss of the exchange product $^{16}O^{18}O$ through its reaction on the catalyst as the concentration of $^{16}O^{18}O$ becomes significant.

For an equi-molar $^{16}O_2/^{18}O_2$ feed, the total rate of oxygen desorption (sum of the desorption rates of $^{16}O_2$, $^{18}O^{16}O$ and $^{18}O_2$) is simply twice the production rate of $^{18}O^{16}O$. It is given by

$$R_d = (n\theta)^2 k_d^{(0)} \exp(-\frac{E_d}{k_B T}),$$

(14)

where $k_d^{(0)}$ and E_d are the pre-exponential factor and activation energy terms of the rate constant k_d, and k_B is the Boltzmann constant. Differentiating the logarithm of R_d with respect to T^{-1} yields the following expression

$$E_d = -k_B \frac{d(\ln R_d)}{dT^{-1}} + 2k_B \frac{d(\ln\theta)}{dT^{-1}}.$$

(15)

The first term to the right of Eq. 15 is the apparent activation energy determined from the Arrhenius plot (48 kcal.mol^{-1}). The second term depends upon the variation of surface oxygen coverage with temperature. The latter was determined to be zero from a pressure dependence study of the isotope exchange rate. In the pressure dependance study, the total oxygen partial pressure was reduced to 20% and 2% of one atmosphere, respectively. The molar ratio of $^{16}O_2$ to $^{18}O_2$ was maintained at 1:1. At conversions less than 10%, the rate of $^{18}O^{16}O$ production in each case was identical to that shown in Fig. 4. This clearly shows that under these conditions, the catalyst is saturated with the surface oxygen, and the desorption of the surface oxygen is the rate-limiting step in the exchange reaction. Thus, the second term in Eq. 15 is zero. As a result, the activation energy of the recombinative desorption of the surface oxygen equals the apparent activation energy of 48 kcal mol^{-1}.

With the surface state saturated ($\theta = 1$), the intercept of the Arrhenius plot in Fig. 4 is equal to $0.5 k_d^{(0)} n^2$, cf. Eq. 14. With $n = 5 \times 10^{14}$ cm^{-2} (the number density at saturation determined in the transient experiments, cf. Sec. 3.1.1.), the value for the pre-exponential factor, $k_d^{(0)}$, is 8×10^{-3} cm^2 s^{-1}. This is a reasonable value for recombinative desorption involving species with moderate surface diffusivity. Thus, the rate coefficient for the recombinative desorption of O(s) is

$$k_d = 8 \times 10^{-3} \exp(-48/k_BT) \ [cm^2 \ s^{-1}]. \tag{16}$$

The rate parameters determined from steady-state measurements gives a value for k_d at 678 K that is roughly a factor of 10 lower than the value obtained in the transient experiments. This is probably reasonable agreement. At the same time, we find that these rate parameters can be used to reproduce the thermal desorption spectrum. The thermal desorption spectrum of the surface oxygen from lithium nickel oxide has been measured by Lambert and coworkers, and is reproduced in left of Fig. 5.[19a] The temperature at the maximum desorption rate is 790 K. Using the rate coefficient k_d determined from our steady-state experiments, and taking a heating rate of 3 K s^{-1}, we can successfully simulate the desorption spectrum. The results are shown on the right side of Fig. 5.

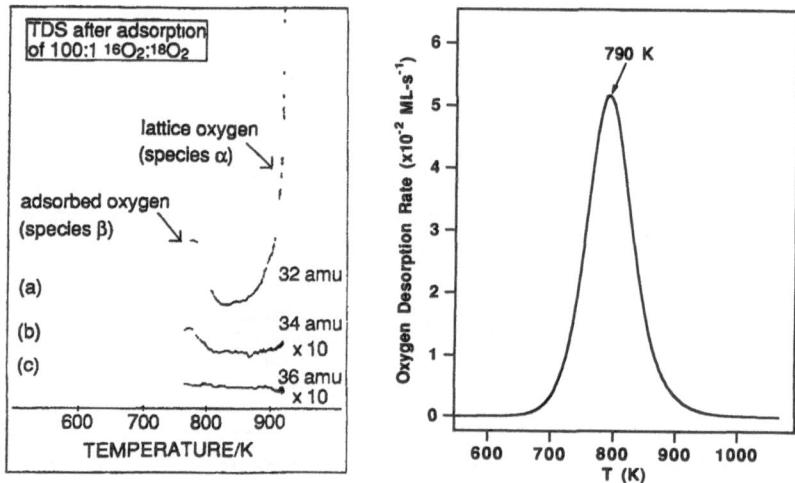

Fig. 5. Left: Thermal desorption spectra of oxygen by Lambert et al.[19a] (Reproduced with permission from *Journal of Catalysis*). Right: Calculated thermal desorption spectrum resulting from recombinative desorption of the surface oxygen, using the rate coefficient expressed by Eq. 16. The heating rate used is 3 K s^{-1}.

3.2. LATTICE OXYGEN STATE

The $^{16}O^{18}O$ transient upon switching from 20% $^{16}O_2$ to 20% $^{18}O_2$ was measured at a number of temperatures above 678 K. The results are presented in Fig. 6. At 678 K, the $^{16}O^{18}O$ transient is substantially longer than the switching time (cf. Sec. 3.1.1.). At 873 K, the O(s) desorption/exchange rate is significantly faster and the FWHM of the transient (~ 2 ms, calculated from k_d) becomes much shorter than the isotope switching time (~1 s). As a result, the rate of $^{16}O^{18}O$ tracks the gas-phase switching rate, i.e., the $^{16}O^{18}O$ intensity is determined by exchange during the switching time when both $^{16}O_2$ and $^{18}O_2$ are still in the reactor. However, there is apparently still no exchange with lattice oxygen since the $^{16}O^{18}O$ signal after the switch decreases rapidly to zero. At 923 K, we start to see $^{16}O^{18}O$ at longer times, which signals the participation of lattice oxygen in the exchange. More and more ^{16}O from the catalyst participates in the exchange as temperature increases above 923 K.

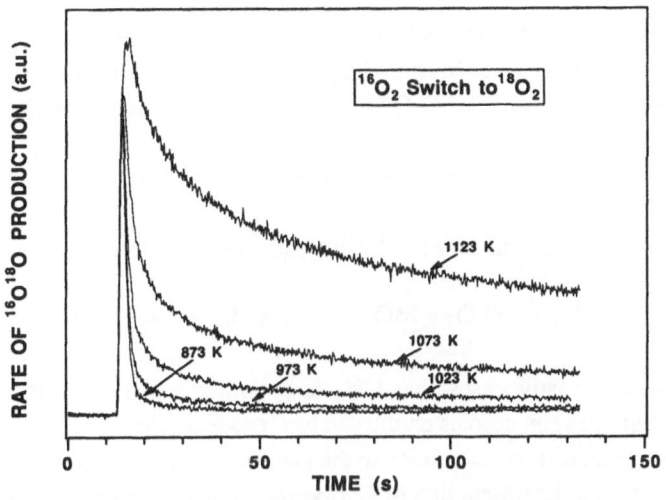

Fig. 6. Transient responses of mass spectrometric intensities of $^{16}O^{18}O$ from 873 to 1125 K upon switching from 20% $^{16}O_2$ in Ar to 20% $^{18}O_2$ in Ar. The flow rate was 40 cc min^{-1}. The catalyst was $Li_{0.44}Ni_{0.56}O$ (weight = 0.36 g).

As lattice oxygen begins to participate in the isotope exchange reaction, ^{18}O populates not only the surface state but also lattice sites, both at the surface and in the near surface region of the catalyst. The depth to which the catalyst becomes labeled depends on the details of the bulk diffusion kinetics, but, in general, the longer the time to which the catalyst is exposed to $^{18}O_2$, the greater the depth of the labelling. Consequently, when the gas-phase oxygen is switched back to $^{16}O_2$, the decay of the $^{16}O^{18}O$ transient that marks the replacement ^{18}O lattice sites will depend on the previous exposure time. This is shown in the right panel of Fig. 2. In the second switch, $^{16}O^{18}O$ is produced with a faster decay in rate and smaller time-integrated area compared to the initial switch from $^{16}O_2$ to $^{18}O_2$. This is due to labelling only to a limited depth, and the competition between replacement by ^{16}O and diffusion further into the bulk. Longer exposure times result in slower decay rates and more ^{18}O showing up the the second transient. This is contrasted by the case where only the surface oxygen is involved in the exchange reaction, shown in the left panel of Fig. 2, and discussed in sec. 3.1.1..

The data in Fig. 6 indicate that temperatures in excess of ~ 900 K are required for the gas-phase oxygen isotope to exchange with the lattice. This is consistent with the observation by Lambert et. al.[19a] that ~ 900 K is needed to cause the lattice oxygen to desorb into the gas phase. The temperature for desorption of the surface oxygen is much lower. Apparently, the activation energy associated with the desorption of the lattice oxygen is greater than that of the surface oxygen (48 kcal mol^{-1}). Although it is not known for $Li_{0.44}Ni_{0.56}O$, the activation energy for the diffusion of oxygen in NiO has been determined to be 57.7 kcal mol^{-1}.[21]

3.3. REACTIONS OF SURFACE AND LATTICE OXYGEN WITH CH$_4$

Oxidative coupling of methane is the reaction of methane with an activated oxygen (active oxygen on a metal oxide catalyst in the present case) to form ethane and water. This reaction can occur in the absence of $O_2(g)$ with many reducible metal oxides [2-6,11,16]

$$2\ CH_4 + MO \longrightarrow C_2H_6 + H_2O + MO_{1-x},\qquad (17)$$

and in the presence of $O_2(g)$ with some reducible and some non-reducible metal oxides [4-10]

$$2\ CH_4 + 1/2\ O_2 + MO \longrightarrow C_2H_6 + H_2O + MO.\qquad (18)$$

It has been shown in many cases that CH_4 is activated on the surface via hydrogen abstraction by certain oxygen species on the catalyst, producing gas-phase methyl radicals.[7b] The methyl radicals can further recombine in the gas phase to produce ethane. Production of ethane competes against the production of CO_2 via deep oxidation of methane, the coupling intermediates, or products. To understand how to design a catalyst that is highly selective for coupling, it is essential to understand the reaction pathways of the different oxygen species on the catalyst.

Reaction of CH_4 with the lattice oxygen of lithium nickel oxide can be studied by reacting CH_4 on $Li_{0.44}Ni_{0.56}O$ in the absence of gas-phase oxygen. In order to determine the intrinsic reactivity and selectivity on a nearly stoichoimetric material, the catalyst was exposed to a small pulse of CH_4 [22]. The duration of the pulse was such that less than three layers of the lattice oxygen were reacted. A typical pulse was 15% CH_4 for 10 s. We find that the lattice oxygen has a rather high selectivity in converting CH_4 to C_2H_6. Between 973 and 1023 K, the selectivity to C_2H_6 was 75%. The remaining 25% was fully oxidized to CO_2, but only 1% remained in the gas phase, 24% remained on the catalyst as carbonate. The amount of carbonate left on the catalyst was determined by oxygen titration following the CH_4 pulse. An initial C_2 selectivity of nearly 100% has been reported.[11d] We think that value is too high, possibly caused by improper counting of the carbonate formed on the surface. So we see that lattice oxygen plays a rather effective role in the coupling pathway. If one extends the duration of the CH_4 pulse, the catalyst maintains the selectivity of 75% for ~ 30 s, and then enters a less selective stage. Eventually, the selectivity decreases precipitously when the catalyst is reduced beyond a critical extent.[12,22]

In order to investigate the role played by the gas-phase and the surface oxygen, pulses of CH_4 and O_2 with fixed CH_4 (13.5%) but different O_2 (0 to 8.2%) partial pressures were reacted on $Li_{0.44}Ni_{0.56}O$ at 1023 K, and conversion of methane and selectivity were determined.[22] With increasing oxygen partial pressure, the production rate of C_2H_6 decreased, accompanied by an increase in the rate of CO_2. Importantly, the conversion of

CH$_4$ did not increased significantly in the presence of gas-phase oxygen. The presence of surface and gas-phase oxygen only reduced the overall selectivity. These results suggest that the surface oxygen does not help substantially the activation of the C-H bond of methane. Rather, if any thing, it leads to complete oxidation.

To further study the reaction between methane and the surface oxygen, we used oxygen isotope labeling to preferentially populate the surface state with ^{18}O. Determining the isotope label of oxygen in carbon dioxide provides us with information on the reactions of methane with the lattice and the surface oxygen. The results at 973 K are shown in Fig. 7.

The reactant-gas composition in the reactor is shown on the top of the diagram. Mass spectrometric intensities of ethane, carbon dioxides, and water (not shown) were monitored continuously as reactants were switched. At first, there was only $^{16}O_2$ in the reactor. Both surface and lattice oxygen had ^{16}O label. When the reactants were switched to $^{18}O_2$ and CH$_4$, the surface oxygen state was rapidly populated by ^{18}O. Fig. 7 shows that reaction of methane on this surface produces both ethane and carbon dioxide. The oxygen of the carbon dioxides, however, is mainly ^{18}O (87% ^{18}O and 13% ^{16}O), i.e. the surface/gas-

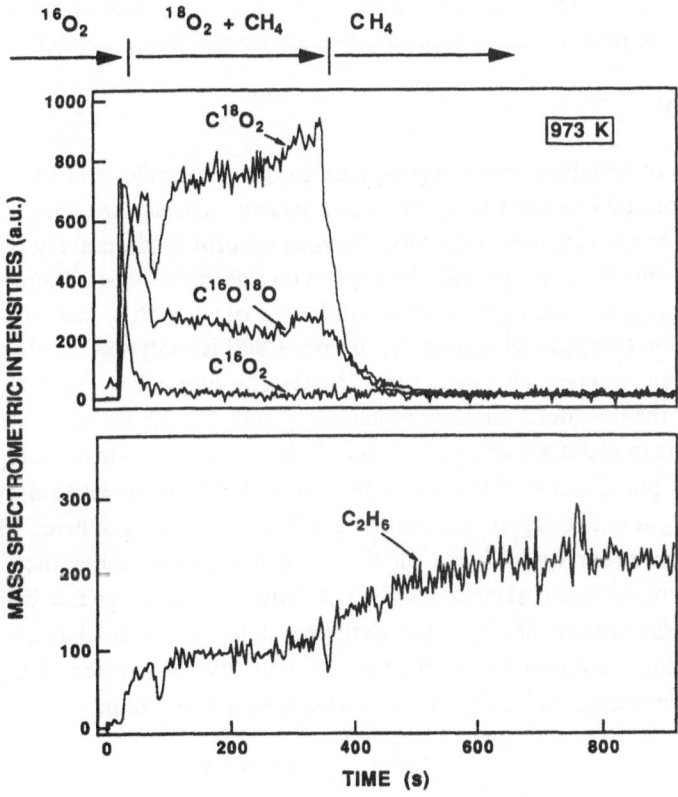

Fig. 7. Reaction of CH$_4$ with the surface ^{16}O and the lattice ^{18}O at 973 K. The reactant composition in the reactor is shown on the top of the figure: 20% $^{16}O_2$ in Ar to 20% CH$_4$ + 10% $^{18}O_2$ in Ar to 20% CH$_4$ in Ar. The low rate was 40 cc min^{-1}. The dips in the mass spectrometric signals at t ~ 80 s are caused by a transient decrease in pressure from the flow controller.

phase oxygen label. No $C^{16}O_2$ was detectable. We estimate that roughly 1/2 the oxidation occurs in the gas phase, and 1/2 is catalyzed by the surface.[22] These results indicate that the surface oxygen leads to complete oxidation, whereas the lattice oxygen is relatively selective to coupling. A couple of factors may complicate the carbon dioxide results, however. First, carbon dioxide can adsorb on the catalyst, producing carbonate by reacting with either a surface ^{18}O or a lattice ^{16}O. When the carbonate decomposes to carbon dioxide, its label can be altered. Second, in addition to the rapid population of the surface oxygen state, the lattice state may also be populated by ^{18}O at 973 K, although at a much slower rate. All these could complicate the interpretation of the carbon dioxide results. We assessed the extent of these processes by reacting a pulse of $C^{16}O_2$ and $^{18}O_2$ on $Li_{0.44}Ni_{0.56}^{16}O$, and monitoring the carbon dioxides. The result shows that ^{18}O can exchange with the oxygen in $C^{16}O_2$ as a result of reactions on the catalyst. Approximately 15% of the carbon dioxide ($C^{16}O_2$) did remain unexchanged. Since no $C^{16}O_2$ was observed in Fig. 7, and since in the absence of the gas-phase oxygen the lattice oxygen converts methane to ethane with rather high selectivity (75%), we conclude that the lattice oxygen does not contribute appreciably to CO_2 formation, whereas the surface oxygen does.

Another important observation in Fig. 7 is that the rate of CO_2 production decreases, accompanied by an increase in the rate of ethane production when the gas-phase oxygen was turned off. This is consistent with the results of the $CH_4 + O_2$ pulse experiments described above, and with the picture that the surface/gas-phase oxygen leads to complete oxidation.

4. SUMMARY

Kinetics of reactions between gas-phase oxygen molecules and $Li_{0.44}Ni_{0.56}O$ have been studied from 673 to 1073 K, using both transient isotopic switching technique and steady-state exchange rate measurements. Oxygen adsorbs dissociatively on the surface, and undergoes isotopic exchange with the oxygen on (in) the catalyst. There are two types of oxygen, namely, surface oxygen O(s) and lattice oxygen O(l), each associated with distinct desorption kinetics and chemistry. Below ~ 900 K, only O(s) exchanges with the gas-phase oxygen, whereas O(l) does not. The density of exchangeable surface oxygen atoms, given by the integrated intensity of the isotopically labeled transient, is 5×10^{14} cm^{-2}. This corresponds to about 1/3 of a monolayer. The steady-state rate of isotopic exchange between the gas-phase and O(s) was measured below 873 K, using an equi-molar mixture of $^{16}O_2$ and $^{18}O_2$ at total oxygen pressures of 2 to 20% of one atmosphere. The exchange rate has a zeroth-order dependence on the oxygen pressure under these conditions, suggesting that the surface is saturated with O(s). Since the exchange rate is limited by the recombinative desorption of O(s), the experimentally determined activation energy corresponds to the activation energy for the recombinative desorption of O(s). The rate coefficient for the recombinative desorption of O(s) is determined to be

$$k_d = 8 \times 10^{-3} \exp(-48/k_B T) \; [\text{cm}^2 \; \text{s}^{-1}].$$

Additional oxygen-isotope tracing experiments reveal that under conditions for the oxidative coupling of methane, O(s) leads mainly to the formation of CO_2, whereas O(l) is responsible for the activation of methane.

ACKNOWLEDGEMENTS

We thank J. G. Chan for helpful discussion.

REFERENCES

1. (a) S. Field, S.C. Nirula and J.G. McCarty, "An Assessment of the Catalytic Conversion of Natural Gas to Liquids", Report of SRI International Project No. 2352, (1987). (b) J.M. Fox, III, T.-P. Chen, and B.D. Degen, *Chem. Eng. Prog.* 42 (1990).
2. G.E. Keller and M.M. Bhasin, *J. Catal.* **73**, 9 (1982).
3. C.A. Jones, J.S. Leonard and J.A. Sofranko, *J. Energy and Fuels* **1**, 12 (1987).
4. M.S. Scurrell, *Appl. Catal.* **32**, 1 (1987).
5. G.J. Hutchings, M.S. Scurrell and J.R. Woodhouse, *Chem. Soc. Rev.* **18**, 251 (1989).
6. Y. Amenomiya, V.I. Birss, M. Goledzinowski, J. Galuszka and A.R. Sanger, *Catal. Rev. - Sci. Eng.* 32, 163 (1990), and references therein.
7. (a) J.H. Lunsford, *Catal. Today* **6**, 235 (1990). (b) D.J. Driscoll, W. Martir, J.-X. Wang and J.H. Lunsford, *J. Am. Chem. Soc.* **107**, 58 (1985). (c) T. Ito and J.H. Lunsford, *Nature* **314**, 721 (1985). (d) C.-H. Lin, T. Ito, J.-X. Wang, and J.H. Lunsford, *J. Am. Chem. Soc.* **109**, 4808 (1987).
8. M. Baerns, *Catal. Today* **1**, 357 (1987).
9. (a) K. Otsuka, K. Jinno and A. Morikawa, *J. Catal.* **100**, 353 (1986).
   (b) K. Otsuka and T. Nakajima, *J. Chem. Soc., Faraday Trans. 1* **83**, 1315 (1987).
10. (a) A. Ekstrom and A. Lapszewicz, *J. Am. Chem. Soc.* **110**, 5256 (1988).
    (b) A. Ekstrom and A. Lapszewicz, *J. Chem. Soc., Chem. Commun.* 797 (1988).
11. (a) J.A. Labinger, S. Mehta, K.C. Otto, H.K. Rockstad and S. Zoumalan, in: *Catalysis*, Ed. J.W. Ward (Elsevier, Amsterdam, 1987) p. 513. (b) J.A. Labinger and K.C. Otto, *J. Phys. Chem.* **91**, 2682 (1987). (c) J.A. Labinger, K.C. Otto, S. Mehta, H.K. Rockstad and S. Zoumalan, *J. Chem. Soc., Chem. Commun.* 543 (1987).
12. (a) K. Otsuka, Q. Lin and A. Morikawa, *Inorg. Chimica Acta* **118**, L23 (1986). (b) H. Hatano and K. Otsuka, *J. Chem. Soc. Faraday Trans. 1* **85**, 199 (1989). (c) K. Otsuka and T. Komatsu, *J. Chem. Soc., Chem. Commun.* 388 (1987). (d) M. Hatano and K. Otsuka, *Inorg. Chimica Acta* **146**, 243 (1988).
13. I.J. Pickering, P.J. Maddox and J.M. Thomas, *Angew. Chem. Int. Ed. Engl. Adv. Mater.* **28**, 808 (1989).
14. R.K. Ungar, X. Zhang and R.M. Lambert, *Appl. Catal.* **42**, L1 (1988).
15. I.J. Pickering, J.T. Lewandowski, A.J. Jacobson and J.A. Goldston, *Solid State Ionics* **53-56**, 405 (1992).
16. W. Li, J.N. Reimers and J.R. Dahn, *Phys. Rev.* **B46**, 3236 (1992).
17. R.B. Hall, Y.-K. Sun, J.T. Lewandowski, G.R. Myers, I.J. Pickering, W.T.A. Harrison and A.J. Jacobson, (to be published).
18. C. A. Mims, R. B. Hall, A. J. Jacobson, J. T. Lewandowski, and G. Myers, in: *Surface Science of Catalysis - in situ Probes and Reaction Kinetics*, Eds: D. J. Dwyer and F. M. Hoffman, ACS Symposium Series 482 (1991).

19. (a) G.D. Moggridge, J.P.S. Badyal and R.M. Lambert, *J. Catal.* **132**, 92 (1991). (b) X. Zhang, R.K. Ungar and R.M. Lambert, *J. Chem Soc., Chem. Commun.* 473 (1989).

20. P.J. Gellings and H.J.M. Bouwmeester, *Catal. Today* **12**, 1 (1992).

21. M. O'Keeffe and W.J. Moore, *J. Phys. Chem.* **65** 1438 (1961).

22. Y.-K, Sun, J.T. Lewandowski, G.R. Myers and R.B. Hall, (to be published).

OXIDATION OF OLEFINS TO ALDEHYDES USING A PALLADIUM–COPPER CATALYST

Timothy T. Wenzel

Union Carbide Corporation

PO Box 8361, South Charleston, WV 25303

INTRODUCTION

The oxidation of terminal olefins with palladium salts usually affords methyl ketones.[1-3] However, in 1986 it was reported that aldehydes could be obtained using a catalyst comprising $(CH_3CN)_2Pd(NO_2)Cl$ and $CuCl_2$ in t-butanol solvent, which was proposed to be bimetallic with the NO_2 group intact.[4] Our studies suggest that this catalyst is best described as a Wacker-like oxidation catalyst modified by an alkyl nitrite, and we report an improved version of this catalyst. Moreover, the application of our system to the oxidation of terminal olefins with allylic substituents has led to some insight as to the potential role of the copper co-catalyst in Wacker-like reactions.

RESULTS AND DISCUSSION

At the time of this report, we were independently working with the same Pd/Cu catalyst but in tetrahydrofuran. We added $CuCl_2$ to $(CH_3CN)_2Pd(NO_2)Cl$,[5] and cationic analogs[6] in order to slow isomerization of the olefinic double bond,[7] but we found that the oxidation rate acquired a positive dependence on $[CuCl_2]$. IR analysis of a mixture of $(CH_3CN)_2Pd(NO_2)Cl$ and $CuCl_2$ in THF revealed that the NO_2 group is immediately transferred to the copper forming $(CH_3CN)_2PdCl_2$ and a mixture of what appear to be two copper nitrate species and a copper nitrosyl species.[*] The identical mixture of copper species can be obtained by treating $CuCl_2$ with one equivalent of $AgNO_2$; treatment with $Ag^{15}NO_2$ proved the bands were N,O stretches. The addition of 1-hexene to the mixture

---

[*]IR data (THF-d_8): Nitrate bands: 1549, 1500, 1300, 1288, 1255 cm^{-1}. Nitrosyl band: 1860 cm^{-1}. The nitrosyl compound could be independently generated by treating $CuCl_2$ with NOCl in THF. This nitrosyl complex is rapidly converted to the mixture of nitrate species on exposure to oxygen.

The Activation of Dioxygen and Homogeneous Catalytic Oxidation,
Edited by D.H.R. Barton *et al.*, Plenum Press, New York, 1993

of copper-NO$_x$ species quantitatively formed 2-chloro-1-nitrohexane (Figure 1). Therefore, no metal nitro catalyst is involved in THF solvent.

We subsequently developed an efficient catalyst for oxidizing terminal olefins to ketones starting with a non-nitro based catalyst. This catalyst comprised PdCl$_2$, CuCl, LiCl, CH$_3$CN and CuCl$_2$ in tetrahydrofuran or sulfolane solvent. No evidence of catalyst deactivation was observed even after >100 turnovers.

At this time, it was reported that the (CH$_3$CN)$_2$Pd(NO$_2$)Cl / CuCl$_2$ catalyst in tBuOH was quite selective for making aldehydes from simple alpha olefins. We decided to examine this catalyst by IR spectroscopy to see if the nitro group was transferred from Pd to the olefin substrate as was the case for our catalyst. In fact we found a much simpler process: when CuCl$_2$ was added, the nitro group was immediately transferred from palladium to t-butanol to give t-butyl nitrite.

We next tested our nitro-free formulation in tBuOH and found that it also gave high yields of aldehyde, along with the corresponding methyl ketone (Figure 2, Table 1). Catalyst lifetimes seem to be very good as we have never seen definite signs of deactivation. The addition of t-butyl nitrite tended to reduce olefin isomerization but it also reduced the aldehyde yield.

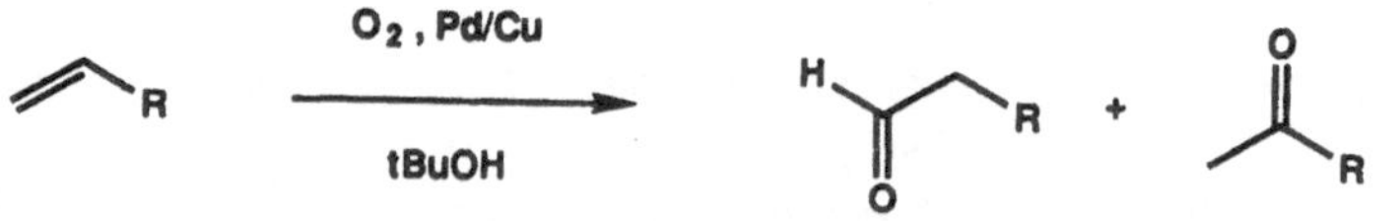

Figure 1. Transfer of the nitro group from Pd to olefin.

Selectivities for the oxidation of simple alpha olefins are significantly higher than in standard Wacker-like reactions where typically no aldehyde is produced. For instance, with only CuCl$_2$ as the copper source, up to 57% aldehyde is initially obtained for 1-octene, but the selectivity decreases as the reaction proceeds (Figure 3). Moreover, an induction period of about 20 minutes is observed. When both CuCl$_2$ and CuCl are used, there is no induction period. However, the aldehyde selectivity for 1-octene is lower (30-35%), but it remains steady throughout most of reaction. Olefin isomerization occurs with both variants and tends to reduce aldehyde selectivity at longer reaction times.

Better aldehyde selectivities are obtained with various allyl derivatives and much less isomerization is observed.[2,3] For instance, up to 90% aldehyde selectivity has been observed with allyl acetate. However, allylic substrates also tend to undergo exchange of the allylic group with the solvent under certain conditions to give t-butoxy aldehydes and ketones. Allyl ethers in particular tend to do this. This can be controlled (or encouraged) to a certain extent by changing the counterion of the auxiliary chloride source. For instance, with allyl acetate and MgCl$_2$ as the chloride source, up to 28% of the product has t-butoxy groups instead of acetoxy groups (Table 1). Switching to tetraalkyl ammonium chlorides completely eliminated any exchange. Allyl acetate tends to undergo several other side reactions that can also be controlled by using tetraalkyl ammonium chlorides.

Table I. Yields and selectivities for the oxidation of various terminal olefins to aldehydes and ketones.

substrate	comments	time (hr)	% conversion	% allylic exchange	aldehyde select.[d]	% yield ald+ket
1-octene[a]		0.5	12	-	30%	8
"		3	49	-	31%	38
"	no CuCl	0.5	4	-	57%	1
"	no CuCl	3	56	-	28%	39
allyl acetate[b]	acetonitrile	1	100	3	75%	60
"	acetonitrile (c)	1	90	5	86%	56
"	p-nitrobenzonitrile	1	100	0	73%	75
"	benzonitrile	1	100	0	69%	63
"	p-methoxybenzonitrile	5	96	0	56%	21
"	MgCl$_2$ (0.25 mmol)[e]	1	93	28	75%	58
"	THF solvent[f]	5	42	0	27%	21
"	10% H$_2$O / DMF[g]	5	100	0	2%	40

[a]0.25 mmol $(CH_3CN)_2PdCl_2$, 1.0 mmol $CuCl_2$, 0.5 mmol CuCl, 0.5 mmol LiCl, 7 mmol 1-octene, 5 mL tBuOH, 60 °C, 40 psi oxygen. [b]0.25 mmol $PdCl_2$, 1.0 mmol nitrile, 0.5 mmol CuCl, 0.5 mmol NaCl, 10 mmol allyl acetate, 5 mL tBuOH, 60 °C, 40 psi oxygen. [c]Same as (b), but 50 °C. [d]Moles aldehyde / (moles aldehyde + moles ketone). [e]In place of NaCl. [f]Tetrahydrofuran with CH_3CN. [g]0.5 mmol $PdCl_2$, 0.5 mmol CuCl, 5.3 mmol allyl acetate, 2.3 mL 10% H_2O / DMF, 20 °C, 40 psi O_2.

The choice of nitrile is also important. In general, nitriles with electron withdrawing substituents, such as p-nitrobenzonitrile, tend to increase aldehyde selectivity and yield (Table 1).

There are several unusual aspects to this catalyst that should be accounted for by any mechanism that we propose. First, with simple alpha olefins like 1-octene, the aldehyde selectivity, and the presence of an induction period, depends on the initial oxidation state of the copper.

Second, both exchange of the allylic substituent and oxidation of the double bond can occur with allylic substrates. The unusual aspect is that *the oxidation reaction can be switched on or off depending on whether oxygen is present, whereas the exchange reaction remains unaffected* . Therefore, the Pd(II) is still present and active under nitrogen, but it will not oxidize the olefin until oxygen is added. This runs counter to the accepted role of copper and oxygen in the Wacker system, which is to simply re-oxidize Pd(0). Figure 4 demonstrates this for the oxidation and exchange of ethyl allyl ether in t-butanol solvent. In this particular case, the amount of t-butoxy aldehyde actually exceeds the amount of ethoxy aldehyde all the way through the reaction, even thought the amount of allyl t-butyl ether is much lower than the amount of allyl ethyl ether. Therefore, most of the t-butoxy aldehyde does not arise by subsequent oxidation of the exchange product (allyl t-butyl ether).

Mechanistic Studies

The superficial mechanism of this reaction appears to be very much like that proposed for the Wacker reaction (Figure 5). However, in this case t-butanol serves as a hindered nucleophile which attacks the palladium-coordinated olefin at the less hindered terminal carbon to give an intermediate sec-alkyl palladium species (1) which either loses HX to give the exchange product or goes on to give aldehyde (Path **b,** Figure 5). Competing attack at the more-hindered internal olefinic carbon leads to the corresponding primary-alkyl palladium species, which leads to methyl ketone. Oxidation product that has exchanged with solvent can either arise by oxidation of the allylically-exchanged product (path **c**) or by partitioning of the intermediate **1** (path **a**). The oxidation pathways most likely occur by β-hydrogen migration followed by further attack on the coordinated vinyl ether by water (as a trace contaminant in the t-butanol which is then regenerated from Pd(0) by CuCl$_2$ / O$_2$ / HCl in the usual way). It is also possible that t-butanol could attack the coordinated vinyl ether to form a di-t-butyl acetal. This is probably less likely given that t-butyl acetals are uncommon and might also be difficult to hydrolyze once formed.

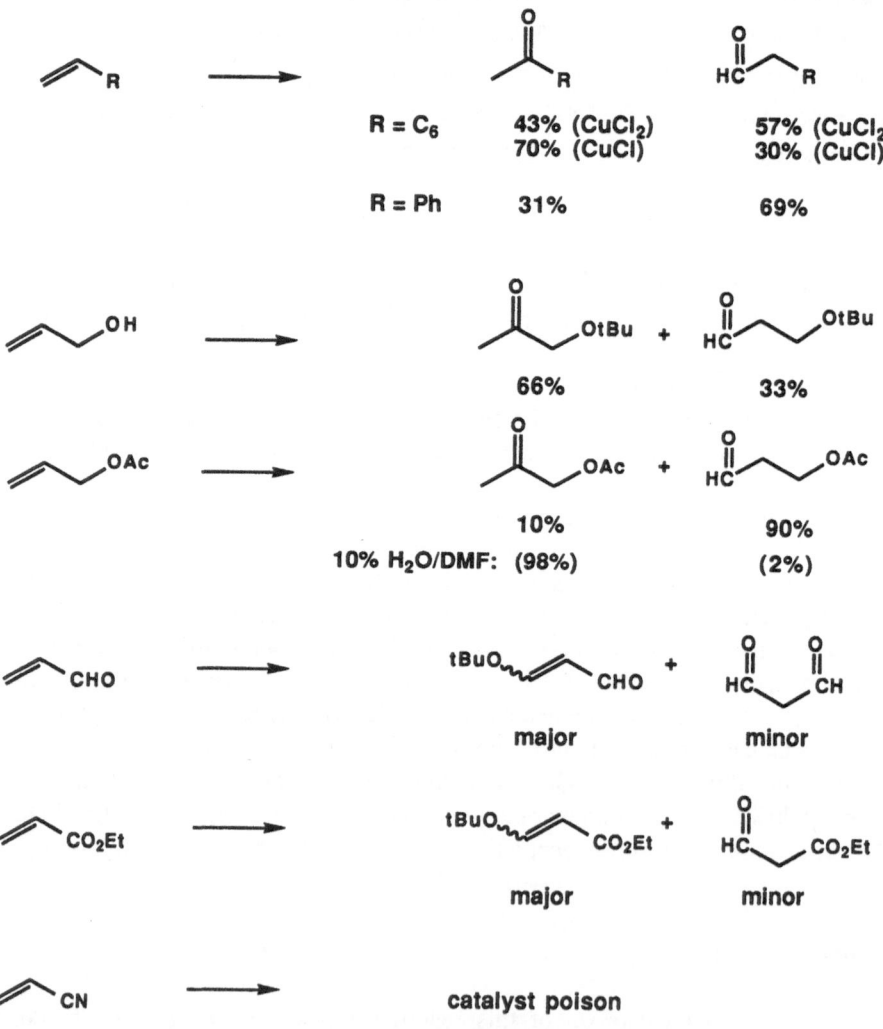

Figure 2. Selectivity for aldehyde vs. ketone for various alkenes.

Evidence for this mechanism is as follows: 1) there is a first order dependence of the initial rate of aldehyde formation on the t-butanol concentration in DMF solvent; 2) the use of n-butanol or s-butanol leads to ketal and acetal products, although those derived from s-butanol readily decompose to the ketone and aldehyde under the reaction conditions;[8] 3) the selectivity for aldehyde increases as: n-butanol < s-butanol < t-butanol; 4) small amounts of water increase the rate, but larger amounts decrease aldehyde selectivity probably due to competing attack by water on the coordinated olefin; and 5) non-protic solvents such as THF give much lower rates and aldehyde selectivities.

The proposed mechanism does not explain the induction period observed when CuCl is omitted nor does it explain how copper and oxygen switch on the oxidation pathway during the exchange of allylic compounds with solvent. These observations are of possible relevance to the Wacker reaction because the influence of copper (and oxygen) on the Wacker reaction has long been debated.[9] Also, little is known about the factors that influence the decomposition of the hydroxy (or alkoxy) palladated intermediate, such as **1**.

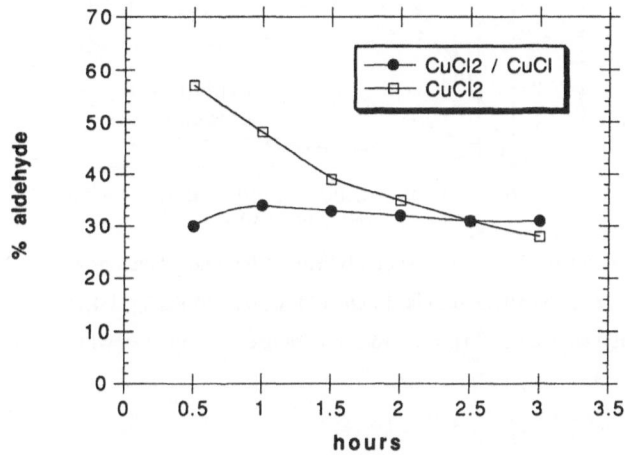

Figure 3. Effect of the oxidation state on the oxidation of 1-octene.

There are at least three possible roles for copper and oxygen in this system: 1) the CuCl serves as an HCl scavenger, to reduce the concentration of HCl formed by spontaneous decomposition of Pd(II) and thus promote oxidation; 2) the possible intermediates involved in the oxidation of CuCl by oxygen, such as copper peroxo species,[12] might function as Lewis acids, oxidants or nucleophiles to promote oxidation; and 3) the reaction might proceed via a Pd-OOH intermediate.[10]

The involvement of a Pd-OOH species, derived from palladium hydride **2** (Figure 2), was excluded because both copper and oxygen are required to initiate the oxidation reaction. Furthermore, no deuterium is incorporated into the aldehyde product when tBuOD / D_2O is used as the solvent, which must occur if the hydride in **2** is converted to a hydroperoxide.

Possibility (1), the formation of an HCl scavenger from CuCl and oxygen, seems likely because stoichiometric experiments with (nBuCN)$_2$PdCl$_2$ and allyl acetate indicate that HCl is required to prevent spontaneous reduction of palladium to Pd(0), while still permitting exchange to occur. This was further confirmed by substituting CuCl$_2$ + LiOH for CuCl$_2$ + CuCl in a catalytic experiment. The two experiments were nearly identical: neither had induction period and the aldehyde selectivity was constant throughout the reaction.

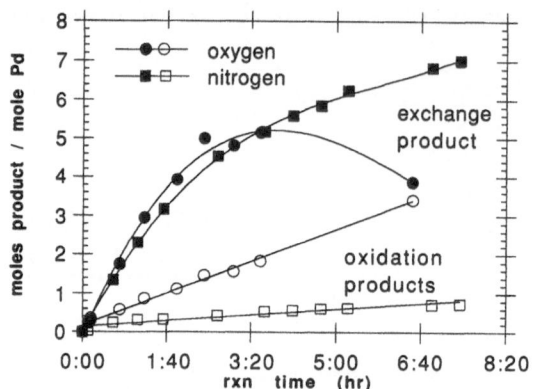

Figure 4. Effect of oxygen on the conversion of allyl ethyl ether to oxidation products (aldehyde + ketone) and exchange product (*t* - butyl allyl ether) in tBuOH (30 °C, 0.05 M PdCl$_2$, 1:4:2:2:4:40 PdCl$_2$:CH$_3$CN: CuCl: NaCl:CuCl$_2$:allyl ethyl ether). Downward curve for exchange product with 80 psi oxygen is due to oxidation of this product.

Possibility (2) was tested by adding oxidants to stoichiometric reactions comprising (nBuCN)$_2$PdCl$_2$ / HCl (1/1) in tBuOH. Ferrocenium, oxygen or CuCl$_2$ neither switched on the oxidation pathway nor affected the exchange rate. However, CuCl / O$_2$, t-butyl nitrite and NO$_2$ all switched on the oxygen pathway without having any effect on the exchange reaction. Since t-butyl nitrite is a known re-oxidant in palladium-based oxidations, this demonstrates that an oxidant can also switch on the oxidation pathway. However, in most cases where it is employed, including these stoichiometric studies, the amount of aldehyde is substantially lowered, which might indicate a change of mechanism. Further experiments will be required to distinguish between the two possible functions listed above for CuCl/O$_2$.

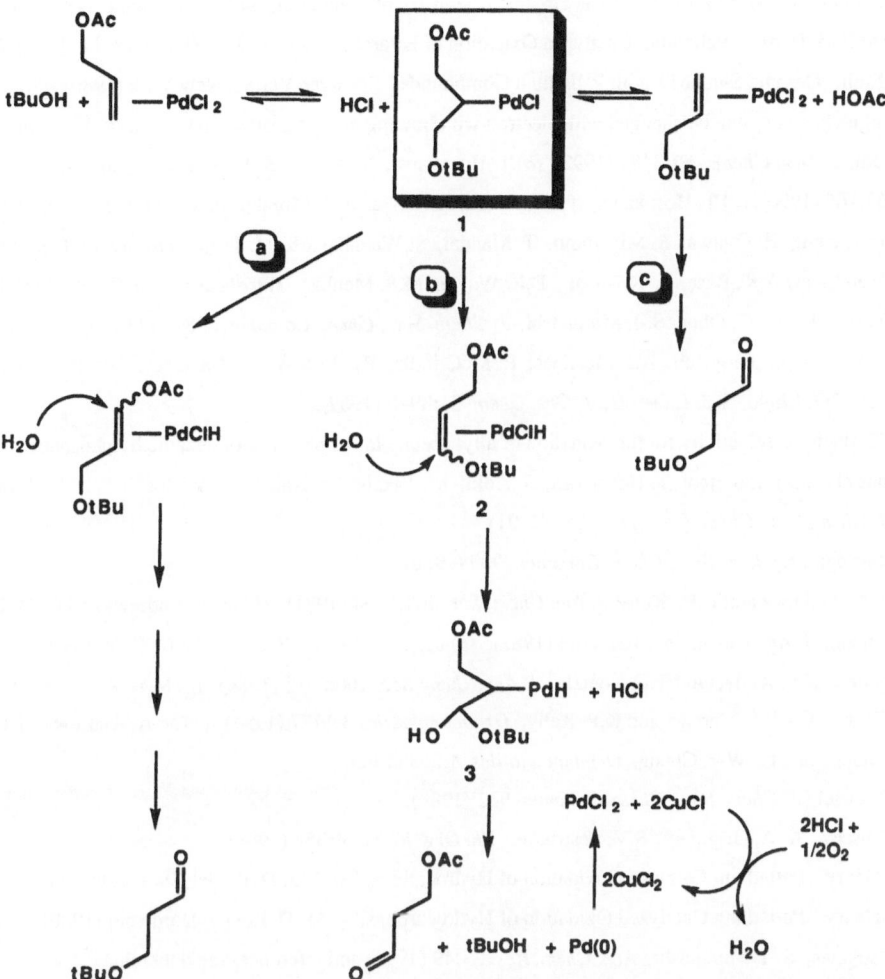

Figure 5. Proposed mechanism for the oxidation and solvent exchange of allylic substrates.

Acknowledgments

I would like to thank D.W. Butler, L.K. Clagg, and B.S. Kagen for their expert technical assistance and Union Carbide for permission to publish this work.

REFERENCES

1. For reviews see: (a) P.M. Maitlis. "The Organic Chemistry of Palladium," vol. 2, Academic, New York (1971). (b) P.M. Henry. "Palladium Catalyzed Oxidation of Hydrocarbons," D. Reidel, Dordrecht (1980). (c) J. Tsuji. "Organic Synthesis with Palladium Compounds," Springer-Verlag, New York (1980).

2. The regiochemistry can be changed with electron withdrawing groups nearby: *(a)* J.-y. Lai, X.-x. Shi, L.-x. Dai, *J. Org. Chem.*. 57:3485 (1992). *(b)* T. Hosokawa, Y. Ataka, S.-I. Murahashi, *Bull. Chem. Soc. Jpn.* 63:166 (1990). *(c)* T. Hosokawa, T. Shinohara, Y. Ooka, S.-I. Murahashi, *Chem. Lett.* 2001 (1989). *(d)* J. Nogami, H. Ogawa, S. Miyamoto, T. Mandai, S. Wakabayashi, J. Tsuji, *Tetrahedron Lett.* 29:5181 (1988). *(e)* A.K. Base, L. Krishnan, D.R. Wagle, M.S. Menhas, *Tetrahedron Lett.* 27:5955 (1986). *(f)* T. Hosokawa, T. Ohta, S.-I. Murahashi, *J. Chem. Soc., Chem. Commun.* 848 (1983). *(g)* E.C. Alyea, S.A. Dias, G. Ferguson, A.J. McAlees, R. McCrindle, P.J. Roberts, *J. Am. Chem. Soc.* 99:4985 (1977). *(h)* W.G. Lloyd, B.J. Luberoff, *J. Org. Chem.* 34:3949 (1969).

3. A 65% aldehyde selectivity for the oxidation of allyl acetate in the presence of hexamethylphosphoric triamide has also been reported: T. Hosokawa, S. Aoki, M. Takano, T. Nakahira, Y. Yoshida, S.-I. Murahashi, *J. Chem. Soc., Chem. Commun.* 1559 (1991).

4. B.L. Feringa, *J. Chem. Soc., Chem. Commun.,* 909 (1986).

5. *(a)* M.A. Andrews and K.P. Kelly, *J. Am. Chem. Soc.* 103:2894 (1981). *(b)* M.A. Andrews and C.-W.F. Cheng, *J. Am. Chem. Soc.* 104:4268 (1982). *(c)* M.A. Andrews, T.C.-T. Chang, C.-W.F. Cheng, T.J. Emge, K.P. Kelly, and T.F. Koetzle, *J. Am. Chem. Soc.* 106:5913 (1984). *(d)* M.A. Andrews, T.C.-T. Chang, C.-W.F. Cheng, and K.P. Kelly, *Organometallics* 3:1777 (1984). *(e)* M.A. Andrews, T.C.-T. Chang, and C.-W.F. Cheng, *Organometallics* 4:268 (1985).

6. T.T. Wenzel, *J. Chem. Soc., Chem. Commun.* 932 (1989).

7. I.I. Moiseev, A. A. Grigor'ev, S.V. Pestrikov, *Zh. Org. Khim..* 4:354 (1968).

8. P.M. Henry. "Palladium Catalyzed Oxidation of Hydrocarbons," p. 133, D. Reidel, Dordrecht (1980).

9. P.M. Henry. "Palladium Catalyzed Oxidation of Hydrocarbons," p 43, D. Reidel, Dordrecht (1980).

10. T. Hosokawa, S.-I. Murahashi, *Acc. Chem. Res.* 23:49 (1990) and references cited therein.

11. Cu(II)-O-O-Cu(II) species: (see, for instance, Z. Tyeklar, K.D. Karlin, *Acc. Chem. Res.,* 22:241 (1989) or Cu(III)-O species: (see, for instance, N. Kitajima, T. Koda, Y. Iwata, Y. Moro-oka, *J. Am. Chem. Soc.* 112:8833 (1990).

NEW METAL COMPLEX OXYGEN ABSORBENTS

FOR THE RECOVERY OF OXYGEN

Dorai Ramprasad*, Andrew G. Gilicinski, Thomas J. Markley and Guido P. Pez

Corporate Research Group, Air Products and Chemicals, Inc.
Allentown, PA 18195-1501 (USA)

INTRODUCTION

Emerging non-cryogenic technologies for the separation of air use zeolites and microporous "molecular sieve" carbons as moderately selective nitrogen and oxygen adsorbents, respectively.[1,2] While the zeolites have a thermodynamic affinity for N_2, use of carbons relies on a kinetic selectivity for the passage of oxygen into the micropores. It is well known that certain coordination compounds of cobalt and iron reversibly react with oxygen under near ambient conditions.[3,4] Since this is a chemical rather than a physical interaction as is seen with zeolites and carbons, it should be possible to use such metal complexes as O_2 equilibrium sorbents for air separation. We have been conducting a long term research effort to prepare such metal complex oxygen carriers for use in future generation non-cryogenic air separation devices.[5] The primary interest in such complexes is in their use in pressure or temperature swing processes for the production of inert gas (N_2,Ar) and oxygen.[6,7] For these applications, the oxygen complex could either be used as a circulating liquid or as a solid sorbent. In order to be useful in a commercial process an oxygen complex has to satisfy several requirements. It must (a) bind O_2 rapidly and reversibly, (b) have a high stability (>1 year lifetime), and (c) be accessible via simple synthetic techniques at minimal cost.

The most difficult problem that has precluded this use of oxygen complexes in a commercial process is that of limited operational lifetime. All known oxygen complexes including hemoglobin and myoglobin in living systems degrade with time.[3,4] With cobalt(II) oxygen complexes there are three general degradation mechanisms: (1) an irreversible formation of 2:1 $Co:O_2$ peroxo-bridged dimers by carriers that otherwise function reversibly as 1:1 $Co:O_2$ carriers, (2) irreversible ligand oxidation, and (3) central metal atom oxidation.[8] The design of ligands that sterically inhibit the formation of the 2:1 peroxo dimers has been an exciting area of fundamental research. Two classical examples are the "picket fence" porphyrin[9] and the lacunar or "dry cave" complexes.[8] However, the problem of ligand oxidation is generally regarded as being the most harmful since it is essentially irreversible.

The Activation of Dioxygen and Homogeneous Catalytic Oxidation,
Edited by D.H.R. Barton *et al.*, Plenum Press, New York, 1993

Martell and coworkers[10] elucidated the degradation mechanisms of a number of polyamine ligand Co(II) complexes which form 2:1 peroxo dimers with oxygen. Their results can be summarized as follows:

(a) Conversion to the O_2-inert Co(III) complexes of the original ligand with release of H_2O_2:

$$LCo^{3+}\text{-O-O-}Co^{3+}L + 2H^+ \rightarrow 2CoL^{3+} + H_2O_2 \qquad (1)$$

(b) Oxidative dehydrogenation of the coordinated ligands:

$$LCo^{3+}\text{-O-O-}Co^{3+}L \rightarrow 2H_2O + 2CoH_{-2}L^{2+} \qquad (2)$$

(c) Oxygen insertion into a coordinated ligand:

$$LCo^{3+}\text{-O-O-}Co^{3+}L \rightarrow 2CoLO^{2+} \qquad (3)$$

As an example of the first mechanism, the 2:1 μ-peroxo complex: $Co_2O_2(OH)(bipy)_4{}^{3+}$ (bipy = 2,2'-bipyridyl) was shown to degrade in aqueous solutions to a Co(III) species and H_2O_2 with the relatively oxidation resistant bipyridyl ligand remaining unchanged.[11]

In other work with polypyridine ligands Huchital and Martell et al[12] prepared $[Co(terpy)(bipy)]^{2+}$ and $[Co(terpy)(phen)]^{2+}$ (terpy = 2,2'; 6',2"-terpyridine; phen = 1,10-phenanthroline) complexes in aqueous solutions which were identified *in situ** by potentiometric titration methods.# Solutions of the former absorbed O_2 at pH 3.0 giving a 2:1 O_2 complex having a half-life of 88 min at room temperature. Since in these $[Co(II)$ polypyridine$]^{2+}$-O_2 systems degradation results (at least in part) from a further reaction of the bound O_2 to yield hydrogen peroxide (mechanism (a)), we sought to prepare $[Co(terpy)(bipy)]^{2+}$ and related polypyridine complexes as discrete compositions and study their O_2 reactivity in aprotic media.

Synthesis and Characterization

Our initial synthetic strategy for generating a $[Co(terpy)(bipy)]^{2+}$ complex in acetonitrile is shown in Equation 4.

$$Co(terpy)Cl_2 + 2AgPF_6 + bipy \rightarrow [Co(terpy)(bipy)](PF6)_2 + 2AgCl\downarrow \qquad (4)$$

The reaction was conducted under an N_2 atmosphere and the silver chloride precipitate was removed by filtration. Attempts were made to isolate the complex $[Co(terpy)(bipy)](PF_6)_2$ from the remaining acetonitrile solutions by precipitating it via addition of ether under nitrogen. However, the orange solids isolated by this procedure did not show any oxygen activity when redissolved in acetonitrile. Fortuitously, we discovered that the same reactions when done in the presence of air generated solids

---

*By reacting these solutions with NH_4PF_6 a solid precipitated which on the basis of chemical analysis for N alone was postulated to be $[Co(terpy)(phen)H_2O]$ $(PF_6)_2$. No O_2 reactivity properties were reported for this material.

#Lunsford et al[13] reported on the synthesis of a $[Co(terpy)(bipy)]^{2+}$ species entrapped in the cage of a Y zeolite. It was shown by EPR spectroscopy to yield with oxygen a mononuclear, 1:1 Co:O_2 reversible oxygen complex.

which appeared to be O_2 active when taken up in acetonitrile. The electrochemistry of solutions of these solids showed the presence of an O_2 active species as the minor component, and an O_2 inactive product. Cyclic voltammograms of these solutions under N_2 and O_2 are reproduced in Figures 1(a) and (b), respectively. The cyclic voltammograms under N_2 show two oxidation waves: one at $E_{1/2}$ = -0.02V (major species) and another at $E_{1/2}$ = 0.21V (minor species).* On introducing oxygen the current due to the minor species at -0.02V diminished and a new oxidation wave appeared at $E_{1/2}$ = 0.84V due to an oxygenated species. The first cited wave at -0.040V was unchanged indicating the major species to be O_2 inactive.

Various experiments were conducted in order to maximize the concentration of the "active" $[Co(terpy)(bipy)]^{2+}$ species. Two synthetic methods proved to be successful (Equations 5 and 6).

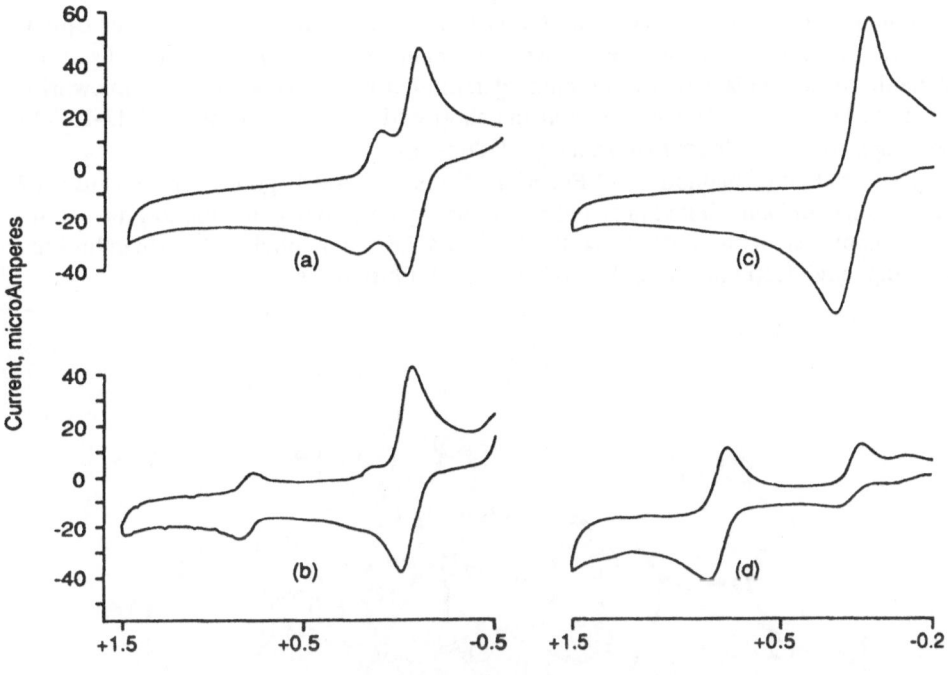

Potential, Volts vs. Ag/0.010 M AgNO$_3$

Figure 1. Cyclic voltammetry results for a 0.001 M solution of Co(terpy)(bipy)]$^{+2}$ complex in acetonitrile with 0.10 M tetrabutylammonium hexafluorophosphate as supporting electrolyte. Scan rate was 0.10 Volts/sec in all experiments. Potentials are referenced vs. silver/0.010 M silver nitrate in acetonitrile. (a) Impure complex under nitrogen, (b) impure complex under oxygen, (c) purified complex under nitrogen, and (d) purified complex under oxygen.

---

*$E_{1/2}$ is taken as the mean of the oxidation and reduction peak potentials for the reversible couple. All potentials reported here are referenced to the Ag/0.01M AgNO$_3$ couple measured at +0.292V versus SCE (saturated calomel electrode).

$$Co(terpy)Cl_2 + 2AgPF_6 + bipy \xrightarrow[\text{acetone}]{\text{air}}$$

$$1/2[(terpy)(bipy)Co\text{-}O\text{-}O\text{-}Co(terpy)(bipy)](PF_6)_4 + 2AgCl\downarrow \qquad (5)$$

$$Co(terpy)Cl_2 + 2NH_4PF_6 + bipy \xrightarrow[\text{air}]{\text{methanol}}$$

$$1/2[(terpy)(bipy)Co\text{-}O\text{-}O\text{-}Co(terpy)(bipy)](PF_6)_4 + 2NH_4Cl \qquad (6)$$

The pure "active" $[Co(terpy)(bipy)]^{2+}$ complex was thus isolated as its oxygen adduct; its identity and purity were established by elemental analyses, a crystal structure determination, and electrochemistry. Figures 1(c) and (d) show cyclic voltammograms of a pure sample of the complex in its free and oxygenated forms, respectively. In contrast to the impure sample (Figures 1(a) and (b)), the species at $E_{1/2} = 0.21V$ is now the major component. With the oxygenated solution a new wave appears at $E_{1/2} = 0.84V$ which indicates the presence of the μ-peroxo dimer.

Single crystals of the above O_2 adduct (as an acetone solvate) were grown from acetone/pentane solutions and the crystal structure determined by x-ray crystallography.[14] Figure 2 shows an ORTEP diagram of the μ-peroxo dimer cation. The cobalt coordination geometry is distorted octahedral, with most of the distortion attributable to the chelate ring dimensions determined by the rigid ligand frameworks. The peroxo bond is 1.419(7)Å and is at the short end of the usual range (1.41-1.49Å) suggesting a considerable π^* to d orbital back donation.

The synthetic methods of Equation 5 also proved good for the use of phenanthroline and substituted phenanthrolines as ligands. We were able to prepare and isolate complexes with 1,10-phenanthroline, 3,4,7,8-tetramethyl-1,10-phenanthroline, and 4,7-diphenyl-1,10-phenanthroline substituting for bipyridine.

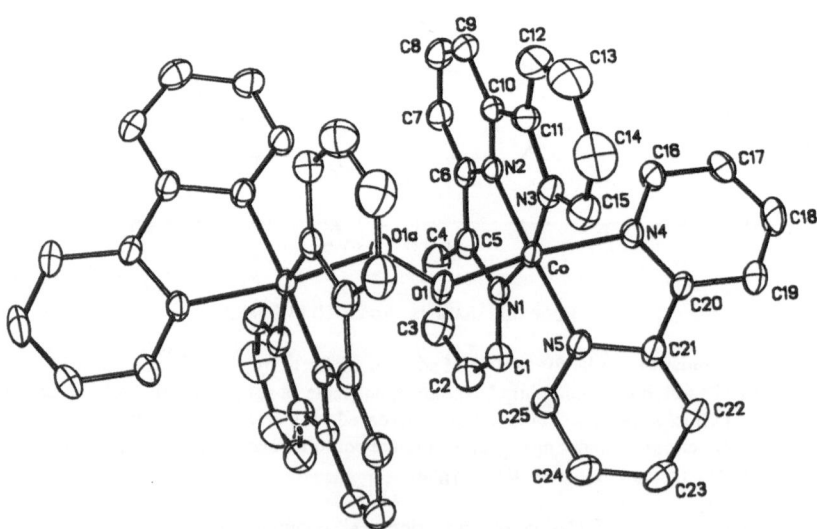

Figure 2. ORTEP view of the $[Co(terpy)(bipy)]_2O_2^{4+}$ ion of $[Co(terpy)(bipy)]_2O_2(PF_6)_4\cdot2CH_3COCH_3$.

Oxygen Chemistry of [Co(terpy)(bipy)]$^{2+}$ Complexes in Acetonitrile

The [Co(terpy)(bipy)]$^{2+}$ complexes were found to bind O_2 reversibly in acetonitrile as shown:

$$2[Co(terpy)(bipy)]^{2+} + O_2 \leftrightarrow [(terpy)(bipy)Co-O-O-Co(terpy)bipy]^{4+}$$

The UV/visible spectrum of a 10^{-3} molar solution of [Co(terpy)(bipy)]$^{2+}$ in acetonitrile showed dramatic differences under N_2 and O_2. A characteristic band was found to develop at ~670 nm upon exposure of the solutions to O_2. The spectral changes could be reversed either by purging with N_2 or by warming. However, these acetonitrile solutions after standing for a week were found to lose much of their O_2 binding capability. Since this change was seen either in the presence of O_2 or N_2, we concluded that the deactivation process did not involve oxidation. Our conclusions were validated by electrochemical measurements done over time with a solution of [Co(terpy)(bipy)](PF$_6$)$_2$ in acetonitrile. A cyclic voltammogram of a pure sample of [Co(terpy)(bipy)]$^{2+}$ under N_2 was recorded in Figure 1(c). Over a few days it was found that the current due to the O_2-active species at $E_{1/2} = 0.21$V decreased and the O_2-inactive species grew in at $E_{1/2} = -0.02$V. The estimated rate of decay was roughly 7% of the active complex per day. These results proved that the O_2-inactive component was a Co(II) species, and that the deactivation was caused by a rearrangement of the [Co(terpy)(bipy)]$^{2+}$ complex and not to oxidation.

Solvent Screening and Synthesis of Triflate Complexes

The oxygen binding studies done on the [Co(terpy)(bipy)]$^{2+}$ complex in acetonitrile showed that the oxygen active form of the complex rearranged slowly to an O_2 inactive form. We felt that a possible reason was that acetonitrile as a coordinating solvent might aid in the rearrangement of the active complex via displacement of the bipy or terpy ligands. Several other solvents were screened, including N,N-dimethylformamide, N-methyl pyrrolidone, dimethylsulfoxide, propylene carbonate, sulfolane and nitrobenzene. Of these solvents nitrobenzene is the least coordinating but still has a high dielectric constant. Indeed, a solution of [Co(terpy)(bipy)]$_2$O$_2$(PF$_6$)$_4$ in nitrobenzene gave a long-lived O_2-adduct but the oxygen was so tightly bound that purging with N_2 did not give the deoxygenated form. We now had a stable oxygen adduct but at the expense of reversibility.

We felt that it would be easier for the oxygen to come off the complex if there were a weakly coordinating anion to replace it. The anion of our choice was CF$_3$SO$_3^-$ (trifluoromethane sulfonate, or triflate). The oxygen adducts of several [Co(terpy)(polypyridine)]$^{2+}$ triflate complexes were prepared using analogous methods to those used for the synthesis of the PF$_6^-$ salts (Equation 5). In contrast to the latter, solutions of the triflate complexes in nitrobenzene could be deoxygenated by purging with N_2.

The presence (in the solid state) of a coordinated triflate on the sixth site was definitively proven by an x-ray crystal structure determination[14] of one complex, [Co(terpy)(3,4,7,8-tetramethyl-1,10-phenanthroline) SO$_3$CF$_3$]SO$_3$CF$_3$. The structure of the cation is shown in Figure 3. The Co(II) center is six coordinate with a highly distorted ligand geometry, caused by the coordination requirements of the terpy ligands. The planes of the two neutral ligands are nearly perpendicular. The Co-N(1) distance, 2.103(4)Å is slightly longer than the Co-N(2), 2.08(4)Å, reflecting the *trans* influence of the one triflate group which is clearly bonded to cobalt with D(Co-O), 2.165(3)Å. All the S-O distances of the coordinated triflate are longer than in the free ion, as expected, with the longest S-O distance being the one coordinated to cobalt.

The utilization of complexes with triflate replacing hexafluorophosphate indeed gave complexes which in nitrobenzene now displayed reversible O_2 affinity, ie. the O_2 could be removed by purging the solutions with nitrogen. This is seen quantitatively in Table 1. For complexes [Co(terpy)(L)X]Y (as defined in the Table) log K_{O_2} values of the order of ca. 4.8 or lower correspond to the above described observable O_2 reversibility. This is realized either in acetonitrile solutions, or in the less coordinating solvent nitrobenzene in the presence of anions of varying donor character. Specifically,

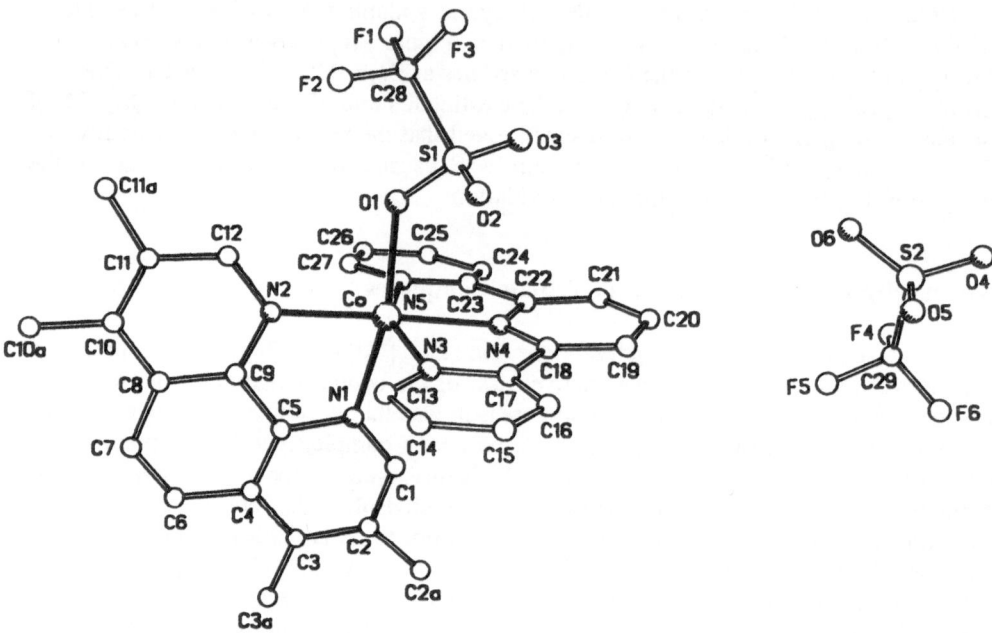

Figure 3. The structure of [Co(terpy)(3,4,7,8-tetramethyl-1,10-phenanthroline)triflate]triflate.

the O_2 binding for [Co(terpy)(bipy)]$(PF_6)_2$ in nitrobenzene was too strong to measure. However, in the presence of the increasingly coordinating anions, triflate or 4-chloro, 3-nitrobenzene sulfonate the O_2-affinity (log K_{O_2}) decreased dramatically from 4.8 to 1.5 respectively. We also investigated other anions as alternatives to triflate and found that NO_3^-, $CH_3SO_3^-$, $C_6H_5SO_3^-$ were all too strongly coordinating to cobalt resulting in non-O_2 active complexes.

Stability and Regeneration of Complexes

The stability (operational lifetime) of the new oxygen complexes was determined by two techniques: a) electrochemical monitoring, b) UV/visible spectroscopy.

Table 1. Dioxygen Binding Constants (log K_{O_2})* for Selected [Co(terpy)(L)X]Y Complexes in Solution

Ligand (L)	Anions (X,Y)	Solvent	log K_{O_2} (25°C) (M^{-1} atm^{-1})
bipy	CH3CN, *1*	CH3CN	4.4
bipy	*1*	PhNO2	*3*
bipy	CF3SO3⁻, *2*	PhNO2	4.8
bipy	4Cl,-3NO2-C6H3SO3⁻, *2*	PhNO2	1.5
1,10-phenanthroline	CH3CN, *2*	CH3CN	3.8
4,7-diphenyl 1,10-phenanthroline	CH3CN, *2*	CH3CN	4.1
4,7-diphenyl 1,10-phenanthroline	CF3SO3⁻, *2*	PhNO2	4.7

*1*Y = PF6⁻   *2*Y = CF3SO3⁻   *3*Too high to measure.

*   The oxygen binding constant log K_{O_2} was defined for the equilibrium, according to the formula:

2 complex + O_2 $\leftrightarrow$ dimer

$$K_{O_2}=[dimer]/[complex]^2 P_{O_2}$$

where P is the pressure of O_2. Concentrations of the bound and unbound forms of the complexes were calculated from cyclic voltammetry data by conversion of the peak current values as described below. Characterization of the current response in the cyclic voltammetry experiment is done using the following equation:[15]

$$i_{peak} \text{ (reversible)} = \text{(constant)} \, n^{2/3} A D^{1/2} v^{1/2} C$$

where i_{peak} is the peak current measured in the voltammetry experiment, n is the number of electrons transferred, A is the electrode area, D is the diffusion coefficient of the reacting species, v is the scan rate of the potential sweep, and C is the concentration of the reacting species. C is determined by measuring the peak current and applying known values to the remaining variables. Of these, n is 1 for the oxidation of the complex, A is determined by the electrode used, D is measured in a separate chronocoulometry experiment, and v is known. Chronocoulometry was used to obtain diffusion coefficients for the concentration calculations, using the relationship between charge and time for diffusion limited current after a potential step:[15]

$$Q = (2nFAD^{1/2}Ct^{1/2})/\pi^{1/2}$$

where F is the Faraday constant, t is time, and the other variables are as defined above. A typical diffusion coefficient was 2.0×10^{-5} cm^2/sec for [Co(terpy)(bipy)](PF6)2 in acetonitrile.

The stability of [Co(terpy)(4,7-diphenyl-1,10-phenanthroline)](triflate)$_2$ under oxygen was evaluated first using cyclic voltammetry. Activity, defined as measured concentration of active complex versus original concentration, was 80% over 30 days under O$_2$ and 91% for the same period under N$_2$, both under near anhydrous conditions. These compared to 35% retained activity under "wet" conditions (contact with lab atmosphere). As far as we are aware, these results represent the best stability yet achieved for solution based equilibrium oxygen complexes, with the possible exception of the Co(II) BISTREN complex.[16]

Lifetime studies were also conducted using UV/visible spectroscopy. The complexes in dry nitrobenzene were stored under O$_2$ for several months. It was found that [Co(terpy)(4,7-diphenyl-1,10-phenanthroline)O]$_2$(triflate)$_4$ as a 0.001 molar solution retained 65% of its activity after two months. Since water had been shown to have a detrimental effect on the stability of the complexes, we felt that increasing the concentration of the complex in solution relative to any adventitious water might improve the stability. Indeed, this was found to be the case and a 0.1 molar solution showed 90% activity after sixty days at 25°C. This corresponds to a projected half life of 8 months.

Finally, we attempted to regenerate oxygen complexes which had lost activity over time. Heating a solution of nearly spent (<5% active) [Co(terpy)(bipy)](triflate)$_2$ in nitrobenzene to 150°C under N$_2$ followed by cooling in an air purge resulted in a partial regeneration of the complex to 46% activity. This may be due to either the removal of water from a coordination site or a rearrangement of the complex to the active form.

CONCLUSIONS

Cobalt(II) polypyridine complexes of general formula [Co(terpy)(L)X]Y where terpy = 2,2';6',2"-terpyridine, L = 2,2'-bipyridine or 1,10 phenanthroline and substituted derivatives thereof and X,Y are anions, were prepared for the first time as discrete crystalline (non-hydrated) compounds. Solutions of the complexes in aprotic solvents were found to in many cases reversibly absorb oxygen. The resulting O$_2$-binding equilibria were characterized by a combination of spectroscopic and electrochemical methods. Oxygen affinities could be rationally altered by the use of solvents or anions of varying coordinating power. Several of the complexes were shown to exhibit a remarkable stability (ie. longevity) in solution. One representative sample retained 90% of its oxygen-binding activity after 60 days at room temperature. Furthermore, a simple method was found to partially regenerate complexes that had lost oxygen activity over time. These results represent a major step towards the goal of utilizing metal complex absorbents for the recovery of oxygen.

ACKNOWLEDGEMENTS

We are grateful to D. Krause for technical assistance, and to Dr. G. Johnson for spectroscopy studies. We also acknowledge the support and continued encouragement of Dr. J. Roth, and thank Air Products and Chemicals for permission to publish this work. Special thanks are due to K. Lakatosh for preparing this manuscript.

REFERENCES

1) H. J. Schroter and H. Jungten, Gas Separation by Pressure Swing Adsorption Using Carbon Molecular Sieves, in: "Adsorption Science and Technology, NATO ASI

Series E," A. E. Rodrigues et al, Ed., Kluwer Academic Publishers, the Netherlands (1989).

2) S. Sircar, Pressure Swing Adsorption Technology in: "Adsorption Science and Technology, NATO ASI Series E," A. E. Rodrigues et al, Ed., Kluwer Academic Publishers, The Netherlands (1989).

3) R. D. Jones, D. A. Summerville, and F. Basolo, Chem. Rev., 79:139 (1979).

4) E. C. Niederhoffer, J. H. Timmons, and A. E. Martell, Chem. Rev., 84:137 (1984).

5) J. A. T. Norman, G. P. Pez, and D. A. Roberts, Reversible Complexes for the Recovery of Dioxygen, in: "Oxygen Complexes and Oxygen Activation by Transition Metals," A. E. Martell et al, Ed., Plenum Press, New York (1988).

6) S. P. Nandi and P. L. Walker, Sep. Sci., 11:441 (1976).

7) H. Jungten, K. Knoblauch, and K. Hardner, Fuel, 60:817 (1981).

8) D. H. Busch, Synthetic Dioxygen Carriers for Dioxygen Transport, in: "Oxygen Complexes and Oxygen Activation by Transition Metals," A. E. Martell et al, Ed., Plenum Press, New York (1988).

9) J. P. Collman, T. R. Halpert, and K. S. Suslick, O_2 Binding to Heme Proteins and Their Synthetic Analogs, in: "Metal Ion Activation of Dioxygen," T. G. Spiro, Ed., John Wiley, New York (1980).

10) A. E. Martell, A. K. Basak, and C. J. Raleigh, Pure & Appl. Chem., 60:1325 (1988).

11) R. F. Bogucki, G. McLendon, and A. E. Martell, J. Am. Chem. Soc., 98:3202 (1976).

12) D. H. Huchital and A. E. Martell, Inorg. Chem., 13:2966 (1974).

13) S. Imamura and J. H. Lunsford, Langmuir, 1:326 (1985).

14) A. Rheingold, University of Delaware, Newark, Delaware 19716.

15) A. J. Bard, L. R. Faulkner, "Electrochemical Methods: Fundamentals and Applications," Wiley, New York (1980).

16) R. J. Motekaitis and A. E. Martell, J. Am. Chem. Soc., 110:7715 (1988).

OXYGENATION OF OLEFINS WITH MOLECULAR OXYGEN CATALYZED BY LOW VALENT METAL COMPLEXES

Teruaki Mukaiyama

Department of Applied Chemistry,
Faculty of Science, Science University of Tokyo
Kagurazaka, Shinjuku-ku, Tokyo 162, Japan

INTRODUCTION

Molecular oxygen is often expected to be the easily available oxidant, but its utilization in organic synthesis has been restricted to several reactions because of the difficulties in carrying a desired reaction out exclusively. Molecular oxygen is caputured and activated by transition-metal complexes coordinated by organic ligands whose stereochmical and electrochemical properties could be tuned by modofication of its ligand systems. Several new and efficient synthetic oxidation reactions of olefins using molecular oxygen were recently developed based on the above concept.

This work has been done at Basic Research Laboratories for Organic Synthesis, Mitsui Petrochemical Industries, Ltd.

COBALT(II) COMPLEX CATALYZED OXYGENATION

Oxygenation of carbon-carbon double bond with molecular oxygen or air is one of the most important reactions in organic synthesis and much effort has been made to develop practical procedure of oxygenation by the use of molecular oxygen. Several oxygenation reactions with molecular oxygen by the combined use of transition-metal complexes and reducing agents have been studied by a number of groups; for example, (tetraphenyl-porphyrinate)manganese(III) complex/NaBH4 (or colloidal Pt-H2)[1], (tetraphenyl-porphyrinate)cobalt(II) complex/Et4NBH4[2] or NaBH4[3] or [bis(salicylidene-γ-iminopropyl) methylamine]cobalt(II) complex/primary or secondary alcohol[4] were shown to be effective catalytic systems for oxygenations of olefins to form ketones as major products. However, there have been only a few reports concerning the selective oxygenation of olefins into the corresponding alcohols by use of molecular oxygen.

Catalytic Oxidation-Reduction Hydration

Because it was reported that bis(acetylacetonato)cobalt(II) readily absorbed molecular oxygen in the coexistence of base, such as pyridine,[5] cobalt(II) complex was first chosen as a transition-metal catalyst. When 4-phenyl-1-butene was treated with an atmospheric pressure of molecular oxygen in 2-propanol in the presence of a catalytic amount of bis(acetylacetonato)cobalt(II), the corresponding oxygenated product, 4-phenyl-2-butanol, was clearly detected on thin-layer chromatography. As shown in Table 1, the reaction smoothly proceeded to afford 4-phenyl-2-butanol as a major product along with 1-phenylbutane and 4-phenyl-2-butanone when a catalytic amount of bis(acetylacetonato)-

The Activation of Dioxygen and Homogeneous Catalytic Oxidation,
Edited by D.H.R. Barton *et al.*, Plenum Press, New York, 1993

cobalt(II) was used in secondary alcohol such as 2-propanol or cyclopentanol.[6] On the contrary, no reaction took place when primary or tertiary alcohol such as ethanol or *tert*-butylalcohol was used as a solvent. Also, cobalt(II)salen complex is not employed as effective catalyst at all.

Table 1. Cobalt(II) complexes and solvent for Oxidation-Reduction Hydration

Entry[a]	Co(II) Complex	Solvent	Conversion / %	Yields / %		
				Alcohol	Ketone	Alkane
1	Co(acac)$_2$	⟩–OH	100	**46**	8	17
2	Co(acac)$_2$	⬠–OH	100	**45**	9	16
3	Co(acac)$_2$	EtOH	No Reaction			
4	Co(acac)$_2$	✕–OH	No Reaction			
5	Co(salen)	⟩–OH	No Reaction			

a) Reaction conditions; 4-phenyl-1-butene (2.0 mmol) and Co(II) complex (0.4 mmol) were heated in 10 ml of solvent at 75 °C for 1 h under O$_2$ atmosphere.

Co(acac)$_2$=   Co(salen)=

It is interesting to point out that the present reaction affords the corresponding hydrated product, alcohol, as a major product, directly from olefin *via* simultaneous transfer of oxygen and hydrogen even under mild oxidative conditions. Since both oxidation (oxygenation) and reduction (hydrogenation) occurred at the same time in the present hydration reaction, it was thus named as "Oxidation-Reduction Hydration".

Table 2. Selectivity for alcohols in Oxidation-Reduction Hydration

Entry[a]	Ligand (LH)	Conversion / %	Yields / %[b]		
			Alcohol	Ketone	Alkane
1	(Ph)	No Reaction			
2	(Hacac)	100	**45**	7	22
3	(Hecbo)	100	**72**	14	2
4	(Ph, CO$_2$Et)	100	**59**	17	15
5	(OEt)	87	**65**	10	2
6	(Hmodp)	94	**74**	7	5
7	(tBu, CF$_3$)	100	**81**	9	4
8	(Htfa)	100	**81**	13	2
9	(Hhfa)	No Reaction			

a) Reaction conditions; 1-decene (2.0 mmol) and Co(II) complex (0.4 mmol) were heated in 10 ml of 2-propanol at 75 °C under O$_2$ atmosphere. b) Determined by GC analysis.

Catalytic activities of several cobalt(II) complexes having various 1,3-diketone-type ligands were next examined. It was interesting to find that the ratio of the hydrated product (alcohol) to the oxidized product (ketone), and to the reduced product (alkane) was influenced by the structure of the ligand system.[7] As shown in Table 2, the selectivity toward the hydration was increased when cobalt(II) complex, having been coordinated by the ligand with electron-withdrawing group, was employed as a catalyst. The yield of alcohol was improved up to 81% by using bis(trifluoroacetylacetonato)cobalt(II). In order to clarify possible catalytic activity, the redox potentials of cobalt(II) complexes were measured (Figure 1).[8] Catalytically active compelxes were characterized by their redox potentials between Co^{2+} and Co^{3+}. It was found that the complexes ranging from 0.0 V to +0.5 V in their redox potentials showed the catalytic activities in the present hydration reaction. No catalytic activities were shown at all by complexes having higher or lower redox potentials other than in the above-mentioned range.

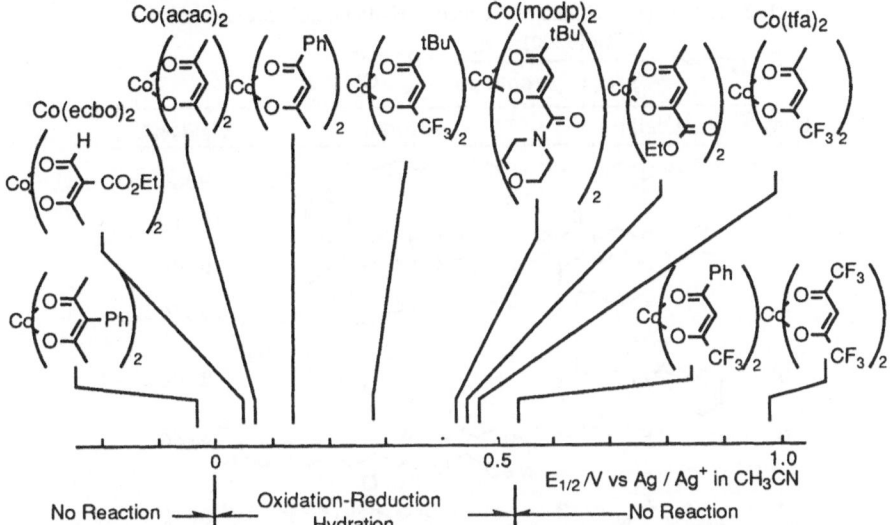

Figure 1. Relationship of redox potentials and catalytic activities of Co(II) complexes

The relationship between structure of the ligands and catalytic activity of cobalt(II) complex can be explained as follows: In the case of cobalt(II) complexes coordinated by ligands with the electron-donating groups such as 3-phenyl-2,4-pentanedione (Entry 1 in Table 2), the redox potential indicates that the complex itself is readily electrically-oxidized and is also oxidized by molecular oxygen, and the oxidized complex, therefore, shows no catalytic activities any longer in the present hydration reaction. On the contrary, in the cases of cobalt(II) complexes coordinated by ligands with strongly electron-withdrawing groups such as hexafluoroacetylacetone (Entry 9 in Table 2), it is reasonable to assume that the caputure of molecular oxygen by cobalt(II) complex is difficult .

As mentioned above, it was considered that one oxygen atom of molecular oxygen (Oxidant) was introduced into olefin to form the corresponding alcohol while another oxygen atom was reduced by two hydrogen atoms from secondary alcohol (Reductant) to form water (Scheme 1). Then, the amount of water formed during the reaction was measured by taking the hydration of 1-decene in 2-propanol catalyzed by Co(ecbo)$_2$. It was observed then that the amount of water increased as the hydration proceeded and nearly two moles of water were formed along with one molar of hydrated product, 2-decanol.

Scheme 1. Oxidation-Reduction Hydration

The addition of water extremely decreased the yield of the hydrated product based on cobalt(II) catalyst because of deactivation of cobalt(II) catalyst. Therefore, it was expected that the yield based on the catalyst would be improved by removal of water formed during the hydration. Then, several methods for removal of water from reaction system were examined. By addition of Molecular Sieves 4A into the reation mixture, the yields of alcohol based on the catalyst increased up to 8500%. Furthermore, azeotropic removal of water was found to be more effective and convenient to afford alcohol in 9140% yield based on cobalt(II) catalyst. It is noted that azeotropic procedure is apparently effective to prevent deactivation of cobalt(II) catalyst and to lead to the improvements of yields of alcohols based on the catalyst.

The present procedure was successfully applied to Oxidation-Reduction Hydration of various olefins. As shown in Table 3, both of acyclic and cyclic olefins were hydrated in high yields based on cobalt(II) catalyst. *Exo*- and trisubstituted olefins were converted into the corresponding tertiary alcohols in more than 10000% yield based on the catalyst (Entries 2 and 4 in Table 3). Also, olefinic compounds having functional groups such as ester, acetal and amide groups were hydrated into the corresponding alcohols in high yields without any decomposition.

Table 3. The effective Oxidation-Reduction Hydration of various olefins

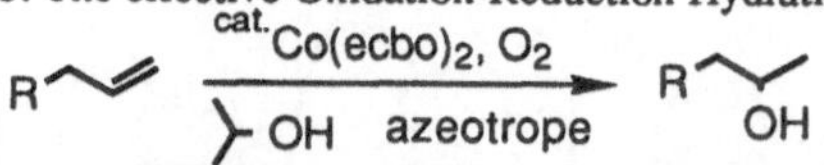

Entry[a]	Olefin	Alcohol	Yield /%[b]
1		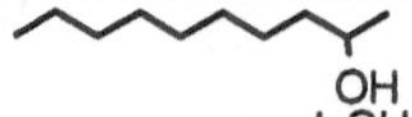	9080
2		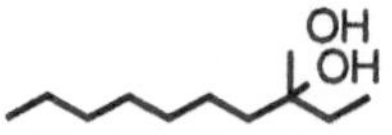	10110
3		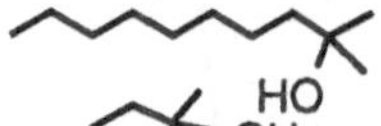	9780
4	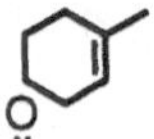	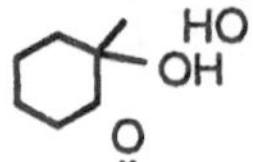	10370
5	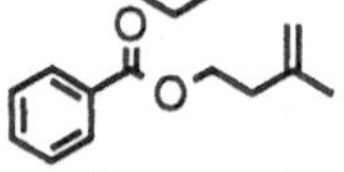	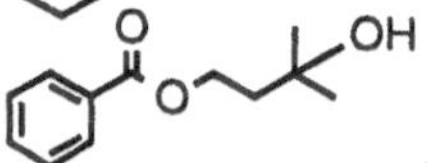	8700
6		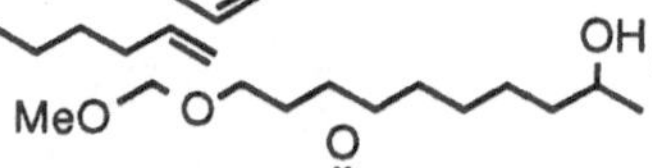	9340
7	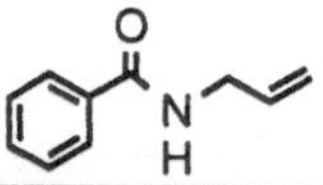	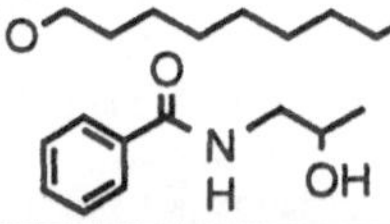	8340

a) Mixture of 5.5 mmol of olefin, 0.043 mmol of Co(ecbo)$_2$ and 25 ml of 2-propanol was gently refluxed under O$_2$. b) Yield based on Co(ecbo)$_2$ and determined by GC analysis.

The present oxygenation reaction catalyzed by cobalt(II) complex with the combined use of molecular oxygen (Oxidant) and 2-propanol (Reductant) was also applied to several oxidation reactions as secondary alcohols into ketones,[9] the direct preparation of ketones from vinylsilanes,[10] and stereoselective oxidative cyclization of 5-hydroxyalkenes into tetrahydrofuran derivatives.[11]

Peroxygenation Using Silanes as Reductant

Silane is expected to be one of the most reliable reductants in organic synthesis.[12] Thus, the empolyment of triethylsilane in place of 2-propanol in Oxidation-Reduction Hydration was tried and was found that triethylsilane also behaved as an effective reductant (hydrogen donor) in the hydration of olefin when bis(1,3-diketonato)cobalt(II) complex was employed as a catalyst.[13] Futhermore, an unexpected peroxygenated product, triethylsilyldioxy derivative, was obtained when the above reaction was carried out at room temperature.[14] For example, 4-phenyl-1-butene reacted with molecular oxygen and triethylsilane at room temperature in the presence of a catalytic amount of bis(acetylacetonato)cobalt(II) complex to give the corresponding 1-phenyl-3-triethylsilyldioxybutane in a good yield (Scheme 2).

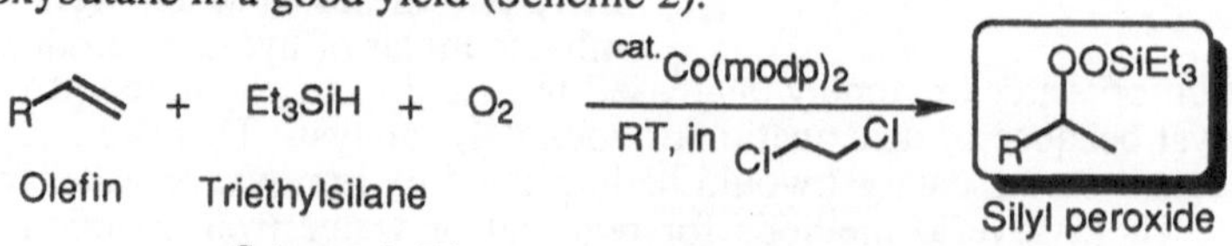

Scheme 2. Direct peroxygenation of olefins

Then, the effects of ligands of cobalt(II) complexes in the above reaction was examined and was found that various bis(1,3-diketonato)cobalt(II), especially Co(modp)$_2$, were effectively employed as catalysts. Preparation of a peroxy compound directly from olefin has been considered to be difficult because of its instability. The present peroxygenation reaction, however, provides a facile and efficient method for the direct introduction of dioxygen function into the carbon-carbon double bond of olefinic compounds under mild conditions (Table 4). In addition, the triethylsilyldioxy derivative here is expected to be a potentially useful synthetic intermediates.

Table 4. Peroxygenation of various olefins with molecular oxygen and triethylsilane

Entry[a]	Olefin	Silyl Peroxide	Yield / %[b]
1	Ph (isopropenyl)	Ph-C(OOSiEt$_3$)(OOSiEt$_3$)	95
2	long-chain terminal olefin	long-chain with OOSiEt$_3$	80
3	Ph-C(=O)-O-CH$_2$-CH=C(CH$_3$)$_2$	Ph-C(=O)-O-CH$_2$-CH$_2$-C(CH$_3$)$_2$OOSiEt$_3$	99
4	Ph-C(=O)-NH-CH$_2$-CH=CH$_2$	Ph-C(=O)-NH-CH$_2$-CH(OOSiEt$_3$)-CH$_2$OOSiEt$_3$	80
5	Ph-CH=CH-C(=O)-Ph	Ph-CH$_2$-CH(OOSiEt$_3$)-C(=O)-Ph	75

a) All reactions were carried out by treating olefin (1 mmol) with triethylsilane (2 mmol) and 0.05 mmol of Co(modp)$_2$ under O$_2$ atmosphere in 5 ml of 1,2-dichloroethane.
b) Isolated yield.

H modp = ^{t}Bu-C(=O)-CH$_2$-C(=O)-C(=O)-N(morpholine)

Besides peroxygenation of simple carbon-carbon double bond of olefins, it was found that several α,β-unsaturated esters were also peroxygenated according to the present procedure to produce the corresponding triethylsilyldioxy derivatives.[15] The introduction of heteroatoms onto α-carbon of α,β-unsaturated carbonyl compound is useful for synthesis of a wide variety of natural products such as amino acids. Therefore, the preparation of α-hydroxy esters from α,β-unsaturated esters was examined in order to demonstarte the synthetic utility of the present peroxygenation reaction. Peroxygenation of α,β-unsaturated ester with molecular oxygen catalyzed by cobalt(II) complex was acheived by the addition of a catalytic amount of *tert*-butylhydroperoxide yielding the corresponding triethylsilyldioxy derivatives in high yields. The triethylsilyldioxy derivative was smoothly desilylated in acidic methanol to give α-hydroperoxycarboxylate whose structure was confirmed by the analysis of ^{1}H NMR spectra after isolation. The hydroperoxide was subsequently converted into α-hydroxy carboxylate in a high yield by reduction with an aqueous solution of Na$_2$S$_2$O$_3$ at room temperature (Scheme 3).

$$R^1\text{-CH=CH-C(=O)OR}^2 \xrightarrow[\text{O}_2, \text{ Et}_3\text{SiH, RT}]{^{cat}\text{Co(acac)}_2, \ ^{cat}\text{ }t\text{-BuOOH}} \left[R^1\text{-CH(OOSiEt}_3)\text{-CH}_2\text{-C(=O)OR}^2 \right]$$

$$\xrightarrow[\text{MeOH, RT}]{H^+} \left[R^1\text{-CH(OOH)-CH}_2\text{-C(=O)OR}^2 \right] \xrightarrow[\text{MeOH/H}_2\text{O}]{\text{Na}_2\text{S}_2\text{O}_3 \text{ aq.}} R^1\text{-CH(OH)-CH}_2\text{-C(=O)OR}^2$$

55 - 88 % yield

Scheme 3. Hydration of α,β-unsaturated carboxylic acid ester

During the course of our continuing study, our interests have been focused on the direct preparation of α-hydroxycarboxylate from α,β-unsaturated esters. In the case of employing phenylsilane[16] as a reductant in the above oxygenation reaction, it was found that the reduction of silyldioxy intermediates, derived by the peroxygenation of α,β-unsaturated ester, proceeded successively to afford the corresponding α-hydroxy carboxylic acid ester in high yield. And manganese(II) complex, bis(dipivaloylmethanato)manganese(II) (Mn(dpm)$_2$), was more effective than cobalt(II) complex for the direct hydration of α,β-unsaturated esters.[17] The reaction proceeded smoothly and the corresponding α-hydroxycarboxylates were obtained regioselectively in high yields (Table 5).

137

Table 5. Manganese(II) catalyzed hydration of various α,β-unsaturated esters

$$R^1\underset{R^2}{\overset{R^3}{=}}CO_2R^4 \quad \xrightarrow[\text{0 °C, 2-propanol}]{\textbf{cat Mn(dpm)}_2,\ O_2,\ PhSiH_3} \quad R^1\underset{R^2}{\overset{R^3\ \ OH}{\diagup}}CO_2R^4$$

Entry[a]	α,β-Unsaturated Ester	α-Hydroxy Ester [c,d]	Yield /% [b]
1	$\diagup\!=\!CO_2Bn$	OH, CO_2Bn	91
2	$\diagdown\!\diagup\!=\!CO_2Bn$	OH, CO_2Bn	91
3	$\diagdown\!\diagup\!\diagdown\!\diagup\!CO_2Bn$ [e]	OH, CO_2Bn	86
4	$\diagdown\!\diagdown\!\diagup\!\diagdown\!\diagup\!CO_2Me$ [e]	OH, CO_2Me	94
5	$\diagdown\!\diagdown\!\diagdown\!\diagup\!\diagdown\!\diagup\!CO_2Me$ [e]	OH, CO_2Me	84
6	$\diagup\!\overset{\ }{\diagup}\!CO_2Bn$	OH, CO_2Bn	92
7	$\diagdown\!\diagup\!\overset{\ }{\diagup}\!CO_2Bn$	OH, CO_2Bn	94
8	$EtO_2C\!\diagup\!=\!\diagup\!CO_2Et$ [f]	$EtO_2C\!\diagup\!\overset{OH}{\diagdown}\!CO_2Et$	82
9	$MeO_2C\!\diagdown\!\diagup\!CO_2Me$ [f]	$MeO_2C\!\diagdown\!\overset{OH}{\diagup}\!CO_2Me$	78
10	$MeO_2C\!\diagup\!=\!\diagup\!CO_2Me$ [g]	$MeO_2C\!\overset{HO\ \ OH}{\diagup}\!CO_2Me$	76

a) Reaction conditions, substrate 2 0 mmol, Mn(dpm)$_2$ 0 04mmol PhSiH$_3$ 4 0 mmol, 2 propanol 10 ml 0 °C, 1 atm O$_2$ b) Isolated yield   c) Entries 1 to 8, no β hydroxycarboxylic acid ester was formed   d) All products gave satisfactory ^{1}H NMR and IR spectra   e) PhSiH$_3$ 3 0 mmol 2 propanol 2 ml   f) Solvent ethanol 5 ml   g) Solvent, ethanol 3 5 ml and 1 2 dichloroethane 1 5 ml

$^tBu\!\diagup\!\diagdown\!\diagup\!^tBu$, O O , Hdpm

Bn = CH$_2$Ph

The present system (phenylsilane and cobalt(II) complex) was applied to the coupling reaction of α,β-unsaturated carbonyl compounds with aldehydes [18]

NICKEL(II) COMPLEX CATALYZED AEROBIC EPOXIDATION

Epoxides are one of the most useful synthetic intermediates for the preparation of oxygen-containing natural products or the production of epoxy resins, etc   Much effort has been made to develop the direct and selective epoxidation of olefins by use of molecular oxygen   However, it is still difficult to control the reaction because of over-oxidations or side-reactions under conventional severe reaction conditions such as high pressure of oxygen or high reaction temperature   Then, it is desired to search for a milder reaction in order to develop an efficient epoxidation method   Only several effective catalysts have been reported for the epoxidatation with molecular oxygen under milder reaction conditions, i e (tetramesitylporphinato)ruthenium(II),[19] or oxoethoxo(tetra-p-tolylporphinato)molybdenum (V),[20] for example   However, practical synthetic methods are not yet successfully established because of the complicated reaction systems

Nickel(II) Catalyzed Epoxidation in Primary Alcohol

"Oxidation-Reduction Hydration" of olefins into the corresponding hydrated compounds using molecular oxygen (Oxidant) and secondary alcohol (Reductant) in the presence of a catalytic amount of bis(1,3-diketonato)cobalt(II) was described in the prior section   Based on the detailed observation on the hydration reaction, it was revealed that one oxygen atom from molecular oxygen and two hydrogen atoms from 2-propanol were simultaneously introduced into olefin to afford the hydrated product, and here, secondary alcohol behaved as most effective reductant. Therefore, in the epoxidation with molecular oxygen, it is postulated that secondary alcohol would also behave as a reliable reductant to accomplish a catalytic cycle.

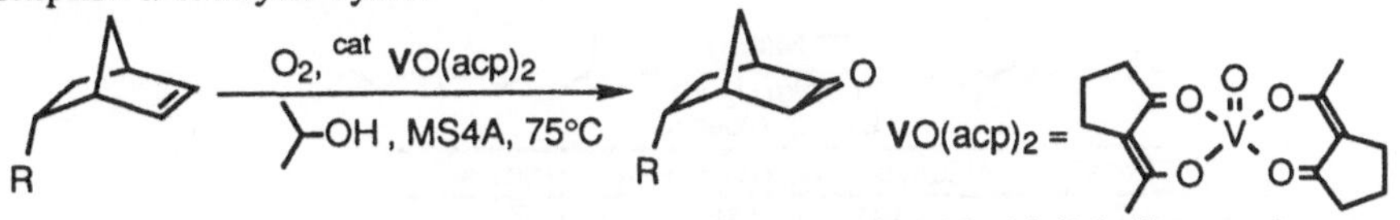

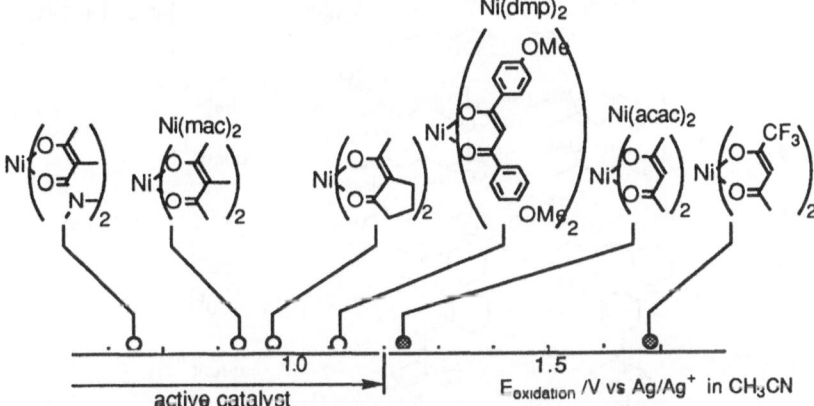

Scheme 4. Epoxidation with molecular oxygen and alcohol catalyzed by bis(1,3-diketonato)oxovanadium(IV)

In the presence of a catalytic amount of bis(2-alkyl-1,3-diketonato)oxovanadium(IV), norbornene analogues were monooxygenated by the combined use of molecular oxygen and 2-propanol to afford the corresponding epoxides in good yields (Scheme 4).[21] Our continuous study revealed that nickel(II) complexes having electron-donating 1,3-diketone ligands were excellent catalysts for epoxidation of olefins with molecular oxygen. In the case of using primary alcohol (1-butanol), it was found that epoxide was formed in higher yield compared with the case of using secondary alcohol (2-propanol) or in absence of alcohol (Scheme 5).[22]

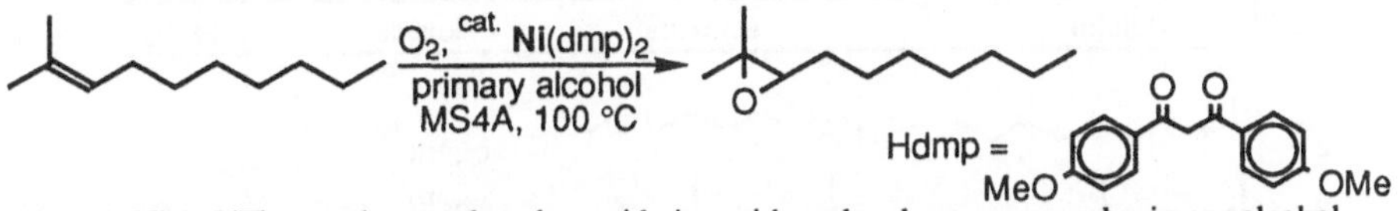

Scheme 5. Nickel(II) complex catalyzed epoxidation with molecular oxygen and primary alcohol

Several 1,3-diketone ligands were examined for epoxidation with molecular oxygen and 1-butanol, and was found that nickel(II) complexes having electron-donating ligands behaved more effectively as catalysts as shown in Figure 2. In the case of using bis[1,3-di(p-methoxyphenyl)-1,3-propanedionato]nickel(II)(Ni(dmp)2), the desired epoxidation proceeded smoothly even under lower oxygen pressure.

Figure 2. Relationship of oxidation potentials and catalytic activities in aerobic epoxidation catalyzed by nickel(II) complexes and primary alcohol

Highly Efficient Aerobic Epoxidation Catalyzed by Nickel(II) Complex

After screening several reductants in the aerobic epoxidation of olefins catalyzed by nickel(II) complexes, it was found that an aldehyde acts as an excellent reductant when treated under an atmospheric pressure of molecular oxygen at room temperature (Scheme 6).[23] Similar reactions have been reported in the patents. Propylene was monooxygenated into propylene oxide with molecular oxygen in the coexistence of metal complexes and aldehyde such as acetaldehyde[24] or crotonaldehyde,[25] but the conversion of olefin and the selectivity of epoxide were never reached satisfactory levels. Recently, praseodymium(III) acetate was also shown to be an effective catalyst for the aerobic epoxidation of olefins in the presence of aldehyde.[26]

First, several aldehydes (reductants) were screened by taking the expoxidation of 2-methyl-2-decene catalyzed by Ni(dmp)$_2$ (Table 6). In case of employing butyraldehyde, both conversion of olefin and yield of the epoxide were low. On the contrary, the corresponding

Table 6. Epoxidation of 2-methyl-2-decene by using several aldehyde

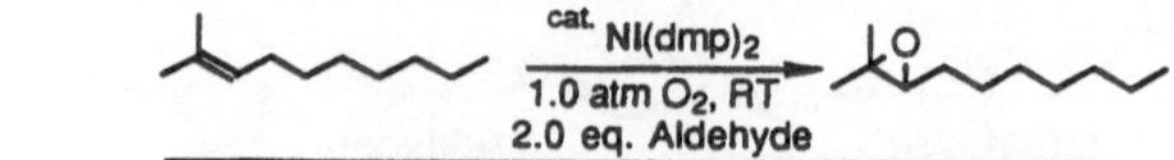

cat. Ni(dmp)$_2$
1.0 atm O$_2$, RT
2.0 eq. Aldehyde

Entry[a]	Aldehyde	Conversion / %[b]	Yield / %[b]
1	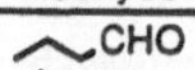 CHO	10	7
2	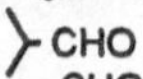 CHO	100	quant.
3	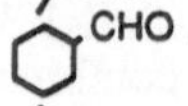 CHO	100	quant.
4	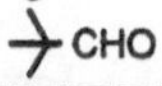 CHO	100	quant.

a) Reaction conditions; 2-Methyl-2-decene 2.5 mmol in 2.0 ml of 1,2-dichloroethane, 1.0 atm of O$_2$ for 12 h. b) Determined by GC.

Table 7. Epoxidation of various olefins(**(A) Standard Method**)[a]

Entry[a]	Olefin	Epoxide	Yield/%	Note
1	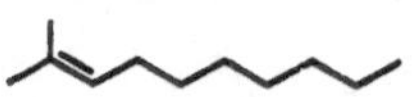	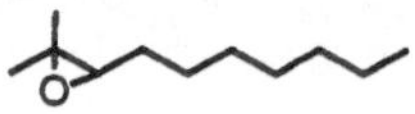	quant.[b]	
2[d]			quant.[b]	
3	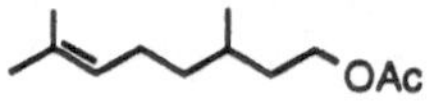 OAc	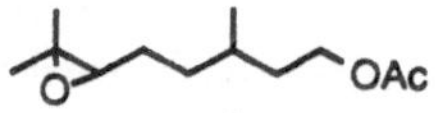 OAc	95[b]	
4	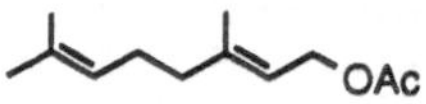 OAc	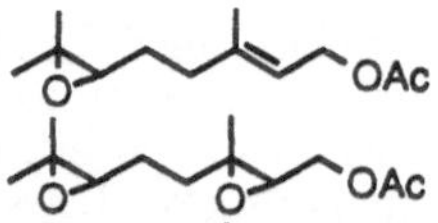 OAc / OAc	68[c] / 19[c]	total yield 87%
5	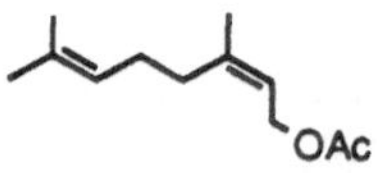 OAc	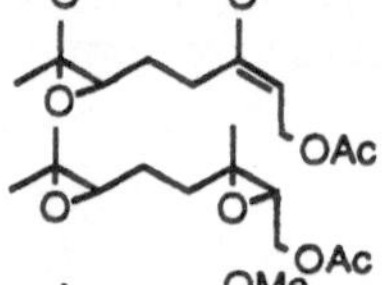 OAc / OAc	61[c] / 13[c]	total yield 74%
6	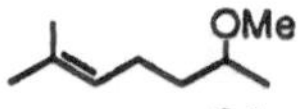 OMe	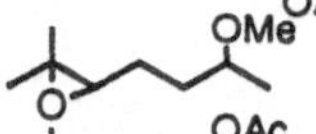 OMe	95[b]	
7	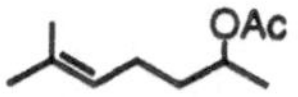 OAc	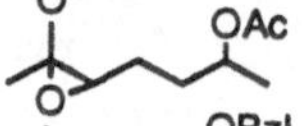 OAc	quant.[b]	
8	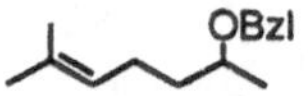 OBzl	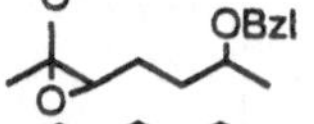 OBzl	95[c]	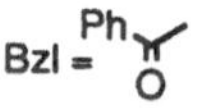 Bzl = Ph—C(=O)
9	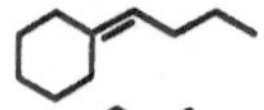	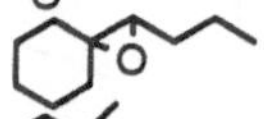	89[b]	
10	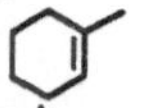	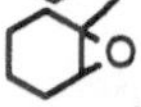	80[b]	
11	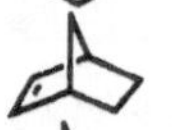	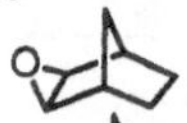	quant.[b]	
12	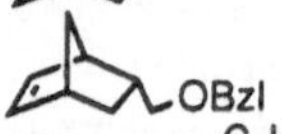 OBzl	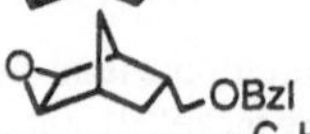 OBzl	84[c]	
13	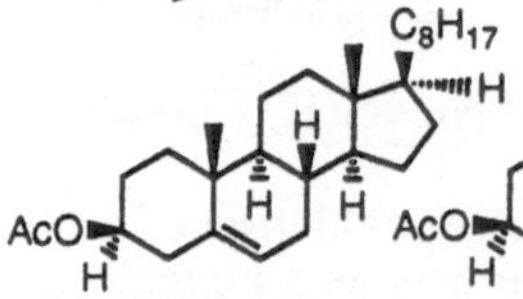 C$_8$H$_{17}$ ... AcO	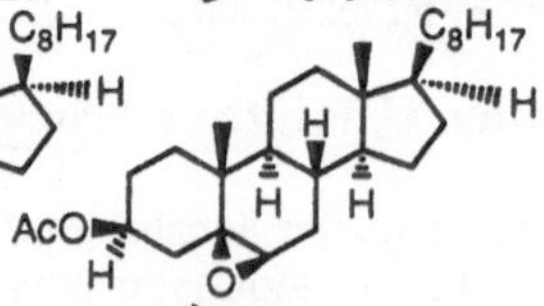 C$_8$H$_{17}$ ... AcO	84[c]	α:β 24:76[e]
14		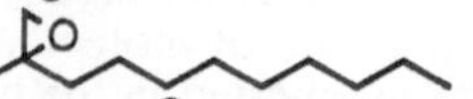	93[b]	
15	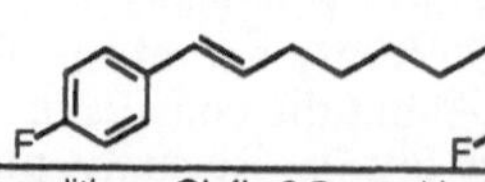 F—C$_6$H$_4$	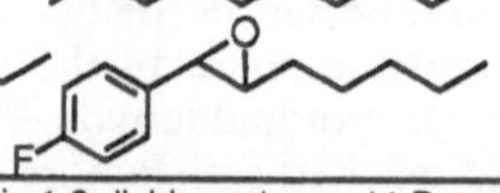 F—C$_6$H$_4$	85[b]	

a) Reaction conditions; Olefin 2.5 mmol in 1,2-dichloroethane. b) Determined by GC analysis.
c) Isolated yield. d) Under an atmospheric pressure of air. e) Diastereomer ratio was determined by NMR.

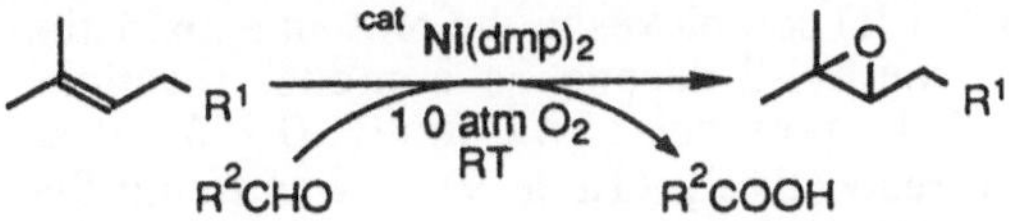

Scheme 6. Epoxidation by Nickel(II)-aldehyde method

epoxide was obtained in quantitative yields when aldehydes having secondary or tertiary carbon next to the carbonyl carbon such as isobutyraldehyde, cyclohexanecarbaldehyde or pival-aldehyde were employed, respectively. According to the above procedure (***Method (A) Standard Method***), various trisubstituted or *exo*-terminal olefins and norbornene analogues were smoothly monooxygenated into the corresponding epoxides in high to quantitative yields under an atmospheric pressure of oxygen at room temperature (Table 7). Here it should be pointed out that, in every case, no over-oxidation at allylic position nor cleavage reaction of carbon-carbon double bonds took place to any extent. *Exo*-terminal olefin and styrene derivative were also monooxidized into corresponding epoxides in high yields, respectively (Entries 14 and 15). In the case of the epoxidation of trisubstituted olefin, the corresponding epoxide was obtained in quantitative yield even under air (Entry 2).

For the aerobic epoxidation of 1,2-disubstituted olefins, the use of smaller amount of nickel(II) complex was effective to improve yield of the corresponding epoxide (***(B) Dilution Method***). Isovaleraldehyde was found to be remarkably effective for the epoxidation of terminal olefin, and 1,2-epoxyalkane was obtained in good to high yield (***(C) Isovaleraldehyde Method***).[27]

Various 1,2-disubstituted olefins were monooxygenated into the corresponding epoxides in high yields by ***Method (B)***, and also in the case of terminal olefin by ***Method (C)***, respectively (Table 8).

Table 8. Epoxidation of various olefins

Entry	Olefin	Epoxide	Yield/%	Note
(B) Dilution Method				
1		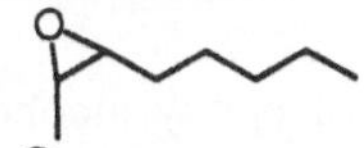	97[a c]	*cis trans*=51 49
2		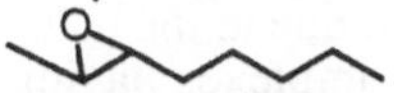	92[a c]	*cis trans*=13 87
3	 Ph	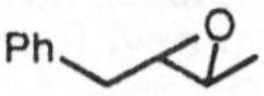 Ph	75[a c]	
4	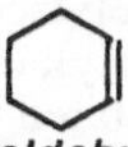	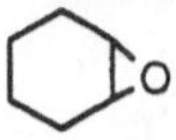	84[a c]	
(C) Isovaleraldehyde Method				
5		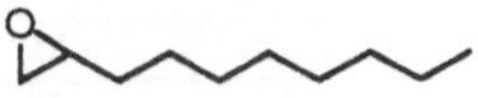	75[b c]	
6		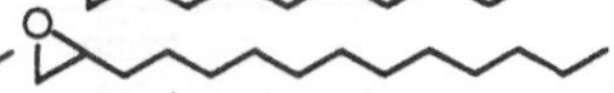	89[b d]	
7	 OBzl	 OBzl	97[b d]	Bzl 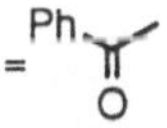 = Ph-C(=O)
8	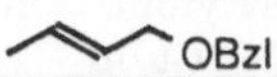 OBzl	 OBzl	62[b d]	

a) ***(B) Dilution method***, Reaction conditions, Olefin 2 5 mmol, isobutyraldehyde 3 0 equiv , Ni(dmp)$_2$ 0 3 mol%, in 1,2-dichloroethane 10 0 ml, 1 0 atm O$_2$,RT b) ***(C) Isovaleraldehyde method*** Reactio conditions, Olefin 3 0 mmol, isovaleraldehyde 6 0 equiv , Ni(dmp)$_2$ 0 3 mol%, in 1,2-dichloroethane 10 0 ml, 1 0 atm O$_2$,RT c) Determined by GC analysis d) Isolated yield

Table 9. Highly efficient epoxidation catalyzed by nickel(II) complex

Entry	Amount of catalyst / mol%	Yield / %[b]	Yield based on catalyst / %[b]
1	4 0	100 0	2,500
2	0 256	100 0	39,000
3	0 0096[a]	98 1	1,020,000

a) Reaction conditions, 2-Methyl-2-decene10 0 mmol, isobutyraldehyde 20 0 mmol, Ni(dmp)$_2$ 0 6 mg in 6 0 ml of 1,2-dichloroethane 1 0 atm O$_2$ for 12 h
b) Determined by GC

The efficiency of nickel(II) complexes in the present epoxidation was demonstrated by taking epoxidation of 2-methyl-2-decene as a model reaction. Through the three experiments carried out in the presence of 4.0 mol%, 0.256 mol%, and 0.0096 mol% of Ni(dmp)$_2$ against olefin, respectively (Table 9), it was found that, even in the case of employing only 0.0096 mol% of Ni(dmp)$_2$, the epoxidation proceeded smoothly and the corresponding epoxide was obtained in 1020000% yield based on Ni(dmp)$_2$.

When the present efficient epoxidation catalyzed by nickel(II) complex was applied to the oxidation of enolates, α-siloxy carbonyl compounds were obtained via possible silyl rearrangement of siloxy epoxides.[28] And α-hydroxy carbonyl compounds were obatined in good yields by successive desilylation with potassium fluoride (Scheme 7).

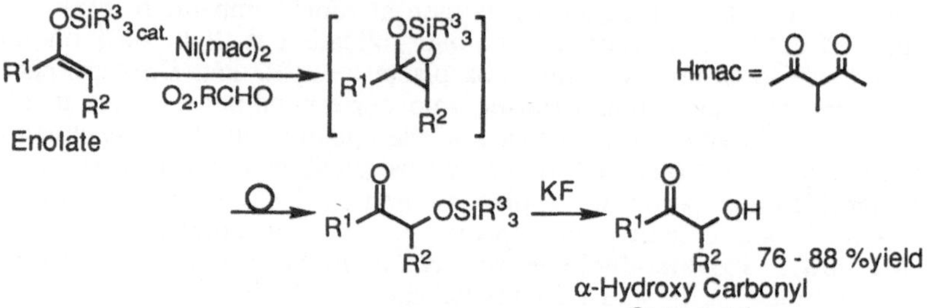

Scheme 7. Synthesis of α-hydroxy carbonyl compounds

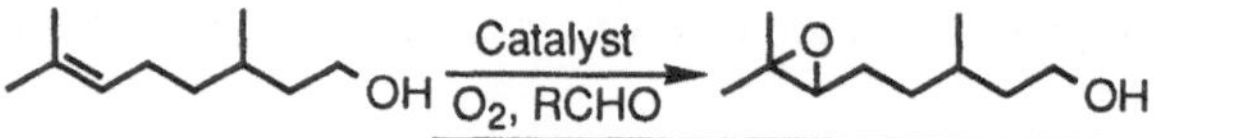

Scheme 8. Epoxidation of citronellol

Catalyst	Conversion / %	Yield / %
Ni(dmp)$_2$	67	57
Fe(dmp)$_3$	100	quant.

As shown in Scheme 8, epoxy alcohol was obtained in quantitative yield without any over-oxidation of hydroxyl group by using iron(III) complex as a catalyst, whereas nickel(II) complex-catalyzed epoxidation of citronellol stopped half-way and the yield of epoxy alcohol was moderate.[29] It was assumed that nickel(II) complex was deactivated due to the undesirable coordination by hydroxyl group of olefinic alcohol, while iron(III) complexes having three 1,3-diketone ligands might not be preferable for the coordination by hydroxyl function.

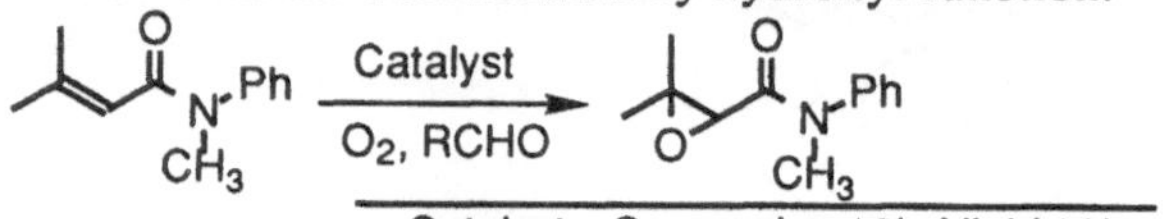

Scheme 9. Epoxidation of α,β-unsaturated carboxamide

Catalyst	Conversion / %	Yield / %
Ni(acac)$_2$	—	trace
Fe(acac)$_3$	33	26
VO(acac)$_2$	81	71
VO(dpm)$_2$	92	87

Bis(acetylacetonato)nickel(II) or tris(acetylacetonato)iron(III), which exhibited excellent catalytic activity for epoxidation of aliphatic or aromatic olefin, was not suitable for the oxygenation of α,β-unsaturated carboxamide and the corresponding epoxide was formed in low yield. On the other hand, it was found that bis(acetyl-acetonato)oxovanadium(IV) is effectively employed as a catalyst, and α,β-unsaturated carboxamide was oxygenated with molecular oxygen and isovaleraldehyde catalyzed by bis(dipivaloylmethanato)oxovanadium(IV)(VO(dpm)$_2$) to afford the corresponding epoxide in 87% yield.[30] It was postulated that the epoxidation of α,β-unsaturated carboxamide would be promoted by oxovanadium(IV) complex which could interact with amide group of α,β-unsaturated carboxamide to afford the corresponding epoxide.

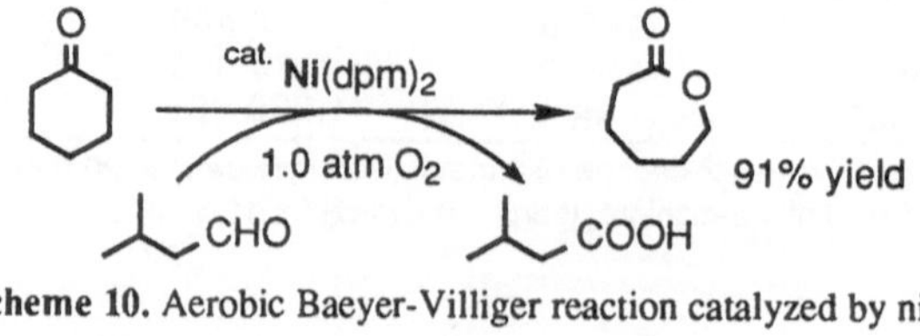

Scheme 10. Aerobic Baeyer-Villiger reaction catalyzed by nickel(II) complex

The present oxidation reaction catalyzed by nickel(II) complex with the combined use of molecular oxygen and aldehyde[31,32] was applied to the aerobic oxidation of aldehyde into the corresponding carboxylic acid,[33] and the aerobic Baeyer-Villiger reaction (Scheme 10)[34], respectively.

MANGANESE(III) CATALYZED ENANTIOSELECTIVE EPOXIDATION

Optically active epoxides have attracted much attention as versatile intermediates[35] for the synthesis of a wide variety of chiral compounds such as biologically active compounds and ferroelectric liquid crystals,[36] *etc.* Sharpless and Katsuki developed the efficient titanium-catalyzed epoxidation of allylic alcohols by using *tert*-butyl hydroperoxide as an oxidant affording optically active 2,3-epoxy alcohols in good yields with very high enantiomeric excesses,[37] which has been successfully applied to the synthesis of a number of natural products. Much effort has been made to develop an efficient and widely applicable enantioselective epoxidation of simple olefins, and several enzymatic systems for terminal alkenes[38] have been reported. In the nonenzymatic systems, artificial metal-porphyrins have been designed as cytochrome P-450 modeling systems for enantioselective epoxidation of styrene analogues as the catalysts.[39] Recently, Jacobsen[40] and Katsuki[41] independently reported that Mn(III)-salen complexes are effective catalysts for enantioselective epoxidation of unfunctionalized olefins by using terminal oxidants such as iodosylbenzene[39-41] or sodium hypochlorite.[42] Except for artificial-bleomycin catalyzed epoxidation,[43] few have been reported on the utilization of molecular oxygen for enantioselective epoxidation of simple olefins.

β-Selective Epoxidation of Cholesterol Derivatives Catalyzed by Manganese(II) Complex

Stereochemistries of the epoxidation of 5,6-double bond in cholesteryl benzoate were examined with molecular oxygen and isobutyraldehyde using several metal complexes coordinated by 1,3-diketones as catalysts (Table 10). It was found that epoxidation catalyzed by metal complexes, such as nickel(II), iron(III), or manganese(II) complexes coordinated with 1,3-diketones (acetylacetone), afforded the *hindered* 5,6 β-epoxide as a major isomer.[44] In the case of employing Mn(dpm)$_2$ as a catalyst, stereoselectivity of 5,6 β-epoxide was improved up to 80% (Entry 5). On the contrary, it was reported that by using peracids such as *m*CPBA or MMPP (magnesium monoperphthalate hexahydrate),[45] cholesteryl benzoate was converted into the corresponding mixture of 5,6 α- and 5,6 β-epoxides in the ratio of 71 to 29 (*m*CPBA, Entry 1),[46] and 85 to 15 (MMPP),[45] respectively. It is interesting to point out that the *less-hindered* 5,6 α-epoxide was obtained as a major product when a peracid was used as an oxidant. These remarkable reverse stereoselectivities in the epoxidation of cholesteryl benzoate obviously indicate that the active oxidant of the present metal complex catalyzed epoxidation is not a simple peroxycarboxylic acid generated from an aldehyde with autoxidation manner, but that an oxygenated metal complex is tentatively considered as the reactive intermediate of the present epoxidation.

Table 10. Stereoselective epoxidation catalyzed by manganese complex

Entry	Epoxidation reagent	α-Epoxide : β-Epoxide[b]
1	*m*CPBA	71 : 29
2[a]	cat. Ni(dmp)$_2$, O$_2$, $\rangle$CHO	31 : 69
3[a]	cat. Fe(acac)$_3$	33 : 67
4[a]	cat. Mn(acac)$_3$	23 : 77
5[a]	cat. Mn(dpm)$_2$	20 : 80

a) Cholesteryl Benzoate 245 mg (0.5 mmol), isobutyraldehyde (2.0 mmol), catalyst (0.0047 mmol, 0.94 mol%) in 1,2-dichloroethane (5.0 ml), RT, 1.0 atm O$_2$, 2.0 h. b) Determined by HPLC analysis.

143

Enantioselective Aerobic Epoxidation of Unfunctionalized Olefins

Stereochemical observations in the epoxidation of cholesterol derivatives mentioned above suggest that the manganese complex participates directly in the oxidation step and the enantioselective aerobic epoxidation should be realized by employing optically active manganese complexes as catalysts.

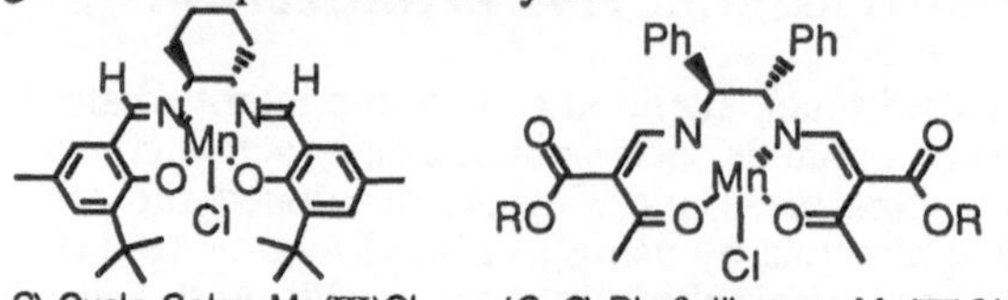

(S, S)-Cyclo-Salen-Mn(III)Cl   (S, S)-Ph-β-diketone-Mn(III)Cl

Figure 3. Optically active manganese(III) catalyst

Optically active manganese (III) complex, Cyclo-Salen-Mn (III)Cl, was prepared by the reported method[40] and purified with column chromatography on silica-gel or washing its benzene solution with aqueous lithium chloride solution.[47] Ph-β-diketone-Mn(III)Cl complex was synthesized from the corresponding alkyl(R) acetoacetate in 5 steps and purified by a similar procedure[48](Figure 3).

As the results of the screening of various aldehydes, pivalaldehyde worked quite effectively for the enantioselective aerobic epoxidation of unfunctionalized olefins such as 1,2-dihydronaphthlene derivatives (Scheme 11). It is noted that i) in the case of epoxidation catalyzed by Salen-type complex, addition of a catalytic amount of N-methylimidazole effectively improved the optical yield of epoxide and ii) bulkiness in ester moiety of β-diketone-type catalyst also influenced the optical yield to higher values.

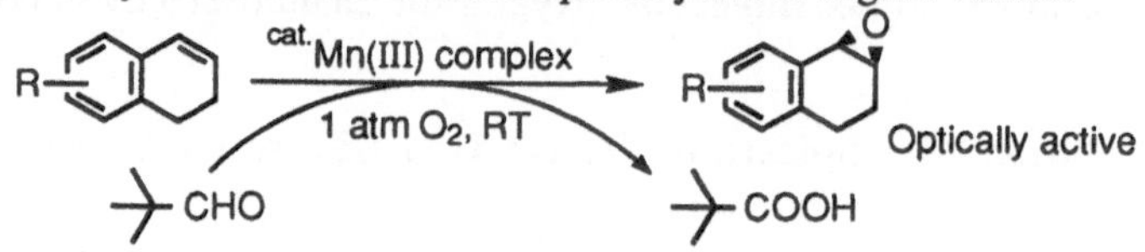

Scheme 11. Enantioselective epoxidation of unfunctionalized oledins with molecular oxygen and aldehyde catalyzed by optically active manganese(III) complexes

The present system was applied to the enantioselective epoxidations of various simple olefines. 1,2-Dihydronaphthalenes which contained no function groups were converted into the corresponding optically active epoxides in good yields with good enantioselectivities (52-72% ee, Entries 1-4). The enantioselective aerobic epoxidation of 1,2-benzo-1,3-

Table 11. Enantioselective epoxidation of various unfunctionalized olefins

Entry	Olefin	Optical yield / %ee[b] (Yield / %[a])	
		Salen-Mn(III)	β-Diketone-Mn(III)
1	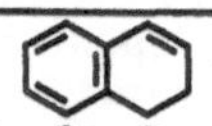	63(78)	64(70)
2	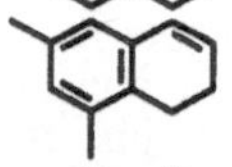	52(73)	53(13)
3	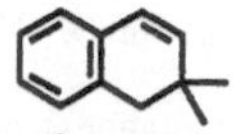	72(80)	53(40)
4	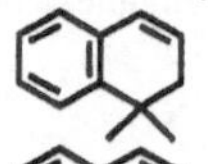	63(35)	70(43)
5	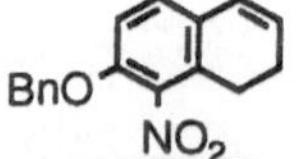 BnO	66[c](38)	43[c](73)
6	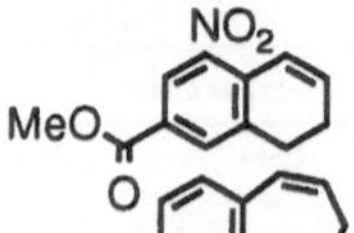 MeO	57(38)	59(67)
7	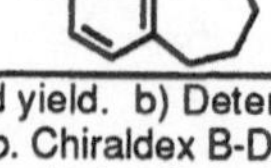	83(52)	84(52)

a) Isolated yield. b) Determined by GC analysis unless otherwise stated. ASTEC Co. Chiraldex B-DA (20 m x 0.25 mm ID x 0.125 μ film). c) Determined by HPLC analysis. Daicel OD(+) (Hexane : 2-propanol).

Further studies on the elucidation of reactive intermediates and stereochemistry in the present asymmetric reaction are under active investigations.

cycloheptadiene afforded the corresponding epoxide with high enantiomeric excess in the coexistence of Mn(III) complex (Cyclo-Salen-Mn(III)Cl: 83%ee, Ph-β-diketone-Mn(III)Cl: 84%ee, Entry 7).

EPOXIDATION OF OLEFINS UNDER NEUTRAL CONDITIONS

Several efficient oxidation reactions with molecular oxygen were developed using transition-metal complexes coordinated by variuos ligands in combination with appropriate reductants. Recently, it was found that cyclic ketones such as 2-methylcyclohexanone[49] and acetals of aldehyde such as propionaldehyde diethyl acetal[50] were effectively employed in aerobic epoxidation of olefins catalyzed by cobalt(II) complexes. In the latter case, ethyl propionate and ethanol were just detected in nearly stoichiometric manner as coproducts (Scheme 12), therefore the reaction system is kept under neutral conditions during the epoxidation.

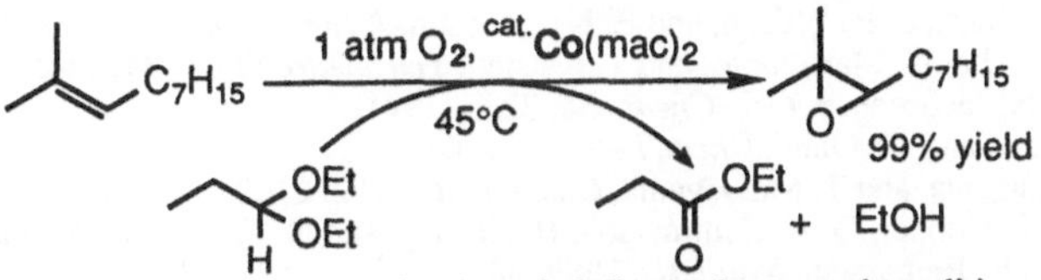

Scheme 12. The aerobic epoxidation of olefins under neutral conditions

ACKNOWLEDGEMENT

The author would like to thank Dr. Shigeru Isayama, Dr. Satoshi Inoki, Dr. Koji Kato, Dr. Tohru Yamada, Mr. Toshihiro Takai, Mr. Kiyomi Imagawa, Mr. Kiyotaka Yorozu, Mr. Ei-ichiro Hata, Mr. Takushi Nagata, Mr. Katsuya Takahashi, and Mr. Oliver Rohde, members of Basic Research Laboratories for Organic Synthesis, Mitsui Petrochemical Industires, Ltd.

REFERENCES

1. I. Tabushi and N. Koga, *J. Am. Chem. Soc.*, **101**, 6456 (1979); I. Tabushi and A. Yamazaki, *ibid.*, **103**, 2884 (1983).
2. T. Okamoto and S. Oka, *J. Org. Chem.*, **49**, 1589 (1984).
3. S. Inoue, Y. Ohkatsu, M. Ohno, and T. Ooi, *Nippon Kgaku Kaishi*, **1985**, 387.
4. A. Zombeck, D. E. Hamilton, and R. S. Drago, *J. Am. Chem. Soc.*, **104**, 6782 (1982); D. E. Hamilton, R. S. Drago, and A. Zombeck, *ibid.*, **109**, 374 (1987); A. Nishinaga, H. Yamato, T. Abe, K. Maruyama, and T. Matsuura, *Tetrahedron Lett.*, **29**, 6309 (1988).
5. E. P. Talsi, Y. S. Zimin, and V. M. Nekipelov, *React. Kinet. Catal. Lett.*, **27**, 361 (1985).
6. T. Mukaiyama, S. Isayama, S. Inoki, K. Kato, T. Yamada, and T. Takai, *Chem. Lett.*, **1989**, 449.
7. S. Inoki, K. Kato, T. Takai, S. Isayama, T. Yamada, and T. Mukaiyama, *Chem. Lett.*, **1989**, 515.
8. K. Kato, T. Yamada, T. Takai, S. Inoki, and S. Isayama, *Bull. Chem. Soc. Jpn.*, **63**, 179 (1990).
9. T. Yamada and T. Mukaiyama, *Chem. Lett.*, **1989**, 519.
10. K. Kato and T. Mukaiyama, *Chem. Lett.*, **1989**, 2233.
11. S. Inoki and T. Mukaiyama, *Chem. Lett.*, **1990**, 67.
12. J. D. Citron, *J. Org. Chem.*, **36**, 2547 (1971).
13. S. Isayama and T. Mukaiyama, *Chem. Lett.*, **1989**, 569.
14. S. Isayama and T. Mukaiyama, *Chem. Lett.*, **1989**, 573.
15. S. Isayama, *Bull. Chem. Soc. Jpn.*, **63**, 1305 (1990).
16. S. Isayama and T. Mukaiyama, *Chem. Lett.*, **1989**, 1071.
17. S. Inoki, K. Kato, S. Isayama, and T. Mukaiyama, *Chem. Lett.*, **1990**, 1869.
18. S. Isayama and T. Mukaiyama, *Chem. Lett.*, **1989**, 2005.
19. M. Tavarès, R. Ramasseul, J.-C. Marchon, B. Bachet, C. Brassy, and J.-P. Mornon, *J. Chem. Soc., Perkin Trans. 2*, **1992**, 1321.
20. Y. Matsuda, H. Koshima, K. Nakamura, and Y. Murakami, *Chem. Lett.*, **1988**, 625.
21. T. Takai, T. Yamada, and T. Mukaiyama, *Chem. Lett.*, **1990**, 1657.
22. T. Mukaiyama, T. Takai, T. Yamada, and O. Rhode, *Chem. Lett.*, **1990**, 1661.
23. T. Yamada, T. Takai, O. Rhode, and T. Mukaiyama, *Chem. Lett.*, **1991**, 1.
24. Nippon Soda Co., Ltd., JP Patent Tokkaisho 46-26063 (1971).
25. Y. Maeda, M. Ai, and S. Suzuki, *Kogyo Kagaku Zasshi*, **73**, 99 (1970).
26. Shell International Research, JP Patent Tokkaisho 59-231077 (1984).

27. T. Yamada, T. Takai, O. Rhode, and T. Mukaiyama, *Bull. Chem. Soc. Jpn.*, **64**, 2109 (1991).
28. T. Takai, T. Yamada, O. Rhode, and T. Mukaiyama, *Chem. Lett.*, **1991**, 281.
29. T. Takai, E. Hata, T. Yamada, and T. Mukaiyama, *Bull. Chem. Soc. Jpn.*, **64**, 2513 (1991).
30. S. Inoki, T. Takai, T. Yamada, and T. Mukaiyama, *Chem. Lett.*, **1991**, 941.
31. T. Takai, T. Yamada, and T. Mukaiyama, *Chem. Lett.*, **1991**, 1499.
32. T. Mukaiyama, K. Imagawa, T. Yamada, and T. Takai, *Chem. Lett.*, **1992**, 231.
33. T. Yamada, O. Rhode, T. Takai, and T. Mukaiyama, *Chem. Lett.*, **1991**, 5.
34. T. Yamada, K. Takahashi, K. Kato, T. Takai, S. Inoki, and T. Mukaiyama, *Chem. Lett.*, **1991**, 641.
35. B. E. Rossiter, "Synthetic Aspects and Applications of Asymmetric Epoxidation," in "Asymmetric Synthesis," ed by J. D. Morrison, Academic Press, Inc., New York (1985), Vol.5, Chap.7, pp.194.
36. H. Nohira, S. Nakamura, and M. Kamei, *Mol. Cryst. Liq. Cryst.*, **180B**, 379 (1990).
37. T. Katsuki and K. B. Sharpless, *J. Am. Chem. Soc.*, **102**, 5974 (1980).
38. S. W. May and R. D. Schwartz, *J. Am. Chem. Soc.*, **96**, 4031 (1974); H. Ohta and H. Tetsukawa, *J. Chem. Soc., Chem. Commun.*, **1978**, 849; K. Furuhashi and M. Takagi, *Appl. Microbiol. Biotechnol.*, **20**, 6 (1984).
39. J. T. Groves and R. S. Myers, *J. Am. Chem. Soc.*, **105**, 5791 (1983); Y. Naruta, F. Tani, N. Ishihara, and K. Maruyama, *ibid.*, **113**, 6865 (1991).
40. W.Zhang, J. L. Loebach, S. R. Wilson, and E. N. Jacobsen, *J. Am. Chem. Soc.*, **112**, 2801 (1990).
41. R. Irie, K. Noda, Y. Ito, N. Matsumoto, and T. Katsuki, *Tetrahedron Lett.*, **31**, 7345 (1990).
42. W. Zhang and E. N. Jacobsen, *J. Org. Chem.*, **56**, 2296 (1991).
43. Y. Kaku, M. Otsuka, and M. Ohno, *Chem. Lett.*, **1989**, 611.
44. T. Yamada, K. Imagawa, and T. Mukaiyama, *Chem. Lett.*, **1992**, 2109.
45. P. Brougham, M. S. Cooper, D. A. Cummerson, H. Heaney, and N. Thompson, *Synthesis*, **1987**, 1015.
46. J.-C. Marchon and R. Ramasseul, *Synthesis*, **1989**, 389.
47. T. Yamada, K. Imagawa, T. Nagata, and T. Mukaiyama, *Chem. Lett.*, **1992**, 2231.
48. T. Mukaiyama, T. Yamada, T. Nagata, and K. Imagawa, *Chem. Lett.*, **1993**, 327.
49. T. Takai, E. Hata, K. Yorozu, and T. Mukaiyama, *Chem. Lett.*, **1992**, 2077.
50. T. Mukaiyama, K. Yorozu, T. Takai, and T. Yamada, *Chem. Lett.*, **1993**, 439.

FORMATION AND ALKENE EPOXIDATION BY HIGHER-VALENT OXO-METALLOPORPHYRINS

Ramesh D. Arasasingham and Thomas C. Bruice

Department of Chemistry
University of California at Santa Barbara
Santa Barbara, CA 93106

1. Introduction

The mechanisms by which an "oxene equivalent" is transferred from an oxygen donor to metalloporphyrins to provide higher valent metallo-oxo porphyrins and the oxygenation reactions of the latter have been the subject of considerable attention. Interest in these reactions stem from their relevance to the biochemical reactions of peroxidases with alkyl hydroperoxides,[1] catalases with hydrogen peroxides[2] and the metabolic reactions catalyzed by cytochrome P-450 enzymes. The cytochrome P-450 enzymes catalyze, among other reactions, the epoxidation of alkenes.[3] Studies on these heme containing monooxygenases reveal that one of the key steps during the reaction is the formation of an enzyme bound hypervalent iron-oxo porphyrin. From this, ensued investigations directed toward the elucidation of the mechanisms of formation and reactions of such species.

We describe here a summary of our studies using iron, manganese and chromium *meso*-tetraarylporphyrins. An understanding of the mechanisms of these simple cases have begun to provide a conceptual basis for the understanding of the enzymatic pathways. Consideration is given to two problems: (1) the mechanisms of oxidation of metalloporphyrins by hydroperoxides in aqueous solution; and (2) the mechanism of alkene epoxidation by higher valent metallo-oxo porphyrins.

2. Reactions of Hydroperoxides with Metallo-Tetraarylporphyrins in Aqueous Solutions[4]

The reactions of hydroperoxides with *meso*-tetraarylporphyrins were carried out in aqueous solution because it is in water that conditions (ionic strength, acidity, ligand species concentration) are best controlled and the data (kinetic, electrochemical, etc.) interpretable. A

The Activation of Dioxygen and Homogeneous Catalytic Oxidation,
Edited by D.H.R. Barton *et al.*, Plenum Press, New York, 1993

description of studies carried out in organic solvents has appeared elsewhere.[5] The water soluble *meso*-tetrakis(2,6-dimethyl-3-sulfonatophenyl)porphyrin, **(1)**H$_2$, and *meso*-tetrakis-(2,6-dichloro-3-sulfonatophenyl)porphyrin, **(2)**H$_2$, (Chart I) were prepared and metallated

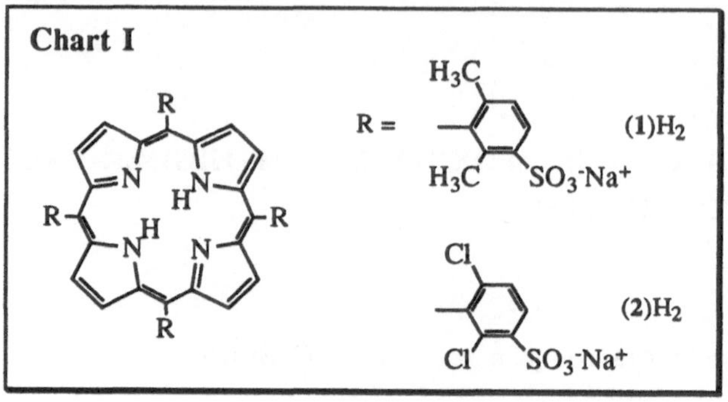

Chart I

with Fe(III) and Mn(III) to provide (1)Fe^{III}(X)$_2$, (2)Fe^{III}(X)$_2$, (1)Mn^{III}(X)$_2$, or (2)Mn^{III}(X)$_2$ (where X = H$_2$O or HO$^-$ axial ligands). The bulky aryl substituents on these metallo-porphyrins prevent the formation of µ-oxo dimers and reduce their tendency to aggregate in solution. Due to the presence of the four *m*-sulfonate substituents (1)Fe^{III}(X)$_2$, (2)Fe^{III}(X)$_2$, (1)Mn^{III}(X)$_2$, and (2)Mn^{III}(X)$_2$ exist as mixtures of four atropisomers. The electronic environments of the metal centers of a set of atropisomers have been shown to be comparable and are kinetically indistinguishable.

2.1. Electrochemical Studies on the Redox Chemistry and Acid-Base Behavior of (1)Fe^{III}(X)$_2$ and (1)Mn^{III}(X)$_2$ in Water[6-8]

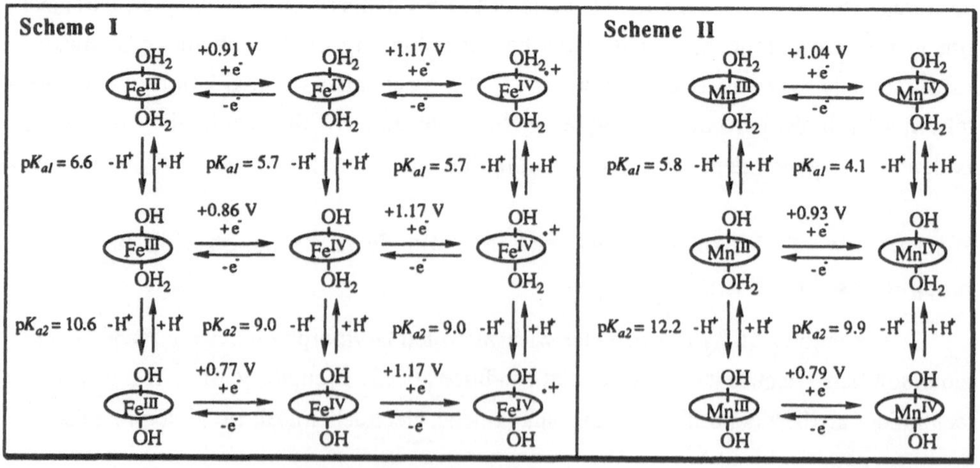

The *pKa* and *E°'* values for (1)Fe^{III}(X)$_2$ and (1)Mn^{III}(X)$_2$ were determined from the the appropriate Nernst-Clark plots of the acidity dependences of the potentials for stepwise 1e$^-$ oxidation and reduction. The results are summarized in Schemes I and II for (1)Fe^{III}(X)$_2$ and (1)Mn^{III}(X)$_2$ respectively. Inspection of Scheme I shows that: (1) the *pKa1* and *pKa2* values for the iron(IV) species are lower than those of the corresponding iron(III) species due to the greater electropositive character of Fe(IV) metal; (2) the $E^{o'}$ potentials for 1e⁻ oxidation of the three Fe(IV) species are essentially identical; and (3) the pK_{a1} values for $(1)Fe^{IV}(H_2O)_2$ and $(\cdot{+}1)Fe^{IV}(H_2O)_2$ are essentially the same as the pK_{a2} values for $(1)Fe^{IV}(HO)(H_2O)$ and $(\cdot{+}1)Fe^{IV}(HO)(H_2O)$. These findings indicate that the electron density of the Fe(IV) moieties in Fe(IV) porphyrin and Fe(IV) porphyrin π-radical cation states are very similar for identical H_2O and HO⁻ ligation. For $(1)Mn^{III}(X)_2$, the decrease in pK_{a1} and pK_{a2} values with increase in oxidation state from Mn(III) to Mn(IV) are qualitatively the same as those for $(1)Fe^{III}(X)_2$.

2.2. Kinetics of Reactions of $(1)Fe^{III}(X)_2$ and $(2)Fe^{III}(X)_2$ with Organic Hydroperoxides[9-15] and Hydrogen Peroxide[16-18]

Kinetics of the reactions were followed by the use of the sodium salt of 2,2'-azinobis-(3-ethylbenzthiazoline-6-sulfonic acid) (ABTS) as an easily oxidizable trap for the reactive higher valent iron-porphyrin intermediates and any radical products derived from YOOH (Y = alkyl or H). The one electron oxidation of ABTS provides ABTS·+ (λ_{max} = 660 nm). Regardless of whether YO-OH bond scission is homolytic (eq 1a) or heterolytic (eq 2a), a

$$(Porph)Fe^{III}(X)_2 + YOOH \rightarrow (Porph)Fe^{IV}(X)_2 + YO\cdot \qquad (1a)$$
$$(Porph)Fe^{IV}(X)_2 + ABTS \rightarrow (Porph)Fe^{III}(X)_2 + ABTS\cdot{+} \qquad (1b)$$
$$YO\cdot + ABTS \rightarrow YOH + ABTS\cdot{+} \qquad (1c)$$

$$(Porph)Fe^{III}(X)_2 + YOOH \rightarrow (\cdot{+}Porph)Fe^{IV}(X)_2 + YOH \qquad (2a)$$
$$(\cdot{+}Porph)Fe^{IV}(X)_2 + 2ABTS \rightarrow (Porph)Fe^{III}(X)_2 + 2ABTS\cdot{+} \qquad (2b)$$

single turnover of catalyst provides two ABTS·+ species (eq 1b,c vs. 2b). Reactions are first order in $(Porph)Fe^{III}(X)_2$ and YOOH and zero-order in ABTS between pH 2 and 13 (eq 3).

$$-d[YOOH]/dt = k_{ly}[(Porph)Fe^{III}(X)_2][YOOH] \qquad (3)$$

The pH dependencies of the second-order rate constants (k_{ly}) for the bimolecular reactions of $(1)Fe^{III}(X)_2$ with three alkyl hydroperoxides are shown in Figure 1, while the pH dependence of k_{ly} for the reaction of $(2)Fe^{III}(X)_2$ with H_2O_2 is shown in Figure 2. The results for the reactions of $(2)Fe^{III}(X)_2$ with alkyl hydroperoxides and $(1)Fe^{III}(X)_2$ with H_2O_2 are much the same. Though the plots of Figures 1 and 2 differ in shape, the kinetics of the reaction display pH dependence consistent with a scheme involving the critical formation of three hydroperoxide-coordinated iron(III) intermediates, IIH_2^+, IIH, and II⁻, at different pH values. The kinetic expressions were derived from Scheme III with a steady state assumption in IIH_2^+, IIH, and II⁻ for alkyl hydroperoxides and a preequilibrium assumption in the same for H_2O_2.

Examination of the pH dependence of k_{ly} for the reactions with alkyl hydroperoxides (Figure 1) shows that at low pH, k_{ly} is independent of acidity as described by eq 4. Near neutrality it displays a "bell-shaped" dependence resulting from a superimposition of the pH log k_{rate} profiles of the reactions of eqs 5 and 6. At high pH, it displays a second "bell-

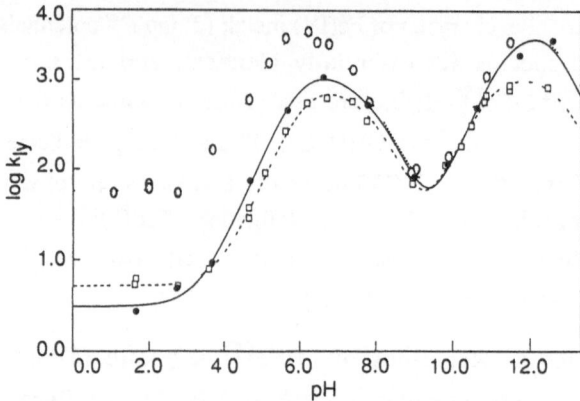

Figure 1. Plot of log k_{ly} vs pH for the reaction of $(1)Fe^{III}(X)_2$ with $(Ph)_2(CO_2Me)COOH$ ($\cdots$); $(Ph)(Me)_2COOH$ (—); and t-BuOOH (---). The points are experimental (ref 15), and the lines were generated from an equation (refs 13 and 14) derived from Scheme III with an assumption of steady state in the intermediates IIH_2^+, IIH, and II^-.

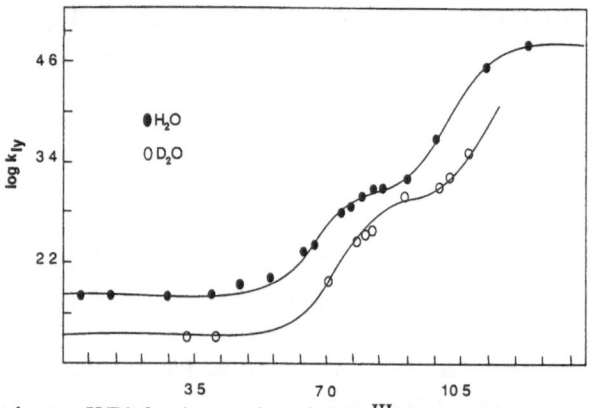

Figure 2. Plot of log k_{ly} vs pH(D) for the reaction of $(2)Fe^{III}(X)_2$ with hydrogen peroxide in H_2O and D_2O. The points are experimental (ref 18), and the lines were generated from an equation that assumes Scheme III and preequilibrium between starting states and the intermediates IIH_2^+, IIH, and II^-. Values of solvent kinetic deuterium isotope effects (k_{ly}^H/k_{ly}^D) can be appreciated by a simple comparison of k_{ly} in H_2O and D_2O in the regions where k_{ly} is independent of pH(D).

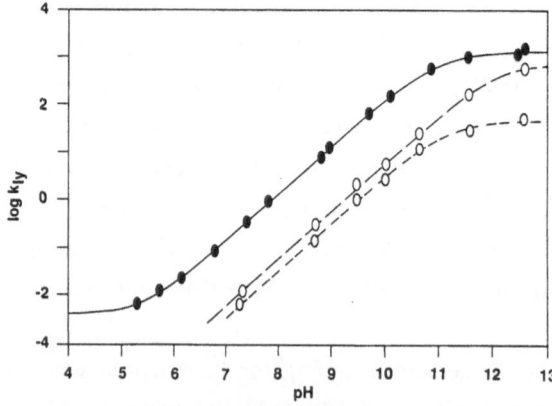

Figure 3. Plot of log k_{ly} vs pH for the reaction of $(1)Mn^{III}(X)_2$ with $(Ph)_2(CO_2Me)COOH$ (solid oval); $(Ph)(Me)_2COOH$ (open oval); and t-BuOOH (hatched oval). The points are experimental (ref 25), and the lines were generated from an equation (refs 24 and 25) derived from eq 16 with an assumption of steady state in the intermediates IIH, and II^-.

Scheme III

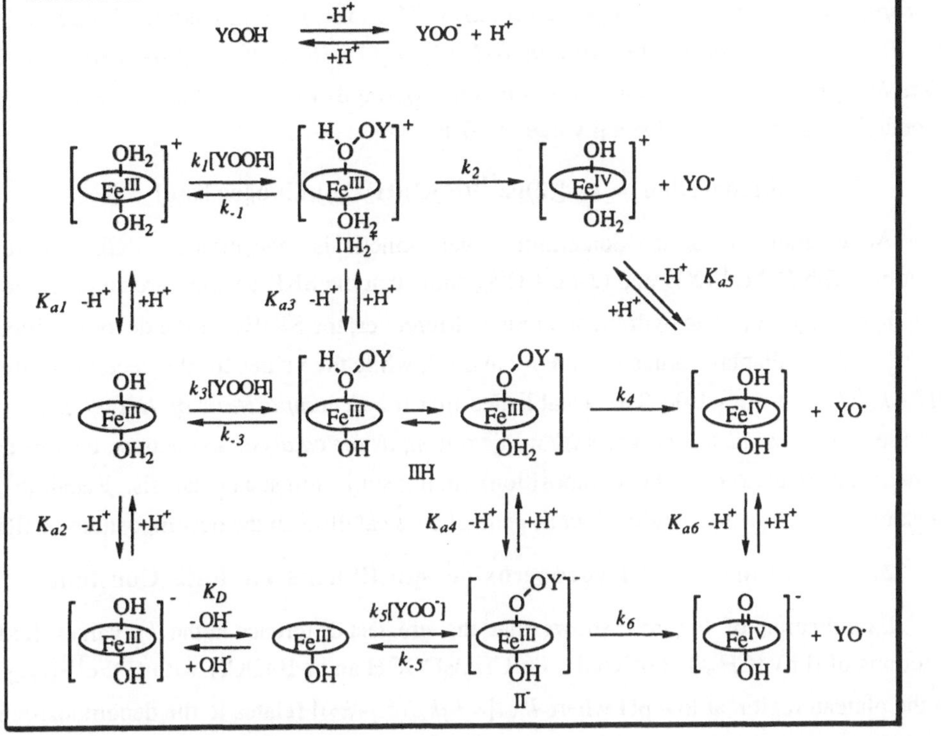

$$YOOH \underset{+H^+}{\overset{K_b, -H^+}{\rightleftharpoons}} YOO^- + H^+$$

$$\begin{bmatrix} OH_2 \\ Fe^{III} \\ OH_2 \end{bmatrix}^+ \xrightarrow[k_{-1}]{k_1[YOOH]} \begin{bmatrix} H_{\diagdown}O^{\diagup}OY \\ Fe^{III} \\ OH_2 \end{bmatrix}^+ \xrightarrow{k_2} \begin{bmatrix} OH \\ Fe^{IV} \\ OH_2 \end{bmatrix}^+ + YO^{\cdot}$$

IIH$_2^{\ddagger}$

K_{a1} $-H^+ \big\updownarrow +H^+$ 　　 K_{a3} $-H^+ \big\updownarrow +H^+$ 　　 $-H^+ K_{a5}$, $+H^+$

$$\begin{bmatrix} OH \\ Fe^{III} \\ OH_2 \end{bmatrix} \xrightarrow[k_{-3}]{k_3[YOOH]} \begin{bmatrix} H_{\diagdown}O^{\diagup}OY & O^{\diagdown}OY \\ Fe^{III} & Fe^{III} \\ OH & OH_2 \end{bmatrix} \xrightarrow{k_4} \begin{bmatrix} OH \\ Fe^{IV} \\ OH \end{bmatrix} + YO^{\cdot}$$

IIH

K_{a2} $-H^+ \big\updownarrow +H^+$ 　　 K_{a4} $-H^+ \big\updownarrow +H^+$ 　　 K_{a6} $-H^+ \big\updownarrow +H^+$

$$\begin{bmatrix} OH \\ Fe^{III} \\ OH \end{bmatrix}^- \underset{+OH^-}{\overset{K_D, -OH^-}{\rightleftharpoons}} \begin{matrix} Fe^{III} \\ OH \end{matrix} \xrightarrow[k_{-5}]{k_5[YOO^-]} \begin{bmatrix} O^{\diagdown}OY \\ Fe^{III} \\ OH \end{bmatrix}^- \xrightarrow{k_6} \begin{bmatrix} O \\ Fe^{IV} \\ OH \end{bmatrix}^- + YO^{\cdot}$$

II$^-$

shaped" dependency described by eq 7. The ascending portions of the "bell-shaped" plots are brought about by the increase of a reactant or intermediate concentration with increase in pH [IIH, eq 5; [(1)Fe^{III}(OH)(H$_2$O))]$^+$, eq 6; (1)Fe^{III}(HO)$_2^-$+ ROO$^-$, eq 7], while the descending portions are due to the concentrations of a reactant or intermediate {[(1)Fe^{III}(H$_2$O)$_2$]$^+$, eq 5; IIH, eq 6; (1)Fe^{III}(OH), eq 7} becoming limited by increase in pH. For the pH *vs* log k_{ly} profiles of the reaction of (2)Fe^{III}(X)$_2$ with H$_2$O$_2$ (Figure 2), the plateau regions at low, intermediate and high pH relate to the rate determining conversion of IIH$_2^+$, IIH, and II$^-$ to products.

$$[(1)Fe^{III}(H_2O)_2]^+ \underset{k_{-1}}{\overset{k_1[YOOH]}{\rightleftharpoons}} IIH_2^+ \xrightarrow{k_2} products \qquad (4)$$

$$[(1)Fe^{III}(H_2O)_2]^+ \underset{k_{-1}}{\overset{k_1[YOOH]}{\rightleftharpoons}} IIH_2^+ \underset{+H^+}{\overset{K_{a3}, -H^+}{\rightleftharpoons}} IIH \xrightarrow{k_4} products \qquad (5)$$

$$[(1)Fe^{III}(OH)(H_2O)] \underset{k_{-3}}{\overset{k_3[YOOH]}{\rightleftharpoons}} IIH \xrightarrow{k_4} products \qquad (6)$$

$$[(1)Fe^{III}(OH)_2]^- \underset{+HO^-}{\overset{K_D, -HO^-}{\rightleftharpoons}} (1)Fe^{III}(OH) \underset{k_{-5}}{\overset{k_5[YOO^-]}{\rightleftharpoons}} II^- \xrightarrow{k_6} products \qquad (7)$$

The reactions of (1)Fe^{III}(X)$_2$ and (2)Fe^{III}(X)$_2$ with alkyl hydroperoxides and H$_2$O$_2$ are not *subject to general-acid or general-base catalysis in the presence of oxygen acids*

/oxygen bases. *Additionally, no catalysis is observed for the reactions for alkyl hydroperoxides with buffer nitrogen acids/bases. However, general catalysis is observed for the decomposition of H_2O_2 through IIH_2^+ (eq 4) with buffers of 4-substituted 2,6-dimethylpyridines.* The Brønsted plot of log k_{gb} (eq 8) vs pK_a of the 4-substituted 2,6-dimethylpyridines·H^+ provide a β value of ~0.1.

$$-d[product]/dt = k_{gb}[(2)Fe^{III}(H_2O)(H_2O_2)][\text{nitrogen base}] \quad (8)$$

An examination of the deuterium solvent kinetic isotope effects (SKIEs) for the reactions with $(1)Fe^{III}(X)_2$ and $(2)Fe^{III}(X)_2$ show that the SKIEs values are comparable for alkyl hydroperoxides and hydrogenperoxide. Moreover, the SKIEs for the decomposition of IH_2^+ (k_2^H/k_2^D) display values that are >2 and <3, while the values for the decomposition of IIH (k_4^H/k_4^D) are ~1.0. This establishes that the *decomposition of IIH_2^+ at low pH represents general-base catalysis with water solvent as the catalyst.* In contrast, the reactions at intermediate pH (i.e., IIH decomposition) are not subject to such catalysis. Recall that 4-substituted 2,6-dimethylpyridines act as general base catalysts in the decomposition of IIH_2^+.

2.3. Influence of Hydroperoxide substituents on Rate Constants

Examination of Figure 1 shows that the greatest difference in the k_{ly} values for the reactions of $(Ph)_2(CH_3OCO)COOH$, $Ph(CH_3)_2COOH$ and t-BuOOH with $(1)Fe^{III}(X)_2$ are in the plateau region at low pH where k_{ly} [$= k_1k_2/(k_{-1}+k_2)$] relates to the decomposition of the IIH_2^+ species (Scheme III). A plot of $k_1k_2/(k_{-1}+k_2)$ vs the pK_a of the substituted alcohols $(Ph)_2(CH_3OCO)COH$, $Ph(CH_3)_2COH$ and t-BuOH displays a slope of β_{1g} ~-0.22.[19] Since complexation of $(1)Fe^{III}(X)_2$ with YOOH to provide the reactive species IIH_2^+ involves the terminal oxygen of the hydroperoxide, and the pK_a's of the most electron deficient $(Ph)_2(CH_3OCO)COOH$ and the least electron deficient t-BuOOH differ by about only 0.4 units, the values of the equilibrium constants (k_1/k_{-1}) for the complexation of the alkyl hydroperoxide with $(1)Fe^{III}(X)_2$ are expected to be much the same for the three alkyl hydroperoxides. Thus, the β_{1g} of -0.22 relates to the breakdown of the IIH_2^+ species. Analysis of the kinetic data for the decomposition of IIH complexes suggests that the β_{1g} value is even smaller. Thus, *the reaction of alkyl hydroperoxides with $(1)Fe^{III}(X)_2$ display a marked insensitivity to polar effects.*

2.4. Mechanism of Iron-Coordinated Peroxide O-O Bond Cleavage as Shown by Products and Stoichiometries

The products of the reaction of $(1)Fe^{III}(X)_2$ with t-BuOOH in the absence of the trapping agent ABTS are $(CH_3)_2C=O$ (90%), CH_3OH (90%), and t-BuOH (15%). Thus, *90% of t-BuOOH is converted to t-BuO·, which fragments to provide $(CH_3)_2C=O$ and CH_3·*. The rate constant for fragmentation of t-BuO· radicals is appreciable in water ($k = 1.4 \times 10^6$ s^{-1}).[20] With increasing [ABTS], the yields of $(CH_3)_2C=O$ and t-BuOH approach 15% and 84%, respectively. Thus, *~15% of the t-BuO· intermediate cannot be trapped by ABTS, regardless of ABTS concentration.* Moreover, the product yields are insensitive to pH (~4-11.5), buffer, ionic strength (with $NaNO_3$), or presence of O_2. Within limits of detection, neither CH_4, C_2H_6, $(t\text{-BuO})_2$ nor O_2 is a product.

The lack of $(t\text{-BuO})_2$ establishes that eqs 9 and 10 are unimportant at the low [t-BuOOH] used. Although $t\text{-BuOO}\cdot$ was detected by spin trapping, the bimolecular reaction of eq 11 is also of little consequence since it is not competitive with unimolecular fragmentation of $t\text{-BuO}\cdot$ at low [t-BuOOH]. $t\text{-BuOO}\cdot$ is likely to result from the reaction of eq 12.

$$2t\text{-BuO}\cdot \rightarrow (t\text{-BuO})_2 \qquad (9)$$
$$2t\text{-BuOO}\cdot \rightarrow (t\text{-BuOO})_2 \rightarrow (t\text{-BuO})_2 + O_2 \qquad (10)$$
$$t\text{-BuO}\cdot + t\text{-BuOOH} \rightarrow t\text{-BuOH} + t\text{-BuOO}\cdot \qquad (11)$$
$$(1)Fe^{IV}(X)_2 + t\text{-BuOOH} \rightarrow (1)Fe^{III}(X)_2 + t\text{-BuOO}\cdot \qquad (12)$$

The 90% yield of $(CH_3)_2C=O$ with $t\text{-BuOOH}$ is not in accord with a mechanism involving the heterolytic cleavage of the O-O bond of iron-coordinated hydroperoxide. A heterolytic mechanism (Scheme IV) can provide only 50% $(CH_3)_2C=O$ while homolysis

---

Scheme IV

$$(1)Fe^{III}(X)_2 + t\text{-BuOOH} \rightarrow (\cdot^+1)Fe^{IV}(X)_2 + t\text{-BuOH}$$
$$(1^{\cdot+})Fe^{IV}(X)_2 + t\text{-BuOOH} \rightarrow (1)Fe^{IV}(X)_2 + t\text{-BuOO}\cdot$$
$$2t\text{-BuOO}\cdot \rightarrow 2t\text{-BuO}\cdot + O_2$$
$$t\text{-BuO}\cdot \rightarrow (CH_3)_2CO + CH_3\cdot$$
$$(1)Fe^{IV}(X)_2 + CH_3\cdot \rightarrow (1)Fe^{III}(X)_2 + CH_3OH$$

Scheme V

$$(1)Fe^{III}(X)_2 + t\text{-BuOOH} \rightarrow (1)Fe^{IV}(X)_2 + t\text{-BuO}\cdot$$
$$t\text{-BuO}\cdot \rightarrow (CH_3)_2CO + CH_3\cdot$$
$$(1)Fe^{IV}(X)_2 + CH_3\cdot \rightarrow (1)Fe^{III}(X)_2 + CH_3OH$$

(Scheme V) provides $(CH_3)_2C=O$ in 100% yield. Thus, *only a mechanism involving homolytic O-O scission can explain the yields of products in the reaction of $(1)Fe^{III}(X)_2$ with t-BuOOH*. Scheme VI accommodates these findings for the reactions of $(1)Fe^{III}(X)_2$ and $(2)Fe^{III}(X)_2$ with $t\text{-BuOOH}$. The fragmentation of $t\text{-BuO}\cdot$ within the intimate pair for $[(1)Fe^{III}(X)_2t\text{-BuO}\cdot]$ is proposed to account for the 15% of $t\text{-BuO}\cdot$ that cannot be trapped by ABTS.

In the reaction of $(1)Fe^{III}(X)_2$ with H_2O_2, rate controlling O-O bond homolysis can be

$$(1)Fe^{III}(H_2O)(H_2O_2) \rightarrow [(1)Fe^{IV}(H_2O)(OH) \cdot OH]$$
$$\rightarrow (\cdot^+1)Fe^{IV}(H_2O)(OH) + HO^- \qquad (13)$$

$$(1)Fe^{IV}(H_2O)(OH) \rightarrow (\cdot^+1)Fe^{IV}(H_2O)(OH) \quad (+1.17V \text{ SCE}) \quad (14a)$$
$$HO^- \rightarrow HO\cdot \qquad\qquad\qquad\qquad (+1.65V \text{ SCE}) \quad (14b)$$

Scheme VI

Association

$(1)Fe^{III}(X)_2 + (CH_3)_3COOH \rightarrow (1)Fe^{III}(X)((CH_3)_3COOH)$

Caged Reactions

$(1)Fe^{III}(X)((CH_3)_3COOH) \rightarrow \{(1)Fe^{IV}(O)(X), \cdot OC(CH_3)_2\}$

$(1)Fe^{III}(O)(X), \cdot OC(CH_3)_2 \rightarrow \{(1)Fe^{IV}(O)(X), CH_3\cdot, (CH_3)_2CO\}$

$\{(1)Fe^{IV}(O)(X), CH_3\cdot, (CH_3)_2CO\} \rightarrow (1)Fe^{III}(H_2O)(X) + CH_3OH + (CH_3)_2CO$

Non-Caged Reactions

$(1)Fe^{III}(X)((CH_3)_3COOH) \rightarrow (1)Fe^{IV}(X)_2 + (CH_3)_3CO\cdot$

$(CH_3)_3CO\cdot \rightarrow (CH_3)_2CO + CH_3\cdot$

$(1)Fe^{IV}(O)(X) + CH_3\cdot \rightarrow (1)Fe^{III}(H_2O)(X) + CH_3OH$

Following Reactions

$(1)Fe^{III}(H_2O)(X) + CH_3\cdot \rightarrow (1)Fe^{II}(H_2O)(X) + CH_3OH$

$2 (1)Fe^{II}(H_2O)(X) + O_2 \rightarrow (1)Fe^{III}(X)-O-O-Fe^{III}(1)(X)$

$(1)Fe^{III}(X)-O-O-Fe^{III}(1)(X) \rightarrow 2 (1)Fe^{IV}(O)(X)$

$(1)Fe^{II}(H_2O)(X) + (1)Fe^{IV}(O)(X) \rightarrow 2 (1)Fe^{III}(H_2O)(X)$

Peroxide Oxidation

$(1)Fe^{IV}(O)(X) + (CH_3)_3COOH \rightarrow (1)Fe^{III}(H_2O)(X) + (CH_3)_3COO\cdot + H^+$

$(CH_3)_3COO\cdot \rightarrow (CH_3)_3CO\cdot + 1/2 O_2$

$(CH_3)_3CO\cdot \rightarrow (CH_3)_2CO + CH_3\cdot$

followed by a second $1e^-$ transfer (eq 13). The oxidation of the Fe(IV) species by $HO\cdot$, within the solvent cage, is thermodynamically favored on the basis of reduction potentials (eq 14).[21] This is not true when $HO\cdot$ is replaced by alkyl-$O\cdot$.

2.5. Summation of Findings for the Reactions of $(1)Fe^{III}(X)_2$ and $(2)Fe^{III}(X)_2$ with Hydroperoxides

The reactions of alkyl hydroperoxides with $(1)Fe^{III}(X)_2$ and $(2)Fe^{III}(X)_2$, (i) are not subject to general-acid or general base catalysis at any pH by either oxygen or nitrogen acids or bases; (ii) display a deuterium SKIE > 2 for the rate-determining decomposition of intermediate IIH_2^+ (Scheme II) but not for IIH or II^-; (iii) show a marked insensitivity of rate constants to polar effects; (iv) display products and product stoichiometry that are pH independent; and (v) produce a product stoichiometry that can only be explained *via* homolytic O-O bond cleavage (eq 1a). The deuterium SKIE at low pH indicates solvent general-base catalysis. We propose a preassociation of H_2O with the iron(III) coordinated hydroperoxide (Chart II).

For the reactions of H_2O_2 with $(1)Fe^{III}(X)_2$ and $(2)Fe^{III}(X)_2$: (i) the deuterium SKIE associated with the decomposition of IIH_2^+ at low pH(D) is > 2; (ii) no catalysis by oxygen bases or acids is observed; and (iii) general base catalysis of decomposition of IIH_2^+ by 2,6-dimethylpyridines displays a Brønsted $\beta = 0.1$. Features i and ii are similar to alkyl hydro-peroxides. A $\beta = 0.1$ indicates an early transition state and little dependence of k_{gb} on the basicity of the catalyst. This would explain the weak base H_2O (at 55M) acting as a general-

Chart II

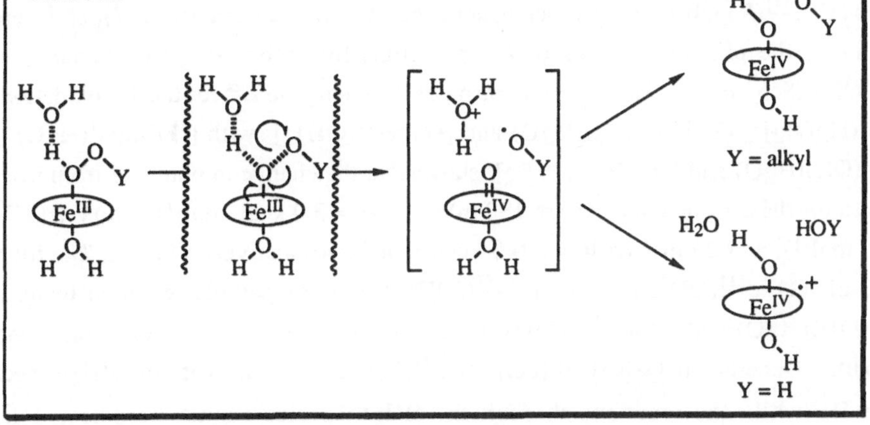

base catalyst (Chart II). The lack of catalysis by oxyanion bases may be explained by eq 15, where the value of C may be greater for 2,6-dimethylpyridines than for carboxylates due to

$$\log k_{gb} = \beta pK_a + C \qquad (15)$$

the greater solvation of carboxylates and/or to the electrostatic repulsion of the negatively charged carboxylates. Shielding by an electrostatic effect is a real possibility. The relationship of pK_{a1} of $(1)Fe^{III}(X)_2$ to ionic strength (Guntelberg-Debye-Huckel expression)[23] suggests that the m-sulfonate substituents on the porphyrin invokes a charge of -2 at the iron(III) center.

2.6. Comparison of the Reactions of $(1)Fe^{III}(X)_2$ and $(1)Mn^{III}(X)_2$ with Hydroperoxides[24-26]

The pH dependencies of k_{ly} for the bimolecular reactions of $(1)Mn^{III}(X)_2$ with three alkyl hydroperoxides are shown in Figure 3. For $(1)Mn^{III}(X)_2$, the kinetics of the reactions display pH dependence consistent with a scheme involving the critical formation of two hydroperoxide-coordinated manganese(III) intermediates at intermediate and high pH values. At low pH the the reaction of $(1)Mn^{III}(X)_2$ with hydroperoxides could not be determined over spontaneous decomposition of hydroperoxide. The kinetic expression was derived from the pathway shown in eq 16 with a steady state assumption in intermediates IIH and II$^-$.

$$(1)Mn^{III}(OH)(H_2O) \xrightleftharpoons[k_{-1}]{k_1[YOOH]} (1)Mn^{III}(OOY)(H_2O)$$
IIH

$$\xrightleftharpoons[+H^+]{\substack{K_{a3} \\ -H^+}} [(1)Mn^{III}(OOY)(OH)]^- \xrightarrow{k_2} \text{products} \qquad (16)$$
II$^-$

A comparison of the log k_{ly} vs. pH profiles for the reaction of $(1)Mn^{III}(X)_2$ and $(1)Fe^{III}(X)_2$ with $(Ph)_2(CH_3OCO)COOH$, $Ph(CH_3)_2COOH$, and t-BuOOH (Figures 1 & 3)

155

shows that the values of k_{ly} are comparable only at high pH (i.e. II⁻ decomposition). As pH decreases $(1)Fe^{III}(X)_2$, becomes a much better catalyst. These differences in k_{ly} at different pH's relate to the pH dependences of $E^{o'}$ potentials for $(1)Fe^{III}(X)_2$ (Scheme I) and $(1)Mn^{III}(X)_2$ (Scheme II). Thus, a comparison of the change in $E^{o'}$ for the 1e⁻ oxidation of $[(1)Fe^{III}(H_2O)_2]^+$, $(1)Fe^{III}(OH)(H_2O)$ and $[(1)Fe^{III}(OH)_2]^-$ with $[(1)Mn^{III}(H_2O)_2]^+$, $(1)Mn^{III}(OH)(H_2O)$ and $[(1)Mn^{III}(OH)_2]^-$ shows that the change in potential from iron to manganese for the different species are 130 mV ($\Delta\Delta G^{\neq} = 3.0$ kcal. mol⁻¹), 70 mV ($\Delta\Delta G^{\neq} = 1.6$ kcal. mol⁻¹), and 20 mV ($\Delta\Delta G^{\neq} = 0.5$ kcal. mol⁻¹), respectively. Hence, k_{ly} for the reactions of $[(1)Fe^{III}(OH)_2]^-$ and $[(1)Mn^{III}(OH)_2]^-$ are comparable, while at lower pH $(1)Fe^{III}(OH)(H_2O)$ and $[(1)Fe^{III}(H_2O)_2]^+$ are far more reactive than their manganese counterparts. General catalysis with $(Porph)Fe^{III}(X)_2$ is seen only with the IIH_2^+ species (Scheme II), while the reactions through the IIH_2^+ species cannot be detected with $(Porph)Mn^{III}(X)_2$. Thus, it is not surprising to find that the reactions of hydroperoxides with $(1)Mn^{III}(X)_2$ and $(2)Mn^{III}(X)_2$ are not subject to general-acid/base catalysis by H_2O, oxygen acids/oxygen bases, or nitrogen bases/nitrogen acids at any pH.

The relationship between the rate constants and the pK_a of $(Ph)_2(CH_3OCO)COH$, $Ph(CH_3)_2COH$ and t-BuOH are comparable for $(1)Mn^{III}(X)_2$ and $(1)Fe^{III}(X)_2$, suggesting homolytic O-O bond cleavage of manganese(III)-coordinated alkyl hydroperoxide. This is supported by the products obtained from the reaction of t-BuOOH with $(1)Mn^{III}(X)_2$. Product analysis in the absence of the ABTS trapping agent provide $(CH_3)_2C=O$ (60-70%), t-BuOH (12%), t-BuOOCH₃ (22-25%), $(t$-BuO)$_2$, CH_3OH and HCHO, while in its presence $(CH_3)_2C=O$ (5%) and t-BuOH (89%) are formed. The product distribution showed no dependence on the pH of the reaction solutions.

With imidazole, $(1)Mn^{III}(X)_2$ forms a monoligated species which reacts with alkyl hydroperoxides (eq 17) with a rate constant, k_{Im}, exceeding k_{ly} by ~4 to 10-fold for t-BuOOH. The product distribution for the reactions in the presence of imidazole showed

$$-d[YOOH]/dt = k_{Im}[(1)Mn^{III}(X)("imidazole")][YOOH] \qquad (17)$$

significant dependence on the pH of the reaction mixtures. At intermediate pH values, the product profiles are consistent with a homolytic mechanism for O-O bond cleavage, where the major product was $(CH_3)_2C=O$ (63-67%), with remainder being t-BuOH (19%), t-BuOOCH₃ (13-16%), $(t$-BuO)$_2$, CH_3OH and HCHO. At high pH (~12), the yield of t-BuOH (63%) increased dramatically with concomitant decreases in the yields of $(CH_3)_2C=O$ (34%), t-BuOOCH₃ (4%), $(t$-BuO)$_2$, CH_3OH and HCHO. The change in product distribution at high pH finds explanation in a change in mechanism of O-O bond cleavage from homolysis to heterolysis as a result of the deprotonation of the manganese(III)-coordinated imidazole (eq 18).[25] A change in the basicity of the axially ligated imidazole ring

$$(1)Mn^{III}(OOY)(ImH) \rightarrow [(1)Mn^{III}(OOY)(Im)]^- + H^+ \quad (pK_a = 11.5) \qquad (18)$$

by proton dissociation represents a mechanism whereby the reactivity at the metal center can be altered.

3. Mechanism of Alkene Epoxidation by Higher Valent Metallo-Oxo Porphyrins[27]

Investigations directed toward the elucidation of the mechanisms of oxidation of alkenes by higher valent iron, chromium and manganese-oxo tetraarylporphyrins have resulted in a number of intermediates being proposed along the reaction path. These include the formation of intermediates such as a metallaoxetane, **I**,[28] an alkene-derived π-radial cation, **II**,[29-31] a carbocation, **III**,[30-34] and a carbon radical, **IV**,[35-39] together with a mechanism involving the concerted insertion of an "oxene" into the alkene double bond, **V**,[40-42] (Chart III). These intermediates have been implicated on the basis of the stereo-

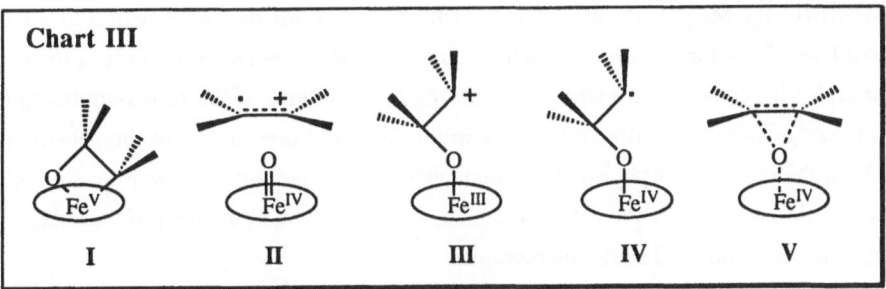

Chart III

I   II   III   IV   V

chemistry of the product epoxides and the nature of other products that are produced. While the stereochemistry of the epoxide products relate directly to the mechanism of epoxidation, the nature of the other products that are formed may not be relevant if these products arise from reactions which are parallel or competing with the epoxidation reaction. Thus, the structures of products, other than epoxide, may or may not contain information concerning the mechanism of epoxidation. For example, the catalytic oxidation of cis-stilbene by $(Cl_8TPP)Mn^{III}(OH)$[43] and C_6F_5IO provide cis-stilbene oxide, trans-stilbene oxide, benzaldehyde, diphenylacetaldehyde, and deoxybenzoin as products. The numerous products can be accommodated by a step-wise mechanism where all products arise from intermediates on the way to epoxide, or by mechanisms which are parallel and competing with epoxidation.

3.1. Metallaoxetanes, I, Are Not Required Intermediates in Epoxidation[44-46]

Epoxidation via metallaoxetane intermediates, **I**, have been proposed to occur by a 2a+2s cycloaddition of the alkene and metallo-oxo species followed by a concerted reductive elimination.[28] The proposed requirement of **I** as an intermediate may be assessed by determining whether epoxidation occurs with metalloporphyrin catalysts that sterically prohibit the formation of **I**. Thus, when the sterically encumbered $(Br_8TPP)Fe^{III}(Cl)$ was used as a catalyst with C_6F_5IO oxidant to generate the $(^{+.}Br_8TPP)Fe^{IV}(O)$ epoxidizing species, all cis-alkenes, including terminal alkenes were cleanly epoxidized. Although quantitative yields of epoxide were obtained, approach of the alkene to the iron(IV)-oxo reaction center is restricted by the bulk of the eight o-bromo substituents.[44,45]

We turned to molecular modeling to determine whether formation of **I** was possible for the reaction of $(^{+\cdot}Br_8TPP)Fe^{IV}(O)$ with alkenes. *o*-Bromo substituents were appended, with a fixed Br-C bond distance of 1.85 Å, on the phenyl rings of the X-ray structures of both the planar $(TPP)Fe^{III}(Cl)$ (S = 5/2)[47] and the extremely convexed $[(^{+\cdot}TTP)Fe^{III}(Cl)](SbCl_6^-)$ (S = 2),[48] and the phenyl rings were rotated to energy minima. In the convexed structure, the iron was 0.46 Å out of the pyrrole nitrogen plane and 0.99 Å out of the pyrrole β-carbon plane. For the construction of the putative metallaoxetane ring, the template X-ray structures of an iridium azametallacyclobutane, $[Cp^*(PMe_3)Ir(CH_2CMe_2NH_2^+)]$,[49] and a platinum oxametallacyclobutane, $[(As(C_6H_5)_3)_2Pt(C_2(CN)_4O)]$,[50] were considered. Both structures displayed similar bond lengths and bond angles for the four-membered rings, and $Cp^*(PMe_3)Ir(CH_2CMe_2NH_2^+)$ was chosen as the template for the construction of the iron metallaoxetane. Structures of the metallaoxetane ring systems derived from *cis*-stilbene and 2,3-dimethyl-2-butene were constructed by fixing the structure of the four-membered ring according to the X-ray structure and energy minimizing the orientation of the substituents. Assembly of the putative porphyrin metallaoxetanes was carried out by superimposing the Fe of the porphyrin and the Ir of the metallaoxetane, followed by readjustment of the substituent positions to provide minimal steric interactions.

Examination of the computer constructed *cis*-diphenylmetallaoxetanes and the tetramethylmetallaoxetanes showed very unfavorable steric interactions between the phenyl and methyl substituents and both the porphyrin ring and the *o*-bromo substituents. These severe steric effects do not depend on the porphyrin being planar or convexed. In order to avoid these steric interactions, the Fe-C and F-O bond lengths of the metallaoxetane rings must be extended to unreasonable values of >3.2 Å. It was concluded that metallaoxetanes could not be required intermediates in alkene epoxidation by hypervalent metal-oxo porphyrins.

3.2. Rate-Limiting Formation of an Alkene-Derived π-Radical Cation, II, Cannot Be a Required Intermediate in Epoxidation[51,52]

The feasibility of the rate-limiting formation of a π-radical cation intermediate, **II**, can be assessed by examining the linear free-energy correlation between the second-order rate constants (k_2) for the bimolecular reaction of alkenes with the relatively stable Cr(V)-oxo tetraarylporphyrins and the potentials for the 1e⁻ oxidation $(AE_{1/2})$ of the alkenes. The linear correlation between $\log k_2$ and $AE_{1/2}$ for the reaction of 16 alkenes with $(Br_8TPP)Cr^V(O)(X)$ is shown in eq 19 and Figure 4. Also, the bimolecular reactions between five variously

$$\log k_2 = -2.99(AE_{1/2}) + 4.02 \qquad (19)$$

substituted Cr(V)-oxo tetraarylporphyrins and norbornene display a linear correlation between k_2 and the 1e⁻ reduction potentials $(CE_{1/2})$ of the Cr(V)-oxo tetraarylporphyrins as shown by eq 20. The electrochemical emf values for the 1e⁻ oxidation of alkenes by

$$\log k_2 = -9.03(CE_{1/2}) + 8.57 \qquad (20)$$

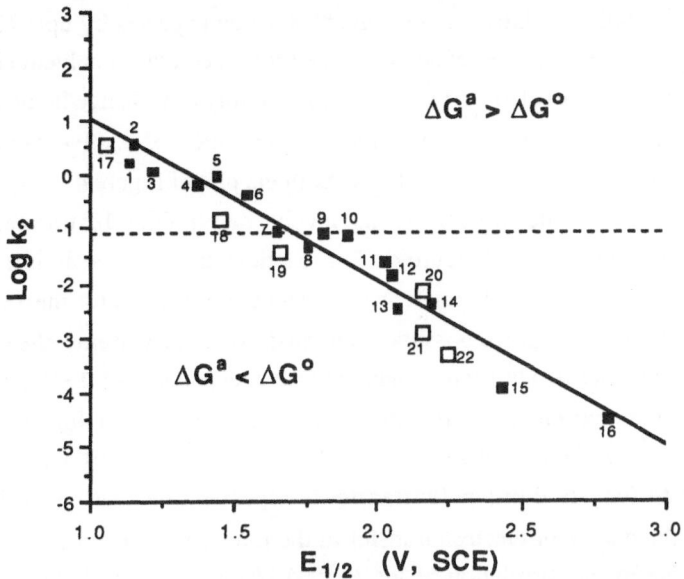

Figure 4. Plot of the log of the second-order rate constants for the reactions of a series of alkenes with $(Br_8TPP)Cr^V(O)(X)$ vs $E_{1/2}$ for $1e^-$ oxidation of the alkenes. The alkenes are as follows: (1) 1,4-diphenyl-1,3-butadiene; (2) 4-methoxystyrene; (3) 1,1-diphenylethylene; (4) 4-methylstyrene; (5) 2,3-dimethyl-2-butene; (6) *cis*-stilbene; (7) styrene; (8) 4-acetoxystyrene; (9) cyclohexene; (10) norbornene; (11) *cis*-cyclooctene; (12) 4-cyanostyrene; (13) *cis*-2-pentene; (14) cyclopentene; (15) 1-hexene; (16) 1-octene; (17) *trans-p,p'*-dimethoxystilbene; (18) (Z)-1,2-bis(*trans*-2,*trans*-3-diphenylcyclopropyl)ethene; (19) *trans*-β-methylstyrene; (20) *trans*-5-decene; (21) *trans*-2-pentene; (22) *trans*-2-hexenyl acetate. The log k_2 values for *trans*-stilbene and *trans-p,p'*-dicyanostilbene exhibit large negative deviations due to steric effects (discussed in ref 71).

$(Br_8TPP)Cr^V(O)(X)$ $(CE_{1/2} = 0.88$ V vs SCE)(eq 22) is given by $CE_{1/2} - AE_{1/2}$ (eqs 21a,b). From the emf values, the standard free-energies ($\Delta G°$) for the equilibria of eq 22 may be calculated and compared with the free-energies of activation ($\Delta G^{\neq}$) obtained from the k_2 values by use of the Eyring equation.[53]

$$(Br_8TPP)Cr^V(O)(X) + e^- \rightleftharpoons (Br_8TPP)Cr^{IV}(O) + X \qquad CE_{1/2} \qquad (21a)$$

$$\text{alkene}^{\cdot+} + e^- \rightleftharpoons \text{alkene} \qquad AE_{1/2} \qquad (21b)$$

$$(Br_8TPP)Cr^V(O)(X) + \text{alkene} \rightleftharpoons (Br_8TPP)Cr^{IV}(O) + \text{alkene}^{\cdot+} \qquad (22)$$

A comparison of $\Delta G°$ and $\Delta G^{\neq}$ values for the alkenes shown in Figure 4, indicates that those alkenes located below the dashed-line display $\Delta G^{\neq} < \Delta G°$ (i.e., free energy of **II** exceeds that of the transition state for the rate-limiting step) while those located above the dashed line display $\Delta G^{\neq} > \Delta G°$ (i.e., free energy of transition state exceeds that of **II**). When $\Delta G^{\neq} < \Delta G°$, the redox reaction of eq 24 cannot be involved and the π-radical cation, **II**, cannot be an intermediate nor can it be a product formed from a competing reaction. About half the alkenes examined displayed $\Delta G^{\neq} < \Delta G°$. Examination of the reaction of $(Br_8TPP)Cr^V(O)(X)$ with 1-octene, shows that $\Delta G°$ exceeds $\Delta G^{\neq}$ by ~21 kcal mol^{-1}. For **II** to exist as an intermediate solvent caged species, $[(Br_8TPP)Cr^{IV}(O)(X) + 1$-octene π-radical cation], it must be stabilized relative to the solvent separated species by upto (21+Y) kcal mol^{-1} [where $Y = \Delta G^{\neq}$ for reversion of solvent-caged intermediates to solvent separated 1-octene π-radical cation and $(Br_8TPP)Cr^{IV}(O)$]. The possibility of a change in mechanism for the epoxidation from one involving the intermediate formation of a solvent caged pair to another is considered unlikely since there is no break in the plot of Figure 4.

The conclusion that the mechanism does not involve rate-limiting 1e$^-$ transfer to form **II** is further supported by the slope of Figure 4. For reactions that involve an initial electron transfer, the relationship between the rate constants and the potentials for the 1e$^-$ oxidation $(\Delta E_{1/2})$ of the alkenes can take three possible scenarios: (i) at one extreme, the diffusion of reactants toward each other is rate-limiting and log k_2 is independent of $\Delta\Delta E_{1/2}$ (slope = 0); (ii) while at the other extreme, 1e$^-$ transfer is so endothermic that log k_2 is directly proportional to $\Delta\Delta E_{1/2}$ for the overall 1e$^-$ transfer (slope = -16.6 V^{-1}); (iii) or an intermediate region, where log k_2 for rate-limiting 1e$^-$ transfer is dependent on $\Delta\Delta E_{1/2}$, with the slope depending on the the degree of electron transfer in the transition state, changing from 0 to -16.6 V^{-1}.[54,55] The linear correlation of log k_2 vs $\Delta\Delta E_{1/2}$ of Figure 4 shows a slope of -2.99 V^{-1}, which corresponds to a Brønsted α of -0.18 in a conventional Brønsted plot of log k_2 vs log K_{eq}. Thus, in the reaction of $(Br_8TPP)Cr^V(O)(X)$ with alkenes, the transition state must be very early such that there is very little electron transfer. Such a situation is equivalent to rate-determining formation of a charge-transfer (CT) complex.

The differences in the slopes of the linear free energy plots of log k_2 vs $AE_{1/2}$ (alkene is varied, eq 21) and log k_2 vs $CE_{1/2}$ ((Porph)Cr^V(O)(X) is varied, eq 22) indicates the presence of a perpendicular (Thornton) contribution to the overall Brønsted slope α. On a scale of 0 (dissociative transition state) to 2 (associative transition state), the Thornton contribution to α can be calculated to be 0.63, and the Leffler-Hammond contribution to α (measure of a progress along the reaction coordinate) can be calculated to be 0.37.[56] Presuming a 1e$^-$ transfer mechanism, the transition state is located ca. 37% along the reaction coordinate and it is somewhat dissociative. This approximation of transition-state structure is not consistent with a full electron transfer (which is endothermic for about one-half of the alkenes studied and therefore requires a late transition state) but is reasonable for rate-limiting formation of a CT complex. An initially formed CT complex could collapse to give $(Br_8TPP)Cr^{III}(X)$ and epoxide (Scheme VII).

The formation of **II** can occur on oxidation of alkenes with low oxidation potentials. That the formation of **II** did not occur at the *rate-determining step* was shown by examination of the oxidation of (Z)-1,2-bis(*trans-2, trans-3*-diphenylcyclopropyl)ethene, **VI-Z**, by $(Br_8TPP)Cr^V(O)(BF_4^-)$ and $(Br_8TPP)Cr^V(O)(ClO_4^-)$ in CH_2Cl_2 solvent.[29] With $(Br_8TPP)Cr^V(O)(BF_4^-)$, **VI-Z** provides the *trans,trans-* and *trans, cis*-dienes **VIII** as products, while $(Br_8TPP)Cr^V(O)(ClO_4^-)$ provides the *trans-2,trans*-3-diphenylcyclopropane carboxyaldehyde as the major product (Scheme VIII). The second-order rate constant (k_2) for oxidation of **VI-Z** was found to fit the line of Figure 4 and was independent of X$^-$ = BF$_4^-$ or ClO$_4^-$. We propose the initial rate-determining step to be the formation of the CT complex, [CT]$^+$[X]$^-$. When X$^-$ = BF$_4^-$, the CT complex provides a carbocation radical, **VII**,

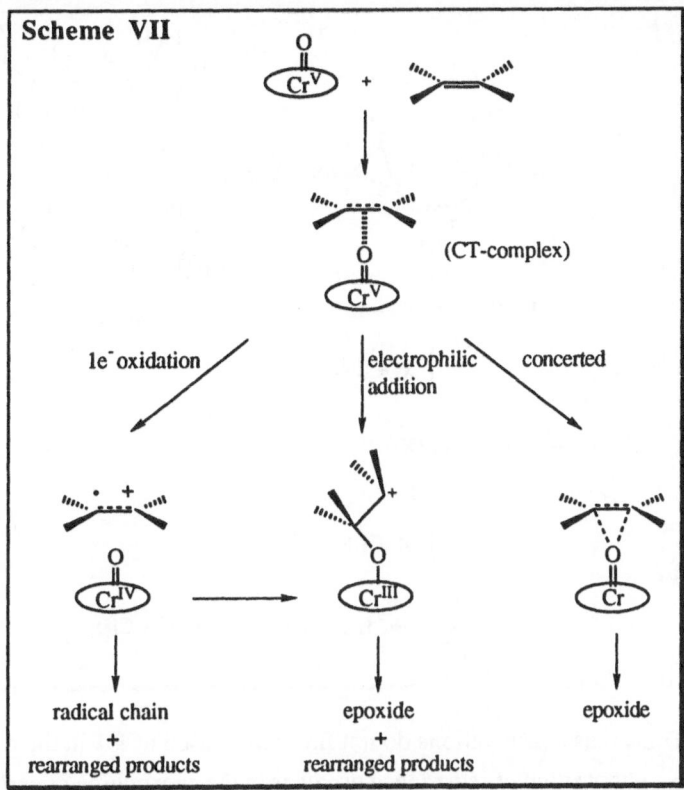

Scheme VII

(CT-complex)

1e⁻ oxidation   electrophilic addition   concerted

radical chain + rearranged products   epoxide + rearranged products   epoxide

which undergoes dual cyclopropylcarbinyl to homoallylcarbinyl radical rearrangement (CPCRR) and cyclopropylcarbinyl to homoallylcarbinyl carbocation rearrangement (CPCCR) and captures the halogen species generated from the solvent to provide the dienes **VIII**. With $X^- = ClO_4^-$, the CT complex breaks down with ClO_4^- as an oxidant to provide *trans-*2,*trans*-3-diphenylcyclopropanecarboxaldehyde. The inclusion of the ClO_4^- cooxidant to provide the [CT]⁺[X]⁻ structure amounts to an enforced oxidation.

3.3. Formation of Carbocation, III, Cannot Be Rate Limiting[52]

Electrophilic addition of the metallo-oxo species to an alkene has been proposed to generate the carbocation intermediate, **III**, which is partially stabilized by the interaction of the positive charge with the electron cloud of the porphyrin nitrogens. The proposed requirement of **III** as an intermediate was based on the rearranged products that are obtained in the reactions with *cis*-stilbene. Thus, the reaction of $(F_{20}TPP)Fe^{III}(Cl)$, $(Cl_8TPP)Fe^{III}(Cl)$ or $(Cl_8TPP)Mn^{III}(Cl)$ catalysts and C_6F_5IO oxidant with *cis*-stilbene produced deoxybenzoin and diphenylacetaldehyde as rearranged carbonyl products.[33]

The plausibility of the rate-limiting formation of **III** can be assessed by examining the linear free-energy relationship of log k_2 and σ^+ for the epoxidation of substituted styrenes by $(Br_8TPP)Cr^V(O)(X)$[51] and $(^{+\cdot}TMP)Fe^{IV}(O)$,[36c] which provide a slope of $\rho^+ = -1.9$. Comparison with known rate-limiting carbocation formations *via* electrophilic additions to substituted styrenes show greater negative ρ^+ values (-3.58 for hydration[57] and -4.8 for

Scheme VIII

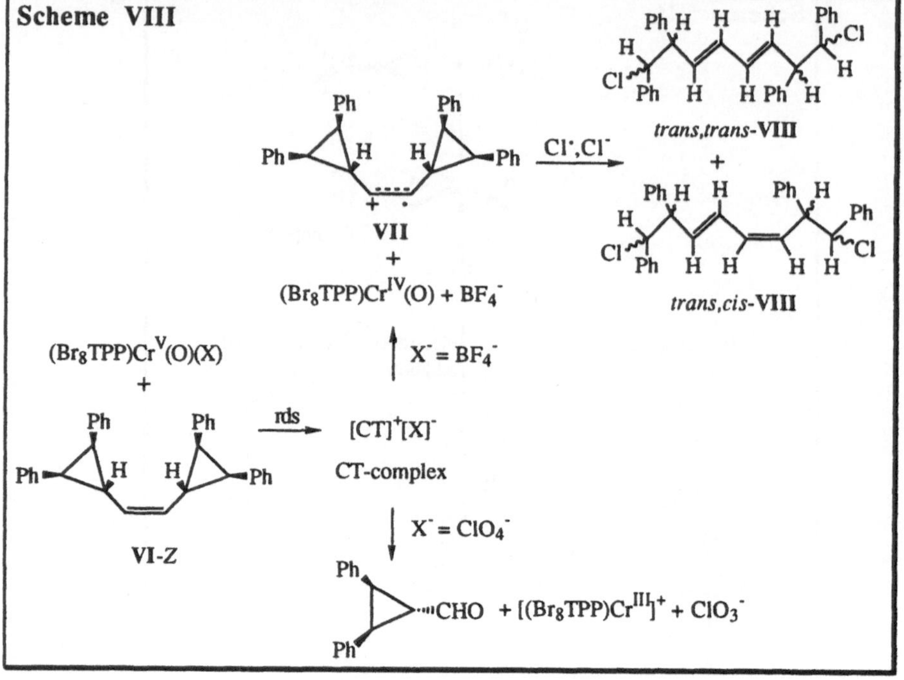

bromination[58]). Thus, these observations do not favor formation of **III** in the rate-limiting step. Moreover, *the observation of rearranged products in the epoxidation of certain alkenes does not necessiate the involvement of an alkene-derived carbocation radical, **II**, or a carbocation, **III**, on the reaction path to epoxide.*

3.4. Radical, IV, Is Not a Required Intermediate in Epoxidation[38]

In a search for radical intermediates, **IV**, the *cis*-olefinic substrate (Z)-1,2-bis(*trans-2, trans-3*-diphenylcyclopropyl)ethene, **VI-Z** (scheme IX), incorporating the hypersensitive radical trapping group, *trans-2, trans-3*-diphenylcyclopropyl, was used in the epoxidation

Scheme IX

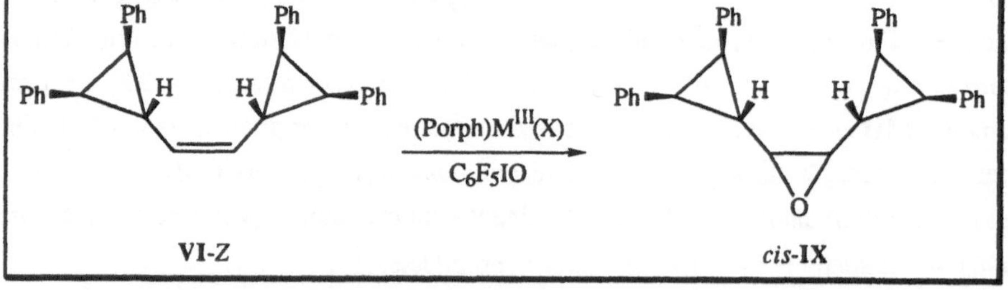

reaction. From competitive studies, it was established that the ring openings of the *phenyl-substituted* cyclopropylcarbinyl radicals by a cyclopropylcarbinyl to homoallylcarbinyl radical rearrangement (CPCRR) is very fast with rate constants $\geq 2 \times 10^{10}$ s^{-1} at room temperature. These radicals have very short lifetimes at room temperature, and are considered to be much faster than diffusional processes and can compete in measurable amounts with even the fastest possible first-order processes.

The reactions of **VI-Z** with $(F_{20}TPP)Fe^{III}(Cl)$, $(Cl_8TPP)Fe^{III}(Cl)$, and $(Cl_8TPP)Mn^{III}(OH)$ catalysts and C_6F_5IO oxidant provided the *cis*-epoxide, **IX**, as the major product (Scheme IX). The yields of the *cis*-epoxide for the various catalysts were 95% for $(F_{20}TPP)Fe^{III}(Cl)$, 90% for $(Cl_8TPP)Fe^{III}(Cl)$, and 84% for $(Cl_8TPP)Mn^{III}(OH)$. No polar products (which could derive from the radical species, **IV**) or any *trans*-alkene or *trans*-epoxide could be detected to 0.1% detection limit. Aside from the *cis*-epoxide, **IX**, very non-polar, non-oxygen containing products were obtained in 5-16% yield.

The test for radical intermediates in the epoxidation reaction is based on the required rate constant for partitioning of the putative radical intermediate to epoxide as compared to other products. Given that the rate constant for CPCRR of the (*trans-2, trans-3*-diphenyl cyclopropyl)carbinyl radical is $\geq 2 \times 10^{10}$ s^{-1} and that the yield of products derived from the radical species **IV** are <0.1%, the conversion of radical to *cis*- **IX** would have a rate constant of $>10^{12}$ s^{-1}. The neutral radical species **IV** cannot be a discrete intermediate in the epoxidation reaction.

Products which could be derived from the carbocation radical **VII** are obtained in 5-16% yields. It follows that the rate constant for conversion of **VII** to epoxide must be $\geq$ (1-2) $\times 10^{11}$ s^{-1}. A rate constant of 10^{11} s^{-1} for the recombination of a solvent-caged pair involving **VII** to provide epoxide is not unreasonable. Recall our conclusion that the formation of **VII** cannot be rate determining. The nonpolar products could have arisen from a reaction that is parallel to epoxidation.

3.5. Rate Limiting Formation of a Radical Intermediate in Epoxidation by Mn(IV)-Oxo porphyrins[59]

The manganese porphyrin systems display alkene oxidation products that can be attributed to a radical type reaction which can be traced to the presence of the Mn(IV)-oxo porphyrins. For example, the catalytic oxidation of *cis*-stilbene by $(Cl_8TPP)Mn^{III}(OH)$ and C_6F_5IO provide both *cis*- and *trans*-stilbene oxides together with benzaldehyde, diphenylacetaldehyde, and deoxybenzoin. The loss of stereochemistry during the epoxidation has been explained by the formation of a carbon radical intermediate. Additionally, the *cis*-stilbene conversion to products was found to exceed the equivalents of C_6F_5IO employed, further suggesting radical condensation processes. The plausibility of the rate limiting formation of a $(Porph)Mn^{III}$-O-C-C· radical intermediate was assessed by examining the linear free-energy correlation between the second-order rate constants (k_2) for the bimolecular reaction of alkenes with $(Cl_8TPP)Mn^{IV}(O)$ and the potentials for the 1e$^-$ oxidation ($AE_{1/2}$) of the alkenes. The linear correlation of log k_2 vs $AE_{1/2}$ is given eq 23

$$\log k_2 = -0.89(AE_{1/2}) + 1.14 \qquad (23)$$

which provides a slope of -0.89 V^{-1}. This corresponds to a Brønsted α of -0.05 in a conventional Brønsted plot of log k_2 vs log K_{eq}. Moreover, the linear free-energy relationship of log k_2 vs σ^+ for the reaction with substituted styrenes provides a slope of $\rho^+ = -0.99$. These observations support a transition state involving very little charge separation, in accord with the rate-determining formation of a charge-transfer complex or a neutral carbon radical intermediate. Product yields determined for the reaction of $(Cl_8TPP)Mn^{IV}(O)$ with various alkenes are in accord with a mechanism involving the formation of a neutral radical intermediate. Thus, the products of cis-stilbene oxidation under aerobic conditions provide cis-stilbene oxide (7%), trans-stilbene oxide (5%) and benzaldehyde (3%) as products. The reactivities of the alkenes showed that the trans-alkenes are slightly more reactive than their cis-isomers and that electron releasing substituents slightly favor the reaction. In a search for the radical intermediate the the cis-olefinic substrate (Z)-1,2-bis(trans-2, trans-3-diphenylcyclopropyl)ethene, **VI**, was used as a radical trap. While no epoxide products were found, *a polar oxygen-containing product resulting from the opening of the cyclopropyl radical to homoallylcarbinyl radical* was detected supporting the formation of a carbon radical species.

3.6. A Concerted Mechanism For Epoxidation Can Be Favored

The following findings support a mechanism involving the concerted insertion of an "oxene" equivalent into the alkene double bond: (i) Epoxidation of alkenes by higher valent oxo-metalloporphyrins is stereospecific for the vast majority of alkenes. In the case of manganese porphyrins, the formation of trans-epoxides from cis-alkenes may be traced to contamination of the manganese(V)-oxo porphyrins by manganese(IV)-oxo porphyrin. (ii) The ρ^+ values reported for the epoxidation of a series of substituted styrenes by $(TPP)Fe^{III}(Cl)$ and C_6H_5IO is -0.93. This value is similar to the ρ^+ values reported for known concerted processes such as the insertion of carbenes into double bonds (-0.62 to -1.61)[60] and the epoxidation of alkenes by perbenzoic acid (-1.2).[61] (iii) The most efficient and selective catalyst for epoxidation is the sterically encumbered $(Br_8TPP)Fe^{III}(Cl)$. The bulky o-bromo substituents can play two roles in favoring a concerted mechanism. The steric bulk could prevent the interaction of the alkenes with the electron density of the porphyrin ring. This would prevent any stabilization of a positive charge on the alkene that could develop in an electrophilic mechanism which would provide the carbocation intermediate, **III**, and the corresponding rearranged products. Additionally, the heavy bromine atoms could facilitate spin-spin interconversions via spin-orbit coupling by creating a favorable environment for concerted oxygen insertion into the alkene double bond and formation of epoxide.[62-65]

Concerted oxygen insertion into the alkene double bond could follow rate-limiting formation of a CT complex if a very fast change of spin states occurs. The iron-bound oxygen in the iron(IV)-oxo π-cation radical species is believed to have a triplet oxenoid character due to the mixing of the p_x and p_y orbitals of oxygen with the d_{xz} and d_{yz} orbitals of iron.[66] Thus, for a concerted mechanism efficient spin inversion from the triplet oxenoid to a singlet oxenoid should occur. Spin inversion does not pose a problem since efficient changes of spin states and the presence of "spin equilibria" are well documented for iron porphyrins.[62-65] It is known that mixing of spin states and spin inversions are accelerated by the presence of other unpaired spins in an unsymmetrical environment.[65] The unpaired electron in the porphyrin π-cation radical might just create such an environment at the point where the electron redistribution within the alkene and the oxidant has commenced. This is supported by the fact that (+·Br$_8$TPP)Fe^{IV}(O)(X), which has a porphyrin centered π-cation radical structure, epoxidizes alkenes in almost quantitative yields, while (Br$_8$TPP)Cr^V(O)(X), which lacks the π-cation radical structure, provides substantial rearranged products.

After rate-determining formation of the CT complex and change of spin-state, the ratio of epoxide to carbocation rearrangement products would be determined by the relative rates of (i) concerted oxygen insertion to give epoxide and (ii) electrophilic attack of metal-bound oxygen on the alkene to provide a carbocation intermediate (Scheme X). The latter would be greatly favored by the overlap of the p orbitals on the alkene and the porphyrin nitrogens.

Scheme X

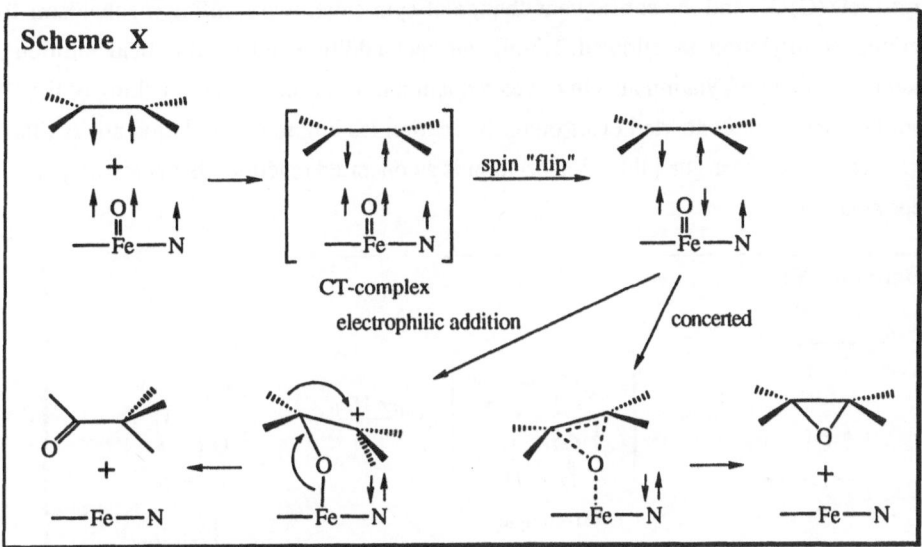

3.7. Porphyrin N-Alkylation Can Be Concerted and Can Compete with Epoxidation

With alkyl-substituted terminal alkenes various degrees of porphyrin N-alkylation accompany epoxidation. The N-alkylation take place exclusively at the unsubstituted terminus of the double bond and could be a concerted or non-concerted reaction. Formation of an intermediate carbocation on the path to N-alkylation can be excluded because it would require the preferential formation of a primary carbocation. Radicals, on the other hand, show much lower preference for substituted *vs* unsubstituted carbon,[67] suggesting that an initial formation of a carbon radical, followed by its collapse to the N-alkyl porphyrin is possible.[41,68] While carbon radicals cannot be discrete intermediates in the epoxidation reaction (*vide supra*), they can be intermediates in N-alkylation, if N-alkylation is a side reaction to epoxidation.

Another possibility is the intermediacy of an alkene-derived π-cation radical **IV**. However, the formation of **IV** from primary alkenes is energetically unfavorable (i.e., the potentials for 1e⁻ oxidation are quite positive for aliphatic primary alkenes), although it is these alkenes which exhibit N-alkylation. Styrene, which has a much lower oxidation potential that would favor the formation of **IV**, does not exhibit N-alkylation.[69] This dichotomy would tend to eliminate carbocation radicals as plausible intermediates in porphyrin N-alkylation.

In the (⁺·Porph)Fe^{IV}(O)(X) oxidant, a partial positive charge resides on the porphyrin nitrogens, and a partial negative charge resides on the iron bound oxygen, thereby making the O-Fe-N moiety similar to many dipolar molecules that are involved in 1,3-dipolar cycloadditions with alkenes.[70] Also, both the iron-bound oxygen and the porphyrin nitrogen possess unpaired electron density, so that the O-Fe-N moiety has a partial 1,3-diradical character. Thus, after initial formation of a CT complex between alkene and (⁺·Porph)Fe^{IV}(O)(X) and the subsequent change of spin-state, two limiting mechanisms for a concerted N-alkylation are allowed, 1,3-dipolar cycloaddition and 1,3-diradical addition to the alkene. While N-alkylation also involves reduction of the iron and the breaking of the Fe-N bond, the means for concerted electron redistribution leading to N-alkylation are available (Scheme XI). It can be argued that N-alkylation is a concerted reaction that is in competition with epoxidation.

Scheme XI

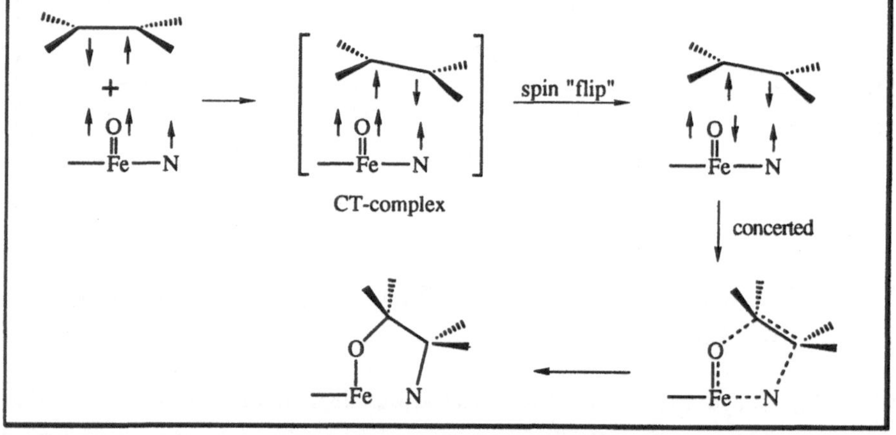

3.8. Summation of Findings on the Mechanism of Epoxidation by Higher Valent Metallo-Oxo Porphyrins

Epoxides are one of several products that are formed in reactions of metallo-oxo porphyrins with alkenes. The nature of the other oxidized products may not relate directly to the mechanism of epoxidation if they arise from reactions which compete with epoxidation. Of the various mechanisms proposed for the epoxidation, the formation of a metallaoxetane intermediate, **I**, along the reaction pathway is sterically impossible as shown for

(+·Porph)Fe^{IV}(O) and (Porph)Cr^V(O)(X). The dependence of the log of the second-order rate constants upon the potential for reduction of alkene-derived π-cation radicals, **II**, (equivalent to ρ_I-σ_I correlation) as well as a comparison of experimentally available values of $\Delta G^{\neq}$ and $\Delta G°$ rules out the required intermediacy of **II** at any step in alkene epoxidation by both (Porph)Cr^V(O)(X) and (+·Porph)Fe^{IV}(O). The ρ_I-σ_I relationship for the epoxidation of styrenes by (Porph)Cr^V(O)(X) and (+·Porph)Fe^{IV}(O) does not support rate-determining carbocation, **III**, formation. Trapping experiments have shown that the existence of an intermediate carbon radical, **IV**, would require a life time $\leq 10^{-12}$ s, indicating that a carbon radical intermediate cannot be formed along the pathway to epoxide for the reactions with (+·Porph)Fe^{IV}(O), (Porph)Cr^V(O)(X) and (Porph)Mn^V(O)(X)). Thus, for the various mechanisms proposed for the epoxidation of alkenes by (+·Porph)Fe^{IV}(O) and (Porph)Cr^V(O)(X) species, those which *require rate-determining formation* of **I**, **II**, **III**, and **IV** may be dismissed. It is most likely the same conclusions can be reached for epoxidations involving (Porph)Mn^V(O)(X). Alkene oxidation products arising from a radical type reaction with manganese porphyrins is attributed to the contamination of the manganese(V)-oxo species by manganese(IV)-oxo species. The oxidation of alkenes by (Porph)Mn^{IV}(O) support the rate-determining formation of a (Porph)Mn^{III}-O-C-C· radical intermediate.

A general picture emerges where the rate-limiting step is the formation of a charge-transfer complex. The reactions following the rate-determining step (epoxidation *vs* rearrangement) are dependent on a number of factors that include the oxidation potentials of the alkenes and the active oxidant, the steric and electronic structures of the reactants (steric bulk and geometry of both species and the metal axial ligand of the metalloporphyrin), as well as the propensity of various substrates to undergo rearrangements.

Acknowledgement. These studies were supported by grants from the National Institutes of Health and the National Science Foundation.

References

1. (a) Sauders, B. C.; Holmes-Siedel, A. G.; Stark, B. P. *Peroxidases*, Butterworths: London, 1964; (b) Dunford, H. B.; Stillman, J. S. *Coord. Chem. Rev.* **1976**, *19*, 187; (c) Sauders, B. C. In *Inorganic Biochemistry*, Eichorn, G. I., Ed.; Elsevier: Amsterdam, The Netherlands, 1973; vol. 2, p.988. (d) Dunford, H. B. *Adv. Inorg. Biochem.* **1982**, *4*, 41.
2. Schonbaum, G. R.; Chance, B. Catalase, In *The Enzymes*, Boyer, P., Ed.; Academic Press: New York, 1976; Vol. 13, p 363.
3. Ortiz de Montellano, P. R. In *Cytochrome P-450; Stucture, Mechanism and Biochemistry*, Ortiz de Montellano, P. R. Ed.; Plenum Press: New York, 1986, Chapter 7, p 217-271.
4. Bruice, T. C. *Acc. Chem. Res.* **1991**, *24*, 243.
5. (a) Bruice, T. C. In *Mechanistic Principles of Enzyme Activity*, Lieberman, J. F., Greenberg, A., Eds.; VCH Publishers: New York, 1988, Chapter 6, p 227; (b) Bruice, T. C. In *Proceedings of the Robert A. Welch Foundation Conference XXXI. Design of Enzymes and Enzyme Models*, 1988, p 37-70.
6. Kaaret, T. W.; Zhang, G.; Bruice, T. C. *J. Am. Chem. Soc.* **1991**, *113*, 4652.
7. Jeon, S.; Bruice, T. C. *Inorg. Chem.* **1991**, *30*, 4311.
8. Jeon, S.; Bruice, T. C. *Inorg. Chem.* **1992**, *31*, 4843.

9. Lindsay Smith, J. R.; Balasubramanian, P. N.; Bruice, T. C. *J. Am. Chem. Soc.* **1988**, *110*, 7411.
10. Bruice, T. C.; Balasubramanian, P. N.; Lee, R. W.; Lindsay Smith, J. R. *J. Am. Chem. Soc.* **1988**, *110*, 7890.
11. Balasubramanian, P. N.; Lindsay Smith, J. R.; Davis, M. J.; Kaaret, T. W.; Bruice, T. C. *J. Am. Chem. Soc.* **1989**, *111*, 1477.
12. Balasubramanian, P. N.; Lee, R. W.; Bruice, T. C. *J. Am. Chem. Soc.* **1989**, *111*, 8714.
13. Murata, K.; Panicucci, R.; Gopinath, E.; Bruice, T. C. *J. Am. Chem. Soc.* **1990**, *112*, 6072.
14. Gopinath, E.; Bruice, T. C. *J. Am. Chem. Soc.* **1991**, *113*, 4657.
15. Gopinath, E.; Bruice, T. C. *J. Am. Chem. Soc.* **1991**, *113*, 6090.
16. Bruice, T. C.; Zipplies, M. F.; Lee, W. A. *Proc. Natl. Acad. Sci. (USA)* **1986**, *83*, 4646.
17. Zipplies, M. F.; Lee, W. A.; Bruice, T. C. *J. Am. Chem. Soc.* **1986**, *108*, 4433.
18. Panicucci, R.; Bruice, T. C. *J. Am. Chem. Soc.* **1990**, *112*, 6063.
19. The calculated pK_a values for (Ph)$_2$(CH$_3$OCO)COH, Ph(CH$_3$)$_2$COH and t-BuOH are 11.1, 15.1, and 16.7, respectively. Discussed in ref 15.
20. Erben-Russ, M.; Michael, C.; Bors, W.; Saran, M. *J. Phys. Chem.* **1987**, *91*, 2362.
21. Sawyer, D. T.; Roberts, J. L., Jr. Acc. Chem. Res. **1988**, *21*, 469.
22. Beck, M. J.; Gopinath, E.; Bruice, T. C. *J. Am. Chem. Soc.* **1993**, *115*, 21.
23. (a) Stumm, W.; Morgan, J. J. In *Aquatic Chemistry*, Halsted Press, John Wiley and Sons, Inc.: New York, 1981, p 136. (b) Robinson, R. A.; Stokes, R.H. In *Electrolyte Solutions*, Butterworths Publications Ltd.: New York, 1959, p 231.
24. Arasasingham, R. D.; Bruice, T. C. *J. Am. Chem. Soc.* **1991**, *113*, 6095.
25. Arasasingham, R. D.; Jeon, S.; Bruice, T. C. *J. Am. Chem. Soc.* **1992**, *114*, 2536.
26. Balasubramanian, P. N.; Schmidt, E. S.; Bruice, T. C. *J. Am. Chem. Soc.* **1987**, *109*, 7865.
27. Ostovic, D.; Bruice, T. C. Acc. Chem. Res. **1992**, *25*, 314.
28. (a) Collman, J. P.; Brauman, J. I.; Meunier, B.; Raybuck, S. A; Kodadek, T. *Proc. Natl. Acad. Sci. U.S.A.* **1984**, *81*, 3245; (b) Collman, J. P.; Brauman, J. I.; Meunier, B.; Hayashi, T.; Kodadek, T.; Raybuck, S. A. *J. Am. Chem. Soc.* **1985**, *107*, 2000. (c) Collman, J. P.; Kodadek, T.; Raybuck, S. A.; Brauman, J. I.; Papazian, L. M. *J. Am. Chem. Soc.* **1985**, *107*, 4343.
29. He, G.-X.; Arasasingham, R. D.; Zhang, G.; Bruice, T. C. *J. Am. Chem. Soc.* **1991**, *113*, 9828.
30. (a) Traylor, T. G.; Nakano, T.; Dunlap, B. E.; Traylor, P. S.; Dolphin, D. *J. Am. Chem. Soc.* **1986**, *108*, 2782; (b) Traylor, T. G.; Mikzstal, A. R. *J. Am. Chem. Soc.* **1987**, *109*, 2770; (c) Traylor, T. G.; Nakano, T; Mikzstal, A. R.; Dunlap, B. E. *J. Am. Chem. Soc.* **1987**, *109*, 3625.
31. Bartolini, O.; Meunier, B. *J. Chem. Soc., Perkin Trans. 2*, **1984**, 1967.
32. Lindsay-Smith, J. R.; Sleath, P. R. *J. Chem. Soc., Perkin Trans. 2*, **1982**, 1009.
33. Castellino, A.; Bruice, T. C. *J. Am. Chem. Soc.* **1988**, *110*, 158.
34. Collman, J. P.; Kodadek, T.; Brauman, J. I. *J. Am. Chem. Soc.* **1986**, *108*, 2588.
35. Guengerich, F. P.; MacDonald, T. L. Acc. Chem. Res. **1984**, *17*, 9.
36. (a) Groves, J. R.; Kruper, W. J., Jr.; Haushalter, R. C. *J. Am. Chem. Soc.* **1980**, *102*, 6375; (b) Groves, J. R.; Myers, R. S. *J. Am. Chem. Soc.* **1983**, *105*, 5791; (c) Groves, J. R.; Watanabe, Y. *J. Am. Chem. Soc.* **1986**, *108*, 507.
37. Fontecave, M.; Mansui, D. *J. Chem. Soc., Chem. Commun.* **1984**, 879.
38. Castellino, A.; Bruice, T. C. *J. Am. Chem. Soc.* **1988**, *110*, 1313.
39. Castellino, A.; Bruice, T. C. *J. Am. Chem. Soc.* **1988**, *110*, 7512.
40. (a) Watabe, T.; Akamatsu, K. *Biochem. Pharmacol.* **1974**, *23*, 1079; (b) Watabe, T.; Ueno, Y.; Imazumi, J. *Biochem. Pharmacol.* **1971**, *20*, 912.
41. Ortiz de Montellano, P. R.; Mangold, B. L. K.; Wheeler, C.; Kunze, K. L.; Reich, N. O. *J. Biol. Chem.* **1983**, *258*, 4208.
42. Hanzlik, R. P.; Shearer, G. O. *Biochem. Pharmacol.* **1978**, *27*, 1441.
43. Abbreviations used: Porph, a generic porphyrin dianion; TPP, dianion of *meso*-tetraphenylporphyrin; TTP, dianion of *meso*-tetra-p-tolylporphyrin; TMP, dianion of *meso*-tetramesitylporphyrin; Cl$_8$TPP, dianion of *meso*-tetrakis(2,6-dichlorophenyl)porphyrin; Br$_8$TPP, dianion of *meso*-tetrakis(2,6-dibromophenyl)porphyrin.
44. Ostovic, D.; Bruice, T. C. *J. Am. Chem. Soc.* **1988**, *110*, 6906.

45. Ostovic, D.; Bruice, T. C. *J. Am. Chem. Soc.* **1989**, *111*, 6511.
46. Lee, R. W.; Nakagaki, P. C.; Balasubramanian, P. N.; Bruice, T. C. *Proc. Natl. Acad. Sci. U.S.A.* **1985**, *85*, 641.
47. Hoard, J. L.; Cohen, G. H.; Glick, M. D. *J. Am. Chem. Soc.* **1967**, *89*, 1992.
48. Gans, P.; Buisson, G.; Duee, E.; Marchon, J.-C.; Erler, B. S.; Scholz, W. F.; Reed, C. A. *J. Am. Chem. Soc.* **1986**, *108*, 1223.
49. Klein, D. P.; Hayes, J. C.; Bergman, R. G. *J. Am. Chem. Soc.* **1988**, *110*, 3704.
50. Schlodder, R.; Ibers, J. A.; Lenarda, M.; Graziani, M. *J. Am. Chem. Soc.* **1974**, *96*, 6893.
51. Garrison, J. M.; Bruice, T. C. *J. Am. Chem. Soc.* **1989**, *111*, 191.
52. Garrison, J. M.; Ostovic, D.; Bruice, T. C. *J. Am. Chem. Soc.* **1989**, *111*, 4960.
53. Eyring, H. J. *J. Chem. Phys.* **1935**, *3*, 107.
54. Rehm, D.; Weller, A. *Isr. J. Chem.* **1970**, *8*, 259.
55. Andrieux, C. P.; Blocman, C.; Dumas-Bouchiat, J.-M.; Saveant, J.-M. *J. Am. Chem. Soc.* **1979**, *101*, 3431.
56. Kreevoy, M. M.; Lee, I.-S. H. *J. Am. Chem. Soc.* **1984**, *106*, 2550.
57. Schubert, W. M.; Keefe, J. R. *J. Am. Chem. Soc.* **1972**, *94*, 559.
58. Yates, K.; McDonald, R. S.; Shapiro, S. A. *J. Org. Chem.* **1973**, *38*, 2460.
59. Arasasingham, R. D.; He, G.-X.; Bruice, T. C. submitted to *J. Am. Chem. Soc.*
60. Moss, R. A. In *Carbenes*, Jones, M. Jr., Moss, R. A., Eds.; John Wiley and Sons: New York, 1973, Vol. 1, p 268-269.
61. Ogata, Y.; Tabushi, I. *J. Am. Chem. Soc.* **1961**, *83*, 3440.
62. Hill, H. A.; Skyte, P. D.; Buchler, J. W.; Lueken, H.; Tonn, M.; Gregson, A. K.; Pellizer, G. *J. Chem. Soc., Chem. Commun.* **1979**, 151.
63. Gregson, A. K. *Inorg. Chem.* **1981**, *20*, 879.
64. Martin, R. L.; White, A. H. *Transition Met. Chem.* **1968**, *4*, 113.
65. Harman, R. A.; Eyring, H. J. *J. Chem. Phys.* **1942**, *10*, 557.
66. Loew, G. H.; Kert, C. J.; Hjemeland, L. M.; Kirchner, R. F. *J. Am. Chem. Soc.* **1977**, *99*, 3534.
67. Tedder, J. M.; Walton, J.C. *Acc. Chem. Res.* **1976**, *9*, 183.
68. Ortiz de Montellano, P. R. In *Bioactivation of Foreign Compounds*, Anders, M. W., Ed.; Academic Press: New York, 1985, Chapter 5.
69. Kunze, K. L.; Beilan, H. S.; Wheeler, C. *Biochemistry* **1982**, *21*, 1331.
70. Huisgen, R.; Grashey, R.; Sauer, J. In *The Chemistry of Alkenes*, Patai, S. Ed.; Interscience Publishers: London, 1964, p 806-878.
71. He, G.-X; Mei, H.-Y.; Bruice, T. C. *J. Am. Chem. Soc.* **1991**, *113*, 5644.

MODELS FOR HORSERADISH PEROXIDASE COMPOUND II:

PHENOL OXIDATION WITH OXOIRON(IV) PORPHYRINS

Nicola Colclough and John R. Lindsay Smith

Department of Chemistry
University of York
York, YO1 5DD, U.K.

INTRODUCTION

The role of iron protoporphyrin (IX), the prosthetic group in most heme proteins, is determined by the local environment of the heme group. Thus in cytochrome b_5 it has two tightly bound ligands and it is involved in electron transport. In hemoglobin, cytochrome P450 monooxygenases and peroxidases, however, the sixth (distal) ligand is either absent or only weakly bound allowing ligation of dioxygen or hydrogen peroxide to the metal centre. The behaviour of the latter group of heme proteins is further defined by the nature of the fifth (proximal) ligand and the polarity of and access to the active site.[1] Much recent research has been directed towards determining these controlling factors since they have important implications in the understanding of biological systems and in the development of new catalytic oxidations.

In horseradish peroxidase (HRP), the most thoroughly studied peroxidase, the iron(III) porphyrin with a proximal histidine ligand is almost entirely surrounded by protein. As a consequence the enzyme is able to restrict and control access to the distal position of the heme group to small alkyl hydroperoxides and H_2O_2.[1] Furthermore by acid/base catalysis and the polar environment of the active site it is able to promote heterolysis of the peroxide O-O bond to give the oxoiron(IV) porphyrin π radical cation, compound 1 [reaction (1)].[2] The protein surrounding the active site also ensures that the subsequent one-electron oxidations of the substrate are restricted to interactions at the edge of the heme molecule and, unlike cytochrome P450, do not involve the oxoiron(IV) group directly. The general reaction sequence for HRP-catalysed oxidations is shown in reactions (1)-(4).

HRP + ROOH	$\longrightarrow$	HRP-I + ROH	(1)
HRP-I + Sub-H	$\longrightarrow$	HRP-II + Sub·	(2)
HRP-II + Sub-H	$\longrightarrow$	HRP + Sub·	(3)
Sub·	$\longrightarrow$	Products	(4)

The Activation of Dioxygen and Homogeneous Catalytic Oxidation,
Edited by D.H.R. Barton *et al.*, Plenum Press, New York, 1993

Peroxidases can catalyse the oxidation of a wide range of substrates. Typically these are phenols and aromatic amines but biopolymers such as lignin are also susceptible to oxidative degradation. The mechanisms of these one-electron oxidations have been suggested to be either an electron transfer, H-atom abstraction or the simultaneous removal of an electron and a proton.[3-8]

Although many high valent oxoiron porphyrins have been prepared there have been very few mechanistic studies of oxidations by these models for HRP compounds I and II.[9,10] This can largely be attributed to the low stability of such species which has restricted investigations to low temperature studies in non-aqueous media. However, in recent studies we[11] and others[12,13] have shown that ionic oxoiron(IV) porphyrins are surprisingly stable in basic aqueous solution (Table 1). This in turn has provided the opportunity to study and compare the oxidations by synthetic oxoiron(IV) porphyrins with those by their enzymic analogues. The results from such a study are presented here.

Table 1. Lifetime of ionic oxoiron(IV) tetraarylporphyrins in aqueous buffer at pH 9.2.[11]

Oxoiron(IV) tetraarylporphyrin[1] Aryl group	Concentration /10^{-6}M	Lifetime /h	% Bleaching
4-carboxyphenyl	9.3	1	50
4-sulphonatophenyl	8.0	1	35
4-trimethylammoniumphenyl	7.9	2	4
4-N-methylpyridyl	9.4	4	3
2-N-methylpyridyl	9.1	13	1
2,6-dichloro-3-sulphonatophenyl	6.8	24	1

[1] Generated from iron(III) porphyrin with 3 equivalents of $^tBuO_2H$ in 0.1 M borate buffer, pH 9.2 at 30°C.

RESULTS AND DISCUSSION

Preliminary Studies

Of the readily available ionic iron porphyrins, we chose the tetra-(N-methylpyridyl) derivatives since their ferryl species are amongst the most stable and their porphyrin rings are relatively resistant to oxidative bleaching. A preliminary series of experiments was designed to classify the types of oxidation brought about by these oxoiron(IV) porphyrins (Figure 1). UV-VIS spectroscopy was used to show that only substrates that are readily oxidised by electron transfer or H-atom abstraction [Figure 2, paths (a) and (b)] are able to reduce the oxidant to iron(III) porphyrin whilst those that might be susceptible to oxygen transfer are inactive [Figure 2, path (c)]. The lack of oxygen transfer from the oxoiron(IV) porphyrin is not unexpected for, although oxoiron(IV) porphyrin π radical cations (compound I analogues) are active mono-oxygen donors, there are few well authenticated examples of oxygen transfer from the corresponding oxoiron(IV) porphyrins (compound II models). Ellis and Lyons[14] have proposed the latter as the active oxidant in the iron porphyrin catalysed autoxidation of alkanes and Weber[15] and his coworkers suggest ferryl porphyrins are

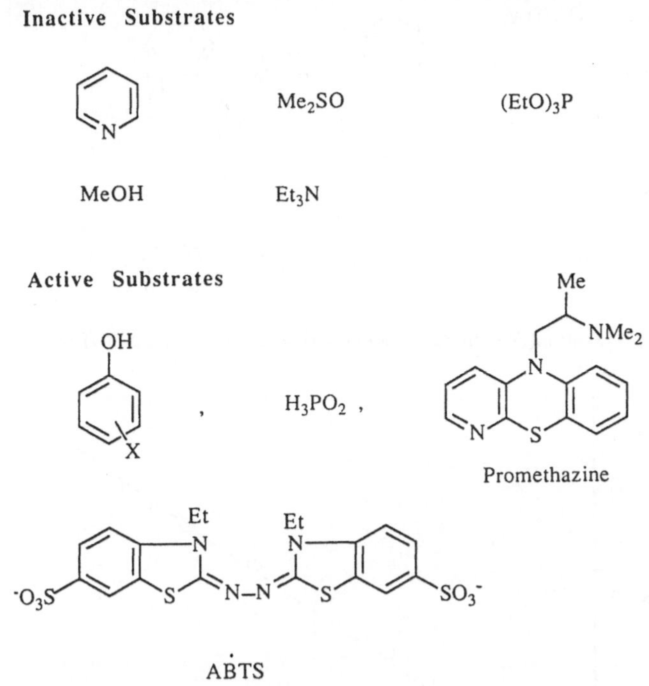

Figure 1. Substrates that are active and inactive reducing agents for OFe(IV)T4MPyP in aqueous buffer at pH 9.2

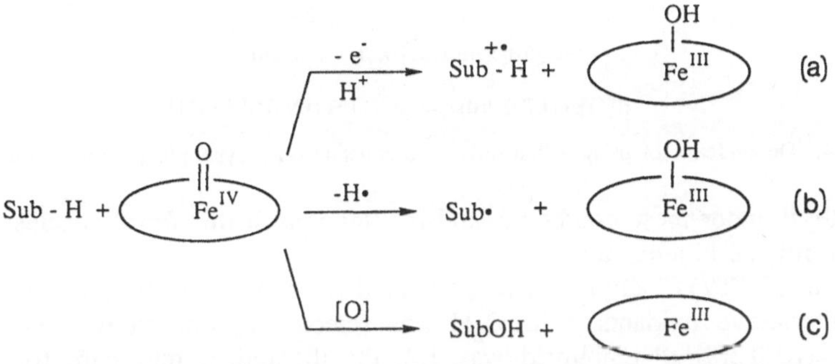

Figure 2. Possible oxidation mechanisms of oxoiron(IV) porphyrins

oxidants generated by the photolysis of iron porphyrin μ-oxo dimers.

Kinetic Study of Phenol Oxidation

Following the preliminary studies, a detailed kinetic investigation of phenol oxidations was undertaken. This was carried out with OFe(IV)T2MPyP under less alkaline conditions (pH 7.7) to allow meaningful comparisons with previous studies with HRP compound II.[5,7] At pH 7.7 OFe(IV)T4MPyP is markedly less stable than it is at pH 9.2 and this leads to experimental complications in studying its reactions. The isomeric OFe(IV)T2MPyP (Figure 3) is, however, ~100 x more stable than

OXIDANT

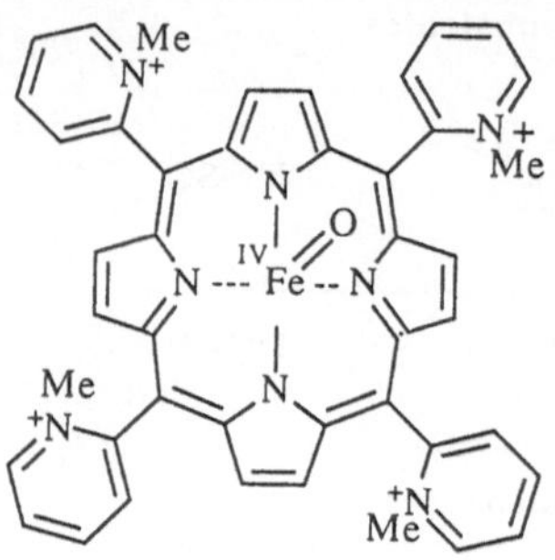

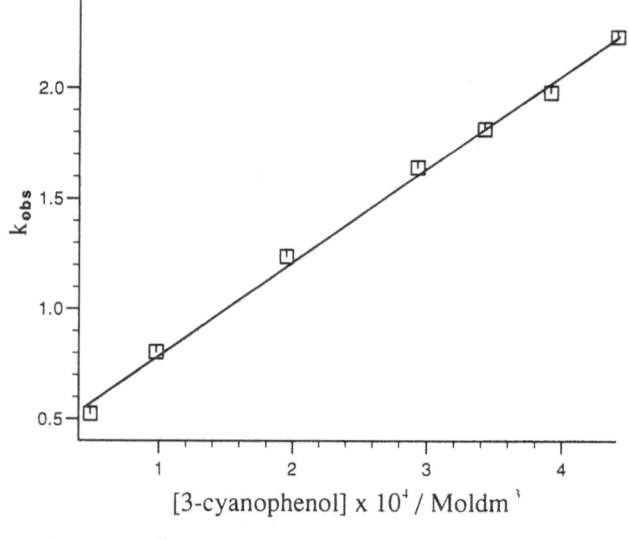

Figure 3. Oxoiron(IV) tetra(2-*N*-methylpyridyl)porphyrin [OFe(IV)T2MPyP]

d[OFe(IV)P]/dt = k_2[OFe(IV)P][ArOH]

Figure 4. Dependence of pseudo first order rate constant on 3-cyanophenol concentration

OFe(IV)T4MPyP under these conditions and for this reason this ferryl species was used as the oxidant for the kinetic studies.

Although OFe(IV)T2MPyP can be generated from the iron(III) porphyrin using a number of alternative oxidants,[11] $^tBuO_2H$ was selected in this study. The ratio of $^tBuO_2H$ to Fe(III)T2MPyP employed was 1:2, the theoretical minimum for complete oxidation of the iron(III). Under these conditions UV-VIS spectroscopy showed that there was a 73% conversion to OFe(IV)T2MPyP. The missing oxidant is consumed in competitive unproductive side reactions of $^tBuO_2H$ and OFe(IV)T2MPyP. An excess of $^tBuO_2H$ over Fe(III)T2MPyP was not employed for the main kinetic studies to avoid possible complications from the recycling of the iron(III) porphyrin during the reaction, although as described and discussed below this precaution was subsequently found to be unnecessary.

The time course of the reactions was monitored, following rapid mixing with a stopped-flow apparatus, by measuring the disappearance of OFe(IV)T2MPyP (λ416mm). The kinetic system was defined using 3-cyanophenol. With an excess of the phenol the reactions follow first-order kinetics for greater than 3 half-lives. The measured pseudo first-order rate constants (k_{obs}) show a linear dependence on the phenol concentration (Figure 4). By contrast, for a constant excess of 3-cyanophenol, the value of k_{obs} was found to be independent of the initial concentration of OFe(IV)T2MPyP. These data

Table 2. Second-order rate constants for oxidation of phenols by OFe(IV)T2MPyP in aqueous phosphate buffer, pH 7.7 at 30°C; μ, 0.2 M^{-1}

X–C$_6$H$_4$OH	pKa	$k_2/10^4$ M^{-1}s^{-1}
3-CN	8.6	0.424
3-F	9.2	1.57
H	10.0	1.78 (1.71)[1]
4-F	9.9	1.92 (1.98)[1]
4-Cl	9.4	2.44 (2.40)[1]
4-Me	10.3	4.05 (4.04)[1]
4-OMe	10.1	19.6 (21.6)[1]

[1] OFe(IV)P from Fe(III)P : ^{t}BuO$_2$H ratio 1 : 1

show that the 3-cyanophenol oxidations follow the simple second-order rate equation; d[OFe(IV)P]/dt = k_2[OFe(IV)P][ArOH].

Linear k_{obs} *versus* [ArOH] plots equivalent to Figure 4 were used to obtain second-order rate constants for six other phenols (Table 2). For five of the phenols the kinetics were also investigated using a 1:1 molar ratio of ^{t}BuO$_2$H : OFe(IV)T2MPyP to generate the oxidant, *i.e.* using an excess of the hydroperoxide. The second-order rate constants obtained from these experiments are within experimental error identical to those where a 1:2 ratio was employed, suggesting that recycling of the excess ^{t}BuO$_2$H under the experimental conditions is unimportant. This conclusion is supported by an examination of the relevant rate constants. For although no second-order rate constant for the reaction of ^{t}BuO$_2$H with Fe(III)T2MPyP at pH 7.7 has been reported, a value of 59 M^{-1}s^{-1} can be calculated for the equivalent reaction with the 4-N-methylpyridyl isomer.[16] Comparison of this value with the measured second-order rate constants from the present study suggests that the rate of regeneration of OFe(IV)T2MPyP from Fe(III)T2MPyP and ^{t}BuO$_2$H under the kinetic conditions would be insignificant by comparison with the rates of reduction of OFe(IV)T2MPyP by the phenols (Figure 5).

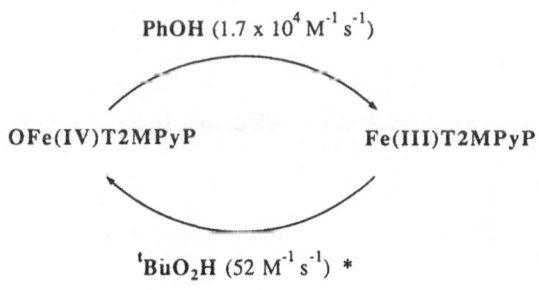

PhOH (1.7 x 10^4 M^{-1} s^{-1})

OFe(IV)T2MPyP   Fe(III)T2MPyP

^{t}BuO$_2$H (52 M^{-1} s^{-1}) *

pH 7.7 , μ = 0.2M ,   [PhOH] = (2.2 - 9.9) x 10^{-5}M

[Fe(III)T2MPyp] = [^{t}BuO$_2$H] = 2.2 x 10^{-6}M

* Value for Fe(III)T4MPyP

Figure 5. Porphyrin cycling in the reaction of OFe(IV)T2MPyP with phenol in aqueous buffer at pH 7.7

EPR Studies

A stopped-flow EPR study was carried out out on the reaction of OFe(IV)T2MPyP with water soluble phenol Trolox C and this gave the phenoxyl radical spectrum in Figure 6. Trolox C was chosen as a substrate for the EPR studies because it is water soluble and its highly substituted phenoxyl radical is relatively stable towards bimolecular radical-radical destruction processes. However, the lifetime of the Trolox C phenoxyl radical was surprisingly short under the reaction conditions (Figure 7). From this we infer that the oxidation of Trolox C follows reactions (5) and (6) and that the self-reaction of the phenoxyl radical [reaction (7)] is unimportant. It is noteworthy that if this conclusion also applies to the oxidation of the phenols in the kinetic study above, the second-order rate constants in Table 2 should all be reduced by a factor of 2.

$$OFe(IV)P \ + \ Trol\text{-}OH \ \longrightarrow \ Fe(III)P \ + \ Trol\text{-}O^{\cdot} \qquad (5)$$
$$OFe(IV)P \ + \ Trol\text{-}O^{\cdot} \ \longrightarrow \ Fe(III)P \ + \ Product \qquad (6)$$
$$2 \ Trol\text{-}O^{\cdot} \ \longrightarrow \ Products \qquad (7)$$

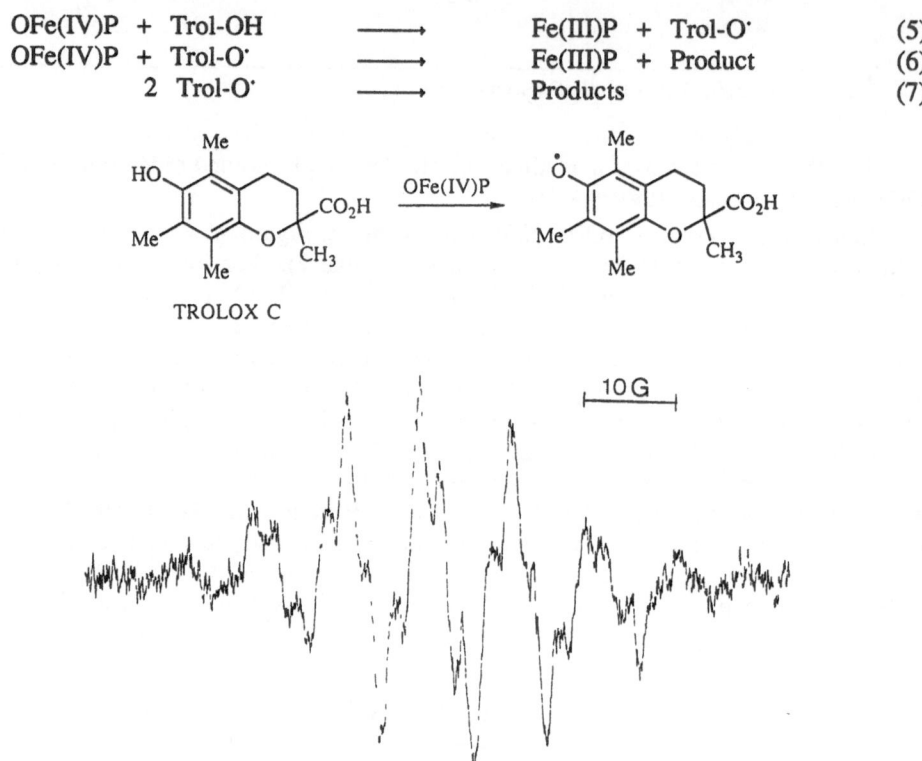

Figure 6. The EPR spectrum of the phenoxyl radical from Trolox C and OFe(IV)T2MPyP at pH 7.7

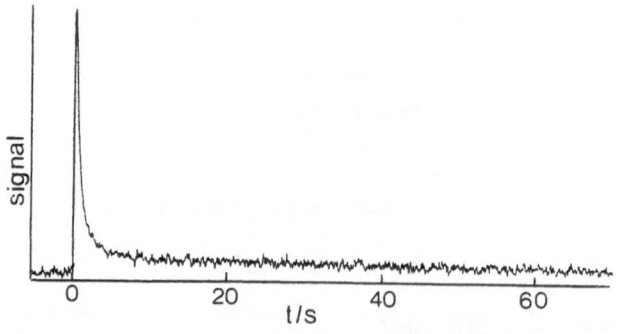

Figure 7. The time dependence of the Trolox C radical from stopped-flow mixing of Trolox C with OFe(IV)T2MPyP at pH 7.7

The Mechanism of Phenol Oxidation by OFe(IV)T2MPyP

Figure 8 identifies three alternative mechanisms for the one-electron oxidation of phenols to phenoxyl radicals. Of these, path (c) the one-electron oxidation of the phenolate ion can be eliminated since the substrates were selected to be predominantly unionised at the pH of the study. Furthermore, the second-order rate constants for each substrate has been corrected for the small proportion of the phenolate present.

To determine whether path (a) or path (b) correctly describes the oxidations, we carried out Hammett and modified Hammett analyses of the rate data (Table 3, Figures 9 and 10). Correlation of the second-order rate constants using a single parameter Hammett equation gives the best linear fit to σ^+. The correlation, however, is markedly improved by the use of the modified dual parameter Hammett equation proposed by Dust and Arnold.[20]

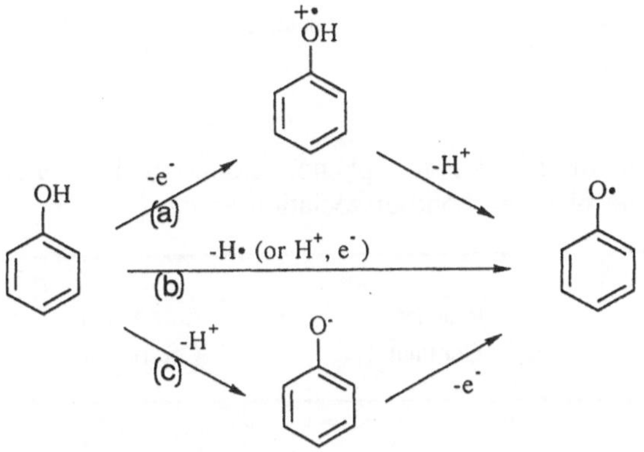

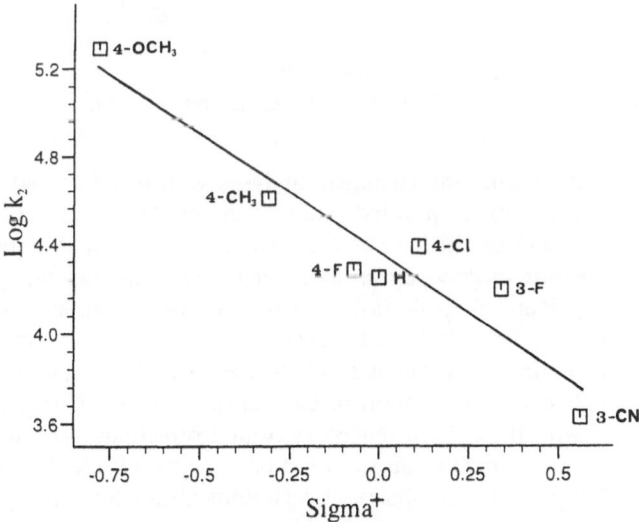

Figure 8. Possible one-electron oxidation mechanisms of phenol to the phenoxyl radical

Figure 9. Correlation of log k_2 *versus* σ^+ for phenol oxidations by OFe(IV)T2MPyP

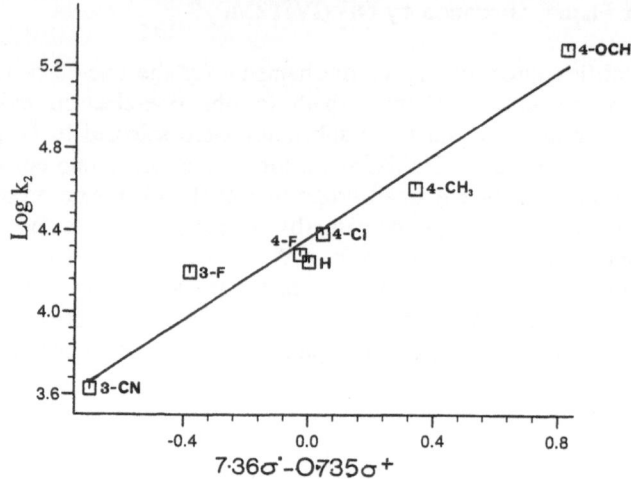

Figure 10. Correlation of log k_2 *versus* dual parameter (σ^+, $\sigma^{\cdot}$) for phenol oxidations by OFe(IV)T2MPyP

Table 3. Correlation of log k_2 for phenol oxidation by OFe(IV)T2MPyP with substituent constants and phenolic bond dissociation energy.

Substituent Constant or BDE	Reaction Constant	Correlation Coefficient	
σ	- 1.52	0.881	1
$\sigma+$	- 1.10	0.956	2
$\sigma^{\cdot}$	18.8	0.921	3
σ^+, $\sigma^{\cdot}$	- 0.735 ($\rho+$) 7.36 ($\rho^{\cdot}$)	0.972	3
$\Delta\Delta H_f$	–	0.989	4

[1] σ values from reference 18; [2] σ^+ values from reference 19; [3] $\sigma^{\cdot}$ values from reference 20; [4] $\Delta\Delta H_f = \Delta H_f ArO^{\cdot} - \Delta H_f ArOH$, reference 17.

The negative ρ values from the Hammett analyses with σ and σ^+ and the improved correlation with the dual parameter plot indicate that the oxidations are assisted by both electron-donating and radical stabilising substituents. At first sight this might be taken to imply that the oxidations proceed by an electron transfer mechanism generating the phenol radical cation [Figure 8, path (a)]. However the Hammett correlations of H-atom abstractions by oxy-radicals have been commonly observed to show significant negative ρ values and to correlate better with σ^+ than σ. This has been attributed to the development of a partial charge separation in the transition state. Comparison of the ρ values from this study with those from related H-atom abstractions and electron transfer oxidations shows that they resemble the former more closely (Table 4). From this we infer that OFe(IV)T2MPyP oxidises phenols by H-atom abstraction. In agreement with this conclusion the log k_2 values show an excellent correlation with the phenolic bond dissociation energies ($\Delta\Delta H_f$) reported by Brewster *et al.*[17]

Table 4. Hammett σ^+ correlations for radical H-atom abstractions and for electron transfer oxidations.

Substrate	Oxidant	Solvent (Temp/°C)	Reaction Constant
H-atom abstraction			
ArOH	$^tBuO \cdot$	PhH(22)	- 0.90 [1]
ArCH$_3$	tBuO$\cdot$	CCl$_4$(40)	- 0.68 [2]
ArCH$_3$	RO$_2 \cdot$	PhCl(30)	- 0.6 [3]
Electron transfer			
ArNMe$_2$	Pb(OAc)$_4$	Ac$_2$O/CHCl$_3$ (38.5)	- 2.4 [4]
ArOH	C$_4$F$_9$O$_2 \cdot$	MeOH(22)	- 2.3 [5]
ArOH	CCl$_3$O$_2 \cdot$	MeOH(22)	- 3.3 [5]

[1] Reference 21; [2] Reference 22; [3] Reference 23; [4] Reference 24 correlation with σ; [5] Reference 25.

Table 5. Dual parameter $(\sigma^+, \sigma \cdot)$ analyses of H-atom abstractions

Reaction	ρ	$\rho \cdot$	$\rho \cdot / \rho$	Correlation Coefficient
ArCH$_3$ with NBS	-1.43	0.672	0.47	0.997 [1]
ArCH$_3$ with $^tBu \cdot$	0.462	1.30	2.81	0.978 [1]
ArCH$_3$ with Fe(III)TPP/PhIO	-1.03	18.7	18.2	0.983 [2]
ArOH with OFe(IV)T2MPyP	-0.735	7.36	10.0	0.972 [3]

[1] Reference 20; [2] Reference 26; [3] This study

The dual nature of some radical reactions has been noted by Dust and Arnold[20] who have attempted to quantify the charge and radical contributions of the reactions. The results from such an analysis of phenol oxidations in the present study are compared with some other radical oxidations in Table 5.

A Comparison of the Oxidation of Phenols by HRP Compound II and OFe(IV)T2MPyP

A comparable analysis of the rate data of Dunford and Adeniran[5] for the oxidation of phenols by HRP compound II at pH 7.0 and 7.6 to that described above reveals some significant similarities and differences between the enzymatic and model oxidations (Table 6). Thus the best single parameter correlation with both systems is with σ^+, however, those with $\sigma \cdot$ and with $\Delta \Delta H_f$ are dramatically worse for compound II than for OFe(IV)T2MPyP. Furthermore, the size of ρ from the σ^+ correlation of the enzymatic oxidation is significantly greater than the value from the present study. A comparison of this ρ value with the values in Table 4 shows that it is similar in magnitude to those from electron transfer oxidations.

Table 6. Correlation of log rate constants for phenol oxidations by HRP compound II with substituent constants and bond dissociation energies.

Substituent Constant or BDE	Reaction Constant	Correlation Coefficient[1]
σ	- 4.70	- 0.930
σ^+	- 2.70	- 0.950
$\sigma^{\cdot}$	(34.2)	0.489
$\Delta\Delta H_f$	–	0.841

[1] Data in reference 5 has been reanalysed using σ values from reference 18, σ^+ values from reference 19, $\sigma^{\cdot}$ values from reference 20 and $\Delta\Delta H_f$ values from reference 17.

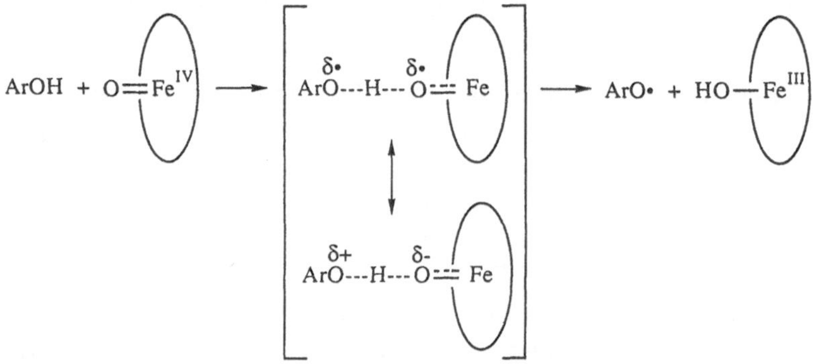

Figure 11. Proposed mechanism for reaction of phenols with OFe(IV)T2MPyP in aqueous buffer at pH 7.7

We conclude that the HRP compound II oxidation of phenols involves a rate determining electron transfer from the phenol. The known steric constraints on access to the iron porphyrin[1] suggest that this oxidation occurs at the edge of the heme ring. The evidence does not support a hydrogen atom abstraction or a simultaneous transfer of an electron and a proton. In contrast, OFe(IV)T2MPyP, without any steric restrictions on reaction at the ferryl oxygen, oxidises phenols by H-atom abstraction. We propose that in this oxidation the ferryl oxygen acts as an electrophilic radical and that the transition state for H-atom abstraction involves significant charge separation (Figure 11).

REFERENCES

1. P.R. Ortiz de Montellano, Control of the catalytic activity of prosthetic heme by the structure of hemoproteins, *Acc. Chem. Res.* 20: 289 (1987)
2. T.L. Poulos, Heme enzyme crystal structures, *Adv. Inorg. Biochem.* 7: 1 (1987)
3. H. Booth and B.C. Saunders, Studies in peroxidase action, Part X. The oxidation of phenols. *J. Chem. Soc.* 940 (1956)
4. D. Job and H.B. Dunford, Substituent effect on the oxidation of phenols and

5. H.B. Dunford and A.J. Adeniran, Hammett $\rho\sigma$ correlation for reactions of horseradish peroxidase compound II with phenols, *Arch. Biochem. Biophys.* 251: 536 (1986)

6. P.R. Ortiz de Montellano, Y.S. Choe, G. DePillis, and C.E. Catalano, Structure-mechanism relationships in hemoproteins. Oxygenations catalysed by chloroperoxidase and horseradish peroxidase, *J. Biol. Chem.* 262: 11641 (1987)

7. J. Sakurada, R. Sekiguchi, K. Sato, and T. Hosoya, Kinetic and molecular orbital studies on the rate of oxidation of monosubstituted phenols and anilines by horseradish peroxidase compound II, *Biochem.* 29: 4093 (1990)

8. B. Meunier, *N*- and *O*-Demethylations catalysed by peroxidases, in Peroxidases in Chemistry and Biology, J. Everse, K.E. Everse, and M.B. Grisham, eds. C.R.C. Press, Boca Raton, 1991, Vol II, 201

9. P. Jones, D. Mantle and I. Wilson, Peroxidase-like activities of ion(III)-porphyrins: kinetics of the reduction of a peroxidatically active derivative of deuteroferriheme by phenols, *J. Inorg. Biochem.* 17: 293 (1982)

10. T.G. Traylor, W.A. Lee, and D.V. Stynes, Model compound studies related to peroxidases II. The chemical reactivity of a high valent protohemin compound *Tetrahedron*, 40: 553 (1984)

11. S.J. Bell, P.R. Cooke, P. Inchley, D.R. Leonard, J.R. Lindsay Smith, and A. Robbins, Oxoiron(IV) porphyrins derived from charged iron(III) tetraarylporphyrins and chemical oxidants in aqueous and methanolic solution, *J. Chem. Soc. Perkin Trans.* 2: 549 (1991)

12. S.-M. Chen and Y.D. Su, Electrochemical and spectral characterisation of stable iron(IV) 5,10,15,20 - (*N*-methyl-4-pyridyl) porphyrin in aqueous solution at room temperature, *J. Chem. Soc. Chem. Comm.* 491 (1990)

13. K.R. Rodgers, R.A. Reed, Y.O. Su, and T.G. Spiro, Resonance Raman and magnetic resonance spectroscopic characterisation of the Fe(I), Fe(II), Fe(III), and Fe(IV) oxidation states of Fe(2-TMPyP)$^{n+}$ aq, *Inorg. Chem.* 31: 2688 (1992)

14. P.E. Ellis and J.E. Lyons, Selective air oxidation of light alkanes catalysed by activated metalloporphyrins - the search for a suprabiotic system, *Coord. Chem. Rev.* 105: 181 (1990)

15 H. Hennig, D. Rehorek, R. Stick, and L. Weber, Photocatalysis induced by light-sensitive coordination compounds, *Pure Appl. Chem* 62: 1489 (1990)

16. J.R. Lindsay Smith and R.J. Lower, The mechanism of the reaction between t-butyl hydroperoxide and 5,10,15,20 tetra (*N*-methyl-4-pyridyl)-porphyrinatoiron(III) pentachlorid in aqueous solution*J. Chem. Soc. Perkin Trans.* 2, 31: (1991)

17. M.E. Brewster, D.R. Doerge, M.-J. Huang, J.J. Kaminski, E. Pop, and N. Bodor, Application of semiempirical molecular orbital techniques to the study of peroxidase-mediated oxidation of phenols, anilines, sulphides and thiobenzamides, *Tetrahedron.* 47: 7525 (1991)

18. D.H. McDaniel and H.C. Brown, An extended table of Hammett substituent constants based on the ionisation of substituted benzoic acids, *J. Org. Chem.* 23: 420 (1958)

19. H.C. Brown and Y. Okamoto, Electrophilic substituent constants, *J. Amer. Chem. Soc.* 80: 4979 (1958)

20. J.M. Dust and D.R. Arnold, Substituent effects on benzyl radical ESR-hyperfine coupling constants. The $\sigma^{\cdot}_{\alpha}$ scale based upon spin delocalisation, *J. Amer. Chem. Soc.* 105: 1221 (1983)

21. P.K. Das, M.V. Encinas, S. Steeken, and J.C. Scaiano, Reaction of tert-butoxy radicals with phenols. Comparison with reactions of carbonyl triplets, *J. Amer. Chem. Soc.* 103: 4162 (1981)

22. B.R. Kennedy and K.U. Ingold, Reactions of alkoxy radicals. 1. Hydrogen atom abstraction from substituted toluenes, *Can. J. Chem.* 44: 2381 (1966).

23. J.A. Howard, K.U. Ingold, and M. Symonds, Absolute rate constants for hydrocarbon oxidation. VIII. The reactions of cumylperoxy radicals, *Can J. Chem.* 46: 1017 (1968)

24. G. Galliani and B. Rindone, Kinetics of the oxidation of some *meta-* and *para-* substituted dimethylanilines with lead tetra-acetate, *J. Chem. Soc. Perkin Trans. 2* . 1803 (1976)

25. G.S. Nahor, P. Neta, and Z.B. Alfassi, Perfluorobutylperoxyl radical as an oxidant in various solvents, *J. Phys. Chem.* 95: 4419 (1991)

26. P. Inchley, J.R. Lindsay Smith and R.J. Lower, Model systems for cytochrome P450 dependent mono-oxygenuses, Part 6. The hydroxylation of saturated C-H bonds with tetraphenylporphyrinatorion(III) chloride and iodosylbenzene, *New J. Chem.* 13: 669 (1989)

NON-PORPHYRIN IRON COMPLEX-CATALYZED
EPOXIDATION OF OLEFINS

Joan Selverstone Valentine,[1] Wonwoo Nam,[2] and Raymond Y. N. Ho[1]

[1]Department of Chemistry and Biochemistry, University of California, Los
Angeles, Los Angeles, California 90024, USA
[2]Department of Science, Hong Ik University, Seoul 121-791, Korea

INTRODUCTION

Elucidation of the mechanisms of oxygen atom transfer reactions catalyzed by monooxygenase metalloenzymes has been a major challenge in the field of oxidation chemistry.[1,2] Such enzymes catalyze aliphatic hydroxylation, olefin epoxidation, aromatic hydroxylation, heteroatom oxidation, and heteroatom dealkylation, with dioxygen as the oxidant. Intensive study of cytochrome P-450 enzymes,[1] which contain heme at their active sites, and of metalloporphyrin model compounds has resulted in a proposed catalytic mechanism which involves a high-valent iron oxo porphyrin complex as a reactive intermediate. By contrast, the active sites of non-heme iron-containing monooxygenase enzymes and their mechanisms are not as fully characterized. In recent years, the chemistry of these latter enzymes has attracted the attention of both bioinorganic and biological chemists who have consequently been characterizing their active sites and attempting to understand the mechanisms of their oxygen atom transfer reactions by studying the enzymes directly as well as model systems.[2,3]

One of the best characterized non-heme iron-containing monooxygenases is methane monooxygenase, which contains a μ-oxo diiron cluster at its active site.[4] Methane monooxygenase catalyzes the oxidation of methane to methanol, and it has been suggested that high-valent iron oxo species are intermediates in its reactions. Examples of other non-heme iron-containing monooxygenases are *Pseudomonas oleovorans* monooxygenase,[5] phenylalanine hydroxylase,[6] squalene epoxidase,[7] and tyrosine hydroxylase.[8] Although iron bleomycin is not an enzyme, it shows a similar reactivity to the monooxygenase enzymes, i.e. catalysis of olefin epoxidation, aromatic hydroxylation, and heteroatom oxidation.[9] Most of the non-heme iron monooxygenase enzymes as well as iron bleomycin have been proposed to involve $Fe^V=O$ or $Fe^{IV}=O$ species as sources of the reactive oxygen atom, although the direct characterization of such high-valent iron oxo intermediates has not yet been achieved for non-heme iron-containing enzymes.

A number of model systems for the non-heme iron-containing monooxygenase enzymes have been developed that give selective oxidation of alkanes using either molecular oxygen or hydrogen peroxide as the oxidant.[10,11] However, the same catalysts typically do not epoxidize olefins but instead give allylic oxidation products. By contrast, synthetic analogues of iron bleomycin have been found to be good catalysts for olefin

The Activation of Dioxygen and Homogeneous Catalytic Oxidation,
Edited by D.H.R. Barton *et al.*, Plenum Press, New York, 1993

epoxidation by hydrogen peroxide giving, for example, predominantly *cis*-stilbene oxide in *cis*-stilbene epoxidation reactions.[12,13] In all of these model systems, high-valent iron oxo species have been proposed as active intermediates by analogy to the metalloporphyrin systems; however, the active oxidizing species in these non-porphyrin iron systems have not yet been definitively characterized.

We have been seeking non-radical types of oxygenation reactions catalyzed by non-porphyrin complexes of iron that may be analogous to oxygen atom transfer reactions occurring in reactions of non-heme iron-containing monooxygenase enzymes or iron bleomycin. Recently, we found that iron complexes of cyclam (1,4,8,11-tetraazacyclotetradecane) and related ligands catalyze the epoxidation of olefins by 30% aqueous hydrogen peroxide in acetonitrile or methanol solutions (eq 1).[14] Epoxides were the predominant products with negligible amounts of allylic oxidation products formed

$$\text{(olefin)} + H_2O_2 \text{ (30\% aqueous)} \xrightarrow[\text{CH}_3\text{OH or CH}_3\text{CN}]{\text{Fe(cyclam)(OTf)}_2} \text{(epoxide)} \qquad (1)$$

under ambient conditions, indicating that the mechanism does not resemble those typical of free radical reactions. The epoxidation of olefins was found to be stereospecific as evidenced by the formation of *cis*-stilbene oxide from *cis*-stilbene and *trans*-stilbene oxide from *trans*-stilbene. We attempted to substitute alkyl hydroperoxides, i.e. *t*-butyl hydroperoxide and ethyl hydroperoxide, for hydrogen peroxide in these reactions, but no epoxide products were found under these conditions. A mechanism was proposed in which complexation of hydrogen peroxide by an iron cyclam complex generates a reactive iron-hydroperoxide species as an intermediate. Molecular modeling of such a species using computer graphics suggested that the preferred conformation of the cyclam ligand would present the HOO⁻ ligand with an axial N-H bond that is well suited for hydrogen bonding.[15] It has been shown that such intramolecular hydrogen-bonding often plays an important role in activating oxygen atom transfer reactions.[16] We suggested that the HOO⁻ complex thus stabilized might react directly with an olefinic substrate prior to O-O bond cleavage. An alternative mechanism is one in which hydrogen bonding activates the bound HOO⁻ ligand and facilitates HO⁻ loss (eq 2). In that case, a high-valent iron oxo species would be expected to be the reactive oxygenating intermediate.

$$
\begin{array}{ccc}
\begin{array}{c}
\text{H} \\ | \\ \text{O} \\ \text{O} \quad \text{H} \\ | \quad | \\ \text{N—Fe}^{n+}\text{—N}
\end{array}
& \longrightarrow &
\begin{array}{c}
\text{H} \\ | \\ \text{O} \\ \text{O}^+ \quad \text{H} \\ | \quad \\ \text{N—Fe}^{n+}\text{—N}^-
\end{array}
& \qquad (2)
\end{array}
$$

RESULTS AND DISCUSSION

We reported previously our observation of iron cyclam-catalyzed epoxidation of olefins by hydrogen peroxide.[14] We have extended this work by examining a variety of ferrous complexes of macrocyclic ligands closely related to the cyclam ligand, as well as iron(II) triflate and iron complexes of BPBH₂ and EDTA ligands, as potential catalysts. In addition to hydrogen peroxide, single oxygen atom donors such as *m*-chloroperbenzoic acid (MCPBA) and iodosylbenzene (PhIO) were investigated in order to obtain information concerning the nature of the oxygenating intermediates. We also studied dioxygen plus an aldehyde as an oxidant in this reaction. Results using the different oxidants for reaction of cyclohexene in the presence of Fe^II(cyclam)(OTf)₂ are given in

Table 1. Cyclohexene oxidations by various oxidants catalyzed by Fe^{II}(cyclam)(OTf)$_2$[a]

oxidants	products (yields, mmol)		
	(epoxide)	OH (cyclohexenol)	O (cyclohexenone)
H$_2$O$_2$	0.17	0.01	0.003
t-BuOOH[b]	trace	0.06	0.09
EtOOH[b]	0	0.07	0.01
PhIO	0.14	0.003	0.007
MCPBA[c]	0.20	trace	0.002
O$_2$/aldehyde[d]	0.6	0	0.04

[a]Except as noted below, reactions were carried out in CH$_3$CN (5 mL) containing Fe^{II}(cyclam)(OTf)$_2$ (0.02 mmol), cyclohexene (1 mmol), and decane (internal standard, 0.01 mmol). Oxidants (0.4 mmol) were added to the solution and the solution was stirred at room temperature for 30 min prior to analysis of the products by GC/MS.
[b]Reactions were run for 120 min with 1 mmol of oxidants.
[c]This reaction was carried out in a solvent mixture of CH$_3$CN (2 mL) and CH$_3$OH (3 mL) containing 3 mmol of cyclohexene for 60 min at -20 °C.
[d]Dioxygen was bubbled through the reaction solution containing isovaleraldehyde (1 mL, 9.3 mmol) for 120 min.

Table 2. Epoxidation of cyclohexene by various oxidants catalyzed by iron complexes and an iron salt.[a]

catalyst	Yield (mmol) of cyclohexene oxide			
	H_2O_2	PhIO	MCPBA[b]	O_2/aldehyde[c]
Fe(OTf)$_2$	0	0.17	0	0 (0)
Fe(cyclam)(OTf)$_2$	0.17	0.14	0.2	0.6 (0.04)
Fe(TIM)(PF$_6$)$_2$	0.030	0.20[d]	0.06	1.5 (0.17)
Fe(HMCD)(OTf)$_2$	0.076	0.15	0.07	0.7 (0.09)
Fe(HMCT)(PF$_6$)$_2$	0.02	0.17	0.03	0.6 (0.10)
Fe(N-Me-cyclam)(OTf)$_2$	0	0.018	0.014	0.5 (0.04)
Fe(BPBH$_2$)(OTf)$_2$	0	0.047[d]	0.01	0.6 (0.03)
NaFe(EDTA)	0	0	0	trace (0.01)

[a]Reactions were run in a solution containing cyclohexene (1 mmol) and the iron catalyst (0.02 mmol) in 5 mL of acetonitrile. Oxidant (0.4 mmol) was added to the reaction solution, and the solution was stirred for 30 min. The yield of products was analyzed by GC/MS.
[b]These reactions were carried out in a solvent mixture of CH$_3$CN (2 mL) and CH$_3$OH (3 mL) containing 3 mmol of cyclohexene for 60 min at -20 °C.
[c]Numbers in parentheses are the amounts of 2-cyclohexen-1-one formed.
[d]Reaction was run for 60 min.

cyclam

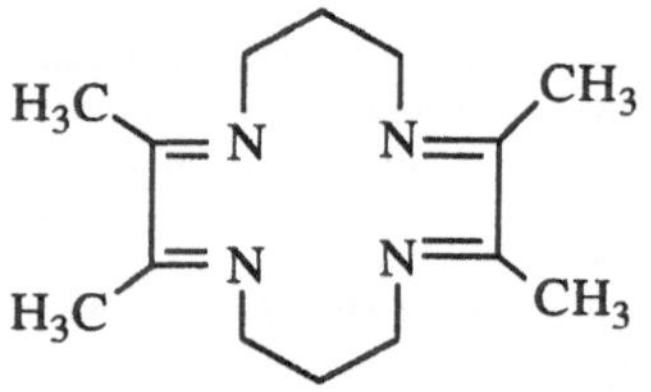

TIM

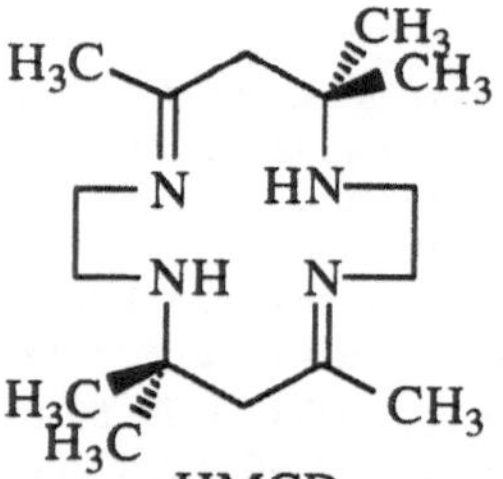

HMCD

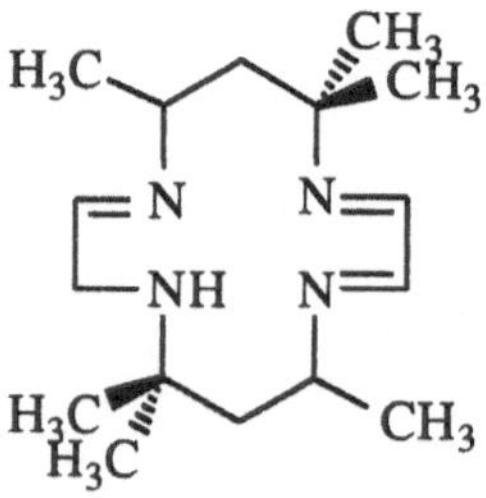

HMCT

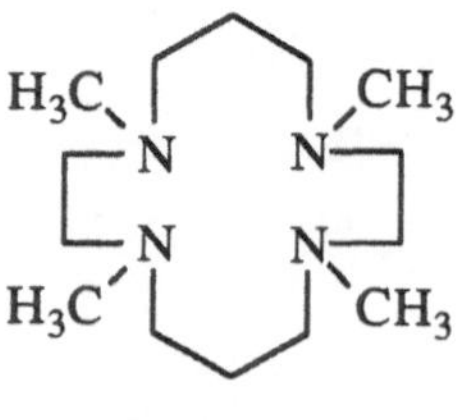

N-Me-cyclam

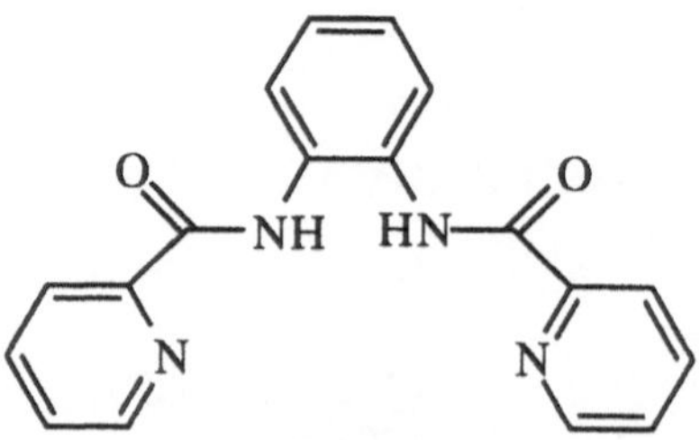

BPBH$_2$

Table 1. The reaction with MCPBA was carried out at -20 ^{0}C since the direct reaction of MCPBA with cyclohexene is very slow at that temperature. We found that with all the oxidants, except for t-BuOOH and EtOOH, the major product of the reactions was epoxide with only small amounts of allylic oxidation products formed, and we conclude that the oxygenating species generated in these reactions are not typical free radicals.

Table 2 gives results for similar reactions using a number of other iron complexes as well as Fe(OTf)$_2$ as catalysts. Of the species tested, we found Fe^{II}(cyclam)$^{2+}$ to be the best catalyst for cyclohexene epoxidation by hydrogen peroxide. Iron complexes of cyclam derivatives such as TIM, HMCD, and HMCT were poor-to-modest catalysts for the hydrogen peroxide reactions. Ferrous triflate, Fe(OTf)$_2$, and the other iron complexes tested were inactive. In those cases, disproportionation of hydrogen peroxide and/or ligand degradation were observed. In contrast to the hydrogen peroxide reactions, the reactivities of Fe(OTf)$_2$ and all of the ferrous cyclam derivatives studied as catalysts for iodosylbenzene reactions were comparable to that of ferrous cyclam itself. In the case of MCPBA as oxidant, in reactions carried out at -20 ^{o}C, only Fe^{II}(cyclam)$^{2+}$ itself gave a good yield of cyclohexene oxide. In the reactions of O$_2$ plus aldehyde, all of the iron complexes except iron triflate salt and Fe^{III}(EDTA) gave cyclohexene oxide as the predominant product, with a small amount of 2-cyclohexen-1-one and no 2-cyclohexen-1-ol. It is of interest to note that the ferrous complex of the N-Me-cyclam ligand was inactive as a catalyst in the reactions of H$_2$O$_2$, PhIO, and MCPBA at low temperature, but gave a high yield in the O$_2$ plus aldehyde reaction. This result suggests that the mechanism of oxygen atom transfer reaction by O$_2$ plus aldehyde is different from those of the other reactions, as discussed below.

The relative reactivities of *para*-substituted styrenes to styrene were determined in iron cyclam complex-catalyzed epoxidations using various oxidants such as H$_2$O$_2$, PhIO, and MCPBA in CH$_3$CN solution. The relative rates for various substituents, X, are listed in Table 3, and a Hammett plot of log k_{rel} vs σ^+ is shown in Figure 1. The correlation with σ^+ was good with ρ = -1.1, -1.6, and -1.6 for the H$_2$O$_2$, PhIO, and MCPBA reactions, respectively. The MCPBA reaction was carried out in CH$_3$CN at room temperature, in CH$_3$CN at -15 ^{o}C, and in a solvent mixture of CH$_3$OH and CH$_3$CN at -15 ^{o}C. All of these reactions gave the same ρ value. Also, the ρ value obtained in styrene epoxidations with MCPBA alone, with no catalyst present, was determined to be -1.0 in CH$_3$CN, similar to results in benzene (ρ= -1.04) and in chlorobenzene (ρ=-0.93) reported previously.[17] The change in ρ that we observe upon addition of the catalyst indicates that the epoxidations with MCPBA in the presence of Fe^{II}(cyclam)$^{2+}$ were iron complex-catalyzed reactions.

Hammett correlations for styrene epoxidations have been reported for a variety of metal-catalyzed oxidations. In iron porphyrin systems, ρ values between -0.83 and -0.94 have been reported for PhIO reactions.[18-20] Groves and Watanabe[21] reported a ρ value of -1.9 for styrene epoxidations where the intermediate is presumed to be the iron(IV) oxo porphyrin cation radical, generated by reaction of Fe(TMP)Cl with MCPBA at -50 ^{o}C. It is noteworthy that ρ values for reactions of iodosylbenzene and hypochlorite catalyzed by Fe(TDCPP)Cl were found by Collman et al. not to be identical.[20] The ρ values for PFIB (pentafluoroiodosylbenzene) and PhIO were -0.86 and -0.91, respectively, whereas the ρ value for LiOCl plus a phase transfer catalyst was -0.57.[20] Those authors suggested that either different oxidants were formed in the iodosylbenzene and hypochlorite reactions or differences in the reaction conditions were responsible for the different ρ values. In our reactions, the smaller ρ value for the H$_2$O$_2$ reaction relative to those for the PhIO and MCPBA reactions suggests that different intermediates are responsible for the oxygenations.

In order to characterize the nature of the oxygenating species formed in the reactions of H$_2$O$_2$, PhIO, MCPBA, and dioxygen plus aldehyde, we studied stereoselectivity in *cis*-stilbene epoxidation, regioselectivity in (+)-limonene and norbornene epoxidations, and intermolecular competitive reactions between cyclohexene

Table 3. Relative reactivities of *para*-substituted styrenes in epoxidation by various oxidants in the presence of $Fe^{II}(cyclam)^{2+}$ [a]

Oxidants	CF$_3$	Cl	H	CH$_3$	OCH$_3$
			p-X-styrene		
H$_2$O$_2$	0.12	0.60	1.0	1.6	5.2
PhIO	0.10	0.42	1.0	2.6	14
MCPBA	0.14	0.62	1.0	3.0	23
MCPBA alone	0.18	0.62	1.0	1.7	5.7

[a]In a typical reaction, a stock solution containing styrene (0.45 mmol), substituted styrene (0.45 mmol), chlorobenzene (0.45 mmol, internal standard), and iron cyclam complex (0.03 mmol) was prepared. The solution was divided into three parts of 5 mL each. Oxidants (0.12 mmol) were added to the reaction solution, and the reaction mixture was stirred for 15 min. The amounts of olefins before and after reactions were determined by GC/MS. The relative reactivities were determined using eq 3,

$$k_x/k_y = \log(X_f/X_i)/\log(Y_f/Y_i) \qquad (3)$$

where X_i and X_f are the initial and final concentrations of substituted styrenes and Y_i and Y_f are the initial and final concentrations of styrene.

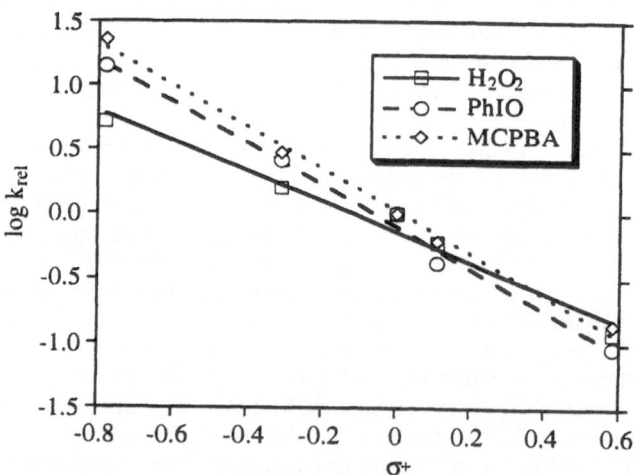

Figure 1. Hammett plot for relative rates of disappearance of *para*-substituted styrenes in reactions of H$_2$O$_2$, PhIO, and MCPBA in the presence of $Fe^{II}(cyclam)^{2+}$.

and 1-hexene and between cyclohexene and norbornene. The results of these experiments are listed in Table 4. In *cis*-stilbene epoxidations, *cis*-stilbene was oxidized to *cis*-stilbene oxide predominantly with only small amounts of *trans*-stilbene oxide and benzaldehyde formed in all of the reactions, except for the O_2 plus aldehyde reaction. In that latter case, the major product was *trans*-stilbene oxide and only a small amount of *cis*-stilbene oxide and benzaldehyde formation were detected. The ratio of *cis*-stilbene oxide to *trans*-stilbene oxide products was 0.26 in the O_2 plus aldehyde reaction, whereas the ratio obtained in the other reactions was greater than 13. This result clearly shows that a different active oxidant was generated in the dioxygen plus aldehyde reaction. The time course for the formation of *cis*-stilbene oxide, *trans*-stilbene oxide, and benzaldehyde in the O_2 plus aldehyde reaction is given in Figure 2. An induction period was observed before product formation. The ratio of *cis*-stilbene oxide to *trans*-stilbene oxide formed was fairly constant during the course of the reaction. Control reactions demonstrated that *cis*-stilbene oxide was stable and that *cis*-stilbene was not isomerized to *trans*-stilbene under the reaction conditions, indicating that the *trans*-stilbene oxide was derived from the oxidation of *cis*-stilbene.

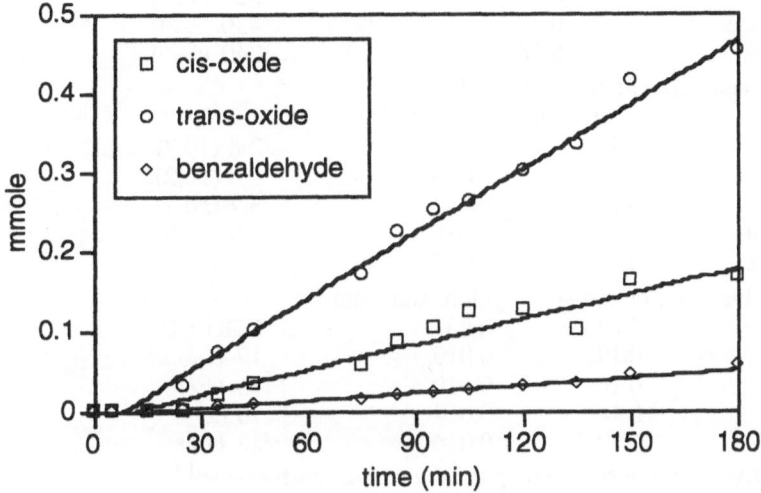

Figure 2. Time course for the formation of *cis*-stilbene oxide, *trans*-stilbene oxide, and benzaldehyde in the oxidation of *cis*-stilbene by molecular oxygen in the presence of an aldehyde and an iron cyclam complex.

Traylor *et al.* studied epoxidation of norbornene in iron porphyrin complex catalyzed-reactions using various oxidants in order to verify the formation of iron(IV) oxo porphyrin cation radical intermediates.[22] They obtained the same ratio of exo- to endo-epoxynorbornane in all of the reactions of the oxidants H_2O_2, PFIB, MCPBA, and ROOH with iron porphyrin catalysts and reached to the conclusion that high-valent iron oxo species was formed via heterolytic O-O bond cleavage of H_2O_2, MCPBA, and ROOH. We have determined the ratio of exo- to endo-1,2-epoxynorbornane products in iron cyclam complex-catalyzed norbornene epoxidations by various oxidants. As shown in Table 4, H_2O_2 and PhIO reactions gave similar ratios of exo to endo isomers, whereas MCPBA reaction gave a lower exo to endo ratio. The ratio of exo to endo isomers

Table 4. Comparison of stereoselectivity, regioselectivity, and competitive reactivities in iron cyclam complex-catalyzed epoxidations by various oxidants.

Oxidant	Product yield (mmol)		Ratio[a]	Yield (%)[b]

1. *cis*-Stilbene epoxidation reaction[c]

Oxidant	*cis*-oxide	*trans*-oxide	*cis*-/*trans*-oxide	
H_2O_2	0.080	0.0043	19	33
PhIO	0.093	0.006	16	40
MCPBA	0.070	0.0052	13	31
MCPBA alone	0.18	0.003	61	51
O_2/Aldehyde	0.034	0.134	0.25	

2. Norbornene epoxidation reaction[d]

Oxidant	1,2-epoxynorbornane	exo-/endo-	Yield
H_2O_2	0.21	100 (±10)	53
PhIO	0.19	120 (±20)	48
MCPBA	0.19	60 (±10)	48
MCPBA alone	0.32	430 (±40)	80
O_2/Aldehyde	0.78	240 (±50)	

3. (+)-Limonene epoxidation reaction[d,e]

Oxidant	1,2-epoxide	8,9-epoxide	1.2-/8,9-	Yield
H_2O_2	0.11	0.040	2.8 (±0.2)	38
PhIO	0.12	0.060	2.0 (±0.2)	45
MCPBA	0.16	0.043	3.7 (±0.2)	51
MCPBA alone	0.32	0.036	8.9 (±0.5)	89
O_2/Aldehyde	0.56	0.087	6.4 (±0.3)	

4. Competitive reaction between cyclohexene and 1-hexene[d,f]

Oxidant	CHO	HO	CHO/HO	Yield
H_2O_2	0.19	0.010	19 (±2)	50
PhIO	0.16	0.010	16 (±2)	43
MCPBA	0.14	0.014	10 (±1)	39
MCPBA alone	0.32	0.014	22 (±1)	84

5. Competitive reaction between cyclohexene and norbornene[d,f]

Oxidant	CHO	NBO	NBO/CHO	Yield
H_2O_2	0.096	0.097	1.0 (±0.2)	48
PhIO	0.084	0.091	1.1 (±0.2)	44
MCPBA	0.082	0.098	1.2 (±0.2)	45
MCPBA alone	0.14	0.20	1.4 (±0.2)	85

[a]Numbers in parentheses are experimental error ranges.

[b]Based on oxidants added.

[c]Reactions were run in a solution containing cis-stilbene (1 mmol) and iron catalyst (0.02 mmol) in 5 mL of acetonitrile. Oxidant (0.4 mmol) was added to the reaction solution, and the solution was stirred for 30 min. The yield of products was analyzed by HPLC.

[d]Reaction procedures were the same as a except that 2 mmol of substrates were used and that GC/MS was used to analyze the yield of products.

[e]The ratio of 1,2-8,9- isomers was determined by combining GC/MS and NMR analyses.

[f]CHO, HO, and NBO stands for cyclohexene oxide, 1,2-epoxyhexane, and 1,2-epoxynorbornane, respectively.

obtained in the reaction of the iron cyclam complex and MCPBA clearly shows that the norbornene epoxidation by MCPBA was iron complex-catalyzed since the exo to endo ratio obtained in the epoxidation of norbornene by MCPBA alone is greater than 400, as reported by Traylor and Miksztal.[23]

In limonene epoxidation reactions, we found the major product to be the 1,2-epoxide derived from the oxidation of the more electron-rich double bond. However, considerably more 8,9-epoxide was observed in these reactions as compared to iron porphyrin systems. It is noteworthy that, in iron porphyrin complex-catalyzed limonene epoxidations, a large preference for the oxidation at the 1,2-position has been observed, e.g. the ratio of the 1,2- to the 8,9-oxides was 19 for the Fe(TPP)Cl-iodosylbenzene system.[24] Suslick and Cook used sterically hindered manganese porphyrin complexes to increase shape selectivity in the epoxidation of limonene.[25,26] Furthermore, Mansuy and coworkers obtained the same ratio of 1,2- to the 8,9-isomers in PhIO, H_2O_2, and O_2 reactions with manganese porphyrin complexes as catalysts, and they concluded that an identical intermediate, i.e. Mn(V)=O, was generated in all of the reactions studied.[27,28] In our reactions, the ratio of 1,2- to 8,9-isomers was different from donor to donor, indicating that different oxidizing intermediates were generated.

Competitions between cyclohexene and 1-hexene and between cyclohexene and norbornene as substrates for epoxidation have been carried out using H_2O_2, PhIO, and MCPBA in the presence of $Fe^{II}(cyclam)^{2+}$ as well as MCPBA alone for comparison. From the reactions between cyclohexene and 1-hexene, a lower preference for cyclohexene epoxidation was observed in the MCPBA reaction as compared to the H_2O_2 and PhIO reactions. The competitive reaction between cyclohexene and norbornene has been used in several laboratories to indicate the nature of the transition state in oxygen atom transfer reactions.[29] Based on those results, the values obtained from the H_2O_2, PhIO, and MCPBA reactions suggest that oxygen atom is transferred to the olefin through a three-membered ring transition state.

ACTIVE OXIDIZING INTERMEDIATES

Possible active oxidants which can be generated in the H_2O_2, PhIO, MCPBA, and O_2 plus aldehyde reactions are shown in Figure 3. Complex **1** is a high-valent iron oxo species similar to those that have been strongly suggested as active intermediates in many oxygen atom transfer reactions. Such intermediates have been directly characterized only in metalloporphyrin systems. Complex **2** is an iron-hydroperoxide species generated by the complexation of hydrogen peroxide and an iron cyclam complex. This species was suggested to be an intermediate in our previous studies.[14,15] Vanadium peroxo complexes, which are very effective reagents for the transformation of olefins to epoxides, have been proposed to contain a similar hydrogen bond.[30] Complex **3** has been suggested as an active intermediate in Lewis acidic metal-catalyzed epoxidations by iodosylbenzene.[31-33] Catalysis of olefin epoxidation by iodosylbenzene has been achieved using non-redox metal catalysts which cannot form high-valent metal oxo species. It has been proposed that the iodine(III) center in the proposed structure will be the electrophilic center responsible for reacting with the olefinic double bond. Complex **4** is the metal-peroxyacid species which is formed by the complexation of MCPBA and an iron cyclam complex. Groves et al.[34] and Watanabe et al.[35] have shown that Mn-peroxyacid and Fe-peroxyacid porphyrin complexes are viable active oxygenation agents for olefin epoxidation at low temperatures. Aldehydes are prone to undergo autoxidation to produce peroxy radicals in the presence of O_2 and metal catalysts.[36] The peroxy radical **5** has been shown to oxidize olefins to give corresponding epoxides.[17,37]

REACTION MECHANISMS

Very recently, Murahashi et al. published the results of study of iron- and ruthenium-catalyzed oxidations of alkanes with dioxygen in the presence of an aldehyde and an acid.[38] In the system studied, olefins were epoxidized as well as alkanes, i.e. cyclohexene was oxidized to give cyclohexene oxide and *cis*-stilbene was oxidized to give *cis*-stilbene oxide and *trans*-stilbene oxide in the ratio of 83:17. Metal-oxo complexes were suggested to be the active oxidizing species. In another system, metalloporphyrins (M = Fe, Mn) were used as catalysts for epoxidation of propylene by molecular oxygen in the presence of an aldehyde.[39] High-valent metal oxo species were proposed as active intermediates for the olefin epoxidations. Metal oxo species were suggested to be formed by the reaction of metal complexes and peroxy acids generated *in*

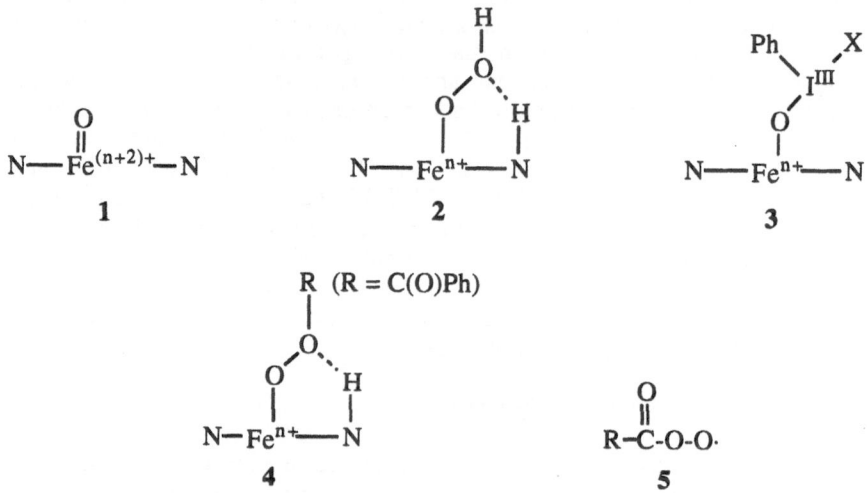

Figure 3. Possible intermediates in metal-catalyzed oxygenation reactions.

situ from the reaction of an aldehyde with dioxygen (Figure 4, pathway A). In our reactions, Fe(cyclam)$^{2+}$ catalyzed the epoxidation of cyclohexene either by MCPBA or by dioxygen plus aldehyde. In the case of *cis*-stilbene, however, the metal-catalyzed MCPBA reaction gave predominantly *cis*-stilbene oxide, whereas the dioxygen plus aldehyde reaction gave more *trans*-stilbene oxide than *cis*-stilbene oxide (see Table 4). These results clearly show that different active intermediates were generated in the MCPBA and O$_2$ plus aldehyde reactions. If a peroxy acid is generated by the reaction of an aldehyde with dioxygen in the presence of iron cyclam complex and if the peroxy acid thus formed reacts with the iron cyclam complex to generate an active intermediate (Figure 4, pathway A), we would expect to see the same ratios of *cis*-stilbene oxide to *trans*-stilbene oxide in both MCPBA and O$_2$ plus aldehyde reactions. However, this is not the case. A possible explanation is that the metal-catalyzed reaction of dioxygen with aldehyde produces an acylperoxy radical, which epoxidizes the olefin directly (Figure 4, pathway B). It is well known that peroxy radicals can epoxidize olefins non-stereospecifically, i.e. giving mixtures of *cis*- and *trans*-stilbene oxide in *cis*-stilbene oxidations.[17,40]

Possible mechanisms for oxygen transfer from H_2O_2, PhIO, and MCPBA to olefins in the presence of an iron cyclam complex are depicted in Figure 5. If the formation of a high-valent iron oxo complex is faster (Figure 5, pathway A and B) than the direct reactions of olefins with intermediates **2, 3**, and **4** (Figure 5, pathway C and D), the

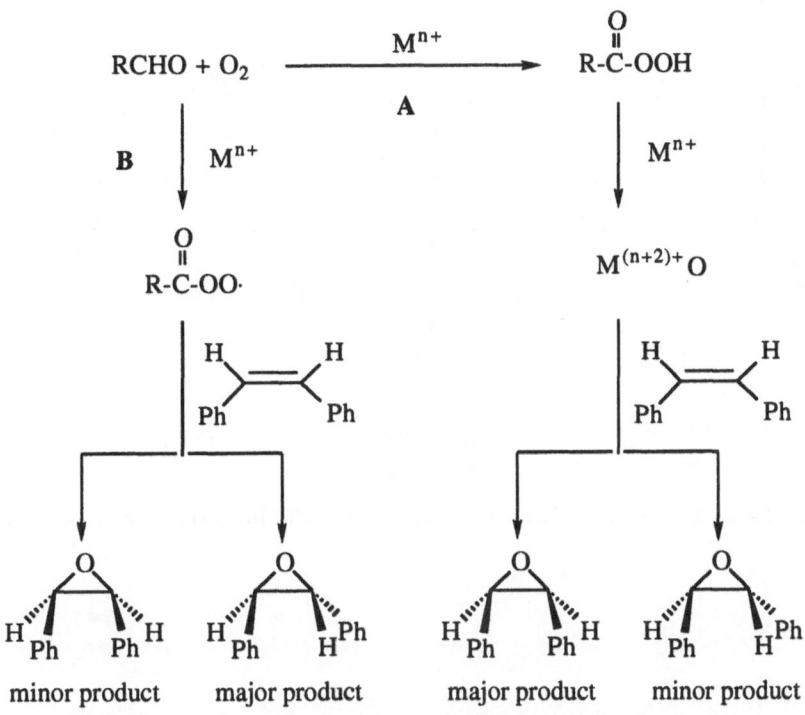

Figure 4. Proposed mechanisms for olefins epoxidation by O_2/aldehyde.

active oxidant will be the high-valent iron oxo complex **1**. If the precursors to the high-valent iron oxo species are reactive enough to transfer their oxygen directly to olefins or if the formation of the iron oxo complex **1** is not feasible due to instability of the species, the oxygen atom will be transferred by the complexes **2, 3**, and **4** (Figure 5, pathway C and D).

It has been widely accepted that the high-valent iron oxo complex **1** is the active oxidant in metalloporphyrin systems and in non-porphyrin iron complex systems as well. A high-valent iron oxo species can be formed via heterolytic O-O bond cleavage of the iron-hydroperoxide **2** or of the iron-peroxyacid species **4** (Figure 5, pathway A). Recently, elegant proof of this heterolytic O-O bond cleavage of hydrogen peroxide, MCPBA, and *t*-butyl hydroperoxide has been provided by Traylor *et al.* in iron porphyrin

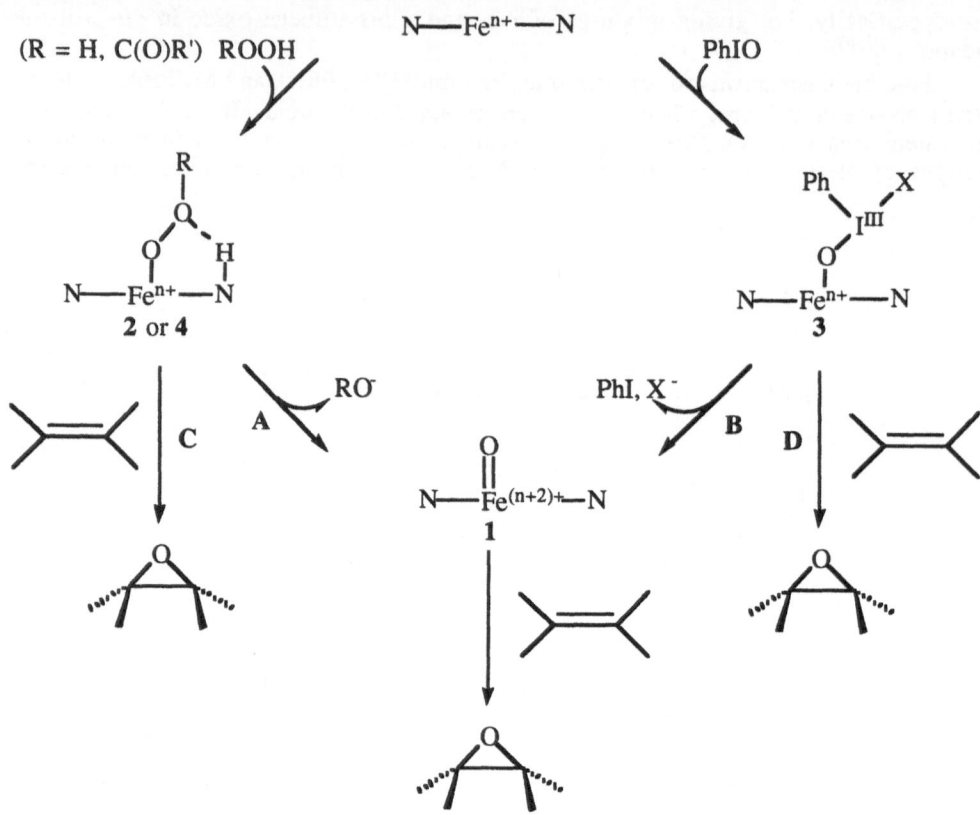

Figure 5. Possible mechanisms for olefin epoxidations by H_2O_2, MCPBA, and PhIO catalyzed by the iron cyclam complex.

complex-catalyzed epoxidation reactions.[41] In non-porphyrin iron complex systems, Bruice and his coworker have concluded that $Fe^{III}(EDTA)$ reacts with peroxy acids or alkyl hydroperoxides by heterolytic O-O bond cleavage accompanied by oxygen atom transfer to $Fe^{III}(EDTA)$.[42,43] The resulting intermediate was proposed to be $(EDTA)Fe^VO$ on the basis of the observation that TBPH (2,4,6-tri-*tert*-butylphenol) was oxidized to the corresponding phenoxy radical. They also compared the rate constants for cyclohexene epoxidation by phenylperacetic acid in the presence and absence of $Fe^{III}(EDTA)$ and reached the conclusion that epoxide formation was 1.5 times faster in the presence of the iron catalyst in methanol. We also studied cyclohexene epoxidation by MCPBA in the presence of $Fe^{III}(EDTA)$ in a solvent mixture of CH_3CN and CH_3OH at a temperature low enough that no direct epoxidation by peracid occurs in the absence of a catalyst. We carried out the same reaction at room temperature for comparison. We found no formation of cyclohexene oxide in the low temperature reaction, but a large amount of cyclohexene oxide was formed at room temperature. In our case, we suspect that the reaction at room temperature occurs by direct reaction of the peracid with the olefin. We also studied the iron cyclam-catalyzed oxidation of TBPH by H_2O_2, PhIO, MCPBA, and *t*-butyl- and ethyl hydroperoxide in acetonitrile and methanol solution. All of these oxidants in the presence of the iron cyclam complex gave oxidation of TBPH to give the highly colored 2,4,6-tri-*tert*-butylphenoxy radical. However, it is important to note that while oxidations of olefins by H_2O_2, PhIO, and MCPBA under these same conditions give epoxides in high yield, those by the alkylhydroperoxides do not.[14] Thus we conclude that the observation of oxidation of TBPH to give the phenoxy radical is not sufficient evidence that a high-valent metal oxo intermediate is formed.

Since it has often been suggested that the observation of substantial levels of ^{18}O incorporation from labeled $H_2{}^{18}O$ into the oxygenated organic products is evidence for the intermediacy of the high-valent iron oxo complex 1,[44-46] we measured the extent of ^{18}O incorporation into the products of iron cyclam complex-catalyzed epoxidations of cyclohexene with H_2O_2, PhIO, and MCPBA when $H_2{}^{18}O$ was added to the reaction mixture.[47] We observed that labeled oxygen was fully incorporated into cyclohexene oxide in the iodosylbenzene reaction (eq 4) but not at all in the H_2O_2 and MCPBA reactions (eq 5).

$$\text{cyclohexene} \; + \; \text{PhI}^{16}\text{O} \; + \; H_2{}^{18}O \quad \xrightarrow{\text{Fe(cyclam)(OTf)}_2} \quad \text{cyclohexene}\,{}^{18}O \qquad (4)$$

$$\text{cyclohexene} \; + \; \text{RO}^{16}\text{OH} \; + \; H_2{}^{18}O \quad \xrightarrow{\text{Fe(cyclam)(OTf)}_2} \quad \text{cyclohexene}\,{}^{16}O \qquad (5)$$

If the assumption that high-valent iron oxo species invariably exchange their oxygen atoms with labeled water at high rates is correct, the active oxidants formed in the iodosylbenzene reaction and in the H_2O_2 and MCPBA reactions would be concluded to be different. However, further studies utilizing iron and manganese porphyrin systems, in which high-valent metal oxo complexes are generally accepted to be the active oxygen atom-transfer reagents, have demonstrated that the assumption is incorrect and that the oxygen atoms in the reactions of iodosylbenzene exchange by another mechanism.[47]

Another approach to determining if the iron-cyclam-catalyzed reactions of H_2O_2, PhIO, and MCPBA occur via a common intermediate is to study intramolecular and intermolecular competitive olefin epoxidations, with the assumption that all three oxidants should give similar reactivity patterns if the active oxygenating intermediates generated from these oxidants are identical. If different active oxidants are generated, then different reactivity patterns would be expected. From the relative reactivity studies of *para*-substituted styrenes, different ρ values for Hammett plot for the H_2O_2 reaction and for the PhIO and MCPBA reactions were obtained, indicating that active oxidant formed in the H_2O_2 reaction was different from that formed in the PhIO and MCPBA reactions. The data obtained from other competitive reactions indicate that the active oxidant formed in the MCPBA reaction was not identical to the active species generated in the H_2O_2 and PhIO reactions. By analyzing results of the competitive reactivity studies, we reached the conclusion that oxygen atom transfer reactions from H_2O_2, PhIO, and MCPBA to olefins in the presence of the iron cyclam complex occur via different intermediates.

ACKNOWLEDGMENTS

This research was supported by the NSF (J.S.V.), NON DIRECTED RESEARCH FUND, Korea Research Foundation, 1992 (W.N.), and the Korea Science and Engineering Foundation, International Cooperative Science Program KOSEF-NSF (W.N.).

REFERENCES

1. P.R. Ortiz de Montellano. "Cytochrome P-450: Structure, Mechanism, and Biochemistry," Plenum, New York (1985).
2. L. Que, Jr. and A.E. True, Dinuclear iron- and manganese-oxo sites in biology, *in*: "Progress in Inorganic Chemistry: Bioinorganic Chemistry," S.J. Lippard, Ed., John Wiley & Sons, Inc., New York (1990).
3. J.B. Vincent, G.L. Olivier-Lilley, and B.A. Averill, Proteins containing oxo-bridged dinuclear iron centers: A bioinorganic perspective, Chem. Rev. 90:1447 (1990).
4. N.D. Priestley, H. G. Floss, W.A. Froland, J.D. Lipscomb, P.G. Williams, and H. Morimoto, Cryptic stereospecificity of methane monooxygenase, *J. Am. Chem. Soc.* 114:7561 (1992).
5. J.E. Colbert, A.G. Katopodis, and S.W. May, Epoxidation of *cis*-1,2-dideuterio-1-octene by *Pseudomonas oleovorans* monooxygenase proceeds without deuterium exchange, *J. Am. Chem. Soc.* 112:3993 (1990).
6. R. Shiman, S.H. Jones, and D.W. Gray, Mechanism of phenylalanine regulation of phenylalanine hydroxylase, *J. Biol. Chem.* 265:11633 (1990).
7. S.E. Sen and G.D. Prestwich, Trisnorsqualene alcohol, a potent inhibitor of vertebrate squalene epoxidase, *J. Am. Chem. Soc.* 111:1508 (1989).
8. T.A. Dix and S.J. Benkovic, Mechanism of oxygen activation by pteridine-dependent monooxygenases, *Acc. Chem. Res.* 21:101 (1988).
9. J. Stubbe and J.W. Kozarich, Mechanisms of bleomycin-induced DNA degradation, *Chem. Rev.* 87:1107 (1987).
10. D.H.R. Barton and M. Ozbalik, Selective functionalization of saturated hydrocarbons by the "Gif" and "Gif-Orsay" systems, *in*: "Activation and Functionalization of Alkanes," C. Hill, ed., John Wiley & Sons, New York (1989).
11. H.-C. Tung, C. Kang, and D.T. Sawyer, Nature of the reactive intermediates from the iron-induced activation of hydrogen peroxide: Agents for the ketonization of methylenic carbons, the monooxygenation of hydrocarbons, and the dioxygenation of arylolefins, *J. Am. Chem. Soc.* 114:3445 (1992).
12. A. Suga, T. Sugiyama, M. Otsuka, M. Ohno, Y. Sugiura, and K. Maeda, Oxidation of alkenes by a chiral non-porphyrinic oxidizing catalyst based on the bleomycin-Fe(II) complex, *Tetrahedron* 47:1191 (1991).
13. D.Y. Dawson, S.E. Hudson, and P.K. Mascharak, Oxygen transfer reactions by synthetic analogues of iron-bleomycin, *J. Inorg. Biochem.* 47:109 (1992).
14. W. Nam, R. Ho, and J.S. Valentine, Iron-cyclam complexes as catalysts for the epoxidation of olefins by 30% aqueous hydrogen peroxide in acetonitrile and methanol, *J. Am. Chem. Soc.* 113:7052 (1991).
15. Y.-D. Wu, K.N. Houk, J.S. Valentine, and W. Nam, Is intramolecular hydrogen-bonding important for bleomycin reactivity? A molecular mechanics study, *Inorg. Chem.* 31:718 (1992).
16. J. Rebek, Jr., Progress in the development of new epoxidation reagents, *Heterocycles* 15:517 (1981).
17. M.A. Brook, J.R. Lindsay Smith, R. Higgins, and D. Lester, Model systems for cytochrome P450 dependent monooxygenases. Part 4. The epoxidation of alkenes by peroxyacids in the presence of cobalt complexes, *J. Chem. Soc., Perkin Trans.* 2 1049 (1985).

18. J.R. Lindsay Smith and P.R. Sleath, Model Systems for cytochrome P450 dependent monooxygenases. Part 1. Oxidation of alkenes and aromatic compounds by tetraphenylporphinatoiron(III) chloride and iodosylbenzene, *J. Chem. Soc., Perkin Trans.* 2 1009 (1982).

19. T.G. Traylor and F. Xu, Model reactions related to cytochrome P-450. Effects of alkene structure on the rates of epoxide formation, *J. Am. Chem. Soc.* 110:1953 (1988).

20. J.P. Collman, P.D. Hampton, and J.I. Brauman, Suicide inactivation of cytochrome P-450 model compounds by terminal olefins. 2. Steric and electronic effects in heme N-alkylation and epoxidation, *J. Am. Chem. Soc.* 112:2986 (1990).

21. J.T. Groves and Y. Watanabe, On the mechanism of olefin epoxidation by oxo-iron porphyrins. Direct observation of an intermediate, *J. Am. Chem. Soc.* 108:507 (1986).

22. T.G. Traylor, W.-P. Fann, and D. Bandyopadhyay, A common heterolytic mechanism for reactions of iodosylbenzenes, peracids, hydroperoxides, and hydrogen peroxide with iron(III) porphyrins, *J. Am. Chem. Soc.* 111:8009 (1989).

23. T.G. Traylor and A.R. Miksztal, Alkene epoxidations catalyzed by iron(III), manganese(III), and chromium(III) porphyrins. Effects of metal and porphyrin substituents on selectivity and regiochemistry of epoxidation, *J. Am. Chem. Soc.* 111:7443 (1989).

24. J.T. Groves and T.E. Nemo, Epoxidation reactions catalyzed by iron porphyrins. Oxygen transfer from iodosylbenzene, *J. Am. Chem. Soc.* 105:5786 (1983).

25. K.S. Suslick and B.R. Cook, Regioselective epoxidations of dienes with manganese(III) porphyrin catalysts, *J. Chem. Soc., Chem. Commun.* 200 (1987).

26. K.S. Suslick and B.R. Cook, Shape selective oxidation as a mechanistic probe, *in:* " Inclusion Phenomena and Molecular Recognition," J.L. Atwood, ed., Plenum, New York (1990).

27. P. Battioni, J.P. Renaud, J.F. Bartoli, M. Reina-Artiles, M. Fort, and D. Mansuy, Monooxygenase-like oxidation of hydrocarbons by H_2O_2 catalyzed by manganese porphyrins and imidazole: Selection of the best catalytic system and nature of the active oxygen species, *J. Am. Chem. Soc.* 110:8462 (1988).

28. W.Y. Lu, J.F. Bartoli, P. Battioni, and D. Mansuy, Selective oxygenation of hydrocarbons and sulfoxidation of thioethers by dioxygen with a Mn-porphyrin-based cytochrome P450 model system using Zn as electron donor, *New J. Chem.* 16:621 (1992).

29. K.B. Sharpless, J.M. Townsend, and D.R. Williams, On the mechanism of epoxidation of olefins by covalent peroxides of molybdenum(VI), *J. Am. Chem. Soc.* 94:295(1972).

30. H. Mimoun, L. Saussine, E. Daire, M. Postel, J. Fischer, and R. Weiss, Vanadium(V) peroxo complexes. New versatile biomimetic reagents for epoxidation of olefins and hydroxylation of alkanes and aromatic hydrocarbons, *J. Am. Chem. Soc.* 105:3101 (1983).

31. W. Nam and J.S. Valentine, Zinc(II) complexes and aluminum(III) porphyrin complexes catalyze the epoxidation of olefins by iodosylbenzene, *J. Am. Chem. Soc.* 112:4987 (1990).

32. Y. Yang, F. Diederich, and J.S. Valentine, Lewis acidic catalysts for olefin epoxidation by iodosylbenzene, *J. Am. Chem. Soc.* 113:7195 (1991).

33. Y. Yang, F. Diederich, and J.S. Valentine, Reaction of cyclohexene with iodosylbenzene catalyzed by non-porphyrin complexes of iron(III) and aluminum(III). Newly discovered products and a new mechanistic proposal, *J. Am. Chem. Soc.* 112:7826 (1990).

34. J.T. Groves, Y. Watanabe, and T.J. McMurry, Oxygen activation by metalloporphyrins. Formation and decomposition of an acylperoxymanganese(III) complex, *J. Am. Chem. Soc.* 105:4489 (1983).

35. Y. Watanabe, K. Yamaguchi, I. Morishima, K. Takehira, M. Shimizu, T. Hayakawa, and H. Orita, Remarkable solvent effect on the shape-selective oxidation of olefins catalyzed by iron(III) porphyrins, *Inorg. Chem.* 30:2581 (1991).

36. R.A. Sheldon and J.K. Kochi. "Metal-Catalyzed Oxidations of Organic Compounds," Academic Press, New York (1981).

37. G.V. Buxton, J.C. Green, R. Higgins, and S. Kanji, Formation of epoxides in the oxidation of β-hydroxyalkyl radicals by copper(II) in aqueous solution, *J. Chem. Soc., Chem. Commun.* 158 (1976).

38. S.-I. Murahashi, Y. Oda, and T. Naota, Iron- and ruthenium-catalyzed oxidations of alkanes with molecular oxygen in the presence of aldehydes and acids, *J. Am. Chem. Soc.* 114:7913 (1992).

39. R. Iwanejko, P. Leduc, T. Mlodnicka, and J. Poltowicz, Metalloporphyrin-catalyzed epoxidation of propylene, in: "Dioxygen Activation and Homogeneous Catalytic Oxidation," L.I. Simandi, ed., Elsevier, Amsterdam (1991).

40. T. Muto, C. Urano, T. Hayashi, T. Miura, and M. Kimura, On the mode of oxygenation with ferric perchlorate-hydrogen peroxide system, *Chem. Pharm. Bull.* 31:1166 (1983).

41. T.G. Traylor, S. Tsuchiya, Y.S. Byun, and C. Kim, Iron(III) porphyrin catalyzed epoxidations with hydrogen peroxide and hydroperoxides, *J. Am. Chem. Soc.* in press.

42. P.N. Balasubramanian and T.C. Bruice, Oxygen transfer involving non-heme iron. The reaction of (EDTA)Fe^{III} with *m*-chloroperbenzoic acid, *J. Am. Chem. Soc.* 108:5495 (1986).

43. P.N. Balasubramanian and T.C. Bruice, Oxygen transfer involving nonheme iron: The influence of leaving group ability on the rate constant for oxygen transfer to (EDTA)Fe(III) from peroxycarboxylic acids and hydroperoxides, *Proc. Natl. Acad.Sci. USA* 84:1734 (1987).

44. J.T. Groves and W.J. Kruper, Jr., Preparation and characterization of an oxoporphinato-chromium(V) complex, *J. Am. Chem. Soc.* 101:7613 (1979).

45. J.T. Groves, W.J. Kruper, Jr., and R.C. Haushalter, Hydrocarbon oxidations with oxometalloporphinates. Isolation and reactions of a (porphinato)manganese(V) complex, *J. Am. Chem. Soc.* 102:6375 (1980).

46. J.T. Groves, R.C. Haushalter, M. Nakamura, T.E. Nemo, and B.J. Evans, High-valent iron-porphyrin complexes related to peroxidase and cytochrome P-450, *J. Am. Chem. Soc.* 103:2884 (1981)

47. W. Nam and J.S. Valentine, Reevaluation of the significance of ^{18}O incorporation in metal complex-catalyzed oxygenation reactions carried out in the presence of $H_2^{18}O$, *J. Am. Chem. Soc.* 115:1772 (1993)

BIOINSPIRED CATALYSIS: HYDROPEROXIDATION OF ALKYLAROMATICS

Jeffrey E. Bond, Sergiu M. Gorun,[*] Robert T. Stibrany, George W. Schriver, and Thomas H. Vanderspurt

Corporate Research Laboratories
Exxon Research and Engineering Co.
Annandale, NJ 08801 USA

INTRODUCTION

Upgrading the value of raw materials by the selective conversion of organic molecules to oxygenated derivatives using air as an oxidant represents a worthy goal and challenge. Industrially, oxygenates are prepared at a rate of 10^9 to 10^{10} lbs/year, mostly using high temperature (above 100°C) processes.[1] Since, in general, the oxygenated products are more reactive then the starting hydrocarbons, secondary reactions which lower the selectivity occur.

Metal-based oxidation catalysts have been described; the chemistries involved are based on both C-H and oxygen activation, the latter category being conceptually subdivided into molecular oxygen (O_2) activation and activation of molecules that already contain a *reduced* form of oxygen (superoxide, peroxide, oxide, etc.). These topics are the subjects of numerous reviews and books.[2]

This paper will describe our use of polynuclear manganese complexes of the type believed to be present at the active site of the oxygen evolving complex (OEC) of photosystem II (PSII)[3] as catalysts for hydrocarbon oxygenation.

FUNCTIONAL ASPECTS OF OEC OF PSII

Four Mn ions are present at the active site of OEC in cyanobacteria, algae and green plants.

The OEC is required for coupling two water molecules and formation of O_2.

$$2H_2O \rightarrow O_2 + 4H^+ + 4e \qquad \text{(Eq. 1)}$$

---

[*] Presenter

The Activation of Dioxygen and Homogeneous Catalytic Oxidation,
Edited by D.H.R. Barton *et al.*, Plenum Press, New York, 1993

199

The polynuclear manganese aggregate of OEC is believed to pass through five oxidation states S_4 to S_0, S_4 denoting the most oxidized.[4] The oxidation state of the Mn ions in the S_0 state is at its lowest level, probably mostly a combination of Mn(II) and (III).

While the primary function of the OEC is to facilitate Equation 1 (a four electron process), in the *absence* of light, PSII exhibits catalatic activity:[5]

$$2H_2O_2 \rightarrow 2H_2O + O_2 \qquad \text{(Eq. 2)}$$

This two electron process is believed to involve only the S_0 and S_2 states.[5]

RESULTS AND DISCUSSION

Structural Studies

We have recently reported the synthesis, and structure of a model for the S_0 state,[6] (BaCa) [Mn$_4$(OHO)(OAc)$_2$L$_2$] **1**, L=1,3-Diamino-2-hydroxypropane-N,N,N',N'-tetraacetic acid. The tetranuclear manganese moiety of this model is represented in Figure 1 along with selected bond distances and angles. The four Mn ions can be "dissected" into two pairs according to the symmetry, Mn(1), Mn(1') and Mn(2), Mn(2'). The oxidation states of the metal centers is II, III, III, III (valence localized), with trivalent Mn(1) and Mn(1') and di- and trivalent Mn(2), Mn(2'). The two Mn pairs are related by a crystallographically imposed mirror plane. Thus, two dinuclear, oxo(hydroxo) and carboxylate bridged Mn units are present. They are linked by an unusual μ_4(O-H-O) bridge, formally formed by deprotonation of two μ_2-H$_2$O bridges.

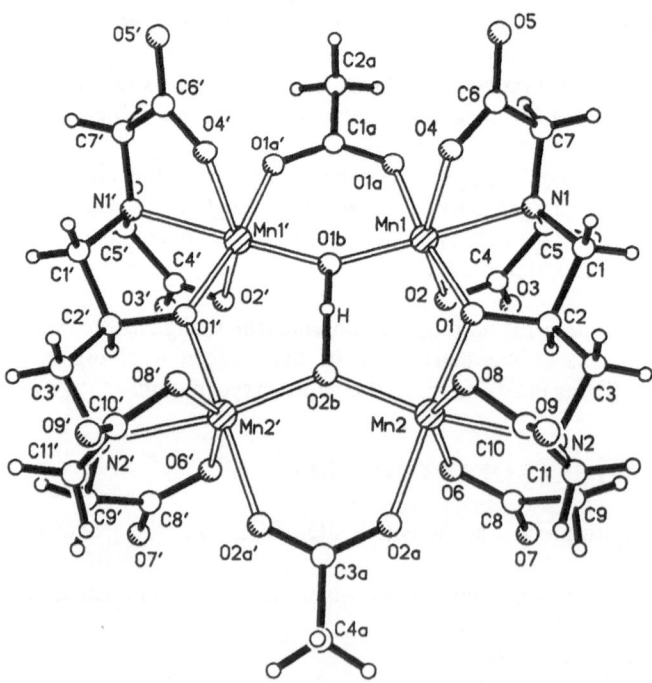

Figure 1. Structure of the anion of **1**. Selected bond lengths [Å] and angles [°] in **1**: Mn1 - Mn1' 3.379(2), Mn2 - Mn2' 3.584(2), Mn1 - Mn2 3.718(1), Mn1 - O1b 1.817(2), Mn1 - O1a 1.961(4), Mn1 - O4 2.217(4), Mn1 - N1 2.120(4), Mn1 - O2 2.152(4), Mn1 - O1 1.954(3), Mn2 - O2b 1.925(3), Mn2 - O2a 2.078(5), Mn2 - O8 2.169(5), Mn2 - N2 2.190(4), Mn2 - O6 2.140(4), O1b - H 1.2(2), O2b - H 1.3(1), Mn1 - O1b - Mn1' 136.8(3), Mn2 - O2b - Mn2' 137.2(4), Mn1 - O1 - Mn2 132.5(2), O1b - H - O2b 156(10).

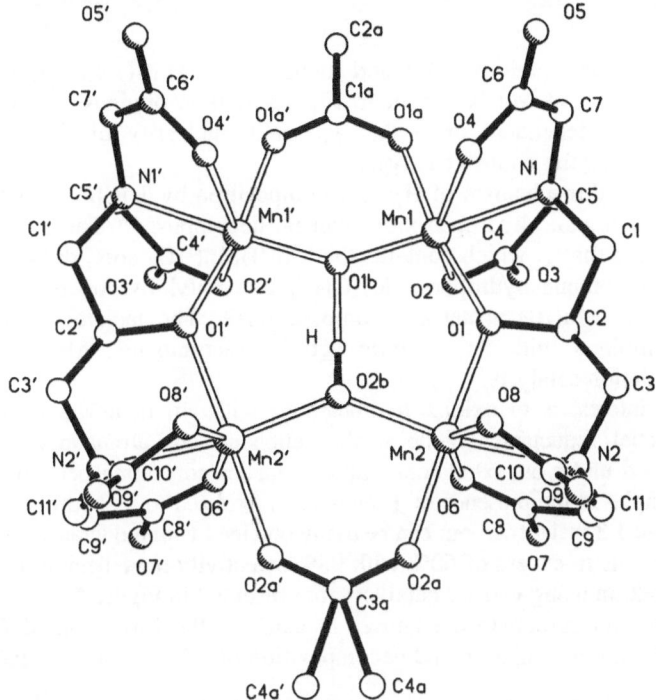

Figure 2. Structure of the anion of **2**. Hydrogen atoms have been omitted for clarity. Selected bond lengths [Å] and angles [°]: Mn1 - Mn1' 3.382(1), Mn2 - Mn2' 3.580(1), Mn1 - Mn2 3.726(1), Mn1 - O1b 1.819(2), Mn1 - O1a 1.971(4), Mn1 - O4 2.183(4), Mn1 - N1 2.121(4), Mn1 - O2 2.151(4), Mn1 - O1 1.942(4), Mn2 - O2b 1.926(3), Mn2 - O2a 2.061(6), Mn2 - O8 2.161(5), Mn2 - N2 2.204(4), Mn2 - O6 2.139(4), O1b - H 1.59(10), O2b - H 0.89(10), Mn1 - O1b - Mn1' 136.7(3), Mn2 - O2b - Mn2' 136.7(4), Mn1 - O1 - Mn2 132.9(2), O1b - H - O2b 180(7).

The dinuclear units are closely related to the dinuclear Mn units believed to be present at the active site of manganese (pseudo)catalases.[7] This view is supported by the previously reported[6] biomimetic (catalytic) activity of **1** (see also below).

Interestingly, the two bridging acetato groups have different degrees of disorder. While the acetate bridging the Mn(1), Mn(1') pair is ordered, the one that bridges the mixed-valence pair Mn(2), Mn(2') is not. Its disorder could reflect true carboxylate lability, be due to the crystallographically imposed averaging of Mn(II) and Mn(III) bond distances and angles and/or other solid state effects. In order to assess these effects, we have replaced Ba by Ca in **1**. The new complex, $Ca_2[Mn_4(OHO)(OAc)_2L_2]$, **2**, is isomorphous with **1**.[6]

The structure was solved;* the tetranuclear manganese moiety is shown in Figure 2. Interestingly, while the acetato group bridging Mn(1), Mn(1') in **2** is ordered (like in **1**), its Mn(2), Mn(2') counterpart is much more disordered (the structures of **1** and **2** have similar degrees of resolution).

Moreover, the H of the O-H-O bridge in **1** seems to be less symmetrically shared by the two μ_2-oxo groups.†

The above results suggest that (i) an open coordination site (or at least a less saturated coordinative environment) can be envisioned at the Mn(II) site, consistent with the higher kinetic lability of Mn(II) vs. Mn(III); (ii) the O-H-O group might accept or lose protons without major structural rearrangement.

---

* R=5.9% Data/parameter ratio 7.9. H of the O-H-O bridge was located in the difference Fourier map and refined. Further details will be reported elsewhere.

† An accurate position of this hydrogen should be obtainable by using neutron diffraction data.

Catalytic Studies

The relative catalatic activity* of $\underline{1}$ and mononuclear Mn(II) and (III) complexes at equimolar concentration in Mn is shown in Figure 3. It is obvious that the rate of O_2 evolution is strongly dependent upon the type and nuclearity of the complexes, the mononuclear ones being the poorest catalysts.

While the detailed mechanism of H_2O_2 decomposition by $\underline{1}$ is not known yet, we are interested if related organic hydroperoxides can be decomposed.† The surprising results, shown in Figure 4, is that $\underline{1}$, which contains the $Mn(II)Mn(III)_3$ core, does not decompose significantly either cumene hydroperoxide (CHP) or t-butyl hydroperoxide (TBHP). In contrast, Mn(II) and Mn(III) acetates decompose CHP. The lack of decomposition of ROOH by $\underline{1}$ combined with its catalytic activity recommend "Mn_4" complexes as candidates for oxidation catalysts.

Our primary interest is to oxidize hydrocarbons with air, at low temperatures, in the absence of sacrificial reductants. While neither aliphatic nor aromatic hydrocarbons are significantly oxidized under our conditions,‡ alkylaromatics containing benzylic hydrogen are oxidized. For example, in the presence of $\underline{1}$, cumene is oxidized to its hydroperoxide; at 65°C, conversions of about 1.3 weight %/hour can be easily obtained with moderate air flows.

Total conversions in excess of 60% with 98% selectivity have been obtained. Relative rates of CHP formation using various catalysts are presented in Figure 5.

Barium oxide and carbonate are known to catalyze the formation of CHP.[8] Cobalt acetate catalyzes both the formation and decomposition of CHP. Other alkylaromatics react

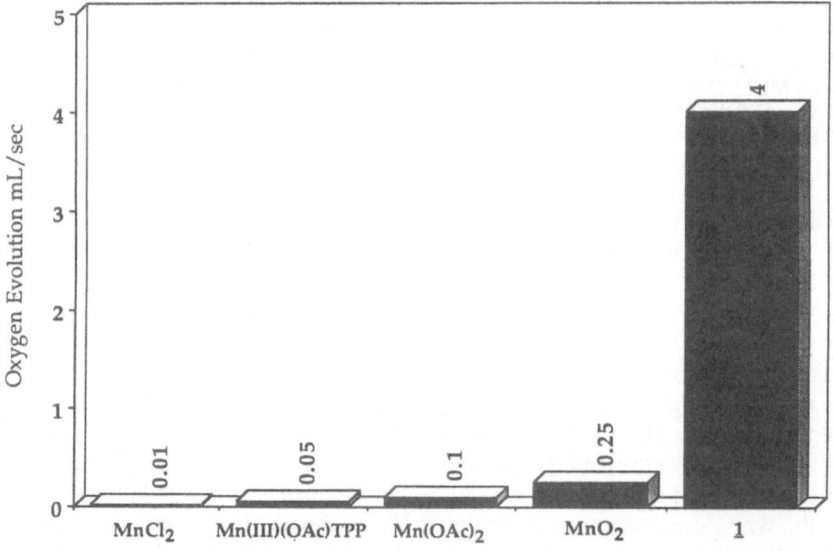

Figure 3. Oxygen evolution from hydrogen peroxide catalyzed by manganese complexes.

---

* Experimental conditions: 30% H_2O_2. O_2 evolution was measured manometrically.

† Experimental conditions: 50-100 mgs solid $\underline{1}$ is suspended in 100 mL solutions of cumene hydroperoxide and t-butyl hydroperoxide (Aldrich), 40% in cumene and 30% in t-butanol, respectively. The reaction mixture was stirred vigorously under inert atmosphere at 65°C and the hydroperoxide content determined by titration. Other metal salts were tested under similar conditions.

‡ Typical experimental conditions: solid $\underline{1}$ is suspended in a liquid hydrocarbon in a round bottom flask equipped with a condenser and stirred vigorously. The temperature is maintained at 65°C (±1°C) and air is passed through the mixture. The O_2 uptake is monitored continuously using a Teledyne oxygen meter. Oxidation products are analyzed by GC/MS, GC and redox titration (for hydroperoxides).

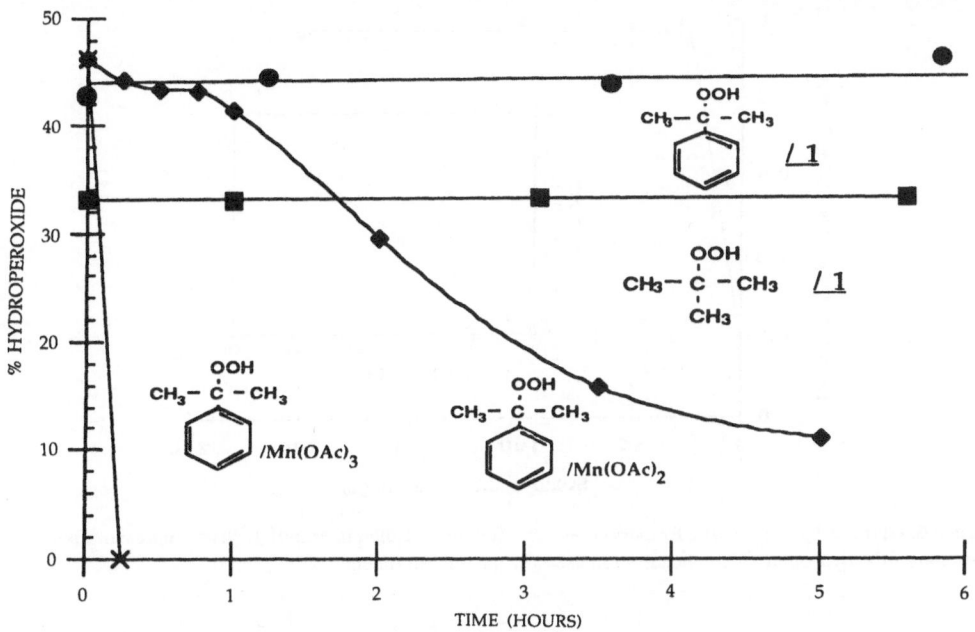

Figure 4. Manganese complexes catalyzed decomposition of organic hydroperoxides.

similarly with reaction rates consistent with the strength of their C-H bonds. The hydroperoxidation of cumene is about an order of magnitude faster than its autoxidation, CHP is obtained in high (practically quantitative) yield with significantly less undesirable acetophenone by-product.

Hydroperoxidation Mechanism: The Role of the Catalyst.

In order to understand the catalytic role of $\underline{1}$, we prepared the Ca_2, $\underline{2}$ (see above), K_4, $\underline{3}$ and Ba_2, $\underline{4}$ analogues. Only the counterions vary in $\underline{1}$, $\underline{2}$, $\underline{3}$ and $\underline{4}$. Since all show the same

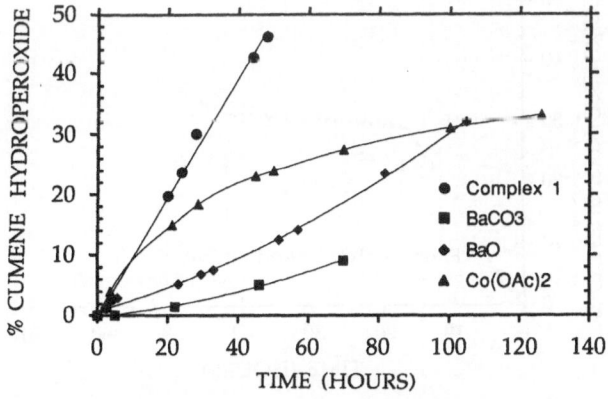

Figure 5. Cumene hydroperoxide formation using neat cumene, air and selected metal compounds at 65°C.

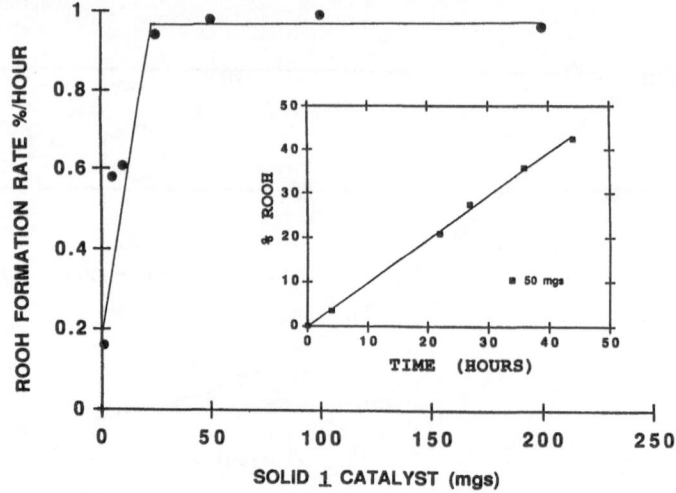

Figure 6. Cumene hydroperoxide formation rate as a function of the amount of 1. Inset: representative time variation of the hydroperoxide concentration used for rate measurement.

catalytic activity, we conclude that the manganese core and not the counterions are responsible for the observed activity. Is the reaction homogeneous or heterogeneous? Samples of hydrocarbons analyzed before and after oxidations contained less than the detection limit (ppb) of Mn.

Moreover, as shown in Figure 6, the reaction rate depends upon the amount of *solid* catalyst (solid catalyst is present at all times during the reaction) and, as shown in Figure 7, removal (filtration) and re-addition of the solid catalyst practically stops and restarts the reaction, respectively. Thus, we conclude that the reaction is heterogeneous.

As far as the reaction mechanism is concerned, the formation of alkyl hydroperoxides suggests alkyl radicals as intermediates. Indeed, when cyclohexylbenzene is used, we observe oxygen insertion not only at the benzylic carbon (1 position) but also at the 2 and 4 positions of the cyclohexyl ring. These products may result if a peroxy radical at the 1 position abstracts a hydrogen atom from the sterically favored 2 or 4 position forming another carbon radical capable of reacting further with oxygen. Further support for the presence of a carbon radical is the classical inhibition pattern observed upon addition of the free radical inhibitor diphenylamine.

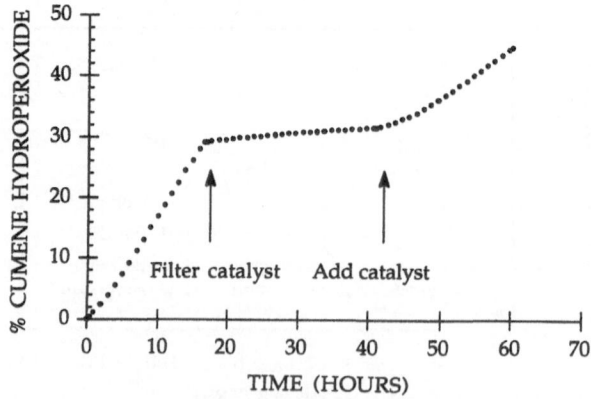

Figure 7. Effect of removal and addition of solid 1 upon the oxygenation rate of cumene.

It is known that manganese salts cause oxidation of hydrocarbons, like cumene, by initiating free radical chain reactions. However, this is normally done by catalytic decomposition of trace amounts of hydroperoxides found in the hydrocarbons.[2b,9] In our case, the catalyst does not seem to decompose CHP, as demonstrated in an independent experiment (see above). If it did, the rate of decomposition should increase in time as the reaction progresses leading to an increase in the autoxidation rate. While we do observe for cumene an initiation period up to the accumulation of 3–5% hydroperoxide, from that point on up to greater than 50% CHP accumulation, the oxidation rate is constant. This initiation period may be due to surface activation of the catalyst.

We can estimate an upper limit for the autoxidation as follows. We assume that despite the fact that we do not observe any catalyzed CHP decomposition even after 120 hours at 65°C, we still decompose it at a rate of 1% over a period of 120 hours.* Using this rate for radical production, the normal solution rate constants for the propagation and termination reactions and the formulae given in Ref. 10, we estimate the rate of CHP production to be 0.54×10^{-5} Ms^{-1}. The observed rate, 2.6×10^{-5} Ms^{-1}, is thus about 5 times larger then the upper limit of the estimated initiated free-radical rate.

Thus, $\underline{1}$ seems to be a true catalyst rather then a new kind of free radical initiator. This behavior is in contrast to the behavior of related manganese complexes. For example, Mn(II) carboxylates are known to decompose CHP during autoxidation of cumene[11]; dinuclear Mn(III) complexes decompose tetralin hydroperoxide during oxidation of tetralin (an inner-sphere Mn-alkyl hydroperoxide intermediate has been proposed)[12]; trinuclear, carboxylate and oxo-bridged complexes containing Mn(II) were found to decompose CHP during the catalyzed oxidation of cumene.[13]

Overall, our results are consistent with a hydrogen rebound mechanism, wherein the catalyst activates oxygen, the oxygenated catalyst abstracts hydrogen from the hydrocarbon which then reacts with O_2 to form a peroxy radical. Recombination of the radicals with the "hydrogenated" catalyst regenerates the catalyst (Mechanism 1):

$$Cat + O_2 \Leftrightarrow Cat\, O_2$$
$$CatO_2 + RH \rightarrow CatO_2H + R\bullet$$
$$R\bullet + O_2 \rightarrow ROO\bullet$$
$$CatO_2H + ROO\bullet \rightarrow CatO_2 + ROOH$$

Alternatively, the catalyst might abstract hydrogen from the hydrocarbon forming a radical which reacts with O_2 to form a hydroperoxide. The catalyst is regenerated by the loss of hydrogen to the hydroperoxide (Mechanism 2):

$$Cat + RH \rightarrow Cat\, H + R\bullet$$
$$R\bullet + O_2 \rightarrow ROO\bullet$$
$$Cat\, H + ROO\bullet \rightarrow Cat + ROOH$$

Both mechanisms assume the formation of a "cage" that prevents ROO• from reacting with RH. This assumption is only a "zeroth order" approximation. Some radicals may escape the cage and start classical free-radical chemistry in solution. It is possible that the hydrocarbon loses a hydrogen atom, but is seems more likely that an electron and a proton are taken up by the manganese catalyst.

Structurally, it is tempting to speculate that if mechanism 1 is valid, the $H^{\oplus}$ can reversibly bind to the central μ_4-O-H-O group while if mechanism 2 is valid, the oxygen binds to the Mn(II) site, which is or could become coordinatively unsaturated due to a labile

---

* This value, 1%, is at least 10 times larger then our CHP titration error.

acetato group (see above). For both mechanisms, however, Mn ions are the "electron binding sites". Two pieces of evidence support this view:

1. We have prepared a complex analog to $\underline{1}$ but in which both Mn(II) and Mn(III) are selectively replaced by the *non-redox* active Zn(II) and Ga(III), respectively. This complex, *isomorphous* with $\underline{1}$, shows no catalytic activity, thus strongly suggesting the requirement of electron transfer for catalysis.

2. Proton coupled electron transfer. Titration of $\underline{1}$ with acetic acid results in the reduction of Mn(III) to Mn(II) as shown in Figure 8. Thus, the transfer of an electron will facilitate the addition of a proton and vice versa. Consequently, the activation energy for catalysis by both above mechanisms is lowered.

It is important to note that even if the proton is attached to the Mn-bound O_2 as in mechanism 1, one cannot exclude its transfer to μ_4-O-H-O bridges present in the bulk of the catalyst. In this way, one surface-bound oxygen can produce (and cap) multiple ROO• radicals.

A very interesting "(hydro)peroxo rebound" mechanism (Mechanism 3) cannot be ruled out:

$$Cat + O_2 \Leftrightarrow CatO_2$$
$$CatO_2 + RH \rightarrow CatO_2H + R• \rightarrow Cat + ROOH$$

This mechanism is reminiscent of the "oxygen rebound" mechanism proposed for cytochrome P450 and which may also be operative in methane monooxyenase. The difference, however, is that in Mechanism 3 *two* (as opposed to one) oxygens are inserted in a C-H bond.

Biological Relevance

A connection between the manganese catalyzed formation of benzyl radicals and tyrosyl radicals can be formally established. It is known that Mn can replace Fe in some ribonucleotide reductases (RR) with retention of activity.[14] Also, a dinuclear Mn active site has been proposed for authentic manganese ribonucleotide reductases (MnRR).[15] Both RR's share the μ-oxo bridged dinuclear manganese motif with $\underline{1}$, except that, as discussed above, $\underline{1}$ comprises *two* such structural units.

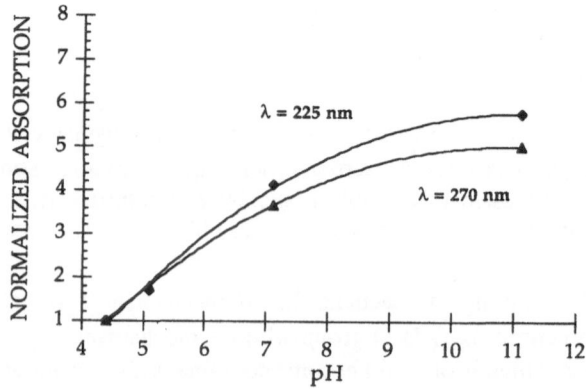

Figure 8. Proton induced reduction of $\underline{1}$. The solution changes color from dark brown to colorless as the pH decreases.

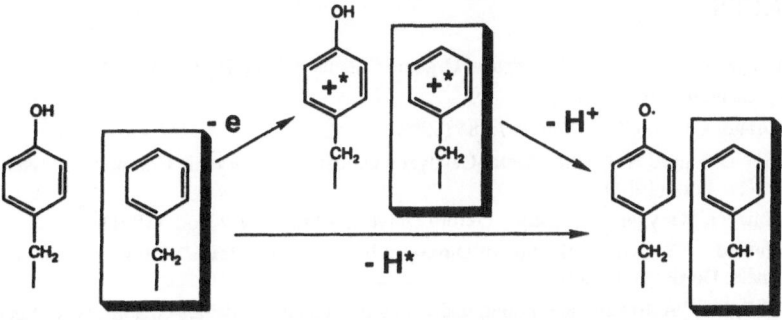

Figure 9. Formal analogy between tyrosyl and benzyl radical formation. Both direct hydrogen atom and proton/electron abstraction mechanisms are shown.

A mechanism by which one proton and one electron are abstracted by a "Mn-oxo" group from the OH of a tyrosine or, in its absence, from the reactive benzylic CH_2 is shown in Figure 9.

It remains to be established which of the two dinuclear moieties of 1 is responsible for catalytic and/or O_2 binding and/or proton/electron abstraction from hydrocarbons. Work in progress, which includes measuring kinetic isotope effects, solvent effects, etc., is aimed at elucidating the oxidation mechanism.

CONCLUSIONS

A novel family of tetranuclear manganese complexes has been synthesized and characterized as structural and functional models for the oxygen evolving complex of Photosystem II. These complexes exhibit biomimetic catalase activity and, interestingly, are also able to function as hydrocarbon air oxidation catalysts. Their activity results in the incorporation of both oxygen atoms of O_2 in organic substrates with no need for a sacrificial reductant. Thus, while they share many features with biological oxidation catalysts, they are *bioinspired* rather than *biomimetic*. The oxygenated products obtained in high yields, alkyl hydroperoxides, are valuable chemicals in their own right and also a source of "activated oxygen" for oxidation of other organic molecules.

Acknowledgements

Drs. A. Walton and K. U. Ingold are gratefully acknowledged for metal analysis and for useful discussions, respectively.

Keywords

Alkyl hydroperoxides, manganese, ribonucleotide reductose, hydrogen peroxide, catalase activity, tetranuclear manganese, PSII, OEC, cumene, cumene hydroperoxide, biomimetic catalysis, bioinspired catalysis, C-H activation (hydrogen activation), oxygen activation, hydroperoxide decomposition, radicals (alkyl radicals and hydroperoxy radicals) and hydrogen rebound (rebound mechanisms).

REFERENCES

1. K. Weissermel and H.-J. Arpe, "Industrial Organic Chemistry", Verlag Chemie, New York (1978).

2. Some representative examples are:

   a) R.S. Drago, *Coord. Chem. Rev.* 117:185 (1992).

   b) R.A. Sheldon and J.K. Kochi, "Metal-Catalyzed Oxidations of Organic Compounds", Academic Press, New York (1981).

   c) D.T. Sawyer, "Oxygen Chemistry", Oxford University Press, New York (1991).

   d) L.I. Simandi, "Catalytic Activation of Dioxygen by Metal Complexes", Kluwer Academic Publishers, Dordrecht (1992).

   e) C.L. Hill, Ed., "Activation and Functionalization of Alkanes", J. Wiley & Sons, New York, (1989).

   f) A.E. Martell and D.T. Sawyer, Eds., "Oxygen Complexes and Oxygen Activation by Transition Metals", Plenum Press, New York (1988).

   g) R.H. Holm, *Chem. Rev.* 87:1401 (1987).

3. a) C.F. Yocum, C.T. Yerkes, R.E. Blankenship, R.R. Sharp, G.T. Babcock, *Proc. Natl. Acad. Sci. USA*, 78:7507 (1981).

   b) J. Amesz, *Biochim. Biophys. Acta*, 726:1 (1983).

   c) G.C. Dismukes in U.L. Schramm, F.C. Welder, Eds., "Manganese in Metabolism and Enzyme Function", Academic Press, London (1986).

   d) G. Renger, *Angew, Chem.*, 99:660 (1987); *Angew, Chem. Int. Ed. Engl.*, 26:643 (1987).

   e) G.W. Brudvig, *J. Bioenerg. Biomemb.*, 19:91 (1987).

   f) V.L. Pecoraro, Ed., "Manganese Redox Enzymes", VCH, New York (1992).

   g) G.N. George, R.C. Prince, S.P. Cramer, *Science*, 243:789 (1989).

4. B. Kok, B. Forbush, M. McGloin, *Photochem. Photobiol.*, 11:457 (1970).

5. W.D. Frasch, R. Mei, *Biochem. Biophys. Acta*, 891:8 (1987).

6. R.T. Stibrany, S.M. Gorun, *Angew. Chem. Int. Ed. Engl.*, 29:1156 (1990). A structurally related complex has been reported by W. H. Armstrong in Ref. 3f.

7. a) V.V. Barynin, A.I. Grebenko, *Dokl. Akad. Nauk S.S.S.R.*, 286:461 (1986).

   b) C.V. Khangulov, N.V. Voevodskaya, V. U. Barynin, A. I. Grebenko, V. R. Melik-Adamyan, *Biofizika*, 32:960 (1987).

8. US Patent 4,153,635 (1979).

9. a) H.S. Blanchard, *J. Am. Chem. Soc.*, 82:4433 (1968).

   b) A.I. Minkov, N.P Keier, *Kinet. i Katal.*, 8:160 (1967).

10. J.A. Howard and K.U. Ingold, *Can. J. Chem.* 46:2655 (1968).

11. Y. Kamiya, *Bull. Chem. Soc. Japan* 43:830 (1969).

12. M. Kashiyae, K. Kashe, S. Yoshitomi, *Nippon Kagaku Kaishi*, 357 (1990).

13. H. Kanai, H. Hayashi, T. Koike, M. Ohsuga, M. Matsumoto, *J. Catal.*, 138:611 (1992).

14. Y. Engström, S. Eriksson, I. Thelan, M. A. Kerman, *Biochemistry*, 18:2941 (1979).

15. A. Willing, H. Follmann, G. Auling, *Eur. J. Biochem.*, 170:603 (1988)

MECHANISM OF DIOXYGEN ADDITION
TO NICKEL-BOUND THIOLATES

Marcetta Y. Darensbourg*, Patrick J. Farmer, Takako Soma,
David H. Russell, Touradj Solouki, and Joseph H. Reibenspies

Department of Chemistry
Texas A&M University
College Station, Texas 77843

INTRODUCTION

Monomeric complexes of Ni^{II} containing terminal (non-bridged) thiolate sulfur donors are relatively rare. The reason for this is explained by the simultaneous strong Ni-S σ-bonds in the presence of four-electron destabilizing interactions of lone-pairs of electrons on thiolate-sulfur with filled metal d-orbitals.[1] Relief of such an anti-bonding interaction is particularly effective by metal electrophiles and hence thiolates serve as

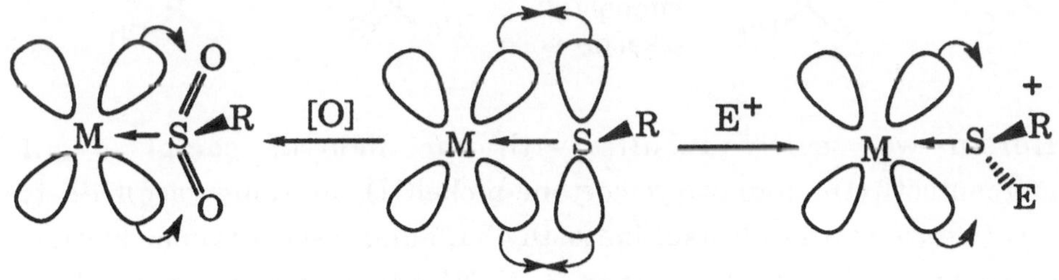

nucleating centers, generating polymetallics, sometimes desired, but more usually a frustration to synthetic or mechanistic chemists.

The Activation of Dioxygen and Homogeneous Catalytic Oxidation,
Edited by D.H.R. Barton *et al.*, Plenum Press, New York, 1993

The air sensitivity of electron-rich or anionic thiolates is also accounted for by this destabilizing π* interaction. Reaction with O₂ is widely believed to be non-specific, generating intractable mixtures of metal oxides, metal sulfides and oxidized ligand products. Nevertheless there have been three recent reports of well-characterized sulfur-oxygenated products resulting from reaction of gaseous oxygen with nickel thiolate complexes.[2-4] Scheme I summarizes the results of Maroney et al., on the reaction of the anionic cyano transdithiolatenickel(II) with O₂ which concludes with a monosulfinato complex, reported to be not further reactive with O₂.[2] In contrast the dianionic, dithiolene complex of Schrauzer and Chadra results in a bissulfinatocomplex with RSO₂⁻ units in cis orientation and shown in the x-ray crystal structure to be tightly ion-paired with a sodium ion, Scheme II.[3]

Scheme I (Maroney, 1989)

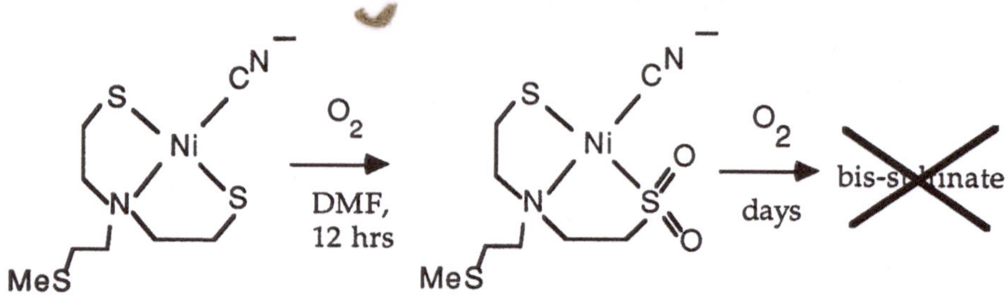

Scheme II (Schrauzer, 1990)

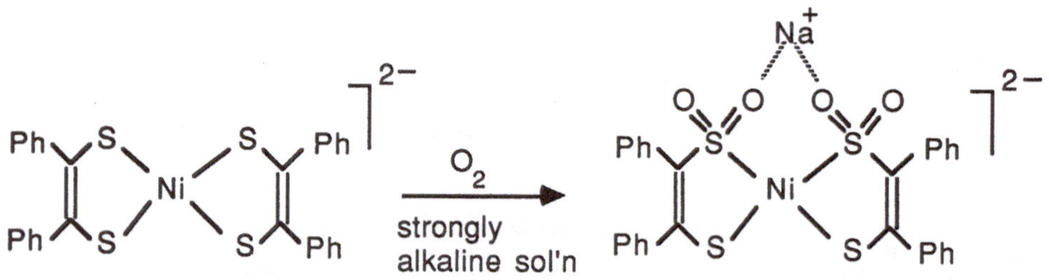

Herein we review results with the neutral complex, N,N'-bis(mercaptoethyl)-1,5-diazacyclooctane-nickel(II), or (bme-daco)Ni^{II}, **1**, to yield N-(mercaptoethyl)-N'-(sulfinatoethyl)-1,5-diazacyclooctanenickel(II), or (mese-daco)Ni^{II}, **2** and N,N'-bis(sulfinatoethyl)-1,5-diazacyclooctanenickel(II), or (bse-daco)Ni^{II}, **3**, Scheme III.[4] The mechanistic implications of isotopic labelling experiments as well as spectroscopically observed intermediates are discussed.

Scheme III (Farmer, Soma, Darensbourg 1992)

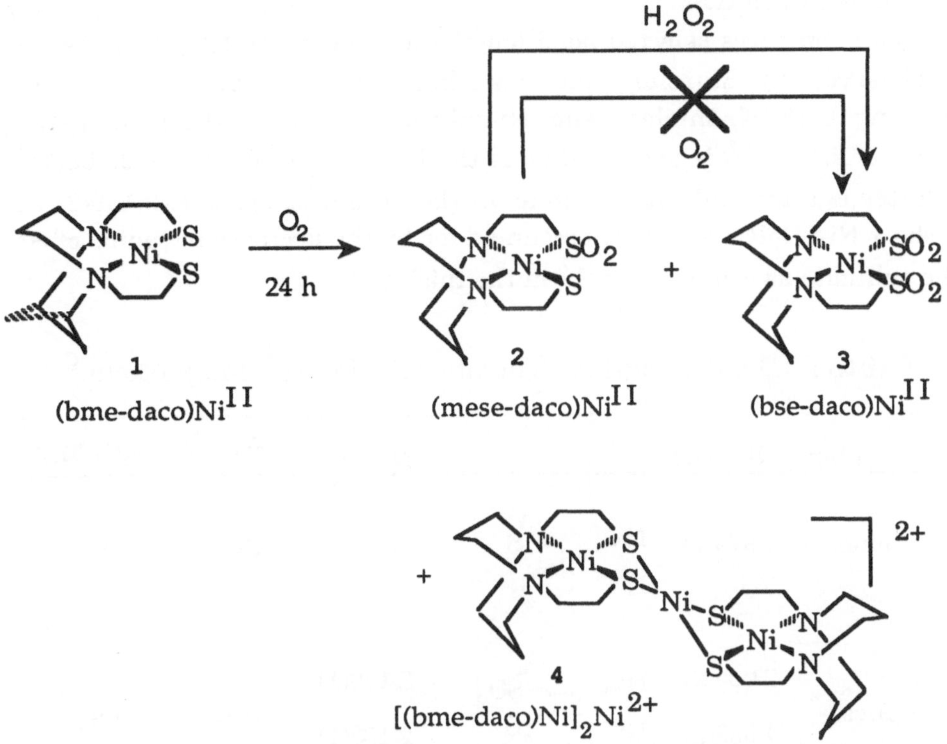

H_2O_2

O_2

O_2

24 h

1

(bme-daco)Ni^{II}

+

2

(mese-daco)Ni^{II}

+

3

(bse-daco)Ni^{II}

+

4

[(bme-daco)Ni]$_2$Ni^{2+}

2+

RESULTS AND DISCUSSION

Characterization of the complexes. Reaction of diamagnetic (bme-daco)Ni^{II} with O_2 in CH_3CN produces the mixture of products shown in Scheme III which may be efficiently separated by flash chromatography on silica gel columns. A typical distribution following a 24 hr reaction at 1 atm O_2 is 50% recovery of (bme-daco)Ni^{II}, 20% and 5% yields of (mese-daco)Ni^{II} and (bse-daco)Ni^{II}, respectively and an undetermined amount (presumably 25% from decomposition of ~8% **1** producing free Ni^{2+}) of the trimetallic remains on the top of the column. All complexes referred to in Scheme III have been characterized by x-ray crystallography. The x-ray crystal structure of starting material, **1**,[5] shows disorder in the daco ring resulting in a 50:50 distribution of chair/chair and chair/boat configurations of the the fused metallodiazacyclohexane rings. The ethylene arms linking N to S are staggered with respect to each other across the NiN$_2$S$_2$ pseudo square plane. The x-ray structures of the sulfinates show them to be four coordinate and approximately square planar with slight tetrahedral twists which possibly account for the observed paramagnetism of the complexes in solution.[4,6]

Interestingly, complex **1** also has a significant Td twist, Table 1, but is diamagnetic in solution.

Table I compares selected bond lengths for complexes **1**, **2**, **3**, and **4**. As was observed by Maroney and coworkers for an octahedral complex containing both Ni-thiolate and Ni-sulfinate groups,[2] the Ni-$S_{sulfinyl}$ distance of 2.140(1)Å is shorter than is the Ni-$S_{thiolate}$ distance, 2.163(1)Å. The latter is essentially the same as in the parent complex **1**, 2.159(2) Å.[5] The short Ni-SO₂R distance is accounted for by the decrease in radius of S^{+2} in the sulfinate as compared to S^{-2} in the thiolate.

Table I. Bond Lengths,[a] Tetrahedral Twist,[b] Magnetism.[c]

Complex	Ni-N(Å)		Ni-S (Å)	Td (°)[b]	μ_{eff}(B.M.)[c]
1 bme-	1.979(7)	N⋯Ni⋯S / N–Ni–S	2.159(3)	13.2	0
2 mese-	2.000(2) / 1.982(2)	N⋯Ni⋯SO₂ / N–Ni–S	2.140(1) / 2.163(1)	18.3	1.33
3 bse-	1.981(9)	N⋯Ni⋯SO₂ / N–Ni–SO₂	2.133(3)	15.9	1.23
4 (Br)₂	1.961(4) / 1.953(4)	[N⋯Ni⋯S–Ni / N–Ni–S]₂	2.155(1) / 2.148(1)		

a) Structural data for **1** from ref. 5, **2** from ref. 4, **3** from ref. 3; b) The Td twist angles are measured as the angle of the intersection of the normals of the S(1)-X-S(2) and N(1)-X-N(2) planes where X is the centroid of the N₂S₂ plane.; c) determined by the Evans' method.

The molecular structure of **4** is the common[7] stair-step structure with avg. Ni-S distances in the central NiS₄ environment (2.200 Å) significantly longer than that in the NiN₂S₂ portion of the complex (2.152 Å). The severe folding angle between the NiS₄ square plane and the best plane calculated for an NiN2S2 unit (Ni is 0.17Å above the N2S2 best plane) is 116.9° resulting in Ni--Ni distances of 2.685(1)Å. This trimetallic is formed on time of mixing of Ni^{2+} with solutions of **1**. It results when salts of Ni^{II} are added to **1** , when **1** is exposed to acid, or when **1** reacts with one-electron oxidants such as cerium(IV), iodine, or NO⁺. Hence it is assumed that the production of **4** in the reaction of **1** with O_2 is a result of an electron transfer path, yielding ligand oxidized product (presumably disulfides) and free nickel(II) which rapidly scavenges its parent complex **1**.

Complexes **2** and **3** were further identified by characteristically intense bands in the $\upsilon(SO)$ region of the IR. For **2**, assignment of the bands at 1182 and 1053 cm⁻¹ to $\upsilon(SO)$ was confirmed by shifts to 1146 and 1018 cm⁻¹ in samples prepared from 98% $^{18}O_2$. Significantly, the UV-vis spectrum of **3** as it eluted off the column showed a highly absorbing contaminant, but the IR spectra in the $\upsilon(SO)$ region was little different from that of the isolated and purified form which has intense bands at 1192, 1180, 1070, and 1031 cm⁻¹.

Ligand Reactions. Thiols and thiolates are known to undergo metal catalyzed O_2 oxidation forming primarily disulfide products.[8] Stirring solutions of both the diprotonated and sodium salt of the bme-daco ligand under O_2 for 2 days produced no intense bands in the $\upsilon(SO)$ region of the IR indicative of S-oxygenates. Likewise, stirring a solution of the dimeric [(bme-daco)Zn]₂ [9] under O_2 for 4 days yielded quantitative recovery of the starting material.

Radical and Oxygen Traps. Reactions carried out in the presence and absence of 10 molar % galvinoxyl as radical trap gave essentially equivalent product yields, ruling out a radical initiated autooxidation (the dithiolate **1** itself should function as an efficient radical trap). Also, in reactions to which Ph₃P or Me₂S were added as activated oxygen traps, products such as Ph₃PO or Me₂SO were not detected.

Kinetic monitor of the oxygenation. Table II lists the electrochemical data derived from cyclic voltammograms of complexes **1** through **4**. Reversible waves assigned to the $Ni^{II/I}$ redox event are more accessible by ca. 300 mV with each conversion of thiolate to sulfinato ligand. Irreversible oxidation events, displaced by ca. 2 V from the $Ni^{II/I}$ couple, are observed for the complexes containing thiolate sulfurs and are ascribed to sulfur radical formation. In the case of the bissulfinate, the oxidation becomes reversible and is assigned to the $Ni^{II/III}$ couple; this assignment is corroborated by the esr spectrum of chemically oxidized **3**.[6]

The good reversibility of the $Ni^{II/I}$ couple in the cyclic voltammograms has permitted a kinetic monitor of the reaction presented in Scheme III

Table II. Cyclic Voltammetric Data.

Complex	Complex Reduction			Complex Oxidation			
	$E_{1/2}$ (mV)	ΔE (mV)	i_{pa}/i_{pc}	E_{pc} (mV) (irrev.)	$E_{1/2}$ (mV) (rev.)	ΔE (mV)	i_{pa}/i_{pc}
1 bme-	- 1944	65	1.01	+ 360			
2 mese-	- 1631	70	0.97	+ 620			
3 bse-	- 1339	72	0.93		+ 847	76	0.92
4 (Cl)₂	-715	63	0.95	+1200			

All potentials scaled to NHE in CH_3CN solutions, ref. 6.

making use of Osteryoung square wave voltammetry. Reactions of 0.1mMol solutions of 1 with O_2 maintained at 1, 2 and 3 atm in various solvents were carried out in a Fischer-Porter tube. Samples were withdrawn at timed intervals and the aliquots degassed with argon prior to electrochemical analysis. The bulk reaction was repressurized and allowed to proceed. A sample analysis showing the disappearance of the starting material, **1**, and

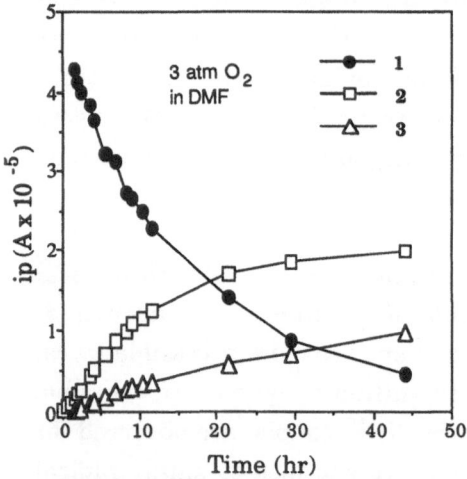

Figure 1. Plot of **1**, **2** and **3** followed over reaction course by OSWV.

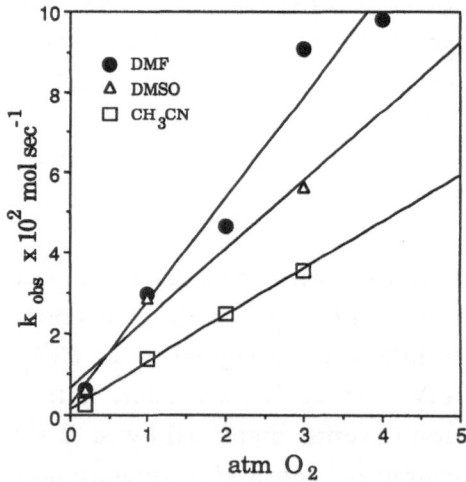

Figure 2. Oxygen dependence in various solvents for reaction of **1** with O_2.

appearance of products, **2** and **3**, is shown in Figure 1. The observed exponential decay of **1** yields linear plots of ln(peak intensity) vs. time. The pseudo-first order rate constants thus obtained show a linear dependence on

O_2 pressure as demonstrated in Figure 2. Hence the kinetic rate expression, determined for the disappearance of **1** is overall second order:

$$rate = -d[bme\text{-}daco)Ni]/dt = k[bme\text{-}daco)Ni][O_2]$$

Notably, the reaction in DMSO and DMF show similar and greater values of of k_{obs} than in the less polar CH_3CN (DMF 2.96, DMSO 2.87, CH_3CN 1.37 x 10^{-2} sec^{-1} at 3 atm oxygen). The protic solvent MeOH completely inhibits reaction while in water, different reaction products are obtained, *vide infra*.

Chemical Oxygenation. Hydrogen peroxide could also be used as the oxygen source and reacted on time of mixing with (bme-daco)Ni^{II} in MeOH to yield the same mixture of products as the O_2 reaction, eq 1. The yield of **2** optimized (70%) at 2 equivalents of H_2O_2, and **3** was the predominant (85%) product for 4 equivalents of H_2O_2 in MeOH or CH_3CN. This stoichiometry suggests that one O-atom is transferred per H_2O_2 molecule used in a sequence as described by eq 1. Further reaction of **2** with additional equivalents of H_2O_2 yields **3**.

Metal bound sulfenates (or metallo-sulfoxides) resulting from sequential O-atom transfer to thiolate sulfur have been spectroscopically identified as intermediates only in the reaction of substitutionally inert Co^{III}-amino-thiolate complexes with hydrogen peroxide.[10] In contrast our work and that of others[11] find the MSR functionality in coordination complexes reacts directly with oxygenating reagents to yield principally metallosulfinates, MSO$_2$R. The propensity to form the MSO$_2$R from MSR extends to single O-atom transfer reagents, such as peracids or dioxiranes.[11a]

Implications. A major difference in O_2 and H_2O_2 reactivity was indicated on the summary in Scheme III. Whereas isolated and purified **2** is **unreactive** with dioxygen under the same conditions as the original reaction (CH_3CN solvent, bubbling O_2, 24 h), reaction of **2** with hydrogen peroxide yields **3** immediately and quantitatively.

We are thus faced with two interesting questions:

1. What is the mechanism of O_2 addition to **1** to yield the monosulfinate complex **2**? Is it O-atom transfer as in the reaction of H_2O_2? Or is a molecular (concerted) process operative?

2. What is the source of the bissulfinate in the original reaction? As indicated in the reaction monitor displayed in Figure 1, the bissulfinate **3** forms early in the reaction and apparently simultaneously with **2**. Do they involve the same or different precursors/intermediates?

Isotopic Labelling Studies.

$$\text{(2)}$$

Label Retention

Label Scrambling

O-atom or Molecular O_2 addition. We addressed the first question in collaboration with Prof. D.H. Russell.[4] Matrix-assisted laser desorption (MALD) ionization was used to obtain Fourier transform ion cyclotron resonance (FT-ICR) mass spectra on Cs^+ doped isotopomers of **2** and **3**. Isotopic labelling experiments ($^{18}O_2/^{16}O_2$ mixtures) demonstrated that both oxygens in the sulfinate ligand of **2** were derived <u>from the same</u> dioxygen molecule. Figures 3a, 3b and 3c display the spectra in the [M+Cs]$^+$ mass region for **2** isolated from reaction of **1** with natural abundance $^{16}O_2$, a 56:44 mixture of $^{16}O_2$:$^{18}O_2$, and $^{18}O_2$(98%), respectively. For the mixed label experiment, Figure 3b, a fit of the data to a combination of the two pathways (O-atom addition vs. O_2 molecular addition) finds that greater than 90% of the product is formed by molecular addition (fit values: 92% retention, 8% scrambling).

Analysis of the [M+Cs]$^+$ mass region for bissulfinate, **3**, isolated from the same reaction mixtures of **1** with natural abundance $^{16}O_2$, a 56:44 mixture of $^{16}O_2$:$^{18}O_2$, and $^{18}O_2$(98%), are seen in Figures 4a, 4b and 4c, respectively. Three mechanistic pathways were considered in analyzing the mixed label experiment. The first, "retention/retention", signifies the product in which one $^{16}O_2$ and one $^{18}O_2$ adds molecularly to **1** at individual sulfur sites, producing the $S^{16}O_2/S^{18}O_2$ isotopomer; or across sulfur sites, producing the

$(S^{16}O^{18}O)_2$ isotopomer (mass spectral analysis of the parent ion can not distinguish the two possibilities, *vida infra*). For the "scrambling", or O-atom addition case, two possibilities exist. The first is that oxygenation of the first sulfur is by molecular addition and the subsequent formation of **3**

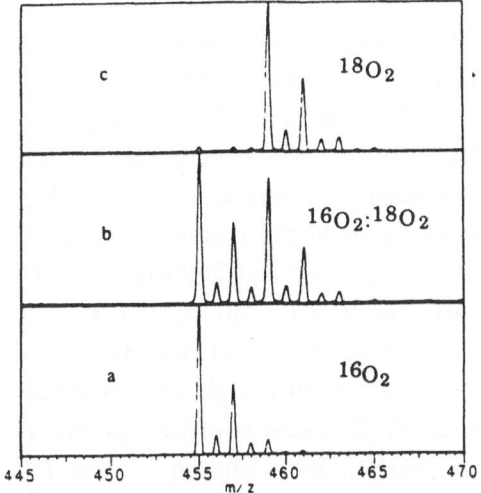

Figure 3. FT-ICR spectra of **2** isotopomers formed from $^{16}O_2/^{18}O_2$ ratios given in text.

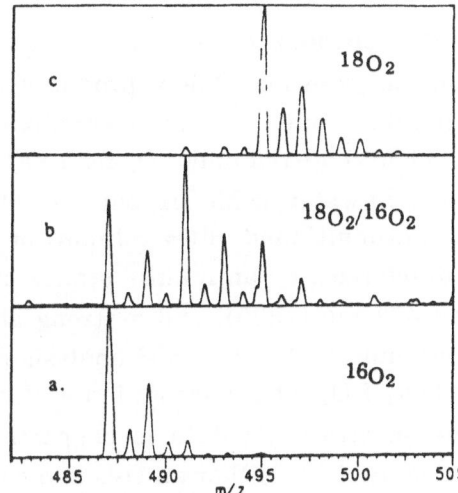

Figure 4. FT-ICR spectra of **3** isotopomers formed from $^{16}O_2/^{18}O_2$ ratios given in text.

scrambles O_2; the second is that both SO_2 units of **3** are formed solely by O-atom addition or a random exchange process. The experimental spectra, Figure 4b, for the mixed label experiment intimate molecular addition of O_2; that is, O-atom or random exchange processes are ruled out based on the lack of isotopomers containing odd numbers of ^{16}O or ^{18}O. A statistical fit finds that > 90% of the product is formed by molecular addition (fit values: 94% retention/retention and 6% retention/scrambling).

Bissulfinate formation. Attempts to limit the mechanistic possibilities for the production of **3**, included an examination of 1) any catalysts in the original reaction of **1** and O_2 that might promote the addition of a second O_2; 2) some intermediate complex which might collapse to **1** or add a second O_2 to yield **3**; 3) further labelling experiments which might shed light on the mechanism of O_2 addition to sulfur-sites prior to formation of **3**.

Experiments were run to test the possibility of the dithiolate, **1**, or the trimetallic, **4**, acting as a catalyst for the production of the bissulfinate. Solutions containing a 5:2 ratio of the monosulfinate, **2**, to the starting dithiolate, **1**, stirred under O_2 for 3 days showed only a slight increase in **3** production from that predicted from **1** alone. A similar experiment using added trimetallic, **4**, produced trace amounts of **2** and **3**, which may rationalized by reaction of an equilibrium concentration of **1** (from the dissociation of **4** into two **1** and one free Ni^{2+}). Thus neither **1** nor **4** are considered candidates for the dioxygenating catalyst.

Precursors to the sulfinate products. The reaction of **1** with O_2 in aqueous solution slowly produces a different product, **X**, which is well separated from **1** by column chromatography. Isolable yield from 0.10 g **1** in 30 ml H_2O stirred under O_2 for 7 days is 40 to 60 mg. The compound is most stable in water or MeOH, but is easily decomposed by trace acid (such as on an unconditioned SiO_2 column) or by heating in organic solvents. It is characterized by an intense orange color [UV-vis max (ε) at 356 nm (>3600) and 473 nm (>380)] and a strong absorbance at 910 to 930 cm^{-1} in its IR spectrum. C, H, N, and O analysis of purified powders matches a dioxygen adduct, **1**·O_2, the monosulfinate, **2**, or a sulfenate with a H_2O of solvation. The spectroscopic data is suggestive of an S=O moiety, either as in a sulfenate [Ni(SO)R] or an O-bound sulfinate [NiO(SO)R], although a Ni or S-bound

peroxidic species (υ(O-O) ~ 900 cm^{-1})[12] is also possible. More importantly, this compound further reacts with O_2 in CH_3CN to give the bissulfinate, **3**. Under typical oxygenation conditions, (0.3 mM in CH_3CN, 1 atm O_2, 1 day) **X** yields 20% **3**, along with 60% recovered **X** and smaller amounts of **2** and **4**.

Although **X** is not observed in the reaction of **1** with O_2 in CH_3CN, a compound **Y** is found as a highly absorbing species (λ_{max} at 365 and 440 nm) which elutes with **3** during workup. It is also found in the mixture of products resulting in reaction of **1** with 3 equiv H_2O_2. Its red-orange color and v(SO) IR spectrum are similar to **X**, but it elutes much more slowly on SiO_2. Concentrated powders of **Y** give C, H, N, and O analysis consistent with a product of intermediate oxidation state such as a mixed sulfinate/sulfenate. It does not react further with O_2.

While difinitive structural characterization of **X** and **Y** elude us, their existence as possible precursors to the sulfinates led to reconsiderations of possible intermediate species in light of the isotopic labelling experiments. Precursors in which the initial O_2 addition adds to the same sulfur site, as in

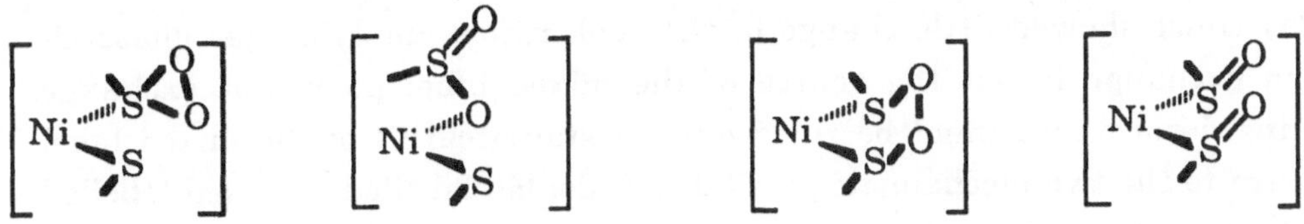

thiadioxirane **A** or the O-bound sulfinate **B**, should produce only dilabelled RSO_2^- in the mixed isotope experiments. If both sulfur sites are involved during the O_2 addition, as in dithiadioxirane **C** or the bissulfenate **D**, some scrambling between RSO_2^- sites may occur in the formation of the bissulfinate.

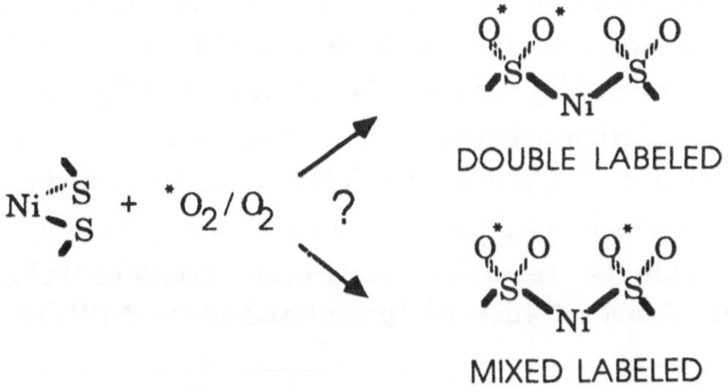

DOUBLE LABELED

MIXED LABELED

Label exchange in the bissulfinate. The possibility of oxygen atom scrambling in the bissulfinate formation (i.e. the formation of $Ni[RS^{16}O^{18}O\text{-}]_2$ from the mixed label experiment, *vida supra*) was checked first by examination of the thermally vaporized FABS MS spectra of **3** from the mixed label experiments above. No parent ion peaks were detected, but SO_2 peaks at m/z ratios 64 ($S^{16}O_2$) and 68 ($S^{18}O_2$) were seen in the spectra from the reaction using natural abundance $^{16}O_2$ and $^{18}O_2(98\%)$, respectively. The mixed label spectra (from a 56:44 mixture of $^{16}O_2{:}^{18}O_2$) displayed an additional peak at m/z ratio 66 which would be assigned to $S^{16}O^{18}O$, the mixed label sulfinate. Both the possibility of exchange over time (~ 6 months elapsed before the FABS analysis) and the presence of variable background ion peaks in the same region (making quantification difficult), led us to repeat the mixed label experiment. FT ICR mass spectra of fresh and purified samples of **3** were analyzed for SO_2^- by using double laser shot desorption in the absence of matrix. Spectra were analyzed from natural abundance $^{16}O_2$, as well as mixtures of $^{16}O_2:^{18}O_2$ in ratios of 21:79 and 53:47. The latter spectra is seen in Figure 5. Both mixed label spectrum clearly show peaks at m/z 66 indicative of mixed label SO_2^-. Variation of the delay times showed little change in the peak ratios, implying gas phase O-atom exchange is not the source of the mixed label peak, but exchange during desorption cannot be ruled out. A statistical fit of the mixed label spectra to the two mechanistic possibilities, double labelled or mixed labelled sulfinates, shows that label mixing is the major pathway (91%).

This result was mechanistically revealing in that the labelling studies show that pairwise addition of dioxygen prevails during the formation of both **2** and **3**, implying the O-O bond is not broken until after both atoms of molecular oxygen are bound to the substrate complex. In **2** a single S-site O_2 addition occurs. In the formation of **3**, the bissulfinate, we now have compelling evidence that each dioxygen molecule is divided between the cis sulfur sites.

Comments on Mechanism: Control experiments have shown that sodium salts of the free bme-daco anion do not add molecular O_2 to sulfur, but rather produce disulfides. Hence the activation of O_2 required for this formally forbidden reaction between a diamagnetic substrate, complex **1**, and $^3O_2$ might be ascribed to a result of thiolate binding to nickel. Dioxygen complexes of nickel are known[13] and have been demonstrated to behave as superoxide-like adducts, resulting in O-atom transfer behavior. For example, the O_2-adduct of nickel^{II} (postulated as O_2^--Ni^{III}) in a tetraaza

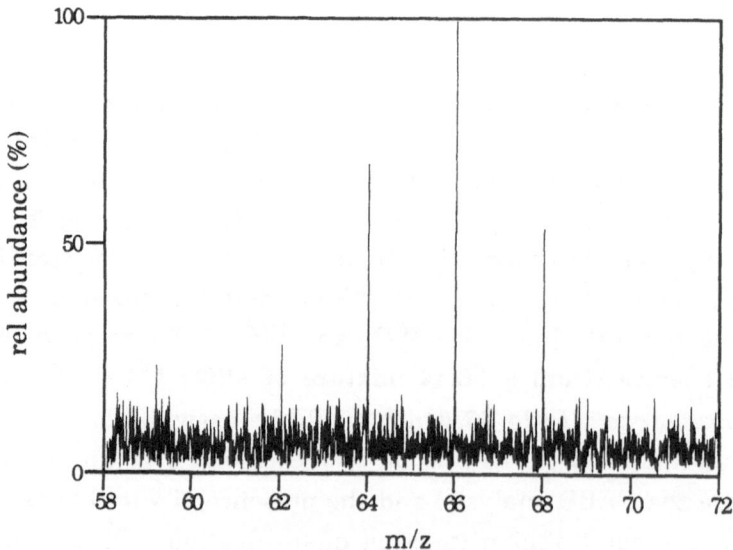

Figure 5. FT-ICR spectra of SO_2^- derived from **3** formed from 53:47 $^{16}O_2:^{18}O_2$ ratio as described in text.

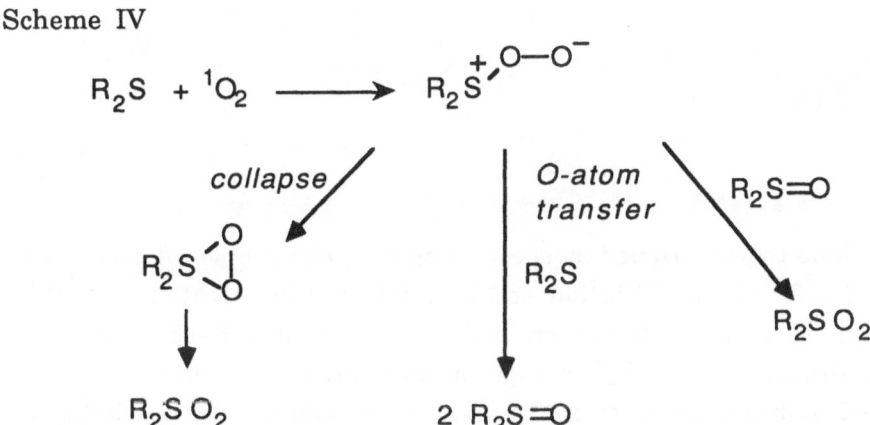

macrocycle hydroxylates activated C-H bonds as in the conversion of benzene to phenol.[13a] The isotopic labelling results described above find no evidence of O-atom chemistry and furthermore there is no evidence for O_2 binding at nickel. Solutions of the nickel complex under O_2 prepared at low temperature show no epr signals nor color changes which might indicate a similar Kimura/Martell type Ni[III]-superoxide complex.

The oxygenation of organic sulfides by singlet oxygen responds to subtituents and solvents in a manner which suggests O-atom transfer from a sulfur-based peroxido intermediate.[14] Sulfoxides are major products with sulfones produced in minor amounts. A recent revisit of this reaction using isotopic $^{18}O_2/^{16}O_2$ mixtures uncovered the alternate molecular route for the sulfone formation as shown Scheme IV.[15] That is, a majority of the sulfone is accounted for by molecular addition of O_2 rather than O-atom transfer.

Scheme IV

The reaction of (bme-daco)Ni[II] with O_2 appears to be similar to the organic sulfides. i.e., sulfur-based. All previous chemistry with 1 suggests great reactivity at sulfur. The mechanism shown in Scheme V invokes an O_2 adduct formation either as a single S-site peroxide or an adjacent double S-site metalladithiadioxirane,[3] which might serve as a common intermediate for the formation of both observed products, the mono- as well as the bissulfinate complexes. Such precursors or intermediates are consistent with the isotopic labelling results of 1) molecular addition of O_2 in both products; 2) label integrity in the SO_2 of the monosulfinate; and 3) mixed label in the SO_2 groups of the bissulfinate.

Scheme V

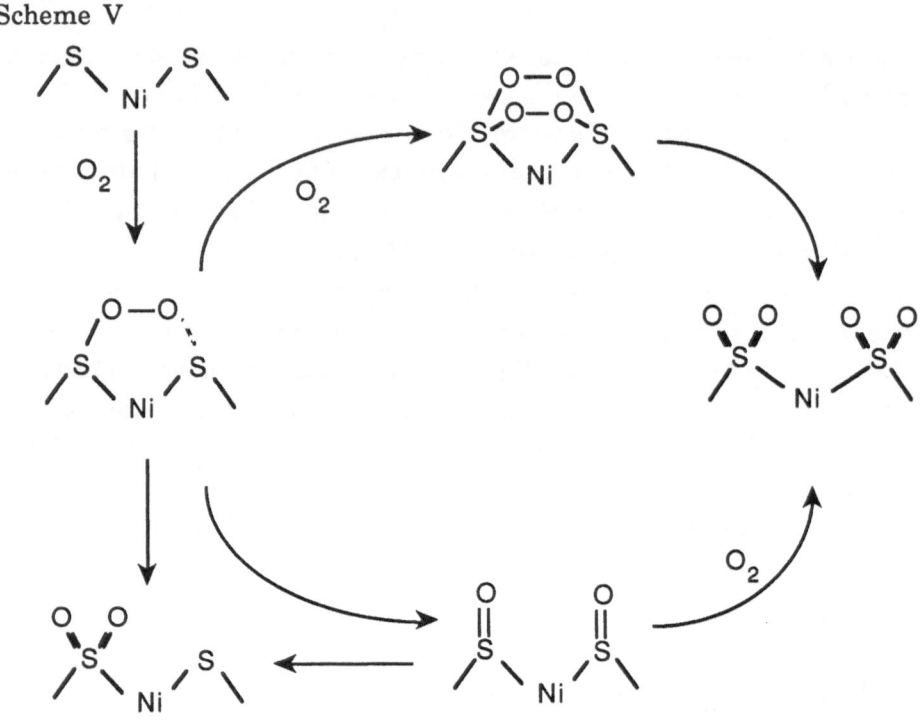

We continue to be intrigued by spectroscopically observable intermediates that have thusfar eluded isolation and complete characterization. To this end, ongoing research is based on a derivative of bme-daco which has replaced hydrogens by methyl groups on the carbons alpha to sulfur.[16] Sterically encumbered sulfur sites are expected to enhance the possibility of isolating partially oxidized products.

Acknowledgment. We thank our TAMU colleagues Professors D.H.R. Barton, A. E. Martell and D.T. Sawyer for their comments and interest in this work. Financial support from the National Science Foundation (CHE 91-09579 to M.Y.D. CHE 88-21780 to D.H.R., and CHE-8513273 for the X-ray diffractometer and crystallographic computing system), and the National Institutes of Health (RO1GM44865-01 to M.Y.D.) is gratefully acknowledged.

References

1. Ashby, M. T.; Enemark, J. H.; Lichtenberger, D. L. *Inorg. Chem.* **1988**, *27*, 191.

2   Kumar, M.; Colpas, G. J.; Day, R. O.; Colpas, G. J.; Maroney, M. J. *J. Am. Chem. Soc.* **1989**, *111*, 8323; Mirza, S.A.; Pressler, M.A.; Kumar, M.; Day, R.O.; Maroney, M. J. *Inorg. Chem.* **1993**, *32*, 977.

3. Schrauzer, G. N.; Zhang, C.; Chadha, R. *Inorg. Chem.* **1990**, *29*, 4104.

4. Farmer, P. J.; Solouki, T.: Mills, D. K.; Soma, T.; Russell, D. H.; Reibenspies, J. H.; Darensbourg, M. Y. *J. Amer. Chem. Soc.* **1992**, *114*, 4601.

5. Mills, D. K.; Reibenspies, J. H.; Darensbourg, M.Y. *Inorg. Chem.* **1990**, *29*, 4364.

6. Farmer, P. J; Reibenspies, J. H.; Lindahl, P. A.; Darensbourg, M. Y. *J. Amer. Chem. Soc.* **1993**, *115*, 4665.

7. a) Wei, C. H.; Dahl, L. F. *Inorg. Chem.* **1970**, *9*, 1878. b) Turner, M. A.; Driessen, W. L.; Reedijk, J. *Inorg. Chem.* **1990**, *29*, 3331. c) Drew, M. G. B.; Rice, D.A.; Richards, K. M. *J. Chem Soc. Dalton.* **1980**, *9*, 2075.

8. a) Cullis, C.F.; Hopton, J.D.; Trimm, D.L. *J. Appl. Chem.* 1968, *18*, 330. b) Cullis, C.F.; Trimm, D.L. *Discuss. Faraday Soc.* 1968, *46*, 144.

9. Tuntulani, T.; Reibenspies, J. H.; Farmer, P. J.; Darensbourg, M. Y. *Inorg. Chem.* **1992**, *31*, 3497.

10. a) Adzamli, I.K.; Lisbon, K.; Lydon, J.D.; Elder, R.C.; Deutsch, E. *Inorg. Chem.* **1979**, *18*, 303. b) Herting, D.L.; Sloan, C. P.; Cabral, A.W.; Krueger, J.H. *Inorg. Chem.* **1978**, *17*, 1649.

11. a) Schenk, W. A.; Frisch, J.; Adam, W.; Prechtl, F. *Inorg. Chem.*, **1992**, *31*, 3329 b) Weinmann, D.J.; Abrahamson, H.B. *Inorg. Chem.* **1987**, *26*, 3034.

12. Otsuka, S.; Nakamura, A.; Tatsuno, Y. *J. Am. Chem. Soc.* **1969**, *91*, 6995.

13. a) Kimura, E.; Sonaka, A.; Machida, R.; Kodama, M. *J. Am. Chem. Soc.* **1982**, *104*, 4255. b) Chen, D.; Motekaitis, R.J.; Martell, A.E. *Inorg. Chem* **1991**, *30*, 1396.

14. a) for example, Foote, C. S.; Peters, J. W. *J. Am. Chem. Soc.* **1971**, *93*, 3795. b) Akasada, T.; Sakurai, A.; Ando, W. *J. Am. Chem. Soc.* **1991**, *113*, 2696.

15. Watanabe, Y.; Kuriki, N.; Ishiguro, K.; Sawaki, Y. *J. Am. Chem. Soc.* **1991**, *113*, 2677.

16. Font, I.; Darensbourg, M.Y. unpublished work.

STUDIES ON THE MECHANISM OF GIF REACTIONS

Gilbert Balavoine[a], Derek H.R.Barton[b],Yurii V.Geletii[a,c,*], David R. Hill[b]

[a]*Institut de Chimie Molèculaire d'Orsay, Université de Paris Sud, Bat. 420, 91405 Orsay, France*
[b]*Department of Chemistry, Texas A&M University, College Station, TX 77843 USA*
[c]*Institute of Chemical Physics in Chernogolovka, 142432, Russia*

INTRODUCTION

The selective functionalization of saturated hydrocarbons which requires the activation of unactivated carbon-hydrogen bonds remains one of the most intriguing subjects of modern chemistry despite the great deal of research interest over the past two decades. The economic and practical functionalization of alkanes is expected to be among the major objectives for the end of this century. Much progress has been achieved in homogeneous and heterogeneous catalysis, in both the liquid and gas phases[1,2]. Catalytic oxidation of alkanes has been explored using various oxidants. Those conducted with dioxygen under mild conditions are especially rewarding goals[2]. Amongst the many approaches to the problem, one is based on mimicking nature. Nature's ability to use dioxygen to oxidize different C-H bonds in a selective manner is attracting chemists from different areas, including bio-, inorganic and organic chemistry. Nature does its job by using enzymes, referred to as oxygenases, which catalyze the oxidation process by directly inserting one or both atoms of dioxygen into the organic substrate giving mainly hydroxyl groups[3].

Oxygenases can be classified into three groups based on the co-factor required for catalytic activity of the enzyme[4]. One is the transition-metal free enzyme containing an organic prosthetic group such as flavin, but oxidation of unactivated C-H bonds by this enzyme has not yet been found. The other two are metalloenzymes containing copper or iron. Despite the fact that these enzymes are largely distributed in nature, the molecular mechanisms of their reactions are known in considerably less detail. Elucidation of this

The Activation of Dioxygen and Homogeneous Catalytic Oxidation,
Edited by D.H.R. Barton *et al.*, Plenum Press, New York, 1993

mechanism may potentially aid in the development of useful transition metal catalysts for mild and selective oxidative transformations and therefore is of current interest and is very important from both a fundamental and industrial chemistry perspective.

Among metallooxygenases the most extensively studied remains cytochrome P-450[5]. Cytochrome P-450 dependent monooxygenase is a family of heme-iron enzymes which catalyze the transfer of one oxygen atom from dioxygen into a wide variety of substrates according to reaction:

$$RH + O_2 + 2e^- + 2H^+ \longrightarrow ROH + H_2O$$

The reaction mechanism of cytochrome P-450 has been for many years the central subject for many chemists and is still in the focus of current interests. The reactive species which delivers oxygen atoms to substrates (at least in the case of alkane hydroxylation) is now generally accepted to be a high-valent porphyrin-iron-oxo intermediate, formally an ^{V}Fe=O species, resulting from the heterolytic cleavage of O-O bond of dioxygen. Many chemical model systems reproducing the chemistry of cytochrome P-450 were proposed, synthetic models for high-valent metal-oxo species being prepared, isolated and well characterized to prove the idea of involvement of such species in the cytochrome P-450 catalyzed reactions[6].

Much less is known about non-heme oxygenases. At the present time, the three well characterized non-heme iron containing enzymes are: methane monooxygenase (MMO) from *Methylococcus capsulatus* (Bath)[7], the hydrocarbon monooxygenase system (POM) from *Pseoudomonas Oleavaranse*[8], and the iron bleomycin complex (FeBLM)[9]. The first two catalyze the hydroxylation of aliphatic hydrocarbons by dioxygen and the third is responsible for oxidative cleavage of DNA. A binuclear iron center was shown to be in the active site of MMO, while POM and FeBLM have mononuclear iron-containing active sites to activate dioxygen.

Phenylalanine hydroxylase catalyzes the conversion of L-phenylalanine to L-tyrosine. This enzyme contains either copper or non-heme iron at the active site and requires a reduced pterin cofactor[10], a metal-peroxo adduct being postulated as an intermediate in the reaction[11].

Over the last two decades many intriguing and valuable non-heme containing catalytic systems have been reported which catalyze alkane hydroxylation by dioxygen and/or peroxides[12]. The simplest models based on the principles of coupled oxidation, are those in biological systems including iron (or other transition metal) salts as a catalysts, an organic ligand and hydrogen peroxide or dioxygen plus an electron source[13]. The cleavage of C-H bonds in the coupled oxidation of alkanes and $SnCl_2$ is performed by hydroxyl radicals[14], the same active species can be suggested in the other similar systems[15], such as that developed by Udenfriend[16].

THE HISTORY AND NOMENCLATURE OF GIF-CHEMISTRY

We have developed a family of chemical systems which we call the Gif-systems for oxidation and functionalization of alkanes under mild conditions. Consideration of the nature of the world after a billion or so years of life under reducing conditions suggested that the arrival of oxygen, generated by the blue-green algae, might have led to a concerted microbial oxidation of iron and of hydrocarbons. To test this idea adamantane was chosen as a key substrate due to its symmetry and simplicity of oxidation product identification, as well as its low volatility thus enabling accurate mass balance determination. Adamantane has 12 equivalent secondary- and 4 equivalent tertiary C-H bonds. The normalized C-H bond activation ratio C^2/C^3 (the ratio of the yields of appropriate oxidation products) serves as a useful mechanistic probe. Indeed, each reactive species exhibits its own regioselectivity ("finger-prints"), and, for example, typical C^2/C^3 ratios for radicals are between 0.05 and 0.15 in accordance with the bond energy of secondary and tertiary C-H bonds, the reactivity order for radical chain autoxidation processes being well established to be tert. C-H > sec. C-H > prim. C-H[17]. Pyridine was chosen as a solvent because metal complexes and hydrocarbons are both readily solubilized in this solvent. Moreover, at the beginning of Gif-chemistry Tabushi *et al.*.[12a] had taken advantage of pyridine as the solvent in a coupled oxidation of 2-mercaptoethanol and adamantane, catalyzed by (FeSalen)$_2$O, and observed in very small yield an unusual value of $C^2/C^3 \approx 1$ for adamantane oxidation but without any mass balance. Taking into account all of the reasons mentioned above, equivalent amounts of iron powder (as a source of electrons and a precursor of a catalytic complex) and acetic acid (as a source of protons) were added to a mixture of pyridine and adamantane. Initially, hydrogen sulfide was also added, but it was subsequently established that this was not needed for the reaction itself. However, it facilitated the start of the reaction by a surface effect. Ten years ago the first paper on Gif-chemistry was published[18] and now Gif-chemistry has its own rich history. In a subsequently developed system we used metallic zinc as a source of electrons and the resulting system (Gif[IV]) was catalytic in iron (up to 2000 turnovers)[18b]. Just after the first publications on Gif-chemistry one of us (YVG) started his research on this subject in Chernogolovka (Russia) and found that iron could be replaced by copper or other metals[19]. An electrochemical version of the Gif system was developed in Orsay, France (Gif-Orsay system)[20]. In this system zinc or iron powder was substituted by a Hg-cathode of an electrochemical cell. At the same time it was found that hydrogen peroxide could be used as an oxidant instead of dioxygen plus electron source, both iron or copper salts being efficient catalysts for alkane oxidation[19b, 21]. After moving to Texas A&M University (Aggieland) it was shown that KO$_2$ under argon could also be used as an oxidant (GoAgg[I] system)[22], however the efficiency (in terms of product yield based on the consumed oxidant) was rather low. Detailed examination of the oxidation by hydrogen peroxide showed that both systems with Fe-based (GoAgg[II])[22] or Cu-based (GoChAgg[I])[23] catalysts share the same characteristics as the other members of the Gif-family. For the former systems it was shown that a catalytic amount of an organic ligand could sharply change the oxidation rate[24], picolinic acid being the best ligand studied[24b-c] (GoAgg^{III}). Recently *tert*-butyl hydroperoxide has been used as an oxidant at elevated temperatures instead of hydrogen peroxide with Fe- or Cu-catalysts[25], the reaction pattern being similar to those of Gif-reactions. Other transition metal (Co, Ru, Ni, etc.) complexes (mainly bipyridine) can also be used in alkane oxidation by hydrogen peroxide in acetonitrile/pyridine solutions[26]. Thus, after ten years of intensive study the Gif-family consists of several members and is being enriched by the addition of new members. The nomenclature used for the Gif-systems is geographically based and is given below.

The Nomenclature of Gif Systems

System	Cat.complex	Reductant	Oxidant
Gif^{III}	Fe(II)/Fe(III)	Fe^0 (powder)	Dioxygen
Gif^{IV}	Fe(II)/Fe(III)	Zn^0 (powder)	Dioxygen
Gif-Orsay	Fe(II)/Fe(III)	Hg-cathode	Dioxygen
GoAgg^I	Fe(II)/Fe(III)		KO$_2$/Ar
GoAgg^{II}	Fe(III)		H$_2$O$_2$/Ar
GoAgg^{III}	Fe(III) picolinic acid		H$_2$O$_2$/Ar
GoChAgg^0	Cu(I)/Cu(II)	Cu^0 (powder)	Dioxygen
GoChAgg^I	Cu(II)		H$_2$O$_2$/Ar

The nomenclature is geographically based: **G** stands for **Gif-sur-Yvette** (France), **O** is for **Orsay** (France), **Agg** is for **Aggieland** (TexasA&M University, USA), **Ch** is for **Chernogolovka** (Russia).

REACTIONS IN GIF-SYSTEMS.

The first transformation discovered in Gif-systems was the oxygenation of saturated hydrocarbons where ketones are the major products of secondary C-H bond oxidation. In the electrochemical (Gif-Orsay) system[27] coulombic yields as high as 60% have been achieved with a hydrocarbon conversion 20-30%. Homogeneous systems utilizing hydrogen peroxide[21-23] as an oxidant are more convenient for mechanistic studies and can give yields higher than 90% based on the oxidant[28]. Ketones are formed without the intermediate formation of alcohols, the latter being oxidized very slowly to the corresponding ketones even at high alcohol concentration[19,29]. It is of great importance, and one of the major peculiarities, that tertiary C-H bonds are not preferentially oxidized under Gif-conditions[29]. At first, a high value for C^2/C^3 = 10-20 was observed for adamantane oxidation, but a more careful study showed that the formation of *tert*-adamantyl-pyridines, but not *sec*-adamantyl-pyridines[30] had occurred *via* the well known Minisci coupling reaction[31] between *tert*-adamantyl radicals and protonated pyridine[32]. When these coupled products were taken into account the C^2/C^3 ratio was lower (around 1)[30,32] than that previously observed but this value is still higher than that for the typical radical oxidation of adamantane (around 0.15). Moreover, adamantane seems to be an exceptional substrate, for other branched alkanes such as 3-ethyl-3-pentane, coupled products and diethylketone were only found in minor quantities[33]. The selectivity order for Gif-chemistry is therefore *tert.* ≤ *prim.* < *sec.* The oxidation of alkyl aromatic compounds led to the corresponding ketones or aldehydes. The secondary C-H bonds were found to be the most reactive[34] and there was no detectable attack on the aromatic ring.

Olefins are not epoxidized in Gif-systems, allylic oxidation to ketone and alcohol is a major pathway in the case of cylohexene, cyclohexene being oxidized at about the same rate as cyclohexane[29]. In the case of 1,1-disubstituted olefins the C=C bond is cleaved giving ketone and formaldehyde[35]. Here it is worth mentioning that alkane oxidation is not suppressed by the presence of other easily oxidizable compounds, such as alcohols, thiols, etc.[21,36,37], this peculiarity is discussed below.

The addition of different trapping reagents diverts the formation of ketone to the appropriate monosubstituted alkyl derivatives: $CBrCl_3$ leads to alkyl bromides, diphenyl diselenide produces (phenylseleno)alkanes, etc. [38-39](see details below). Thus, the most important features of Gif-chemistry are given below.

The Main Peculiarities of Gif-Reactions.

- Oxidation of saturated hydrocarbons under mild conditions with high yields and selectivities where the major products are ketones. Alcohols are not reaction intermediates.
- Addition of different trapping reagents diverts the formation of ketones to the appropriate monosubstituted alkyl derivatives, e.g. $CBrCl_3$ affords RBr in quantitative yield, Ph_2Se_2 produces a quantitative yield of the RSePh, *etc*.
- The presence of an excess of some easily oxidizable compounds (such as alcohols, aldehydes, PhSeH, Ph_3P, $P(OMe)_3$, *etc.*) does not significantly suppress the oxidation of alkanes.
- Olefins are not epoxidized, but instead yield unsaturated ketones, oxidation of some olefins may lead to C=C bond cleavage as well.
- **The selectivity of C-H bond oxidation is *sec.* > *tert.* > *prim.***
- ***sec.* Alkyl free radicals are not reaction intermediates.**

THE THEORY OF TWO INTERMEDIATES

Under Gif-conditions ketones and alcohols are not interconverted. Reducing agents, such as dianisyl telluride, added in excess compared to the alkane transformed did not suppress the total oxidation yield but changed the ketone/alcohol ratio in favor of the alcohol[40], a similar effect being observed for PhSH and Ph_3P[37]. This suggested that a common reaction intermediate B was a precursor of both the ketone and alcohol.

As mentioned above, the addition of different trapping reagents diverts the formation of ketone and alcohol giving other alkyl derivatives in similar yields to that of the oxygenation process[38-39]. This data proves that the reaction pathway includes an intermediate A, which is a precursor of intermediate B[37,41].

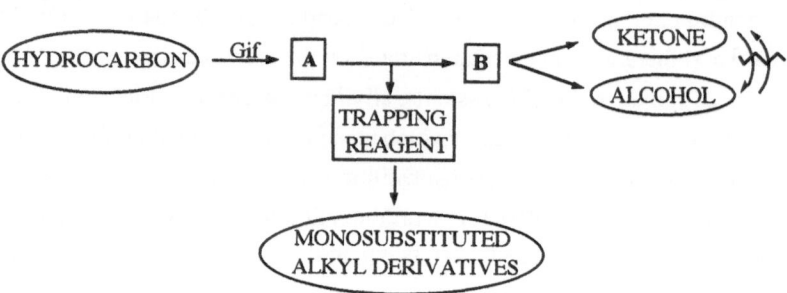

THE IDENTIFICATION OF INTERMEDIATE B[37,41]

Chemical and spectroscopic studies (^{13}C and ^2H nmr) of the GoAgg[II] system have fully proved that intermediate B is an alkylhydroperoxide. Cyclohexyl hydroperoxide was isolated from cyclohexane oxidation in the GoAgg[III] system and was identified by comparison with an authentic specimen. In the Gif[IV] system reducing agents such as thiophenol, benzeneselenol, diphenyl disulfide and diphenyl diselenide strongly decreased the ketone to alcohol ratio without significantly changing the total oxidation yield, proving the intermediacy of the hydroperoxide. Similar data (^{13}C nmr and quenching experiments with triphenylphosphine) has been obtained for GoChAgg reactions[42]. These results can be interpreted unambiguously, namely, that the alkyl hydroperoxide is an intermediate in Gif-type reactions. Analysis of kinetic curves for alkylhydroperoxide, ketone and alcohol formation (see ref. 37, Fig. 7), shows that initial rate of ketone accumulation, was not equal to zero, therefore one could not exclude another (not *via* alkylhydroperoxide) pathway affording ketone: but other explanations are also possible.

THE ORIGIN OF THE OXYGEN ATOMS IN OXIDATION PRODUCTS

During oxygenation under Gif conditions, the oxygen atoms in products may conceivably arise from water, hydrogen peroxide or dioxygen. This dilemma has been resolved by using labeled $H_2^{18}O$ and $^{18}O_2$. The oxygen atom of cyclododecanone, formed from the oxidation of cyclododecane by hydrogen peroxide under GoAgg[II] conditions, is in fact largely derived from dioxygen[37,43]. More detailed studies[42] have also confirmed this result in other Gif-systems. Thus, hydrogen peroxide was not a direct chemical source of the oxygen atoms in ketone and alcohol but, played two different roles: it formed the reactive iron species and produced dioxygen, which later afforded the alkyl hydroperoxide. Moreover, it was proved that all three products (alkyl hydroperoxide, ketone and alcohol) contained oxygen atoms from dioxygen, and not from $H_2^{18}O$, alcohol being formed from the reduction of the hydroperoxide. Consequently, the oxidation pathway in Gif reactions includes interaction of dioxygen with intermediate A to afford intermediate B (the alkyl hydroperoxide).

STUDIES ON THE NATURE OF INTERMEDIATE A:
Why Is It Not a Free Alkyl Radical?

As mentioned above, intermediate A reacts with dioxygen to give the alkyl hydroperoxide. In the presence of different trapping reagents the corresponding monosubstituted alkyl derivatives are formed instead of ketone or alcohol. These trapping reagents: Ph_2Se_2, Ph_2S_2, $CBrCl_3$, CCl_4, TEMPO (2,2,6,6 tetramethyl-1-piperidinyloxy radical) as well as dioxygen itself are well known traps for alkyl radicals and a free-radical pathway has been proposed for the GoAgg^{II} system[43]. Taking these results into account one could propose the following oxidation pathway for Gif-reactions:

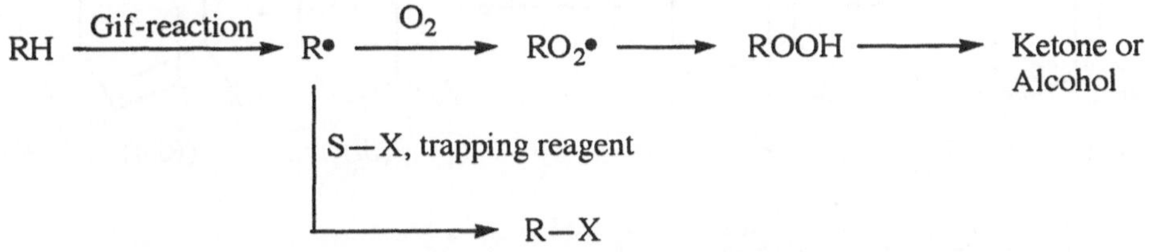

$$RH \xrightarrow{\text{Gif-reaction}} R\bullet \xrightarrow{O_2} RO_2\bullet \longrightarrow ROOH \longrightarrow \text{Ketone or Alcohol}$$

$$R\bullet \xrightarrow{\text{S}-\text{X, trapping reagent}} R-X$$

Why then, is intermediate A not a free alkyl radical? In order to examine the reaction pathway of genuine free alkyl radicals under Gif-conditions, they were generated at room temperature by the photolysis of N-hydroxy-2(*1H*)-thiopyridone derivatives of alkane carboxylic acids[43] and were allowed to react with dioxygen or added trapping reagents (Scheme 2):

In the absence of added trapping reagents genuine alkyl radicals always produced both alkyl pyridine derivatives and oxygenation products. The ratio of these products was strictly dependent on dioxygen partial pressure[32]. However, this is not the case for Gif-systems since *sec*-alkyl pyridines have never been observed[23,30,32]! Adamantane appeared to be an exception since *tert*-adamantyl pyridines were indeed formed under Gif-conditions The ratio of 1-adamantanol to *tert*-adamanatyl pyridines decreased at lower dioxygen pressure, therefore the intermediacy of free *tert*-adamantyl radicals was perceived. Genuine *tert*-adamantyl radicals exhibit a rather particular reactivity towards $CBrCl_3$ unlike that of

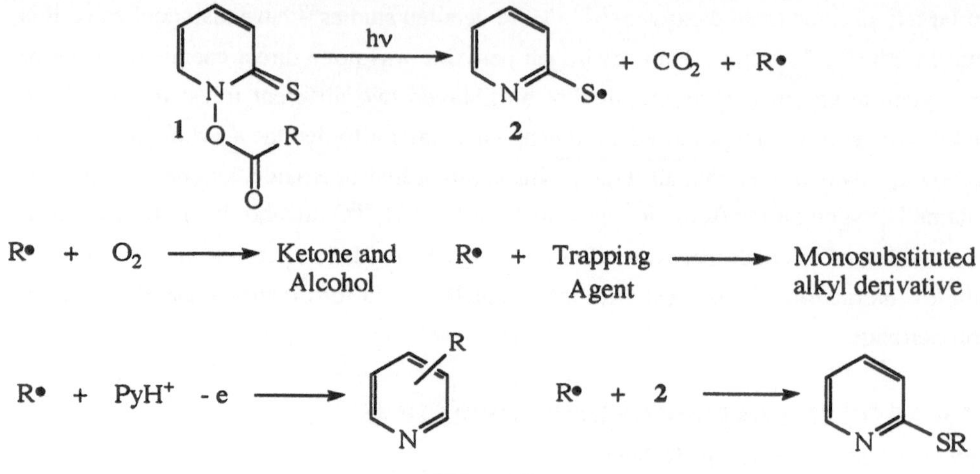

hν

1 → 2 + CO$_2$ + R•

R• + O$_2$ ⟶ Ketone and Alcohol

R• + Trapping Agent ⟶ Monosubstituted alkyl derivative

R• + PyH$^+$ - e ⟶ (pyridine with R)

R• + 2 ⟶ (pyridine-SR)

Scheme 2.

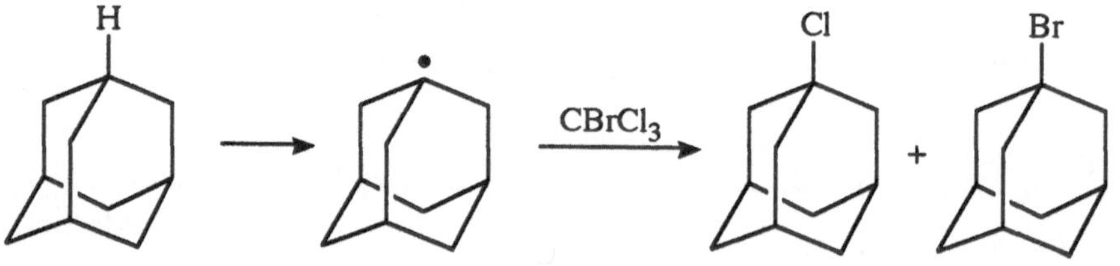

H (adamantane) ⟶ • (adamantyl radical)

CBrCl$_3$ ⟶ Cl (RCl) + Br (RBr)

other alkyl radicals, forming adamantyl chloride together with the expected alkyl bromide[45] and this needs to be studied further.

The ratio of RCl to RBr found in the GoAgg^{III} and GoChAgg^I systems appeared to be significantly lower (0.20 - 0.23) as compared with that of the Fenton reaction (H$_2$O$_2$: Fe^{II} = 1:1) in pyridine (0.59) and that for the reaction induced by photochemical generation of hydroxyl radicals from N-hydroxy-2-thiopyridone (0.4) (for the generation of hydroxyl radicals)[46,47] as well as for radical halogenation of adamantane in CBrCl$_3$ at 130°C (0.59)[45].

Under Gif^{IV} conditions, the addition of diphenyl diselenide diverted the formation of ketone to alkyl phenylselenides[38] with yields similar to the oxygenation, the by-product being mainly (96%) benzene selenol[41a]. If alkyl radicals are formed under these conditions, they should inevitably react with benzene selenol with a rate constant of about 2×10^9 M^{-1}s^{-1} at room temperature[48], but not with dioxygen due to its much lower concentration. This should lead to the reduction of the alkyl radical to the starting hydrocarbon and hence to a significant decrease in the total yield. However, this phenomenon was not observed experimentally.

In the Gif^{IV} reaction of cyclohexane in the presence of P(OMe)$_3$, ketone and a new product were formed instead of the expected alcohol. This new product was identified as cyclohexyl dimethyl phosphate[41a, 49]. If a carbon-centered cyclohexyl radical was an intermediate, one would expect the catalytic conversion of trimethyl phosphite into trimethyl phosphate in the presence of dioxygen. The latter reaction was indeed observed when genuine cyclohexyl radicals were photochemically generated under similar conditions to those employed in the Gif^{IV} reaction[41a,49]. So, once again this demonstrates that Gif-oxidation does not produce free alkyl radicals.

As mentioned above, the normal ketonization process seen in Gif-systems could be intercepted by different brominating reagents (CH$_2$Br$_2$, CBrCl$_3$, etc.) to yield alkyl monobromides[38-39,41a]. To demonstrate once again that free alkyl radicals are not intermediates in Gif-reactions, the reactivity order of different brominating reagents was compared with that for radical chain reaction conditions[39,41a]. Intermediate A in the Gif-reaction was allowed to compete for the brominating reagent and dioxygen, the ratio of alkyl bromide to ketone being the measure of its reactivity. A comparison was then made of the relative reactivities of the same reagents towards radicals generated by the photolysis of N-hydroxy-2-thiopyridone derivatives[44], the alkyl radicals thus formed were allowed to compete for the brominating reagent and a standard radical trap, thiophenol which gave alkanes by a hydrogen atom abstraction process[50]. The ratio of alkyl bromide to alkane served as a measure of relative reactivity. The results are given below:

Reactivity order of brominating reagents towards free cyclohexyl radical (the values in brackets are for methyl radical in gas phase[51]).

CBr$_4$	>	CBr$_2$Cl$_2$	>	CBrCl$_3$	>	(CBrCl$_2$)$_2$	>	CBr$_2$F$_2$
2.81 (4.3)		1.18 (2.3)		1.00 (1.00)		0.79 (-)		0.01 (0.013)

The reactivity order of brominating reagents under GoAggII conditions.

CBrCl$_3$	>	(CBrCl$_2$)$_2$	>	CBr$_4$	>	CBr$_2$Cl$_2$	>	CBr$_2$F$_2$
1.00		0.60		0.20		0.05		0.001

Again, the results obtained under GoAgg^{II} conditions are in agreement with the scarce data in the literature and are not compatible with a carbon radical structure for intermediate A.

WHAT IS THE REACTIVE SPECIES IN GIF-REACTIONS

The aspect of Gif-chemistry which is least understood is the nature of the reactive species involved. In early studies a trinuclear organo-iron carboxylate cluster $Fe^{II}Fe^{III}_2O(OAc)_6(py)_{3.5}$ was isolated after oxidative dissolution of iron powder in pyridine/acetic acid. Its catalytic activity was demonstrated in the Gif^{IV} system[29]. Naturally, the replacement of acetic acid by another carboxylic acid led to other carboxylate complexes, however no strong catalytic effect on the oxidation process was observed[52a-b]. Recently the nature of the Fe^{III} species was studied in the presence of picolinic acid (GoAgg^{III} system) and a single-iron core structure was deduced[52c]. Copper complexes serve as a catalysts in GoChAgg systems[19,21,23] and the complexes of other transition metals are also catalytically active[26]. Such diversity of catalytically active complexes has so far precluded the formulation of any generalization concerning the structure of the catalyst in Gif-systems.

The hydrogen kinetic isotopic effect (KIE) is widely used to establish the mechanism of chemical reactions[54]. The value of KIE being a criterion for understanding the nature of the reactive species involved in the process of C-H bond activation[2a,54]. We have measured[55] the values of KIE in different Gif-systems for oxygenation, bromination, and selenation reactions, as well as the KIE data for the Fenton reaction in pyridine solution and that of hydroxyl radicals photochemically generated from N-hydroxy-2-thiopyridone[46]. The relative reactivity of cyclopentane and cyclohexane (C_5/C_6) is another useful criterion. This ratio depends on whether a carbocation or alkyl radical is involved as an intermediate in C-H bond cleavage[2a,54]. For H-atom abstraction this value is greater than unity. Here it is noteworthy that for reactions involving one or more intermediates, affording several products as a result of intermediate compound fragmentation such as the Gif systems, the ratios of products may not represent the KIE, because the intermediate fragmentation may have its own KIE[55]. Therefore the values of KIE, measured as a ratio of only ketones or only alcohols varied widely[12c, 28a]. Thus, it is beleived that these values are not related to the C-H bond activation step. The data obtained and those available in the literature are shown below.

Hydroxyl radicals are believed to be intermediates in the Fenton system in aqueous solutions[15, 56a]. In pyridine solution, under conditions similar to those in Gif-reactions, the Fenton reaction produced free alkyl radicals, which were successfully trapped by $(PhSe)_2$[57]. The relative C-H bond reactivity was in agreement with values obtained in aqueous solution[58]. In addition, the KIE value was lower than that observed for Gif-reactions.

The values of KIE and (C_5/C_6) obtained are considered together, showing clearly that for all the Gif systems examined the first step of the reaction is similar and involves the cleavage of the hydrocarbon C-H bond in the rate determining step. The mechanism of C-H bond activation differs from hydrogen atom abstraction by hydroxyl radical, the reactive

Kinetic Isotopic Effect and Partial Relative Selectivity (C_5/C_6)

System	Major product	KIE	5/6
Gif[a]	ROH, R'=O	2.0 - 2.3	0.6 - 0.85
Gif/CBrCl$_3$[b]	RBr	2.1 - 2.4	0.57
Fenton/Py	R-Py	1.5	
Fenton/Py/(PhSe)$_2$	R-Se-Ph	1.7	
Fenton/Py/CBrCl$_3$	RBr	1.5	
(HO·)[c]/(PhSe)$_2$	R-Se-Ph	1.7	
Fenton/H$_2$O[d]	Δ(RH)/Δ(RD)	1.1	1.14
HMnO$_4$/H$_2$O[d]	Δ(RH)/Δ(RD)	5.0	0.88
NO$_2^+$/H$_2$SO$_4$[d]	Δ(RH)/Δ(RD)	2.2	0.25

[a] Gif^{IV}, GoAgg^{II}, GoAgg^{III}, GoChAgg^{I}; [b] GoAgg^{II}, GoAgg^{III}, GoChAgg^{I};
[c] photochemically generated from N-hydroxy-2-thiopyridone[46]; [d] see ref. 56.

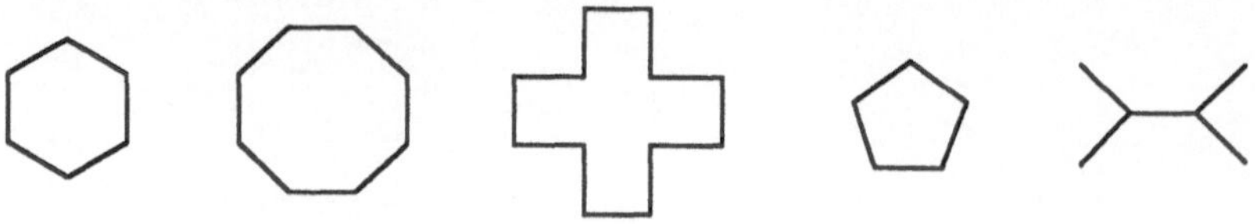

	⬡	⯃	✚	⬠	⤬
Radical Chain Bromination using CBrCl$_3$	1.0	3.3	1.9	1.3	10.2 (tert)
GoAggIII Bromination using CBrCl$_3$	1.0	0.76	0.69	0.63	0.18 (prim) 0.06 (tert)
GoAggII Oxidation reaction	1.0	0.75	0.43	0.85	—
GoChAggI Oxidation reaction	1.0	0.76	0.56	0.80	—

species not being a strong electrophile. This is substatiated by data obtained from the oxidation of 1,1-disubstituted olefins[35].

Recently the relative reactivity of different alkanes in bromination[39] and oxidation reactions[59] was studied under Gif-conditions and compared with the corresponding radical chain processes (competitive oxidation of pairs of alkanes in CBrCl$_3$ at reflux under Ar, initiated by dibenzoyl peroxide). The following hydrocarbons were studied and the data obtained (normalized per one C-H bond) is given below:

Additional proof for the difference between Gif- and radical bromination was obtained from bromination of cyclohexyl bromide. Radical induced hydrogen atom abstraction occurrs at the β-position of bromoalkanes[60] and in accordance with "Skell-Walling effect"[61] results in the formation of the *trans*-1,2-dibromide[60]. This was confirmed qualitatively for the radical bromination of cyclohexyl bromide. In contrast, this was not the case in the GoAgg[III] bromination reaction[39]. The *trans*-1,2-dibromide was found to be only a minor product, while *trans*-1,4- and *cis*-1,3-dibromocyclohexanes were the major products. Thus all this data illustrates the different nature of Gif-reactions and radical reactions. Here it is worth mentioning that tertiary C-H bonds appeared to be the least reactive in the bromination process, as is found in the Gif- oxidation reactions.

Besides the unusual regioselectivity of alkane oxidation (tertiary C-H bonds are less reactive than secondary C-H bonds[62]) another interesting feature of the reactive species in Gif-reactions is the unusual relative reactivity of alkane C-H bonds compared with some other readily oxidizable compounds such as alcohols, thiols and triphenylphosphine etc.[37,38, 41a]. The explanation of such behavior is not clear so far.

IS PYRIDINE THE KEY COMPONENT OF THE SOLVENT FOR GIF-REACTIONS?

Regular Gif-systems use pyridine as a major component of the solvent mixture. However, it can be replaced in part by other solvents with some improvment in the yield[22-23,26,30,52]. Pyridine can also be replaced by other pyridine derivatives[26b,30], but this gives lower yields, thus, pyridine still remains the solvent of choice. Recently it was reported that Gif-reactions can be carried out in t-butanol or aqueous solution at an appropriate pH[52c,63]. The value of pH (which could be adjusted by the varying the ratio of Pyridine and acetic acid used) was shown to be important in regular Gif-systems in terms of the rate of the reaction, the rate being higher, but not the yields, when the concentration of acid was lowered[23].

WHAT KIND OF ENZYMES IS GIF-CHEMISTRY EMULATING?

Some similarities can be perceived between Gif-reactions and the processes catalyzed by methane monooxygenase[64]. The insertion of sulfur into unactivated C-H bonds occurs in Gif-systems[65], modeling penicillin cyclase[66] and biotin synthase[67].

GIF PARADOXES

The term 'Gif-chemistry' is now accepted and widely used in the literature. Herein we would like to define Gif-chemistry. The best way to do this is to emphasize the chemical paradoxes observed in these systems in terms of the major peculiarities of the reactive species and the intermediate A.

Reactive species

• Does not react preferentially with the weakest (tertiary) C-H bond.
• Does not react with some readily oxidizable compounds such as Ph_3P, PhSeH, *etc.*
• Its reactivity does not depend strongly on the nature of the catalyst.

Intermediate A.

• Is not trapped by PyH^+, but is efficiently trapped by other radical traps.
• Shows a different reactivity order towards traps than that of genuine alkyl radicals.

SOME SPECULATIONS ON THE MECHANISM OF GIF-REACTIONS

We have shown very thoroughly that Intermediate A is not a carbon radical. Within the context of Gif chemistry it must be an organometallic species, suitably with an iron-carbon bond. The reagent which attacks the hydrocarbon is formally an Fe^V oxenoid species. However, the only way to understand the selectivity of this species towards saturated hydrocarbons is to postulate that there is another species which is inert until it meets and reacts with the hydrocarbon (the Sleeping Beauty Hypothesis)[68].

Important work on the nature of the pyridine N-oxide radical cation has been published. This was stimulated by consideration of the possible role of such a species in Gif chemistry.[69,70,71]

REFERENCES

1. a) A.E. Shilov. "Activation of Saturated Hydrocarbons by Transition Metal Complexes", Reidel, Dordrecht, (1984); b) "Activation and Functionalization of Alkanes", C.L. Hill, ed., Wiley, NY, (1989); c) A.E. Shilov, and G.B. Shulpin, *Russ. Chem. Rev.*, 56:442 (1987); d) H. Mimoun, *New J. Chem.*, 11:513 (1987); e) E.H. Grigoryan, *Russ. Chem. Rev.*, 53:210 (1984); f) "Selective Hydrocarbon Activation" J.A. Davies, P.L. Watson, A. Greenberg, J.F. Liebman, eds. VCH Publisher, NY (1990).

2. a) E.S. Rudakov. "Reactions of Alkanes with Oxidants, Metallocomplexes and Radicals in Solutions," Naukova Dumka, Kiev (1985) (in Russian); b) "Studies in Surface Sciences and Catalysis, " vol. 66, L.I. Simandi, ed., Elsevier, Amsterdam (1991); c) D.T. Sawyer. "Oxygen Chemistry, " Oxford University Press, NY (1991).

3. a) "Molecular Mechanism of Oxygen Activation," O. Hayashi, ed., Academic Press, NY (1974); b) "Biological Oxidation System", C. Reddy, G.A. Hamilton, K.M. Madyastha, eds., Academic Press, NY (1990).

4. K. Lerch. Chapter 5, *in* "Metal Ions in Biological Systems, " vol. 13, H. Sigel, ed., p. 143, Marcel Dekker, NY (1981).

5. a) "Cytochrome P-450: Structure, Mechanism and Biochemistry," P. Ortiz de Montellano, ed., Plenum Press, NY (1981); b) P.W. White, *Bioorg. Chem.*, 440 (1990); c) T.C. Bruice, *Acc. Chem. Res.* , 24:243 (1991).

6. See, the last comprehensive review, B. Meunier, *Chem. Rev.*, 92:1411 (1992).

7. a) J. Colby, D.I. Stirling, and H. Dalton, *Biochem. J.* , 165:395 (1977); b) J. Green, and H. Dalton, *J. Biol. Chem.* , 264:17698 (1989); c) A. Ericson, B. Hedman, K.O. Hodgson, J. Green, H. Dalton, J.G. Bensten, R.H. Beer, and S.J. Lippard, *J. Am. Chem. Soc.*, 110:2330 (1988); d) B.G.Fox, W.A.Froland, J.Dege, and J.D.Lipscomb, *J. Biol. Chem..*, 264:10023 (1989)

8. a) S.W. May, and B. Abbot, *J. Biol. Chem..*, 248:1725 (1973); b) A.G. Katopodis, K. Wimalasena, J. Lee, and S.W. May, *J. Am. Chem. Soc.* 106:7928 (1984).

9. a) S.M. Hecht, *Acc. Chem. Res.*, 19:383 (1986); b) J. Stubbe, and J.W. Kozarich, *Chem. Rev.* 87:1107 (1987).

10. a) S.O. Pember, J.J. Villafranca, and S.J. Benkovic, *Biochemistry*, 25:661 (1986); b) G. Guroff, M. Levitt, J. Daly, and S. Udenfriend, *Biochem. Biophys. Res. Commun.*, 25:253 (1966).

11. Th.A. Dix, and S.J. Benkovic, *Acc. Chem. Res.*, 21:101 (1988).

12. a) I. Tabushi, T. Nakajima, and K. Seto, *Tetrahedron Lett.*, 21:2565 (1980); b) N. Kitajima, H. Fukui, and Y. Mora-oka, *J. Chem. Soc., Chem. Commun.*, 485 (1988); c) J.B. Vincent, J.C. Huffman, G. Christou, Q. Li, M.A. Nanny, D.N. Hendricson, R.H. Fong, and R.H. Fish, *J. Am. Chem. Soc.*, 110:6898 (1988); R.H. Fish, M.S. Konings, K.J. Oberhausen, R.H. Fong, W.M. Yu, C. Christou, J.B. Vincent, DeA.K. Coggin, and R.M. Buchanan, *Inorg. Chem.*, 30:3002 (1991); d) H. Sugimoto, and D.T. Sawyer, *J. Org. Chem.*, 50:1784 (1985); H. Sugimoto, L. Spencer, and D.T. Sawyer, *Proc. Natl. Acad. Sci. U.S.A.*, 84:1731 (1987); e) J.T. Groves, and M. Van Der Puy, *J. Am. Chem. Soc.*, 96:5274 (1974); f) A.M. Khenkin, and A.E. Shilov, *New J. Chem.*, 13:659 (1989); g) R.A. Leising, R.E. Norman, and L. Que, Jr., *Inorg. Chem.*, 29:2553 (1990); B.A. Brennan, Q. Chen, C. Juarez-Garsia, A.E. True, C.J. O'Connor, and L. Que, Jr., *Inorg. Chem.*, 30:1937 (1991); h) T. Funabiki, K. Kashiba, T. Toyoda, and S. Yoshida, *Chem. Lett.*, 2303 (1992); i) Sh.-I. Murahashi, Y. Oda, T. Naota, and N. Komiya, *J. Chem. Soc., Chem. Commun.*, 139 (1993).

13. a) S. Udenfriend, C.T. Clark, J. Axelrod, and B.B. Brodie, *J. Biol. Chem.*, 208:731 (1954); b) H. Mimoun, I. Seree de Roch, *Tetrahedron*, 31:777 (1975); c) C. Sheu, A. Sobkowiak, S. Jeon, and D.T. Sawyer, *J. Am. Chem. Soc.*, 112:879 (1990).

14. Yu.V. Geletii, E.I. Karasevich, A.P. Moravsky, and A.A. Shteinman, *Kinetika i kataliz*, 22:349 (1981).

15. C. Walling, *Acc. Chem. Res.*, 8:125 (1975).

16. S. Ito, K. Ueno, A. Miterai, and K. Sasaki, *J. Chem. Soc., Perkin Trans. 2*, 255 (1993).

17. J. Fossey, D. Lefort, M. Massoudi, J.-Y. Nedelec, and J. Sorba, *Can. J. Chem.*, 63:678 (1985).

18. D.H.R. Barton, M.J. Gastiger, and W.B. Motherwell, *J. Chem. Soc., Chem. Commun.* 41 (1983); b) D.H.R. Barton, M.J. Gastiger, and W.B. Motherwell, *J. Chem. Soc., Chem. Commun.*, 731 (1983).

19. a) Yu.V. Geletii, G.V. Lubimova, and A.E. Shilov, *Kinetika i Kataliz*, 26:1023 (1985); b) Yu.V. Geletii, V.V. Lavrushko, and A.E. Shilov, *Dokl. Acad. Nauk SSSR*, 288:139 (1986).

20. G. Balavoine, D.H.R. Barton, J. Boivin, A. Gref, N. Ozbalik, and H. Rivière, *Tetrahedron Lett.*, 27:2849 (1986).

21. Yu.V. Geletii, V.V. Lavrushko, and G.V. Lubimova, *J. Chem. Soc., Chem. Commun.*, 936 (1988)

22. D.H.R. Barton, F. Halley, N. Ozbalik, E. Young, G. Balavoine, A. Gref, and J. Boivin, *New J. Chem.*, 13:177 (1989).

23. D.H.R. Barton, E. Csuhai, D. Doller, and Yu.V. Geletii, *Tetrahedron*, 33:47 (1991).

24. a) D.H.R. Barton, E. Csuhai, and N. Ozbalik, *Tetrahedron Lett.*, 31:2817 (1990) b) E. About-Jaudet, D.H.R. Barton, E. Csuhai, and N. Ozbalik, *Tetrahedron Lett.*, 31:1657 (1990); c) C. Sheu, S.A. Richert, P. Cofré, B. Ross, Jr., A. Sobkowiak, D.T. Sawyer, and J.R. Kanofsky, *J. Am. Chem. Soc.*, 112:1936 (1990).

25. a) D.H.R. Barton, S.D. Bévière, W. Chavasiri, D. Doller, and B. Hu, *Tetrahedron Lett.*, 33:5476 (1992); b) D.H.R. Barton, S.D. Bévière, W. Chavasiri, D. Doller, and B. Hu, *Tetrahedron Lett.*, 34:567 (1993); c) A. Sobkowiak, A. Qui, X. Liu, A. Llobet, and D.T. Sawyer, *J. Am. Chem. Soc.*, 115:609 (1993).

26. a) H.-C. Tung, and D.T. Sawyer, *J. Am. Chem. Soc.*, 112:8214 (1990); b) D.H.R. Barton, D. Doller, and Yu.V. Geletii, *Mendeleev Commun.*, 115 (1991); c) H.-C. Tung, C. Kang, and D.T. Sawyer, *J. Am. Chem. Soc.*, 114:3445 (1992).

27. G. Balavoine, D.H.R. Barton, J. Boivin, A. Gref, P. Le Coupanec, N. Ozbalik, J.A.X. Pestana, and H. Rivière, *Tetrahedron*, 44:1091 (1988).

28. a) H.-C. Tung, C. Kang, and D.T. Sawyer, *J. Am. Chem. Soc.*, 114:3445 (1992); b) U. Schuchardt, C.E.Z. Krähembül, and W.A. Carvalho, *New J. Chem.*, 15:955 (1991); c) U. Schuchardt, W.A. Carvalho, P. Pereira, and E.V. Spinacé, this book.

29. D.H.R. Barton, J. Boivin, M.J. Gastiger, J. Morzycki, R.S. Hay-Motherwell, W.B. Motherwell, N. Ozbalik, and K.M. Schwartzentruber, *J. Chem. Soc., Perkin Trans. 1*, 947 (1986).

30. D.H.R. Barton, J. Boivin, W.B. Motherwell, N. Ozbalik, K.M. Schwartzentruber, and K. Jankowski, *Nouv. J. Chem.*, 10:386 (1986).

31. F. Minisci, E. Vismara, and F. Fontana, *Heterocycles*, 28:489 (1988).

32. D.H.R. Barton, F. Halley, N. Ozbalik, M. Schmitt, E. Young, and G. Balavoine, *J. Am. Chem. Soc.*, 111:7144 (1989).

33. D.H.R. Barton, D. Doller, N. Ozbalik, G. Balavoine, A. Gref, and J. Boivin, *Tetrahedron Lett.*, 31:353. (1990)

34. a) D.H.R. Barton, F. Halley, N. Ozbalik, and W. Mehl, *Tetrahedron Lett.*, 30:6615 (1989); b) E. Baciocchi, E. Muraglia, and G. Sleiter, *Tetrahedron Lett.*, 32:2647 (1991).

35. D.H.R. Barton, K. W.Lee, W. Mehl, N. Ozbalik, and L. Zhang, *Tetrahedron*, 46:3753 (1990)

36. D.H.R. Barton, E. Csuhai, and N. Ozbalik, *Tetrahedron*, 46:3743 (1990).

37 D.H.R. Barton, S.D. Bévière, W. Chavasiri, E. Csuhai, D. Doller, and W.-G. Liu, *J. Am. Chem. Soc.*, 14:2147 (1992).

38. G. Balavoine, D.H.R. Barton, J. Boivin, P. Lecoupanec, and P. Lelandais, *New J. Chem.*, 13:691 (1989).

39. D.H.R. Barton, E. Csuhai, and D. Doller, *Tetrahedron* 48:9195 (1992).

40. D.H.R. Barton, E. Csuhai, and N. Ozbalik, *Tetrahedron Lett.*, 31:2817 (1990).

41. a) D.H.R. Barton, and D. Doller, *Acc. Chem. Res.* 25:504 (1992); b) D.H.R. Barton, and D. Doller, *in* "Dioxygen Activation and Homogeneous Catalytic Oxidation", L.I. Simándi, ed., Elsevier, Amsterdam (1991).

42. D.H.R. Barton, S.D. Bévière, W. Chavasiri, E. Csuhai, and D. Doller, *Tetrahedron* 48:2895.(1992).

43. C. Knight, and M.J. Perkins, *J. Chem. Soc., Chem. Commun.* 925 (1991).

44. D.H.R. Barton, *Aldrichimica Acta* 23:3 (1990).

45. I. Tabushi, J. Hamuro, and R. Oda, *J. Am. Chem. Soc.* 89:7127 (1967).

46. J. Boivin, E. Crepon, and S. Zard, *Tetrahedron Lett.*, 31:6869 (1990); D.H.R. Barton, J.Cs. Jaszberenyi, and A.I. Morrel, *Tetrahedron Lett.*, 32:311 (1991).

47. D.H.R. Barton, and Yu.V. Geletii, *unpublished data.*

48. M. Newcomb, and M.B. Manek, *J. Am. Chem. Soc.*, 112:9662 (1990).

49. D.H.R. Barton, S.D. Bévière, and D. Doller, *Tetrahedron Lett.*, 32:4671.(1991).

50. J.A. Franz, B.A. Bushaw, and M.S. Alnajjar, *J. Am. Chem. Soc.*, 111:268 (1989).

51. D.M. Tomkinson, and H.O. Pritchard, *J. Phys. Chem.*, 70:1579 (1966).

52. a) D.H.R. Barton, R.S. Hay-Motherwell, and W.B. Motherwell, *Tetrahedron Lett.*, 24:1979 (1983); b) Yu.V. Geletii, and A.E. Shilov, *in* "Studies in Organic Chemistry", vol.33, p.293, W. Ando, and Y. Moro-oka, eds., Elsevier, Amsterdam (1988); c) D.H.R. Barton, S.D. Bévière, W. Chavasiri, D. Doller, W.-G. Liu, and J.H. Reibenspies, *New J. Chem.*, 16:1019 (1992).

53. L. Melander, and W.H. Saunder, Jr., "Reaction Rates of Isotopic Molecules," Wiley, NY (1979).

54. E.S. Rudakov, *Dokl. Phys. Chem. (Engl.Trans.)*, 263:262 (1982).

55. D.H.R. Barton, D. Doller, and Yu.V. Geletii, *Tetrahedron Lett.*, 32:3811 (1991).

56. a) E.S. Rudakov, L.K. Volkova, V.P. Tret'yakov, and V.V. Zamashchikov, *Kinetics and Catalysis (Eng.Trans.)*, 23:18 (1982); b) E.S. Rudakov, N.A. Tishchenko, and L.K. Volkova, *Kinetics and Catalysis (Eng.Trans.)*, 27:949 (1986); c) E.S. Rudakov, V.V. Zamashchikov, A.I. Lutsyk, and A.P. Yaroshenko, *Dokl. Phys. Chem.. (Eng.Trans.)*, 224:916 (1975).

57. C. Sheu, A. Sobkowiak, L. Zhang, N. Ozbalik, D.H.R. Barton, and D.T. Sawyer, *J. Am. Chem. Soc.*, 111:8030 (1989).

58. A.F. Trotman-Dickenson, *Adv. Free Radical Chem.*, 1:1 (1965).

59. D.H.R. Barton, S.D. Bévière, W. Chavasiri, E. Csuhai, and D. Doller, *Tetrahedron*, 48:2895 (1992).

60. C. Walling, and B.B. Jacknow, *J. Am. Chem. Soc.*, 82:6113 (1960).

61. P.S. Skell, and K.J. Shea, *in* "Free Radicals", vol.2, p. 809, K.J. Kochi, ed., Wiley, NY (1973).

62. D.H.R. Barton, Ph.E. Eaton, and W.-G. Liu, *Tetrahedron Lett.*, 32:6263 (1991).

63. C. Larpent, H. Patin, *J. Mol. Catal.* 72:315 (1992).

64. D.H.R. Barton, E. Csuhai, D. Doller, N. Ozbalik, and G. Balavoine, *Proc. Natl. Acad. Sci.U.S.A..*, 87:3401 (1990).

65. G. Balavoin, D H.R. Barton, A. Gref, and I. Lellouche, *Tetrahedron Lett.*, 32:2351 (1991).

66. J.E. Baldwin, *in* "Recent Advances in the Chemistry of β-Lactam Antibiotics" Royal Soc.Chem. Special Pub., p. 62, A. Brown, ed., S.M. Roberts, London (1985).

67. F. Frappier, G. Guilerm, A.G. Salib, and A. Marquet, *Biochem. Biophys. Res. Commun.* 91:521 (1979); F. Frappier, and A. Marquet, *ibid.*, 103:1288 (1981).

68. D.H.R. Barton and D. Doller, *Acc. Chem. Res.* 25:504 (1992).

69. a) E.M. Koldasheva, Yu.V. Geletii, V.V. Yanilkin, and V.V. Strelets, *Izv. Akad. Nauk, SSSR, Ser. Khim.* 994 (1990); b) E.M. Koldasheva, A.F. Shestakov, Yu.V. Geletii, and A.E. Shilov, *Izv. Akad. Naukk SSSR Ser. Khim.* 845 (1992); c) Yu.V. Geletii, V.V. Strelets, V.Ya. Shafirovich, and A.E. Shilov, *Heterocycles*, 28:677 (1989); d) E.M. Koldasheva, Yu.V. Geletii, V.V. Strelets, and A.E. Shilov, *in.* "Studies in Surface Science and Catalysis," vol. 66, L.I. Simandi, ed., Elsevier, Amsterdam (1991); e) E.M. Koldasheva, Yu.V. Geletii, A.F. Shestakov, A.I. Kulikov, and A.E. Shilov, *New J. Chem.,* (1993) (in press).

70. a) Yu.V. Geletii, V.A. Kuzmin, P.P. Levin, and V.Ya. Shafirovich, *React. Kinet. Catal. Lett.,* 37:307 (1988); b) Yu.V. Geletii, V.E. Zubarev, P.P. Levin, G.V. Lubimova, and V.Ya. Shafirovich, *Kinetika i Kataliz* 31:802 (1990).

71. A.E. Shilov, *React. Kinet. Catal. Lett.* 41:223 (1990); b) A.E. Shilov, *Khim. Fizika*, 10:758 (1991).

CYCLOHEXANE OXIDATION: CAN GIF CHEMISTRY

SUBSTITUTE FOR THE CLASSICAL PROCESS ?

Ulf Schuchardt, Wagner A. Carvalho, Ricardo Pereira
and Estevam V. Spinacé

Instituto de Química, Universidade Estadual de Campinas
Caixa Postal 6154, 13081-970 Campinas, SP (Brazil)

ABSTRACT

Cyclohexane oxidation was studied using the Gif^{IV}, the GoAgg^{II}, the GoAgg^{III} and the GoChAgg systems. The Gif^{IV} system gives a conversion of cyclohexane of 40%, with a concentration of one + ol of 0.120 M, but shows low selectivity and needs large quantities of metallic zinc. The GoAgg^{II} system produces a 0.143 M solution of one + ol with 100% selectivity but is too slow for any industrial use. The GoAgg^{III} system allows accumulation of the oxidation products. A 0.370 M solution of one + ol can be produced in 180 min of reaction time, but large quantities of catalyst are needed and the selectivity of one + ol is only 80%. Addition of HCl during the accumulations allows obtaining a 0.267 M solution of one + ol with 100% selectivity in 60 min of reaction time, compared to a 0.3 M solution of one + ol with only 80% selectivity in 40 min in the industrial oxidation process. On the other hand, the catalyst forms iron (hydr)oxide particles during the reaction and the pyridine has to be substituted by a cheaper and a less toxic solvent. The GoChAgg system is less active and deactivates rapidly during the accumulations. The best result obtained so far is a 0.120 M solution of one + ol with 100% selectivity in 120 min of reaction time. Substituition of pyridine by *tert*-butanol reduces the activity of the GoAgg^{III} and the GoChAgg systems. Precipitation of the catalyst and rapid deactivation are observed. Up to now we could only prepare a 0.040 M solution of one + ol. Even so, we hope to find a solvent/buffer system which avoids the deactivation of the catalyst while maintaining the high activity and selectivity of the process.

INTRODUCTION

In the classical oxidation process only 4% of the cyclohexane is converted in order to minimize over-oxidation of the products.[1] This means that 96% of the cyclohexane

The Activation of Dioxygen and Homogeneous Catalytic Oxidation,
Edited by D.H.R. Barton *et al.*, Plenum Press, New York, 1993

must be separated from the products and recycled. Even at this low conversion the selectivity for cyclohexanone plus cyclohexanol (one + ol) is only 80%, which corresponds to a concentration of one + ol in the reaction mixture of about 0.3 M (mol/L). The remainder consists mostly of n-butyric, n-valeric, succinic, glutaric and adipic acids,[2] formed by ring cleavage of cyclohexanone and/or cyclohexyloxy radicals (Scheme 1). For each ton of one + ol produced, 30 tons of cyclohexane have to be recycled and 0.25 tons of acids neutralized and subjected to disposal which turns the process expensive and its improvement a challenge for chemists working in this field.

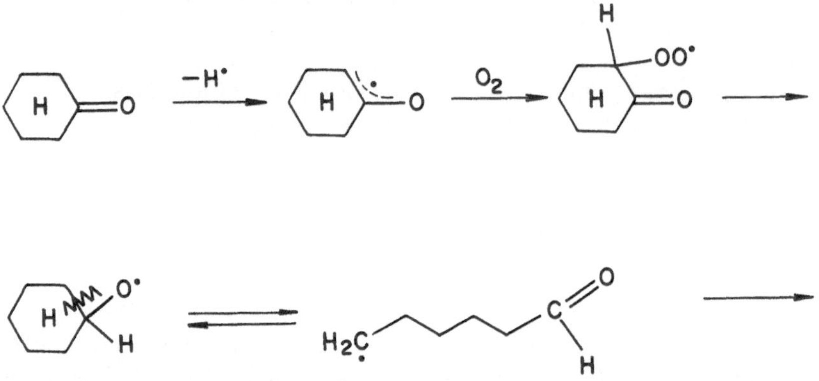

Scheme 1. Ring cleavage reactions in classical cyclohexane oxidation.

To improve the selectivity of the process, the concentration of cyclohexanone in the reaction mixture has to be reduced and the formation of cyclohexyloxy radicals avoided. The formation of cyclohexanone can be reduced by complexing the cyclohexanol and/or its hydroperoxide with boric acid, which then precipitate under reactions conditions.[3] This process allows a selectivity for one + ol of 90% at 10% conversion and is used on an industrial scale by the Institut Français du Pétrole. However, its operational costs are high as large amounts of solids need to be separated and decomposed and the boric acid has to be recycled.[3]

We have studied the possibility of first oxidizing cyclohexane to its hydroperoxide which, in a second step, is selectively decomposed to one + ol in the presence of an appropriate transition metal catalyst. Using a reactor passivated with sodium diphosphate ($Na_4P_2O_7$) and 10% (v/v) *tert*-butanol as a stabilizer, we were able to accumulate cyclohexylhydroperoxide in the reaction mixture. After 100 min at 155°C we obtained a conversion of 10%, giving 7% of cyclohexylhydroperoxide and 2% of one + ol.[4] This represents a total selectivity of 90%, as obtained in the presence of boric acid, but has the advantage of keeping the cyclohexylhydroperoxide in solution.

The hydroperoxide can be rapidly decomposed in the presence of soluble transition metal compounds in the temperature range of 80-100°C. Using a cobalt catalyst, an one:ol ratio of 1:3 was obtained, while an one:ol ratio of 3:1 was observed with a chromium salt.[5] The total one + ol selectivity was always better than 95%. Soluble iron salts are normally not active at this temperature; on the other hand, potassium hexacyanoiron(III) as well as hexacyanoiron(II) showed good activity, giving an one:ol ratio of approximately 4:1.[6]

According to patents,[7] cyclohexylhydroperoxide epoxidizes propylene in the presence of soluble molybdenum catalysts at the same temperature range of 80-100°C. The selectivity for one + ol is reported to be better than 95% and that for propylene oxide at least 70%. This opens the possibility of making a modified process for the classical oxidation of cyclohexane economically interesting. For each ton of one + ol produced, only 8.46 tons of cyclohexane have to be recycled, only 0.15 tons of acids have to be neutralized and in addition 0.3 tons of propylene oxide are obtained (Scheme 2).[4] On the other hand, recycling of *tert*-butanol, used as stabilizer in the process, is difficult as it forms an azeotropic mixture with cyclohexane and the cyclohexylhydroperoxide in the reaction mixture has to be concentrated to 20-30% in order to turn the epoxidation reaction effective.

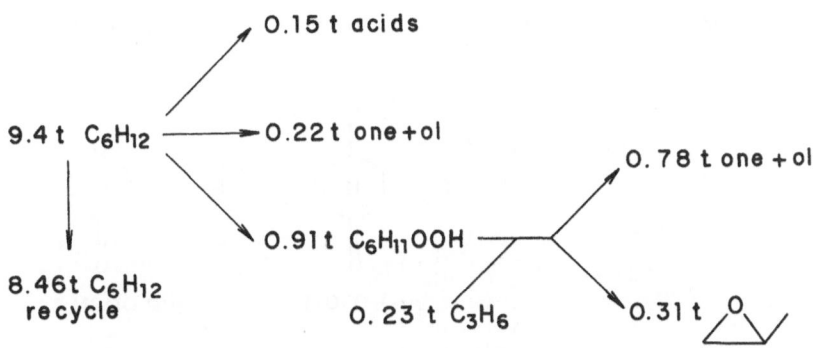

Scheme 2. Product distribution in a modified commercial process.

During the last five years we have investigated the possibility of using the Gif system for industrial cyclohexane oxidation. We first studied the Gif^{IV} system, later the GoAgg^{II}, GoAgg^{III} and the GoChAgg systems.[8] We wish to report here our results and evaluate the possibility of substituting the classical cyclohexane oxidation process by a process based on the Gif system.

Gif^{IV} SYSTEM

Of the Gif systems developed in Gif-sur-Yvette, Barton *et al.*[9] had already shown that the Gif^{IV} system, which uses an iron(II) salt in pyridine-acetic acid with zinc dust-oxygen as an oxidant, is the most efficient. In cyclohexane oxidation they obtained mainly cyclohexanone with a turnover number of 121 in 4 h of reaction. On the other hand, the concentration of one + ol (0.027 M) was small. We optimized the Gif^{IV} system for cyclohexane oxidation and obtained a 0.035 M solution of oxidized products but the selectivity was slightly lower than 100%, as cyclohexylpyridines were also formed.[10]

When we studied the same reaction under an atmosphere of pure oxygen, we could increase the concentration of oxidized products to 0.047 M but found a selectivity for one + ol of only 83%.[11] By addition of new portions of zinc, acetic acid, cyclohexane and catalyst, we were able to accumulate three reactions and obtained a conversion of cyclohexane of approximately 40%, which corresponds to a final concentration of one + ol of 0.120 M, but found a further decrease in selectivity to 70%. Besides cyclohexylpyridines, over-oxidation products such as 1,4-cyclohexadione and 4-hydroxycyclohexanone were formed (Scheme 3).[12] Taking into consideration that large quantities of zinc are needed and that the selectivity for one + ol was not good, we concluded that the Gif^{IV} system is not suitable for cyclohexane oxidation under industrial conditions.

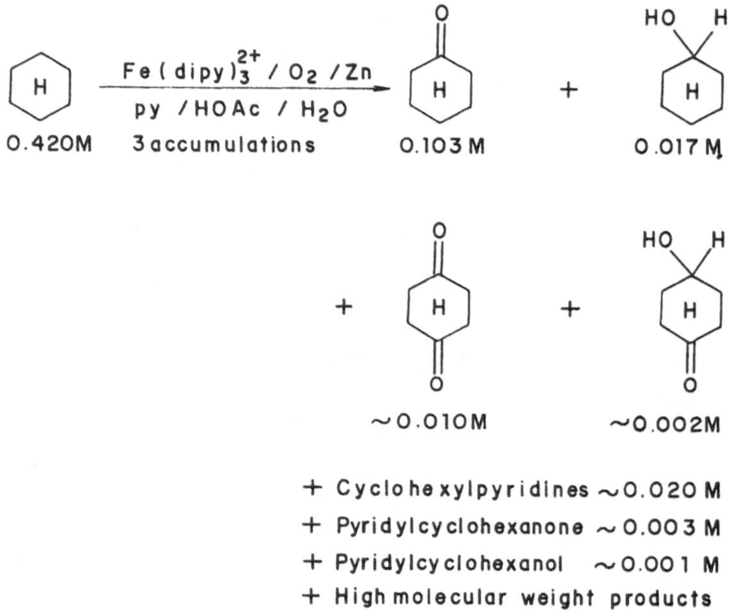

Scheme 3. Oxidation products formed by the Gif^{IV} system in three accumulations.

GoAgg^{II} SYSTEM

The GoAgg^{II} system uses hydrogen peroxide as an active oxidant in the presence of an iron(III) salt, having the advantage of forming a homogeneous mixture of the reactants. In cyclohexane oxidation Barton et al.[13] reported a 0.053 M solution of one + ol after 5 h of reaction time. In open air we obtained a 0.063 M solution of oxidized products after 8 h but the selectivity for one + ol was only 87%.[14] Under a static atmosphere of argon the concentration of one + ol increased to 0.133 M with 100% of selectivity. This result suggests that the presence of molecular oxygen in the GoAgg[II] system causes formation of coupling products with pyridine and reduces the efficiency.[4] A careful kinetic investigation (Fig. 1) showed that the efficiency with respect to hydrogen peroxide can be as high as 91%, giving a 0.143 M solution of one + ol, but the reaction needs 10 h to be complete, which is too slow for any industrial use.[15]

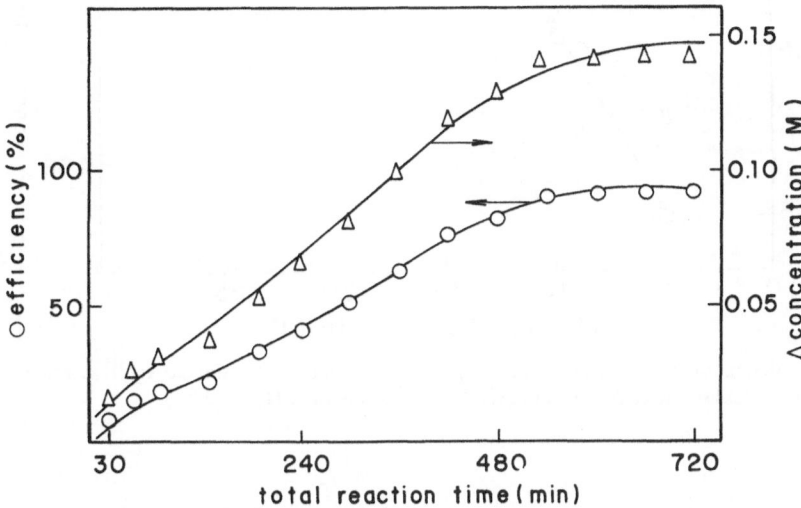

Figure 1. Time dependence of efficiency and concentration of one + ol for the GoAgg[II] system (1 mmol of FeCl₃, 20 mmol of cyclohexane, 10 mmol of H₂O₂, 20°C).

GoAgg[III] SYSTEM

In early 1990 Sawyer *et al.*[16,17] and Barton *et al.*[18] published the results on a modified GoAgg system which uses picolinic acid as a ligand for the iron catalyst (GoAgg[III] system). They found that this system is 40 times faster than the GoAgg[II] system. We obtained with the GoAgg[III] system in only 15 min of reaction time a 0.059 M solution of one + ol with 100% selectivity, but the efficiency with respect to hydrogen peroxide was only 35%.[14] The time dependence of the concentration of one + ol and of the efficiency with respect to hydrogen peroxide are shown in Fig. 2. The efficiency can be increased to 53% by the use of a higher concentration of cyclohexane in the reaction mixture.[14] This corresponds to a 0.088 M solution of one + ol. We used this system for the accumulation of oxidation products. As shown in Fig. 3 we were able to accumulate 12 reactions and reached a final concentration of one + ol of 0.370 M.[19] On the other hand, the efficiency dropped to 20% during the accumulations. The selectivity for one + ol was 100% for the first 90 min of reaction time, giving a 0.260 M solution of one + ol. After this, coupling and over-oxidation products were formed and the selectivity was reduced to 80%.[19]

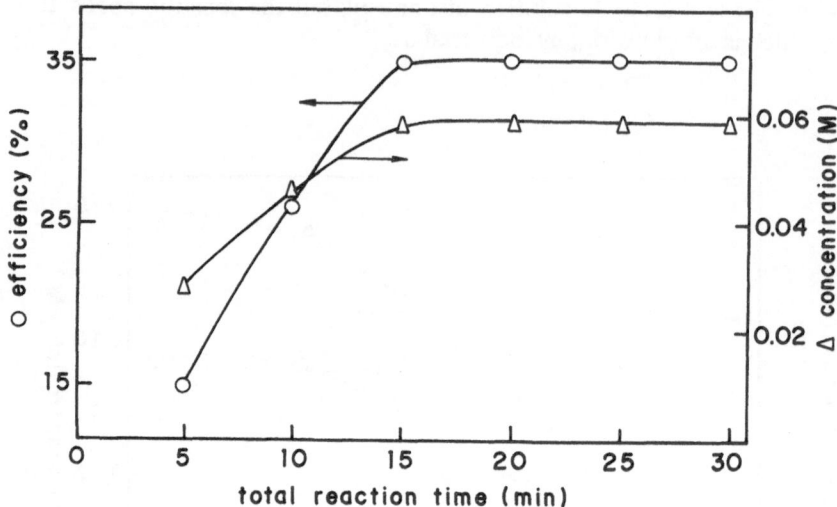

Figure 2. Time dependence of efficiency and concentration of one + ol for the GoAgg[III] system (1 mmol of FeCl₃, 3 mmol of picolinic acid, 20 mmol of cyclohexane, 10 mmol of H₂O₂, 20°C).

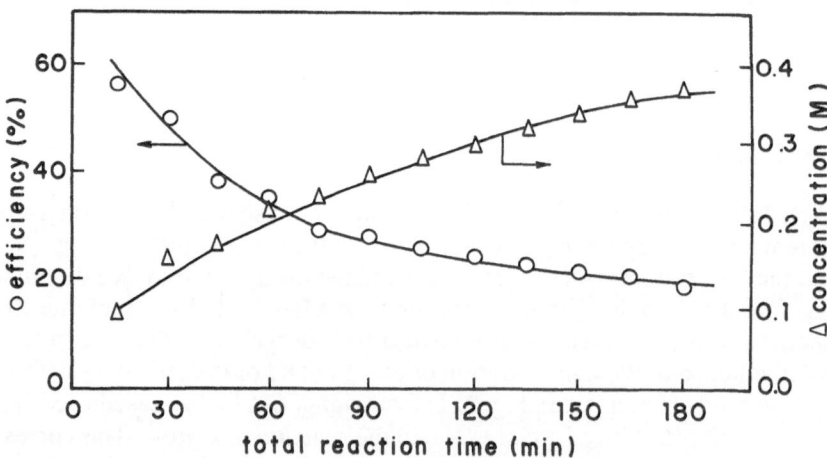

Figure 3. Time dependence of efficiency and concentration of one + ol in the accumulation reactions (3 mmol of picolinic acid, 100 mmol of cyclohexane, 20°C; every 15 min: 1 mmol of FeCl₃ and 10 mmol of H₂O₂).

In the accumulation reactions it was necessary to add a new portion of the catalyst together with the hydrogen peroxide after each reaction as, in its absence, the hydrogen peroxide was only decomposed.[19] In order to understand why it was necessary to add iron(III) chloride after each reaction, we tried to explain the colour change from light yellow to dark brown during the reaction course. Turbidity measurements showed the formation of colloidal particles of iron (hydr)oxide in the nm range.[20] This is expected as the pH of the reaction mixture (5.6) is high enough to deprotonate the hexaaquoiron(III) cations, forming μ-(hydr)oxodiiron(III) complexes which, under reaction conditions, hydrolyse to polynuclear complexes and finally to iron (hydr)oxide particles which are not active in the oxidation of cyclohexane but simply decompose the hydrogen peroxide.

We tried to avoid the hydrolysis of the μ-(hydr)oxodiiron(III) complexes by reduction of the pH of the reaction mixture with 1 M HCl in acetic acid. Adding 1 mmol of HCl together with 10 mmol of hydrogen peroxide after each reaction, we were able to perform 4 accumulations.[20] The results are shown in Fig. 4.

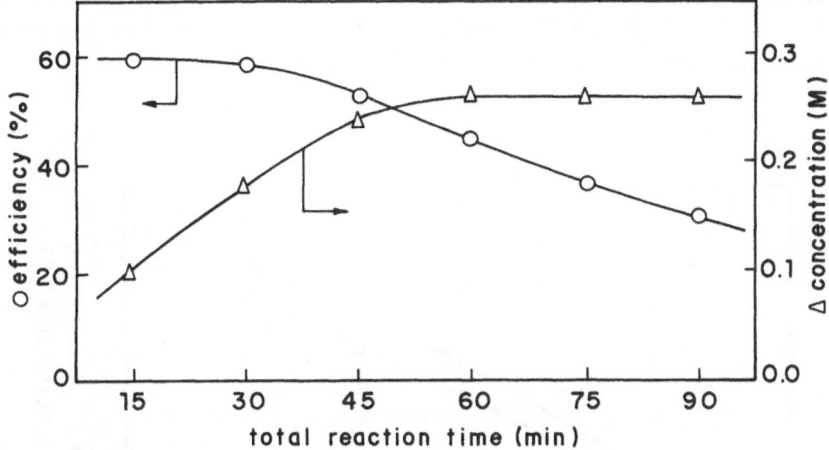

Figure 4. Time dependence of efficiency and concentration of one + ol in the reactions with HCl (1 mmol of FeCl₃, 3 mmol of picolinic acid, 100 mmol of cyclohexane, 20°C; every 15 min: 10 mmol of H₂O₂ and 1 ml of HCl/HOAc).

We obtained a 0.267 M solution of one + ol with 100% of selectivity in only 60 min of reaction time. The overall efficiency with respect to hydrogen peroxide was 44%.[20] After that time the formation of the iron (hydr)oxide colloid was irreversible. Further addition of iron(III) chloride, HCl in acetic acid and hydrogen peroxide caused the formation of two phases and the oxidation reaction did not proceed.

The acidification of the reaction mixture can also be performed with 1 M HClO₄ in acetic acid. The results are slightly inferior, giving a 0.255 M solution of one + ol.[20] On the other hand, substitution of ferric chloride by ferric perchlorate in the absence of hydrochloric acid strongly reduced the efficiency of the system, giving only a 0.128 M solution of one + ol. This corresponds to the results obtained by Nappa and Tolman,[21] who found that tetraphenylporphyrin iron(III) looses its activity for cyclohexane oxidation if the axial chlorine ligand is substituted by perchlorate. We, therefore, believe that chlorine is an essential ligand for the GoAgg[III] system.

Under Gif conditions copper(II) salts show a similar reactivity in the oxidation of saturated hydrocarbons with hydrogen peroxide.[22] In the oxidation of cyclododecane, Geletii *et al.*[23] obtained only a 0.019 M solution of oxidized products which shows that the activity of the copper catalyst is smaller than that of the corresponding iron(III) catalyst. On the other hand, the GoChAgg system, without addition of picolinic acid, is nearly as rapid as the GoAgg[III] system and the addition of water does not reduce its efficiency.[23] There are further differences between the two systems which prove that the metal itself participates actively in the activation process.[24]

As at pH 5-6 no copper (hydr)oxide is formed, we thought that it might be interesting to study cyclohexane oxidation by the GoChAgg system. Using copper(II) chloride or perchlorate under the same conditions as used in the GoAgg[II] system, we obtained a 0.028 M solution of cyclohexanone with 100% selectivity.[25] If the reaction was performed in the absence of acetic acid, copper(II) chloride was shown to be more active than the perchlorate, giving final concentrations of cyclohexanone of 0.056 M and 0.028 M, respectively.[25] These values were obtained under a static argon atmosphere. In open air in the absence of acetic acid, copper chloride was less active. The time dependence of the oxidation in the presence of copper(II) chloride is shown in Fig. 5. The total reaction time was 30 min and the efficiency with respect to hydrogen peroxide was 32%.[25]

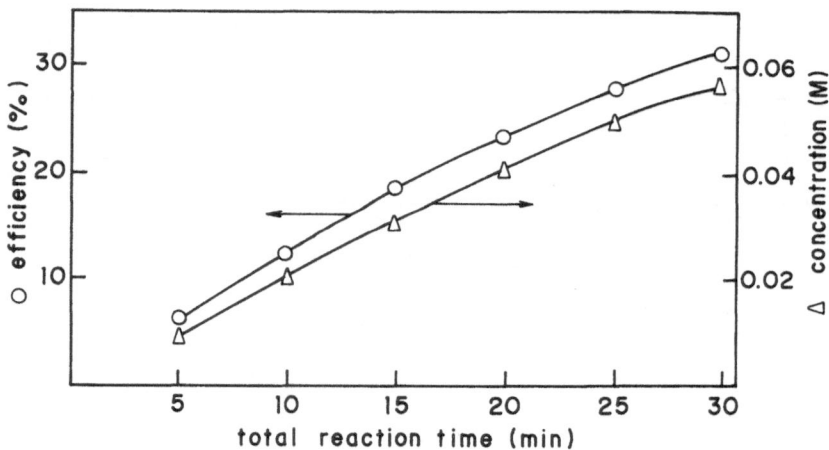

Figure 5. Time dependence of efficiency and concentration of one for the GoChAgg system (1 mmol of CuCl$_2$, 20 mmol of cyclohexane, 10 mmol of H$_2$O$_2$, 28°C).

We tried to accumulate oxidation reactions using copper chloride under these conditions (Fig. 6) but observed that the system had already started to deactivate after the first reaction. Furthermore, the system began to produce cyclohexanol in the accumulation reactions, which was not observed before. After 4 accumulations we obtained a 0.120 M solution of one + ol (8.3:1) with 100% selectivity but only 18% efficiency with respect to hydrogen peroxide.[25] Using copper perchlorate in the absence of acetic acid, we were also able to accumulate one + ol to a concentration of 0.120 M (one:ol = 5.9:1), but observed that the reaction was less efficient at the beginning (Fig. 7). Interestingly, the

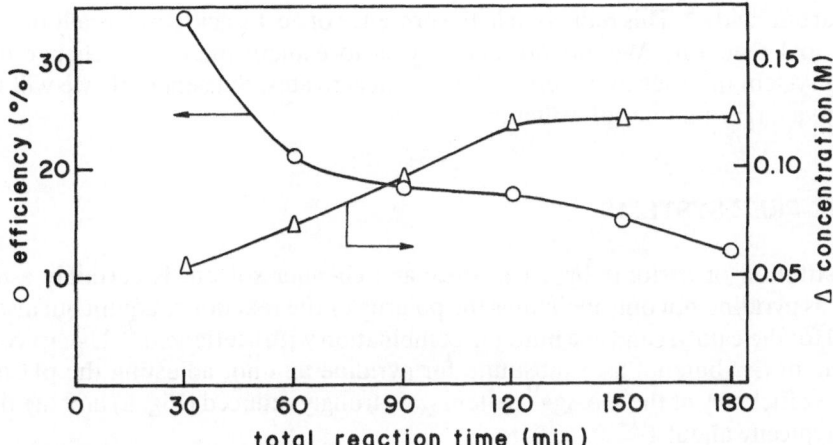

Figure 6. Time dependence of efficiency and concentration of one + ol in the accumulation reactions (1 mmol of $CuCl_2$, 20 mmol of cyclohexane, 28°C; every 30 min: 10 mmol of H_2O_2).

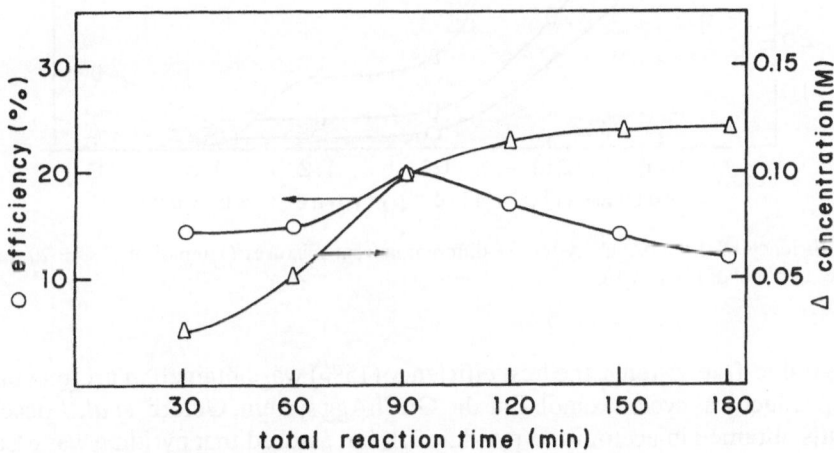

Figure 7. Time dependence of efficiency and concentration of one + ol in the accumulation reactions (1 mmol of $Cu(ClO_4)_2$, 20 mmol of cyclohexane, 28°C; every 30 min: 10 mmol of H_2O_2).

efficiency increased in the first three accumulations, after which the same value observed for copper chloride was obtained.[25]

As no particle formation was observed, the deactivation of the catalyst must be related to the colour change of the copper complex from light green to dark brown. If acetic acid was added in the accumulation reactions, the colour change was slower but, even so, the total concentration of oxidized products was smaller (0.100 M in the presence of 1 ml of acetic acid).[25] This reduction in the presence of acetic acid was already observed in the normal reactions. We are presently trying to explain the colour change of the GoChAgg system in order to understand why it deactivates. Subsequently we will try to accumulate a larger number of oxidation reactions.

PYRIDINE-FREE SYSTEMS

Substitution of pyridine by a less toxic and cheaper solvent is certainly a major challenge as pyridine not only maintains the polarity of the reaction medium but also acts as a ligand for the catalyst and as a buffer in combination with acetic acid.[26] Using acetone, acetonitrile or *tert*-butanol as a substitute for pyridine and not adjusting the pH of the system, the efficiency of the GoAgg[II] system was strongly reduced (Fig. 8) and the one:ol ratio was typically about 1.[14]

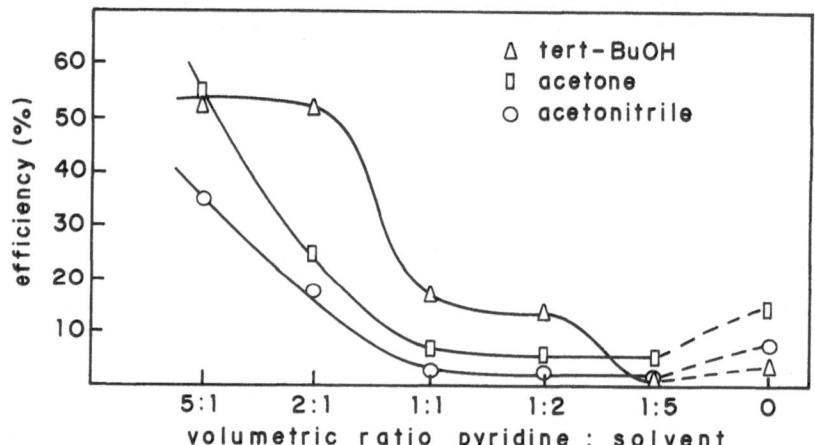

Figure 8. Efficiency of the GoAgg[II] system in different solvent mixtures (1 mmol of FeCl$_3$, 20 mmol of cyclohexane, 10 mmol of H$_2$O$_2$, 20°C).

For the pyridine-free systems, the best efficiency (15%) was obtained in acetone but the principal product was cyclohexanol. For the GoChAgg system, Geletii *et al.*[27] described good results obtained in acetonitrile/pyridine 2:1, but assured that pyridine was essential and must be present in the system. On the other hand, in a very recent publication Barton *et al.*[28] reported that pyridine can be completely replaced by *tert*-butanol with only a small reduction of the total hydrocarbon activation and a slightly reduced one:ol ratio. However, the initial oxidation rate was pH dependent and buffering of the system was necessary. We tried to repeat these experiments in cyclohexane oxidation using the GoAgg[III] and the GoChAgg systems, but had problems in achieving pH values of 6-7 due to the low solubility of sodium acetate in *tert*-butanol/acetic acid. Furthermore we observed the precipitation of an iron or copper compound, which we are presently trying to identify, during the reaction course. Using 10:1 (v/v) *tert*-butanol/acetic acid with iron(III) picolinate as a catalyst, we obtained a 0.026 M solution of one + ol (1:1 ratio) with an efficiency of 13% at a pH of 2.2.[29] By addition of sodium acetate to the reaction mixture we could increase the pH to 3.8, but obtained only a slightly higher efficiency of 15%. Using copper(II) chloride under the same conditions, the concentration of one + ol (11:1 ratio) was only 0.012 M (efficiency of 8%). Addition of 2 ml of water increased the activity of the copper catalyst (0.019 M solution) but reduced the one:ol ratio to 3.8:1. In comparison, Barton *et al.*[28] obtained a concentration of 0.015 M of one + ol with iron(III) chloride/picolinic acid in *tert*-butanol. Accumulation of the oxidation products was difficult (Fig. 9); the best value obtained so far is 0.040 M in one + ol.[29]

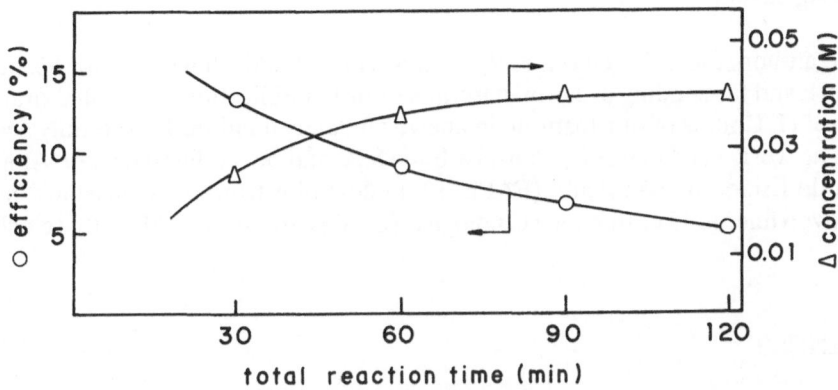

Figure 9. Time dependence of efficiency and concentration of one + ol for the GoAgg[III] system in *tert*-butanol (1 mmol of FeCl$_3$, 20 mmol of cyclohexane, 28°C; every 30 min: 10 mmol of H$_2$O$_2$).

Larpent and Patin[30] studied the oxidation of saturated hydrocarbons using reverse microemulsions of iron(III) chloride and hydrogen peroxide in the hydrocarbon. They obtained only low concentrations of oxidized products (0.030 M of cyclooctanone in cyclooctane) with a low efficiency of 7-8%. We were able to obtain similar results in the cyclohexane oxidation; a 0.020 M solution of one + ol (1:1 ratio), with an efficiency with respect to hydrogen peroxide of 5%, was easily prepared. To increase the concentration of oxidized products in this system seems to be difficult, as the phase diagram changes with higher concentrations of the reagents, which causes further reduction of the efficiency.

CONCLUSIONS

The GoAgg[III] system shows a good potential for substituting for the classical cyclohexane oxidation process. In the presence of hydrochloric acid it produces a 0.267 M solution of one + ol with 100% selectivity in 60 min of reaction time. In comparison with the classical process (0.3 M solution of one + ol with only 80% selectivity in 40 min) these results are totally compatible but have the advantage of being obtained at room temperature without applying pressure. On the other hand, the turnover numbers are low

($\approx$ 10), due to hydrolysis of the catalyst to iron (hydr)oxide particles, and the pyridine has to be substituted by a cheaper and less toxic solvent. Our attempts to overcome these problems have not yet been successful. Copper(II) salts, which do not form (hydr)oxide particles, are less efficient and also deactivate. An appropriate solvent/buffer system still has to be found. We do believe, though, that this will be possible by using other ligands for the catalyst and/or alternative bases which are soluble in the reaction mixture without addition of large amounts of water. Furthermore, *tert*-butanol is probably not an appropriate solvent, as its separation from cyclohexane and its oxidation products is complex. We are working to find a system which shows sufficiently high activity and selectivity at room temperature and, hopefully, one which can substitute for one of the less satisfactory industrial processes.

Acknowledgements

The authors thank Sir Derek H.R. Barton, Texas A&M University, for his interest in our work and for sending us his manuscripts prior to publication. The collaboration of Prof. Carol H. Collins of our Institute in analytical determinations is gratefully acknowledged. The work was financed by Nitrocarbono S.A. and by the Fundação de Amparo à Pesquisa do Estado de São Paulo (FAPESP). Fellowships from the Conselho Nacional de Desenvolvimento Científico e Tecnológico (CNPq) and from FAPESP are acknowledged.

REFERENCES

1. K.U. Ingold, *Aldrichimica Acta* 22:69 (1989).
2. R.A. Sheldon and J.K. Kochi, "Metal-Catalysed Oxidations of Organic Compounds", Academic Press, New York (1981); Chap. 11.
3. S. Ciborowski, *in*: "Studies in Surface Science and Catalysis", vol 66, "Dioxygen Activation and Homogeneous Catalytic Oxidation", L.I Simándi, ed, Elsevier, Amsterdam (1991); p. 623.
4. U. Schuchardt, W.A. Carvalho and E.V. Spinacé, *Synlett*, in press.
5. W.A. Franco Jr. and U. Schuchardt, *in*: "Proceedings of the XI Simpósio Iberoamericano de Catálisis," Instituto Mexicano del Petróleo y Universidad Autonoma Metropolitana, México (1988); p. 1503.
6. U. Schuchardt and D. Mandelli, unpublished results.
7. See for example: M.N. Sheng, J.G. Zajacek and T.N. Baker III (to Atlantic Richfield Co, New York, N.Y.) U.S. Patent 3,862,961 (1976); C.A. 84: P135449.
8. The nomenclature is geographical. Gif stands for Gif-sur-Yvette in France where Gif^I through Gif^{IV} were invented. GoAgg stands for Gif-Orsay-Aggieland, introducing work done at the Université de Paris-Sud and Texas A&M University. Ch comes from Chernogolovka (USSR), where the GoChAgg system was discovered.
9. D.H.R. Barton, J. Boivin, M. Gastinger, J. Morzycki, R.S. Hay-Motherwell, W.B. Motherwell, N. Ozbalik and K.M. Schwartzentruber, *J. Chem. Soc., Perkin Trans. I* 947 (1986).
10. U. Schuchardt and V. Mano, *in*: "Studies in Surface Science and Catalysis", vol 55, "New Developments in Selective Oxidation", G. Centi and F. Trifiró, eds., Elsevier, Amsterdam (1990); p.185.
11. U. Schuchardt, E.V. Spinacé and V. Mano, *in*: "Studies in Surface Science and Catalysis", vol 66, "Dioxygen Activation and Homogeneous Catalytic Oxidation", L.I. Simándi, ed., Elsevier, Amsterdam (1991); p.47.
12. E.V. Spinacé, Master's Thesis, Universidade Estadual de Campinas (1991).
13. D.H.R. Barton, F. Halley, N. Ozbalik, E. Young, G. Balavoine, A. Gref and J. Boivin, *New J. Chem.* 13:177 (1989).
14. W.A. Carvalho, Master's Thesis, Universidade Estadual de Campinas (1992).
15. U. Schuchardt, C.E.Z. Krähembühl and W.A. Carvalho, *New J. Chem.* 15:955 (1991).

16. C. Sheu, S.A. Richert, P. Cofré, B. Ross Jr., A. Sobkowiak, D.T. Sawyer and J.R. Kanofsky, *J. Am. Chem. Soc.* 112:1936 (1990).
17. C. Sheu, A. Sobkowiak, S. Jeon and D.T. Sawyer, *J. Am. Chem. Soc.* 112:879 (1990).
18. G. Balavoine, D.H.R. Barton, J. Boivin and A. Gref, *Tetrahedron Lett.* 31:659 (1990).
19. U. Schuchardt, C.E.Z. Krähembühl and W.A. Carvalho, *in*: "Abstracts of the 8th International Symposium on Homogeneous Catalysis", Amsterdam (1992); P-261.
20. U. Schuchardt and W.A. Carvalho, unpublished results.
21. M.J. Nappa and C.A. Tolman, *Inorg. Chem.* 24:4711 (1985).
22. Yu. V. Geletii, V.V. Lavrushko and G.V. Lubimova, *J. Chem. Soc., Chem. Commun.* 1936 (1988).
23. D.H.R. Barton, E. Csuhai, D. Doller and Yu. V. Geletii, *Tetrahedron* 47:6561 (1991).
24. D.H.R. Barton, S.D. Bévière, W. Chavasiri, E. Csuhai and D. Doller, *Tetrahedron* 48:2895 (1992).
25. U. Schuchardt and R. Pereira, unpublished results.
26. D.H.R. Barton, J. Boivin, W.B. Motherwell, N. Ozbalik and K.M. Schwartzentruber, *New J. Chem.* 10:387 (1986).
27. D.H.R. Barton, D. Doller and Yu. V. Geletii, *Mendeleev Commun.* 115 (1991).
28. D.H.R. Barton, S.D. Bévière, W. Chavasiri, D. Doller, W.G. Liu and J.H. Reibenspies, *New J. Chem.* 16:1019 (1992).
29. U. Schuchardt and A.A. Bellini, unpublished results.
30. C. Larpent and H. Patin, *J. Mol. Catal.* 72:315 (1992).

CATALYTIC CHEMISTRY OF CYTOCHROME P450 AND PEROXIDASES

Paul R. Ortiz de Montellano

Department of Pharmaceutical Chemistry
School of Pharmacy
University of California, San Francisco
San Francisco, CA 94143-0446

INTRODUCTION

Iron protoporphyrin IX and other iron porphyrins have been shown in model studies to catalyze the full range of reactions supported by hemoproteins, albeit with little selectivity.[1] The basic catalytic chemistry of hemoproteins is thus intrinsic to the iron porphyrin moiety. A key function of the protein in hemoproteins is therefore to channel the chemistry of the heme group into a single productive mode, either by accelerating one of the possible reactions or by suppressing the alternatives. This reaction control is mediated by the nature of the fifth iron ligand, by specific interactions of the heme group or its axial ligands with active site residues, by the protein active site architecture, and by the ability of the protein to interact with other redox proteins.

The majority of hemoproteins can be classified as either electron transfer proteins (e.g., cytochrome b5, cytochrome c), oxygen transport proteins (e.g., myoglobin), peroxidases (e.g., horseradish peroxidase, myeloperoxidase), or monooxygenases (e.g., cytochromes P450), although there are additional, more limited, classes of hemoproteins (e.g., catalase, guanylate cyclase). Electron transfer proteins are differentiated from other hemoproteins by the fact that their heme iron atom is firmly hexacoordinated, leaving no coordination site vacant and thus effectively preventing catalysis.[2,3] A water is bound to the sixth coordination site of the iron in the myoglobins and hemoglobins, but the water molecule is readily displaced by oxygen, allowing the proteins to function as reversible oxygen carriers.[4] The globins do not function normally as reaction catalysts but the presence of a vacant iron coordination site conveys on them considerable catalytic potential.[5] Among the proteins that have been evolutionarily optimized as catalysts, a primary distinction is to be made between peroxidases and monooxygenases/peroxygenases. Although catalysis by both classes of proteins has been attributed to related ferryl (formally $Fe^V=O$) species, peroxidases remove electrons one-at-a-time from their substrates with concomitant reduction of the ferryl oxygen to water,[6] while monooxygenases transfer the ferryl oxygen atom to theirs (Figure 1).[7] We have proposed that the key to whether peroxidative or monooxygenative (peroxygenative) chemistry occurs is the relationship between the substrate binding site and the ferryl oxygen: oxygen transfer occurs when the substrate is able to interact with the ferryl oxygen but peroxidation predominates when this interaction is prevented by the protein structure.[6,8]

The Activation of Dioxygen and Homogeneous Catalytic Oxidation,
Edited by D.H.R. Barton *et al.*, Plenum Press, New York, 1993

Studies with horseradish peroxidase provided the first evidence that peroxidase substrates are prevented by the protein structure from interacting with the ferryl oxygen species. These studies demonstrated that reaction with phenylhydrazine results in addition of the phenyl group to the δ-meso carbon or abstraction of the vicinal 8-methyl hydrogen rather than, as with most other hemoproteins, in formation of a phenyl-iron complex.[9] The unique reactivity of the δ-meso carbon has been confirmed by the formation of adducts at that position with alkylhydrazines, azide, cyclopropanone hydrate, and nitromethane.[10-13] Nuclear Overhauser experiments have provided independent evidence that substrates bind to the ferric enzyme in the vicinity of the 8-methyl group.[14] These results suggest that substrates bind near the δ-meso carbon and transfer an electron to the heme edge without coming in contact with the iron (or ferryl oxygen). The substrate radical produced by the electron transfer is therefore able to diffuse away rather than being trapped by combination with the radical-like ferryl oxygen atom.

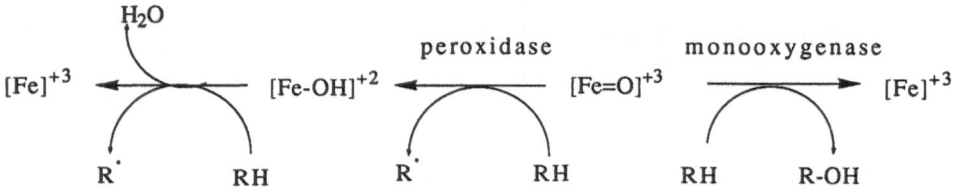

Figure 1. Monooxygenase (peroxygenase) versus peroxidase pathways for reaction of the ferryl species ($[Fe=O]^{+3}$) with a substrate RH.

Differentiation of peroxidases from monooxygenases across the range of such enzymes appears, from recent evidence, to involve more than simple segregation of the substrate and ferryl sites within the protein active site. The nature of structural parameters that determine the reaction outcome has been probed by investigating the oxidation of styrenes and thioanisoles by cytochrome $P450_{cam}$, cytochrome c peroxidase, horseradish peroxidase, and myoglobin. The differences in the oxidation of these substrates by the monooxygenase and the two peroxidases, and by a hemoprotein not designed to catalyze either reaction but which catalyzes both, provide valuable information on the relationship between structure and function in hemoproteins.

EPOXIDATION AND SULFOXIDATION BY CYTOCHROME $P450_{cam}$

Cytochrome $P450_{cam}$ was once thought to be a specific catalyst for camphor 5-hydroxylation but has recently been found to catalyze the oxidation of styrenes,[15] thioanisoles,[16] and other substrates (e.g., nicotine)[17] unrelated to camphor. Studies of the epoxidation mechanism using *trans*-2-phenyl-1-vinylcyclopropane as a probe suggest that the epoxidation reaction does not involve a discrete radical or cation intermediate.[18] This conclusion, which agrees with recent results obtained with metalloporphyrin models,[19] is based on the observation that oxidation of the olefin does not detectably yield cyclopropane ring-opened products (Figure 2). Thioether sulfoxidation, on the other hand, is believed to involve electron transfer to give the sulfur radical cation that combines via a "rebound" mechanism with the ferryl oxygen.[7]

To explore the influence of active site topology on the epoxidation and sulfoxidation reactions, we have determined the absolute stereochemistries of the reaction products. We have compared the results with independent stereochemical predictions made by Dr. Gilda Loew (Molecular Research Institute) and her group based on molecular dynamics studies of the substrates in the crystallographically defined active site of cytochrome P450$_{cam}$. These calculations were carried out with a truncated active site model rather than the whole protein to minimize computation time. Good agreement was found between the experimental and calculated enantiomeric excesses for the epoxidation of styrene, *cis*-β-methylstyrene, and *trans*-β-methylstyrene (Table 1),[15] and reasonable agreement for the sulfoxidation of thioanisole and *p*-methylthioanisole.[16] These results suggest that the reaction outcome can be predicted when the detailed structures of the enzyme active site and the substrate are known, at least for substrates with a single reaction site and few degrees of internal freedom.

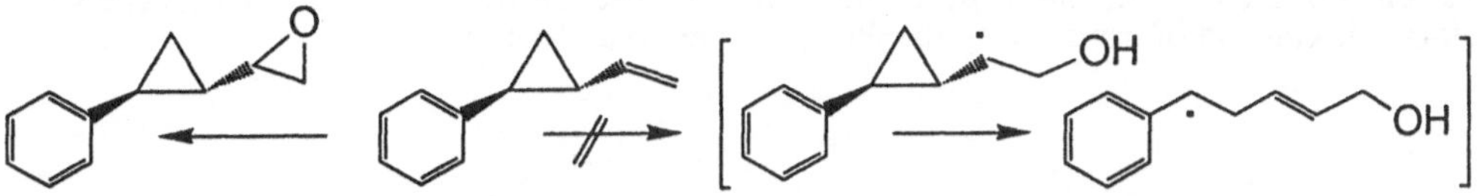

Figure 2. Epoxidation of *trans*-2-phenyl-1-vinylcyclopropane by cytochrome P450$_{cam}$. The epoxide is obtained but the ring opened products expected from a radical mechanism do not appear to be formed.

Table 1. Experimental and calculated absolute stereochemistries of the styrene and *cis*- and *trans*-β-methylstyrene epoxides produced by cytochrome P450cam.

Substrate	Stereoisomers	Experimental	Calculated
Styrene	1*S*:2*R*	83:17	65:35
Cis-β-methylstyrene	1*S*,2*R*:1*R*,2*S*	89:11	84:16
Trans-β-methylstyrene	1*S*,2*S*:1*R*,2*R*	75:25	75:25
Thioanisole	*R*:*S*	72:28	65:35
p-Methylthioanisole	*R*:*S*	48:52	22:78

EPOXIDATION AND SULFOXIDATION BY CYTOCHROME C PEROXIDASE AND HORSERADISH PEROXIDASE

The normal substrate for cytochrome c peroxidase is cytochrome c but the enzyme also oxidizes guaiacol and other classic peroxidase substrates.[20] The crystal structure of cytochrome c peroxidase shows that the heme is buried in the protein but is accessible from the exterior via a narrow channel that disgorges into the active site in the vicinity of the δ-meso carbon atom.[21] Oxidation of cytochrome c requires electron transfer between the heme groups in the complex of the two proteins over an edge-to-edge distance of roughly 16 Å.[22,23] Guaiacol and other small substrates, however, are oxidized at the δ-meso edge of the prosthetic heme group after they migrate down the access channel into the active site (Figure 3). This is clearly shown by the demonstration that reconstitution of cytochrome c peroxidase with δ-meso-alkyl-substituted hemes blocks the oxidation of guaiacol but does not interfere with cytochrome c oxidation.[20] The δ-meso edge therefore appears to be as important for small substrate oxidation by cytochrome c peroxidase as it is for horseradish peroxidase.

Cytochrome c peroxidase, in addition to catalyzing the one-electron oxidation of guaiacol and other peroxidase substrates, has been shown to promote the oxidation of styrenes to styrene oxides.[24] Retention of stereochemistry in the epoxidation of cis-β-methylstyrene and incorporation of oxygen from H_2O_2 into the epoxide product indicate that epoxidation is mediated by a ferryl oxygen transfer mechanism analogous to that of the cytochrome P450 monooxygenation reaction. Cytochrome c peroxidase can therefore function as both a peroxygenase and a peroxidase, in conflict with the simple idea that oxygen transfer reactions are prevented by protein barriers in peroxidases. Cytochrome c peroxidase is an unusual peroxidase, however, in that its normal protein substrate is held at a considerable distance from the prosthetic heme group by protein-protein contacts. It can therefore be argued that there has been little need to refine the active site structure to discriminate between small substrate peroxidation and peroxygenation. The peroxygenase reaction proves, in fact, that a peroxidase, in the absence of interference by active site residues, is capable of transferring the ferryl oxygen to a substrate.

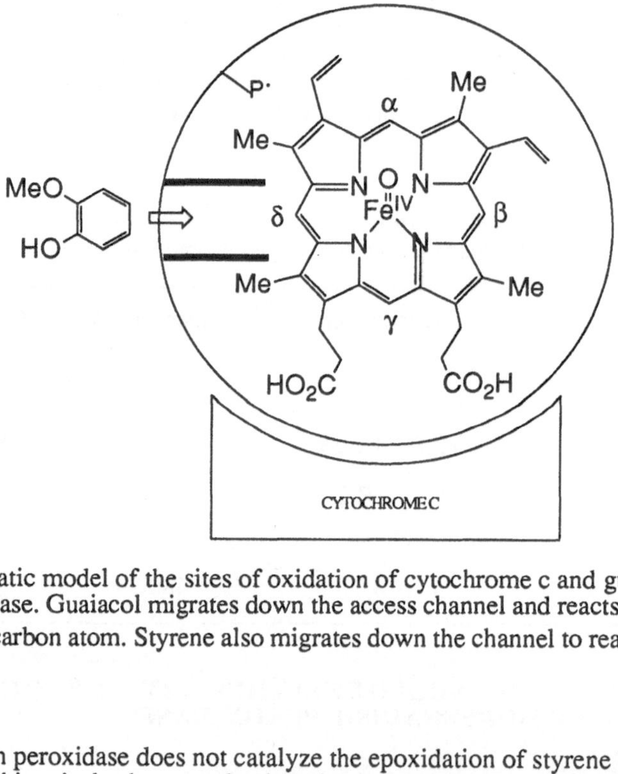

Figure 3. Schematic model of the sites of oxidation of cytochrome c and guaiacol by cytochrome c peroxidase. Guaiacol migrates down the access channel and reacts with the heme near the δ-meso carbon atom. Styrene also migrates down the channel to react with the ferryl oxygen.

Horseradish peroxidase does not catalyze the epoxidation of styrene but catalyzes the sulfoxidation of thioanisoles by a mechanism that incorporates an atom of oxygen from the peroxide into the product.[25,26] Thus, even this archetypical peroxidase is able to promote an oxygen transfer reaction in contradiction of the proposal that oxygen transfer reactions are suppressed in peroxidases by the protein structure. Recent work suggests that there are distinct binding sites in horseradish peroxidase for peroxidative and peroxygenative substrates.[26] Key evidence in this regard is provided by the observation that reconstitution of horseradish peroxidase with δ-meso-ethylheme suppresses the peroxidation of guaiacol but increases the sulfoxidation of thioanisole (Table 2). This agrees with the hypothesis that peroxidation is mediated by electron transfer to the heme edge whereas oxygen transfer requires interaction of the substrate with the ferryl oxygen. In accord with earlier studies, a δ-meso methyl group does not prevent either sulfoxidation or guaiacol peroxidation, providing direct evidence that the effect of the ethyl group is steric rather than electronic in nature. The stereospecificity of the sulfoxidation reaction is decreased by both d-meso-alkyl substituents, however. These results suggest that the binding site for substrates that are oxidized by removal of one electron is distinct from that for substrates to which the ferryl oxygen is transferred. Evidence for at least two substrate sites has been obtained by kinetic and spectroscopic methods.[26] Kinetic studies suggest that inhibition of the oxidation of guaiacol by thioanisole is not strictly competitive and therefore that the two compounds bind at distinct, albeit possibly overlapping, sites. Spectroscopic studies show that the changes in the spectrum of the enzyme caused by thioanisole are not suppressed by even very high concentrations of guaiacol, which causes a different change in the spectrum. These results provide evidence for the existence of multiple binding sites, but do not resolve the question of what determines whether a compound is bound in the peroxidative or peroxygenative site. Peroxidation appears to be mediated by electron transfer to the δ-meso heme edge, but it appears that the active site structure, and the factors that control the reaction outcome, are more complex than a simple protein barrier.

Table 2. Sulfoxidation of thioanisole and peroxidation of guaiacol by horseradish peroxidase (HRP) and modified HRP

Enzyme	Sulfoxide ee	Thioanisole sulfoxidation (%)	Guaiacol peroxidation (%)
Native HRP	70	100	100
Meso-ethyl HRP	26	180	5
Meso-methyl HRP	22	70	165
Heat denatured HRP	4	15	3

EPOXIDATION BY MYOGLOBIN

Myoglobin, unlike cytochrome $P450_{cam}$, cytochrome c peroxidase, and horseradish peroxidase, normally functions as an oxygen transport protein rather than as an enzyme. However, the availability of an iron coordination site in the ferric and deoxy ferrous states makes possible a reaction between the hemoprotein and H_2O_2. This reaction produces a ferryl species and a transient protein radical.[27,28] In the case of sperm whale myoglobin, the protein radical is quenched, in part, by the formation of dityrosine cross-links between myoglobin monomers.[29] The heme group is also crosslinked to the protein in a fraction of the protein molecules, again through covalent bond formation to one of the tyrosine residues.[30] The residues involved in these processes are Tyr-103 and Tyr-151. The cross-linking results and the spectroscopic data clearly indicate that the protein radical is located, to a large extent, on these two tyrosine residues.

Expression of a synthetic sperm whale myoglobin gene in *E. coli* has made possible site specific mutagenesis studies of the radical site.[31] Site specific mutants have been prepared in which the three tyrosines (Tyr-103, Tyr-146, and Tyr-151) have been replaced by phenylalanines individually, two at a time, or all three at once. Reaction of H_2O_2 with each of these mutants, including the triple mutant with no tyrosine residues, gives rise to an EPR-detectable protein radical.[31] The EPR signal is observed in all instances except that of the tetra mutant with no tyrosines and His-64 replaced by a valine, although the shape of the signal differs among the various mutants. His-64 serves as a gate into the heme crevice and forms a hydrogen bond to the water that coordinates to the iron in metmyoglobin. In accord with earlier results, protein cross-linking is only observed when Tyr-151 is retained in the protein. It is clear from the results that the unpaired electron density migrates readily over

Table 3. Product formation and incorporation of oxygen from H_2O_2 in the epoxidation of cis-β-methylstyrene by sperm whale myoglobin and myoglobin mutants.[34]

Myoglobin[a]	Relative yields %		^{18}O incorporation from $H_2^{18}O_2$ %	
	cis-β-methyl styrene oxide	trans-β-methyl styrene oxide	cis-β-methyl styrene oxide	trans-β-methyl styrene oxide
Native	74	26	nd[b]	nd
Wild type	54	46	90	8
Y103F	47	53	98	17
Y146F	52	48	97	20
Y151F	49	51	90	18
Y146F/Y151F	75	25	nd	nd
Y103F/Y151F[c]	72	28	nd	nd
Y103F/Y146F Y151F[c]	48	52	95	15
H64V	92	8	100	nm[d]
H64V/Y103F/ Y146F/Y151F[c]	99	1	102	nm

[a]Y = tyrosine, F = phenyl-alanine, H = histidine. [b]nd = not done. [c]All Tyr-103 -> Phe mutants also carry a Lys-102 -> Gln mutation to stabilize the Tyr-103 mutation. [d]nm = too small to measure.

several residues, including Tyr-103, Tyr-151, His-64, and possibly other non-tyrosine residues.

Metmyoglobin and methemoglobin catalyze the H_2O_2-dependent epoxidation of styrene and other aryl-substituted double bonds.[32,33] Stereochemical and isotope incorporation experiments (Table 3) imply the existence of a plurality of epoxidation mechanisms. Thus, a fraction of the epoxide formed from cis-β-substituted styrenes is formed with retention of the olefin stereochemistry and incorporation of oxygen from the peroxide. These properties of the reaction characterize it as a normal cytochrome P450-like ferryl oxygen transfer process. However, a fraction of the epoxide is formed with loss of olefin stereochemistry and incorporation of oxygen from molecular oxygen or even, to a small extent, from water.[32,33] To explain the incorporation of molecular oxygen with loss of stereochemistry, we proposed that the epoxide is formed by a co-oxidation mechanism in which reaction of oxygen with the protein radical yields a protein-peroxy radical that adds to the double bond of styrene (Figure 4). The evidence for a tyrosine radical makes Tyr-103 or Tyr-151 a strong candidate for the protein residue involved in the co-oxidation reaction .

Analysis of the epoxidation of styrene and cis-β-methylstyrene by the sperm whale myoglobin mutants shows, however, that the cis:trans epoxide product ratio and the incorporation of oxygen from molecular oxygen into the trans epoxide occurs to about the same extent in the presence and absence of Tyr-103 and/or Tyr-151![34] Indeed, only with the His-64 mutants does the epoxidation proceed almost exclusively with retention of stereochemistry and incorporation of an oxygen from the peroxide into the epoxide (Table 3). His-64 thus appears to play a critical role in the epoxidation mechanism. Either it is itself responsible for the co-oxidation process, or its replacement by a valine improves access to the ferryl group to the extent that ferryl oxygen transfer swamps out the co-oxidation reaction. The former mechanism appears more likely because the absolute rate of epoxide formation increases by only approximately 2-fold with the His-64 mutants, a rate increase that does not appear to be high enough to overwhelm the co-oxidation reaction.[34]

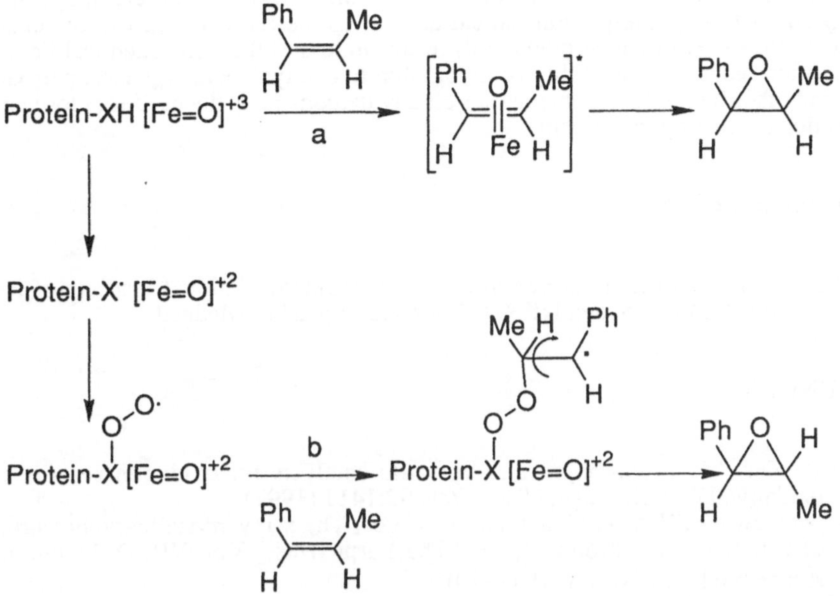

Figure 4. Ferryl transfer versus protein-mediated co-oxidation of styrene proposed for the myoglobin-catalyzed formation of styrene epoxides with, respectively, (a) retention of stereochemistry and incorporation of oxygen from H_2O_2, and (b) loss of stereochemistry and incorporation of oxygen from O_2.

SUMMARY

Styrene epoxidation and thioanisole sulfoxidation are informative probes of the abilities of hemoproteins to catalyze ferryl oxygen transfer reactions. Classical peroxidases with a histidine as the fifth iron ligand do not appear to oxidize styrene but are able to catalyze the sulfoxidation of thioanisole. An exception to this rule is cytochrome c peroxidase, which catalyzes the epoxidation of both styrene and cis-β-methylstyrene. These results show that peroxidases have a limited ability to catalyze ferryl oxygen transfer reactions. If a single substrate binding site is involved, it must allow sufficient interaction between the ferryl oxygen and the thioanisole for the ferryl oxygen to be transferred to the sulfur atom, but does not allow the more intimate interaction of the ferryl oxygen with a double bond required for epoxide formation. Sulfur oxidation requires minimal interaction between the electrons in one of the sulfur orbitals and the ferryl oxygen, whereas epoxidation requires intimate contact between both carbons of the olefin and the ferryl oxygen. Kinetic and spectroscopic evidence suggests that the peroxygenase and peroxidase substrate binding sites of horseradish peroxidase are distinct, but the nature of the two binding sites and the substrate features that determine to which site it is bound remain unclear. The postulate that the protein in a peroxidase minimizes contact between the substrate and the ferryl oxygen appears to be valid, but it is increasingly clear that significant variations of this central theme will occur.

Studies of the epoxidation of styrene and sulfoxidation of thioanisoles by cytochrome P450$_{cam}$, a monooxygenase of known structure, show that the absolute stereochemistry of the reaction can be predicted by molecular dynamics calculations. Extension of this approach to other reactions, and to more complicated substrates, should provide an important step towards the rationalization of oxidative metabolism and the construction of tailor made oxidative catalysts.

The oxidation of styrene by metmyoglobin/H$_2$O$_2$ illustrates the complexity that is possible in catalysis by a hemoprotein not specifically constructed as a catalyst. The ferryl oxygen in this situation is not well stabilized and readily oxidizes the protein, opening the way for participation of the protein radical as well as the ferryl oxygen in the catalytic chemistry. In the case of metmyoglobin, this results in loss of the stereochemical fidelity of the epoxidation reaction. The ability of myoglobin to catalyze peroxygenase, peroxidase, catalase, and other hemoprotein chemistries clearly demonstrates the need for close control of the reaction process by the protein.

Acknowledgments

Support of the studies described in this manuscript by National Institute of Health Grants GM25515, GM32488, and DK30297 is gratefully acknowledged.

REFERENCES

1. B. Meunier, Metalloporphyrins as versatile catalysts for oxidation reactions and oxidative DNA cleavage, *Chem. Rev.* 92:1411 (1992).
2. F.S. Mathews, E.W. Czerwinski, and P. Argos, The x-ray crystallographic structure of calf liver cytochrome b$_5$, *in*: "The Porphyrins," Vol. VII, D. Dolphin, ed., Academic Press, New York (1979).
3. S. Ferguson-Miller, D.L. Brautigan, and E. Margoliash, The electron transfer function of cytochrome c, *in*: "The Porphyrins," Vol. VII, D. Dolphin, ed., Academic Press, New York (1979).
4. R.E. Dickerson and I. Geis. "Hemoglobin: Structure, Function, Evolution, and Pathology," Benjamin/Cummings, Menlo Park (1983)
5. R.J. Mieyal, Monooxygenase activity of hemoglobin and myoglobin, in: "Reviews in Biochemical Toxicology,", Vol. 7, E. Hodgson, J.R. Bend, and R.M. Philpot, eds., Elsevier, New York (1985).
6. P.R. Ortiz de Montellano, Catalytic sites of hemoprotein peroxidases, *Annu. Rev. Pharmacol. Toxicol.* 32:89 (1992).
7. P.R. Ortiz de Montellano, Oxygen activation and transfer, *in*: "Cytochrome P-450: Structure, Mechanism, and Biochemistry," P.R. Ortiz de Montellano, ed., Plenum, New York (1986).
8. P.R. Ortiz de Montellano, Control of the catalytic activity of prosthetic heme by the structure of hemoproteins, *Accts. Chem. Res.* 20:289 (1987).
9. M.A. Ator and P.R. Ortiz de Montellano, Protein control of prosthetic heme reactivity. Reaction of substrates with the heme edge of horseradish peroxidase, *J. Biol. Chem.* 262:1542 (1987).
10. M.A. Ator, S.K. David, and P.R. Ortiz de Montellano, Structure and catalytic mechanism of horseradish peroxidase. Regiospecific *meso* alkylation of the prosthetic heme group by alkylhydrazines, *J. Biol. Chem.* 262:14954 (1987).
11. P.R. Ortiz de Montellano, S.K. David, M.A. Ator, and D. Tew, Mechanism-based inactivation of horseradish peroxidase by sodium azide. Formation of meso-azidoprotoporphyrin IX, *Biochemistry* 27:5470.
12. J.S. Wiseman, J.S. Nichols, and M.X. Kolpak, Mechanism of inhibition of horseradish peroxidase by cyclopropanone hydrate, *J. Biol. Chem.* 257:6328 (1982).

13. D.J.T. Porter and H.J. Bright, The mechanism of oxidation of nitroalkanes by horseradish peroxidase, *J. Biol. Chem.* 258:9913 (1983).
14. J. Sakurada, S. Takahashi and T. Hosoya, Nuclear magnetic resonance studies on the spatial relationship of aromatic donor molecules to the heme iron of horseradish peroxidase , *J. Biol. Chem.* 261:9657 (1986).
15. J. Fruetel, J.R. Collins, D.L. Camper, G.H. Loew, and P.R. Ortiz de Montellano, Calculated and experimental absolute stereochemistry of the styrene and b-methylstyrene epoxides formed by cytochrome P450$_{cam}$, *J. Am. Chem. Soc.* 114:6987 (1992).
16. J.A. Fruetel, Y.-T. Chang, .R. Collins, G. Loew, and P.R. Ortiz de Montellano, Thioanisole sulfoxidation by cytochrome P450$_{cam}$ (CYP101): experimental and calculated absolute stereochemistries, *submitted for publication.*
17. J.P. Jones, W.F. Trager, and T.J. Carlson, The binding and regioselectivity of reaction of (R)- and (S)-nicotine with cytochrome P-450$_{cam}$: parallel experimental and theoretical studies, *J. Am. Chem. Soc.* 115:381 (1993).
18. V.P. Miller, J.A. Fruetel, and P.R. Ortiz de Montellano, Cytochrome P450$_{cam}$-catalyzed oxidation of a hypersensitive radical probe, *Arch. Biochem. Biophys.* 298:697 (1992).
19. D. Ostovic and T.C. Bruice, Mechanism of alkene epoxidation by iron, chromium, and manganese higher valent oxo-metalloporphyrins, *Acc. Chem. Res.* 25:314 (1992).
20. G.D. DePillis, B.P. Sishta, A.G. Mauk, and P.R. Ortiz de Montellano, Small substrates and cytochrome c are oxidized at different sites of cytochrome c peroxidase, *J. Biol. Chem.* 266:19334 (1991).
21. B.C. Finzel, T.L. Poulos, and J. Kraut, Crystal structure of yeast cytochrome c peroxidase refined at 1.7-Å resolution, *J. Biol. Chem.* 259:13027 (1984).
22. T.L. Poulos and J. Kraut, A hypothetical model of the cytochrome c peroxidase.cytochrome c electron transfer complex, *J. Biol. Chem.* 255:10322 (1980).
23. H. Pelletier and J. Kraut, Crystal structure of a complex between electron transfer partners, cytochrome c peroxidase and cytochrome c, *Science* 258:1748 (1992).
24. V.P. Miller, D.G. DePillis, B.P. Sishta, and A.G. Mauk, The monooxygenase activity of cytochrome c peroxidase, *J. Biol. Chem.* 267:8936 (1991).
25. S. Kobayashi, M. Nakano, T. Goto, T. Kimura, and A.P. Schaap, An evidence of the peroxidase-dependent oxygen transfer from hydrogen peroxide to sulfides, *Biochem. Biophys. Res. Commun.* 135:166 (1986).
26. R.Z. Harris, S.L. Newmyer, and P.R. Ortiz de Montellano, Horseradish-peroxidase-catalyzed two-electron oxidations. Oxidation of iodide, thioanisoles, and phenols at distinct sites, *J. Biol. Chem.* 268:1637 (1993).
27. N.K. King, F.D. Looney, and M.E. Winfield, Amino acid free radicals in Oxidised metmyoglobin, *Biochim, Biophys. Acta* 133:65 (1967).
28. Y. Yonetani and H. Schleyer, Studies on cytochrome c peroxidase. IX. The reaction of ferrimyoglobin with hydroperoxides and a comparison of peroxide-induced compounds of ferrimyoglobin and cytochrome c peroxidase, *J. Biol. Chem.* 242:1974 (1967).
29. D. Tew and P.R. Ortiz de Montellano, The myoglobin protein radical. Coupling of Tyr-103 to Tyr-151 in the H$_2$O$_2$-mediated cross-linking of sperm whale myoglobin, *J. Biol. Chem.* 263, 17880 (1988).
30. C.E. Catalano, Y.S. Choe, and P.R. Ortiz de Montellano, Reactions of the protein radical in peroxide-treated myoglobin. Formation of a heme-protein cross-link, *J. Biol. Chem.* 264:10534 (1989).
31. A. Wilks and P.R. Ortiz de Montellano, Intramolecular translocation of the protein radical formed in the reaction of recombinant sperm whale myoglobin with H$_2$O$_2$, *J. Biol. Chem.* 267:8827 (1992).
32. P.R. Ortiz de Montellano and C.E. Catalano, Epoxidation of styrene by hemoglobin and myoglobin. Transfer of oxidizing equivalents to the protein surface, *J. Biol. Chem.* 260:9265 (1985).

33. C.E. Catalano and P.R. Ortiz de Montellano, Oxene-transfer, electron abstraction, and cooxidation in the epoxidation of stilbene and 7,8-dihydroxy-7,8-dihydrobenzo[a]pyrene by hemoglobin, *Biochemistry* 26:8373 (1987).

34. S.I. Rao, A. Wilks, and P.R. Ortiz de Montellano, The roles of His-64, Tyr-103, Tyr-146, and Tyr-151 in the epoxidation of styrene and β-methylstyrene by recombinant sperm whale myoglobin, *J. Biol. Chem.* 268:803 (1993).

OXYGEN-DERIVED TOXINS GENERATED BY NEUTROPHILS

AND THEIR MICROBICIDAL MECHANISMS

James K. Hurst

Department of Chemistry, Biochemistry, & Molecular Biology
Oregon Graduate Institute of Science & Technology
Portland, OR 97291-1000

INTRODUCTION

This review is designed as a short introduction to phagocytic cells (white blood cells) found in the peripheral circulation of mammals. Emphasis will be placed upon the oxidative chemistry and biochemistry of these cells, and will include a discussion of selected current problems in fields relating to these topics. A bibliography of recent general reviews and monographs is given for readers wishing a more encyclopedic coverage[1-6] and pertinent topical reviews will be cited in the subsection headings.

Several types of white blood cells are formed from precursor cells produced in the bone marrow (Figure 1). Predominant among these is the neutrophil, a motile cell whose primary function in host resistance to disease is to recognize and destroy unicellular pathogenic organisms. Among the other white blood cells, the monocyte appears to have very similar biochemical capabilities. Upon leaving the blood stream to reside in host tissues, this cell differentiates further, forming macrophages which lack the peroxidase (termed myeloperoxidase) found in neutrophils and monocytes, but are otherwise similar. The eosinophil, which will be mentioned in a comparative sense, has unique biochemical capabilities and appears to be directed primarily against multicellular parasitic organisms.[7]

Each of these cells, upon recognizing and binding a bacterium or other foreign body, undergoes a progression of biochemical transformations that serve to isolate the organism within an inimical environment. These events are illustrated in stylized form for the neutrophil in Figure 2. Upon binding a particle (here, an antibody-coated bacterium), a dormant electron transport chain located within the plasma membrane is activated, causing respiration to increase about 100-fold over the basal rate. This respiratory chain catalyzes a unique one-electron reduction of O_2 to superoxide ion, which then disproportionates quantitatively (or nearly so), forming hydrogen peroxide. The immediate source of electrons is NADPH; because this reductant is cyclically regenerated by glucose oxidation via the hexose monophosphate pathway, the other

The Activation of Dioxygen and Homogeneous Catalytic Oxidation,
Edited by D.H.R. Barton *et al.*, Plenum Press, New York, 1993

A. Phagocytic cells of host resistance:

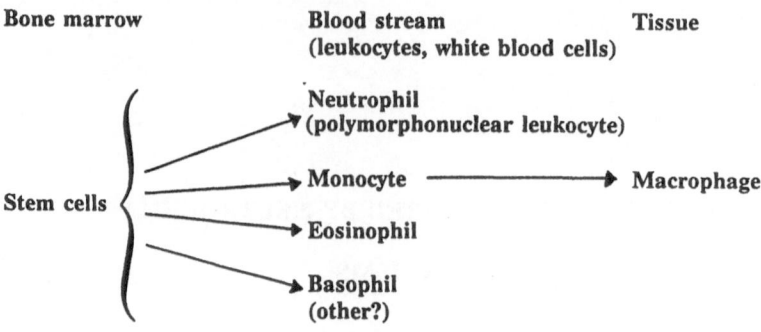

B. Diseases of granulocytes (neutrophils, monocytes):

Chronic granulomatous disease (CGD)
> (inability to undergo stimulated respiration)

Myeloperoxidase (MPO) deficiency
> (neutrophils lack active peroxidase)

Figure 1. A glossary of terms. The three columns in part A refer to intracellular location of the particular cell types, as labeled.

respiratory end product is carbon dioxide. Coincident with stimulation of respiration, phagocytosis is initiated. The membrane invaginates, ultimately surrounding the particle and pinching off to form a sealed vacuole (phagosome) containing the particle. Because the membrane everts during this process, the respiratory chain is now vectorially oriented to produce H_2O_2 within the phagosome from the enzymatic glucose oxidation system located in the cytosol. Neutrophils also contain lysosomal particles called granules that, upon activation of the cell, migrate to the phagosomal and plasma membranes, at which point the membranes fuse and the lysosomal contents are discharged into the vacuole or extracellular space. Granule components include among a variety of digestive enzymes and biopolymers two proteins that are potentially involved with oxidative metabolism, the enzyme myeloperoxidase (MPO) and the iron-binding protein, lactoferrin. The amount of MPO contained within normal neutrophils is truly staggering, comprising up to 5% of the total cell weight. Although lactoferrin containing bound iron might also utilize H_2O_2,[2] the protein appears to be present primarily in its apo, or demetalated form.

Are the oxidative reactions associated with the respiratory burst important to bactericidal mechanisms? Clinical evidence suggests that they are. A set of congenital defects known as chronic granulomatous disease (CGD) which manifests itself as an inability of affected individuals' neutrophils to undergo stimulated respiration, but with otherwise apparently normal phagocytic capabilities, is characterized by persistent life-threatening bacterial infections.[6] In contrast, hereditary myeloperoxidase deficiency is markedly less debilitating, and apparently has only minor clinical manifestations.[8,9] However, as summarized by Klebanoff,[4,8] MPO-deficient neutrophils are considerably less effective than normal neutrophils in cell-free studies of bactericidal potency. This behavior implies the existence of dual, i.e., MPO-dependent and MPO-independent, oxidative bactericidal mechanisms. Further, not all species of bacteria require O_2 for phagocytic killing,[10,11] and the neutrophil granules contain cationic proteins that are

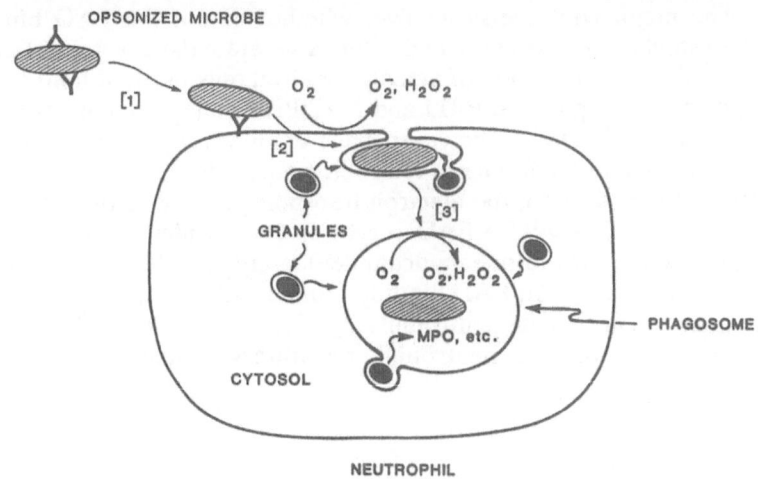

Figure 2. Diagram of phagocytosis by neutrophils. (1) Binding of the antibody-coated bacterium elicits the respiratory burst; (2) the neutrophil membrane invaginates; and (3) ultimately pinches off forming a new vacuole (phagosome). Simultaneous degranulation leads both to extracellular secretion and intraphagosomal compartmentation of granule components. Reproduced from reference 2 by permission of CRC Press.

toxic to bacteria by nonoxidative mechanisms.[12-14] The need for this apparent redundancy is not presently obvious, but may ultimately be found to lie within the diversity of bacterial species with which our bodies must contend.

HOW IS O_2 "ACTIVATED" BY NEUTROPHILS?[15]

Although the mechanistic details are as yet incomplete, activation of the respiratory chains in phagocytic cells appears to involve a secondary messenger cascade triggered by binding at specific receptor sites on the plasma membrane envelope, followed by enzymatic activation, phosphorylation, and translocation of cytoplasmic proteins and their assembly with membrane-localized proteins to form the intact electron transport chain. Specifically, binding of bacteria or other agonists is thought to release a trimeric G-protein from the receptor sites, which in turn activates phospholipases (A_2, C, D), which directly (C), or indirectly (D) generate diacylglycerol or other secondary messengers (A_2) by hydrolyzing phospholipid components of the plasma membrane. These secondary messengers activate protein kinase C, which phosphorylates several cytosolic proteins, including a 47-kDa protein. Phorbol esters and similar surface-active compounds are apparently able to mimic functionally these secondary messengers, bypassing the usual triggering mechanisms to directly activate the enzyme. The phosphorylated 47-kDa protein, together with specific 67-kDa and 21-kDa cytosolic proteins, then complex with the plasma membrane-localized redox proteins to form the electron transport chain. The requirement and sufficiency of these three cytosolic proteins for activity appears firmly established from cell-free reconstitution studies in which the separate proteins purified by recombinant DNA techniques were added to membranous fractions of purified oxidase redox components.[16,17] The minimal membrane protein composition is less well defined, however. General agreement exists that, in addition to an NADPH binding site, the chain contains an FAD molecule and an unusual heteromeric low-potential b-type cytochrome. The major controversy involves whether or not the FAD binding site exists within the cytochrome or is contained within a separate flavoprotein. Two groups have reported[16,17] that reconstitution of activity required only purified lipid-solubilized cytochrome and identified putative FAD and NADPH binding domains from genetic mapping comparisons.[16,18] Others, however, have demonstrated a requirement for a separate membrane-bound flavin enzyme in their preparations.[19]

Irrespective of composition, the electron transport pathway is thought to involve the following sequence: NADPH → FAD → cyt b → O_2, in which the FAD is properly situated to mediate the two-to-one noncomplementary transfer of electrons from NADPH to the cytochrome.[2] An FAD binding protein with the appropriate reduction potential has been identified in neutrophils by ESR spectroscopy.[20] The general features for activation of the neutrophil respiratory oxidase are summarized schematically in Figure 3.

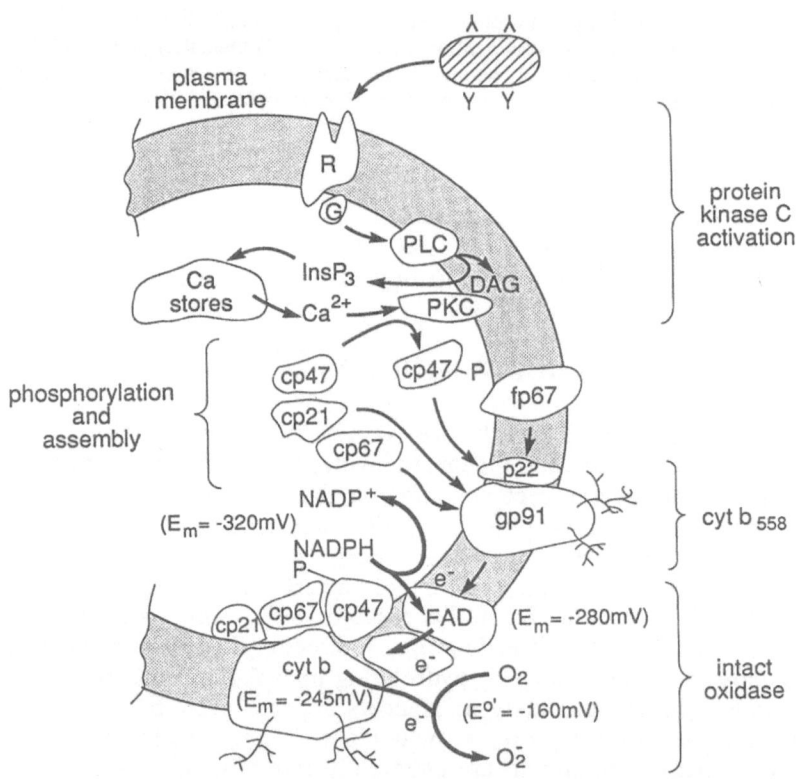

Figure 3. A putative pathway for neutrophil oxidase activation. As discussed in the text, bacterial binding at specific receptor (R) sites releases a trimeric G protein, which activates phospholipase C (PLC), catalyzing formation of inositol-1,4,5-triphosphate ($InsP_3$) and diacylglycerol (DAG). $InsP_3$ binding to Ca storage proteins releases Ca^{2+} ion which, with DAG, activates protein kinase C. The kinase phosphorylates a 47-kDa cytosolic protein (cp47) which initiates assembly of other cytosolic proteins and membrane-localized redox proteins to form the intact oxidase; fp and gp stand for flavoprotein and glycosylprotein, respectively. The structural organization of the complex is unknown, and the issue of the location of flavin adenine nucleotide (FAD) is unresolved. Reduction potentials for the redox centers, as measured for isolated components, are given in parentheses. This figure is an adaptation of drawings from reference 15.

THE ENIGMATIC HEMES OF THE NEUTROPHIL

Cytochrome b_{558} (b_{-245})

The respiratory cytochrome is a heteromeric integral membrane protein complex containing two types of subunits, a 22-kDa protein and a 67- to 91-kDa glycoprotein;[21,22] it is often described as a dimer, although the subunit stoichiometry has not been firmly established. Based upon the optical properties of the purified cytochrome, there appear to be 2-3 hemes bound within the complex.[23] Although denaturation studies have suggested that heme binding is heterogeneous, physical studies give little evidence for heme inequivalence. Thus, the spectroelectrochemically determined reduction potentials and optical, magnetic circular dichroic, and resonance Raman vibrational spectra are all consistent with the presence of a single or equivalent hemes in the oxidase.[24] The only reported physical evidence suggesting inequivalence is the observation that approximately one-half of the ferric heme is easily photoreduced to the ferrous oxidation state, whereas the remainder is extremely resistant to photoreduction.[24] The resonance Raman spectra clearly indicate that the heme is low-spin six-coordinate in both its ferrous and ferric oxidation states. These studies also suggest N-atom ligation at both axial binding sites, most likely from histidyl imidazole substituent groups. This assignment was problematical when originally proposed because earlier structural studies had suggested that heme binding was localized to the small subunit,[25,26] which was known from its genetic sequence to have only one conserved histidine group.[27] However, more recent studies have demonstrated heme binding to both subunits, suggesting an interfacial binding locus.[23] In that case, the second imidazole ligand could be provided by the large subunit, which has five histidine groups. Alternatively, the sixth ligand could be an ϵ-amino substituent from a protein lysine group.

Strong axial ligation on both sides of the heme plane accounts for the inability of the oxidase to bind small anions and for the ferrous cytochrome to only weakly bind carbon monoxide.[2] Consequently, the oxidase is unaffected by addition of general respiratory inhibitors such as azide or cyanide ions. This property also suggests the absence of an oxygen binding site, which has prompted considerable discussion of the appropriateness of the cytochrome as the oxidative terminus of the respiratory chain. However, since formation of O_2^- requires transfer of only one electron to O_2, heme coordination of O_2 is not essential and the reaction could involve outer-sphere electron transfer from the heme periphery. Precedent for this mechanism is the reduction of ferricytochrome c by O_2^-, which is essentially the reverse reaction.[28] In this case, electron transfer is more rapid than the rate of axial ligand substitution on the heme, requiring an outer-sphere mechanism. Imidazole ligation can potentially also account for the unusual ferric-ferrous reduction potential of the heme, which is the lowest reported for a mammalian respiratory heme. (The designation cyt b_{-245} used by some researchers, rather than the more conventional cyt b_{558}, is based upon the measured redox potential of the human cytochrome.) Potentials of similar magnitude are observed in peroxidases, whose proximal imidazole ligand is strongly hydrogen bonded to a second basic amino acid within the peptide backbone. This hydrogen bonding enhances the imidazolate character of the ligand, increasing its σ-electron donating capabilities with consequent stabilization of the ferric oxidation state.[2,29]

One puzzling aspect of the physical properties of the cytochrome is the absence of detectable EPR signals from the ferric heme. Deoxycholate-solubilized oxidase gave no characteristic signals expected for low-spin hemes, the only detectable signals being a high-spin rhombic signal attributable to myeloperoxidase present in the membrane fraction and a complex g = 2.0 organic radical signal (Figure 4).[24] Reports assigning

271

both high-spin[30] and normal low-spin[31] heme signals to the oxidase have appeared, but probably arise from contaminating heme impurities. More recently anomalous low-spin signals at g = 3.3-3.4 were observed for highly purified cyt b_{558} preparations (Figure 4),[18,32] which have the features of highly anomalous low-spin (HALS)-type bis-imidazole hemins.[33] The bands in these spectra are typically very broad and often only the low-field g_z component is clearly resolved. If these EPR spectra represent the native form of the oxidase (purified cyt b_{558} is active only when reconstituted with lipids), then the signals should also have been detectable in the solubilized oxidase preparation. An alternative interpretation is that the heme interacts strongly with a second paramagnetic center, possibly another heme, and is even-spin or diamagnetic, therefore PER-silent, or that relaxation-induced broadening causes the signals to drop below detectable levels. The circular dichroic spectrum of cyt b_{558} exhibits an unusual bilobed band in the Soret region which might be indicative of heme-heme interactions,[26] but the notion of a dimeric heme appears inconsistent with the bis-imidazole ligation suggested from resonance Raman spectroscopy. An additional feature of the EPR spectra is that the g = 2 signal observed in the solubilized oxidase apparently copurifies with the cytochrome (Figure 4). The apparent splitting of this signal and the dependence of its bandshape upon temperature suggest interaction with a second paramagnetic center.[24] One intriguing possibility is that this signal arises from the semiquinone form of the respiratory FAD redox center, although preliminary redox titrations suggest that the signal titrates at a potential 100-200 mV above the redox poise of the respiratory chain. Despite the uncertainties, recent progress in the structural and functional characterization of the oxidase components has been rapid, and we can anticipate resolution of many of the remaining issues in the near future.

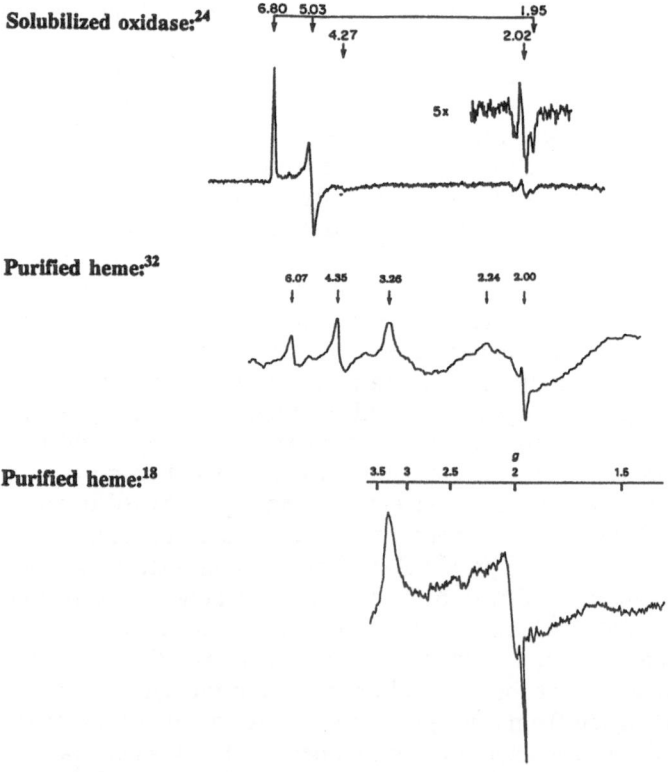

Figure 4. Electron paramagnetic resonance spectra of neutrophil cytochrome b_{558} (b_{-245}). Adapted from references 18, 24, and 32.

As a consequence of having a strong visible absorption band at 570 nm, MPO is green, rather than the usual rust coloration characteristic of other peroxidases. This spectral feature has been attributed alternatively to the presence of formyl or other electron-withdrawing substituents at heme peripheral sites, e.g., as in heme a of cytochrome oxidase, or to symmetry-breaking reduction at one of the β-pyrrolic ring positions, forming a chlorin. Historically, evidence from physical studies has favored assignment of the heme as a chlorin, whereas chemical derivatization studies were more consistent with the presence of formyl substituents. Studies designed to isolate and characterize the heme have been frustrated because the heme is tightly bound, probably covalently, to the protein and the relatively harsh treatments required to isolate it have invariably caused physical and/or chemical modifications.

Resonance Raman spectroscopy has proven exceptionally useful in characterizing the heme in the native enzyme.[34,35] Several additional bands were observed in the ring breathing region when compared to similar peroxidases containing normal heme prosthetic groups such as lactoperoxidase[36] and eosinophil peroxidase.[37] These differences, which could arise from reduction in ring symmetry, were similar to those found in comparisons of chlorin and heme model compounds.[38,39] Furthermore, there was no indication of a carboyl carbon-oxygen stretching band, which should have been a prominent spectral feature if a ring-conjugated formyl substituent were present. To accommodate the chemical derivatization results, we suggested that perhaps the heme was indeed a formylporphyrin, but was attached to the protein by addition of an amino acid substituent (specifically a cysteine thiol group) across the β-carbon double bond of the pyrrole moiety bearing the formyl substituent.[29] In this case, the native MPO would have the physical characteristics of a chlorin since the formyl substituent was not in conjugation with the heterocyclic ring, but its presence would be unmasked upon cleavage of the covalent bond to the protein. This suggestion was based upon structural studies on lactoperoxidase from which it was inferred that the heme was covalently linked to the apoprotein through a disulfide bond between a cysteine residue and a heme β-thiomethyl ring substituent.[40,41] If this were a formyl substituent on the MPO heme, a nearby cysteinyl group could preferentially add across the β-pyrrolic double bond, forming a pseudo-chlorin that is the analog of well-described green sulfhemins.[42] Recent NMR[43] and X-ray crystallographic[44] studies do not support this notion, however. Although these studies could not distinguish between chlorin and heme prosthetic groups, the spectral signature of MPO did not match well the reported spectra of sulfhemins, and the heme binding pocket contained no protein cysteinyl residues. From the crystallographic data, the site of covalent attachment has been tentatively assigned to a substituent on the heme pyrrole ring B and a protein glutamyl residue, although the chemical nature of the linkage remains uncertain. A likely possibility is formation of ester or thioester bonds to a modified protoporphyrin IX-based heme. These modifications might involve alterations of the normal 3-methyl or 4-vinyl β-pyrrolic ring substituents or attachment to hydroxyl groups of a ring-reduced 3,4-dihydroxyprotochlorin IX.[44] Precedent for this type of chlorin structure is the cytochrome d terminal oxidase from *E. coli*, which has been identified as a 5,6-dihydroxyprotochlorin IX,[45] i.e., the analogous structure reduced at the β-carbon-carbon double bond of protoporphyrin IX pyrrole ring C. Similarities noted in the magnetic circular dichroic spectra between MPO and a model system constructed by reconstituting apomyoglobin with 2-formyl-4-vinyl-deuteroheme IX have also been interpreted to indicate that the MPO heme is a formylporphyrin.[46] Mild denaturation of MPO, which was partially reversible, has been shown to cause reduction in the number of characteristic bands in its resonance Raman spectrum, consistent with an increase in chromophore symmetry.[47] Based primarily upon this observation, it was speculated that the anomalous physical properties of native MPO were caused by the presence of an uncompensated negative charge from the protein in the heme binding pocket.[48] Presumably, protein denaturation relaxed this symmetry-breaking perturbing influence by allowing the anionic group to move away from the heme. The heme in the denatured form of MPO was green, suggesting that it is a formylporphyrin. However, a formyl $C=O$ vibrational mode could not be clearly identified in the resonance Raman spectrum of the denatured MPO. Thus, the structure of the MPO heme remains controversial, and will probably be conclusively identified only when X-ray crystallographic data are refined to atomic resolution.[43]

Kinetic characterization of the catalytic reactions of MPO is complicated by many factors, including the reversibility of H_2O_2 binding to form compound I,[49] its catalase[50] and superoxide dismutase activities[51] and possible reaction sequences involving both O_2^- and H_2O_2,[52,53] and formation of an inhibitory complex with the physiological substrate, chloride ion.[29] A schematic summary of these reactions, taken primarily from the studies by Winterbourn, Wever, and their collaborators, is given in Figure 5; compounds I, II and III have all been shown to form in the presence of its substrates, O_2^- and H_2O_2.[51,52,54] The central reaction of importance, however, is between compound I and Cl$^-$ ion, yielding hypochlorous acid. MPO is unique among the peroxidases in its capacity to catalyze two-electron oxidation of Cl$^-$ to a form that is freely diffusible from the enzyme active site.[55] Since the phagosomal milieu contains MPO, chloride ion (at ~0.15 M), and an H_2O_2-generating system, formation of HOCl within that vacuole during stimulated respiration is plausible.

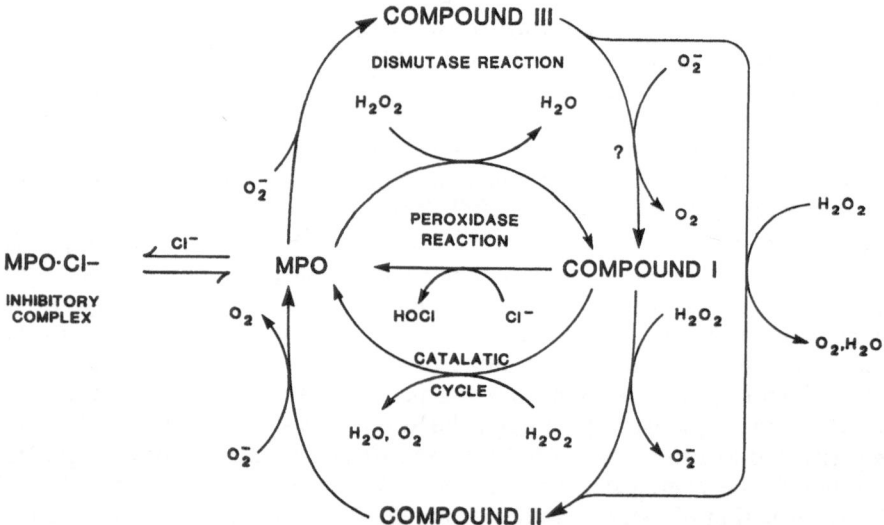

Figure 5. Reactions of neutrophil myeloperoxidase with products of the respiratory burst. The kinetic scheme is a composite of various reaction steps described in references 29 and 49-55.

IS HOCl FORMED IN THE PHAGOSOME?

Early studies using chemical trapping agents for HOCl such as taurine[56] (2-aminoethane sulfonate) and 1,3,5-trimethoxybenzene[57] indicated that stimulated neutrophils produced HOCl, but the topographic location of these reactions in relation to the cell could not be identified. We subsequently used fluorescein covalently attached to yeast cell wall fragments (zymosan) as a phagocytosable probe.[58] Fluorescein is rapidly oxidized by HOCl to mono- and dichloro derivatives that have red-shifted optical absorption and fluorescence bands and reduced fluorescence quantum yields. Changes in the fluorescent properties of the dye upon phagocytosis of these particles by normal neutrophils were consistent with fluorescein ring chlorination, although these reactions were not observed when neutrophils from MPO-deficient or CGD patients were used. Azide ion, an inhibitor of MPO, blocked the reaction in normal neutrophils, and addition of MPO or an H_2O_2-generating enzymatic system to the suspending medium containing MPO-deficient or CGD cells, respectively, elicited spectroscopic changes analogous to those observed for normal neutrophils. Although these changes were consistent with intraphagosomal chlorination, the probe could not be recovered for confirming structural analyses. In part to remedy this deficiency, we have recently constructed artificial particles containing covalent linkages that can be cleaved upon demand, allowing dye recovery. The reaction sequence given below illustrates preparation of one such particle.

Syntheses:

$$\bullet-\overset{\overset{\text{O}}{\|}}{\text{C}}\text{NH}_2 + \text{OH}^- \longrightarrow \bullet-\overset{\overset{\text{O}}{\|}}{\text{C}} - \text{O}^- \quad (I)$$

$$I + \text{NH}_2(\text{CH}_2)_2\text{S-S}(\text{CH}_2)_2\text{NH}_2 \xrightarrow{\text{EDC}} \bullet-\overset{\overset{\text{O}}{\|}}{\text{C}}\text{NH}(\text{CH}_2)_2\text{S-S}(\text{CH}_2)_2\text{NH}_2 \quad (II)$$

$$II + \text{FITC} \longrightarrow \bullet-\overset{\overset{\text{O}}{\|}}{\text{C}}\text{NH}(\text{CH}_2)_2\text{S-S}(\text{CH}_2)_2\text{NH}\overset{\overset{\text{S}}{\|}}{\text{C}}\text{NH-fluorescein} \quad (III)$$

Fluorescein release:

$$III + \text{RSH} \longrightarrow \text{HS}(\text{CH}_2)_2\text{NH}\overset{\overset{\text{S}}{\|}}{\text{C}}\text{NH fluorescein} \quad (IV)$$

Polyacrylamide beads of approximately the same size as bacteria (1-2 μm) were partially hydrolyzed (I), the carboxy groups formed were amidated with cystamine (II) using a carbodiimide coupling reagent (EDC), then fluorescein isothiocyanate (FITC) was attached to the other amino terminus of the linker group through formation of a thiourea bond (III). When serum-opsonized, the particles were avidly phagocytosed by neutrophils, as observed by fluorescence microscopy. Subsequent fluorescence spectral shifts and intensity changes were consistent with intraphagosomal chlorination of the probe. The cells were isolated and lysed, then the fluorescence was released by disulfide exchange with soluble thiol compounds (Figure 6); the dye was recovered in 90-95% yield. Subsequent HPLC analysis revealed peaks that chromatographed with authentic samples of the recovered fragments (IV) that had been directly chlorinated with HOCl. These results leave little doubt that HOCl is indeed generated within the phagosomes of stimulated neutrophils.

MYELOPEROXIDASE-DEPENDENT TOXICITY -- HOW ARE BACTERIA KILLED?

Although HOCl is a strong oxidant, it exhibits remarkable selectivity with respect to its reaction partners.[2] This is probably a consequence of its tendency to act as a two-electron oxidant toward organic compounds. In this case, electron transfer may require nascent bond formation, as in atom transfer reactions. In any event, the

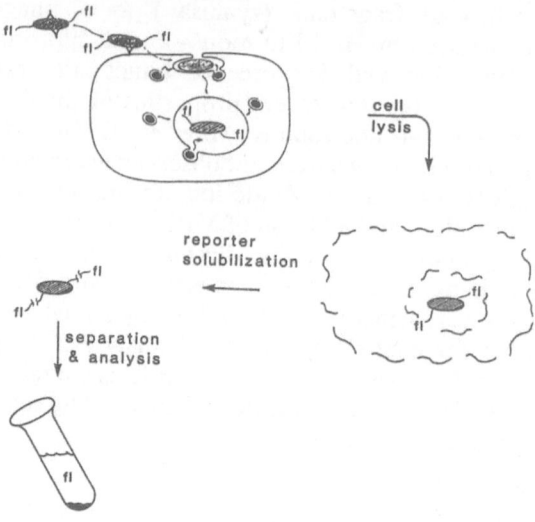

Figure 6. "Smart" probes for timing post-phagocytic events; properties include a reporter group that can be recovered upon demand and the capacity to discriminate between pH changes, chlorination, and other oxidative reactions. The symbol fl refers to the fluorescein moiety.

chlorine atom in HOCl behaves as an electrophile, reacting most rapidly at nucleophilic sites within other molecules. As we have shown, the most reactive biological molecules are those containing nitrogen heterocycles (hemes, nucleotide bases), conjugated polyenes (e.g., carotenes), or other electron-rich π-delocalized centers (e.g., iron-sulfur clusters).[59] Free amines react with HOCl at near-diffusion controlled rates yielding, initially, N-chloramines, but their conjugate acids are unreactive; at physiological pH values, the rates are several orders of magnitude below the diffusion limit.[60] The highly polarizable sulfur atom of cysteinyl groups is also easily oxidized; consequently enzymes and transport proteins containing essential cysteine functional groups are particularly sensitive to oxidative inactivation by HOCl.

These selectivities, which were based upon kinetic studies with simple bio-molecules, have been shown to extend to bacteria themselves. Thomas has shown that oxidation of sulfhydryl substituents and N-chlorination of amines on the bacterial envelope are early events in the progressive oxidation of bacteria by HOCl.[61,62] We have found in representative examples that plasma membrane-localized respiratory components (cytochromes, iron-sulfur containing dehydrogenases) and carotene pigments are also particularly vulnerable to oxidative degradation.[59,63] In fact, the selectivity is sufficiently high that pseudo-titrimetric curves can be constructed that plot loss of susceptible molecules and sites, or loss of biological function, against the amount of added HOCl. These curves are generally sigmoidal in shape, with the initial titrimetric range in which no changes occur in the molecule or function being monitored representing oxidations at other, more reactive sites within the bacterium.

The reactivity patterns that emerge from comparing several sets of these curves can be used to identify the bactericidal mechanisms. Although some species differences exist, the patterns for the various bacteria that we have examined lead to identification of a common cidal mechanism; a typical result is illustrated schematically for *Escherichia coli* in Figure 7. Vulnerable molecules existing within the bacterial cytosol such as sulfhydryl-dependent enzymes and the nucleotide bases, particularly adenine, are not damaged until an amount of HOCl is added that is in large excess over that required to kill the organisms[64,65] (Figure 7, solid line). Similarly, the functional integrity of the bacterial plasma membrane is not lost, as measured by its intrinsic impermeability to small molecules and ions and its ability to maintain an electrochemical gradient (the protonmotive force).[65] Respiratory capabilities are lost after addition of only severalfold HOCl in excess of that required for killing, accompanied approximately by oxidative bleaching of the respiratory cytochromes and destruction of the iron-sulfur redox sites.[63] However, loss of viability coincides with (or in some instances, is preceded by) loss of metabolic energy reserves by massive ATP hydrolysis, inhibition of metabolite transport by all available transport mechanisms, and inhibition of the proton-translocating ATP synthase.[66] Within the framework of the chemiosmotic hypothesis,[67] this means that the cell is functionally dead because its capacity for energy transduction has been destroyed. Specifically, the locus of respiration-driven energy transduction in bacteria is the plasma membrane, which contains the electron transport chain and proteins involved with metabolite and ion transport. Although the plasma membrane is intact and respiratory capabilities are apparently sufficient to maintain a normal protonmotive force at this level of oxidation, the organelle required to couple the electrochemical gradient to intracellular ATP synthesis, the ATP synthase, has been destroyed. Similarly, other metabolite transport systems which couple directly to the electrochemical gradient, i.e., by proton symport, have also been inactivated. Further, loss of ATP inhibits energy-linked metabolite and ion transport, including the phosphoenolpyruvate-dependent glucose phosphotransferase system. Since glucose transport is the first step in substrate-level phosphorylation by the glycolytic pathway, it appears that cells lacking metabolic energy reserves would be unable to generate ATP by this pathway as well. Without a constant supply of ATP, cells could not maintain homeostasis, much less undergo biosynthesis and repair, and would soon expire. These HOCl-inflicted dysfunctions therefore identify a set of reactions that are sufficient to account for the universally bactericidal properties of HOCl. The molecular sites of attack within the transport proteins have not been identified, but are probably cysteine sulfhydryls, since these groups appear to be critical to their proper functioning.[66]

MYELOPEROXIDASE-INDEPENDENT MECHANISMS -- METAL ION-PROMOTED H_2O_2 TOXICITY

General Mechanistic Considerations

Hypochlorous acid is potently toxic to bacteria; typically, less than 10^8 HOCl molecules per bacterium (or 50 μmol HOCl/g) are required to inhibit cellular division of *E. coli* and similar organisms.[68] The demonstration that HOCl forms within the phagosome therefore implies that HOCl is a primary microbicide generated by the respiratory burst. However, as previously discussed, the clinical manifestations of MPO deficiency are very minor,[8,9] suggesting that the phagosome generates other oxidants that are also effective microbicides. The respiratory end products, superoxide ion and hydrogen peroxide, are by themselves not particularly toxic.[2] For example, bacteria are able to survive H_2O_2 quantities as high as 50 mmol H_2O_2/g in metal-free environments (conditions which correspond to dilute cellular suspensions in 0.05-0.1 M H_2O_2!).[69]

When certain redox metal ions and reducing agents are present, H_2O_2 toxicity is markedly potentiated.[69] In model systems, cupric was by far the most effective metal ion, enhancing toxicity over a wide range of medium conditions to levels approaching those measured for HOCl. In contrast, ferric complexes commonly used as catalysts in H_2O_2 oxidation reactions were much less effective, and measurable effects could be demonstrated only under a narrowly defined set of conditions.[70] Mechanisms of toxicity

E. coli cellular reactions with HOCl

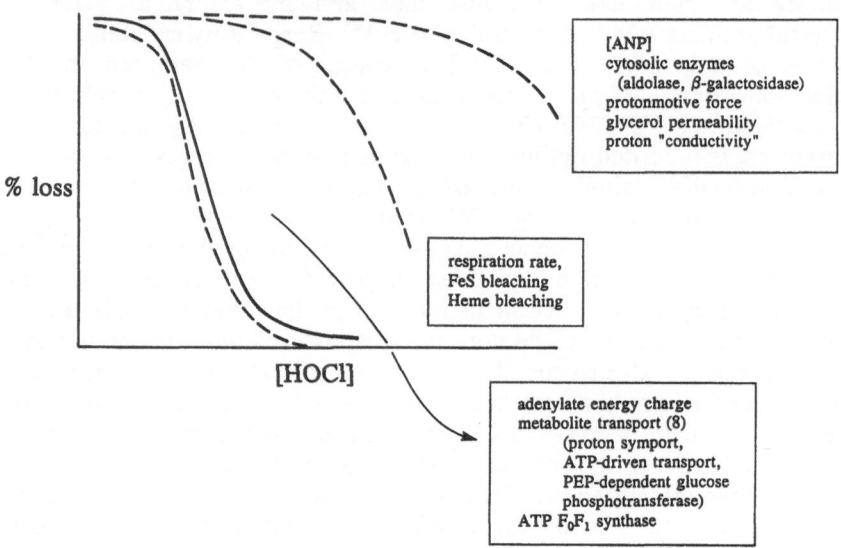

Figure 7. Schematic representation of titrimetric loss of cellular function or biochemical components of *E. coli* with addition of HOCl and/or cell-free MPO-H_2O_2-Cl$^-$ bactericidal systems. Adapted from references 63-66.

are thought to involve cyclic one- or two-electron oxidation of the metal ions in their lower valence states (Cu^I, Fe^{II}) by H_2O_2, followed by reduction to the original oxidation state by the reductant and/or components of the bacterial cells.[71-73] Either the higher-valent ions (Cu^{III}, Fe^{IV}=O) or hydroxyl radical might be the causative agents of cellular damage, although the actual oxidant has not yet been conclusively identified in any physiological system. A conceptual problem arising in considering these reactions is that, given the extremely reactive character of the hydroxyl radical and higher-valent metal ions and the plethora of potential biological reactants, how might selectivity for vulnerable targets within the organism be attained? This problem is particularly acute for phagocytic cells involved with host resistance, where the reagents for H_2O_2 formation, i.e., O_2 and respiratory reductants, are in limited supply. One popular hypothesis is that the metal ions themselves, as catalytic agents for H_2O_2 reduction, confer selectivity on the systems *de facto* by the cellular location of their binding sites.[74] In this site-specific mechanism, the powerful oxidants generated are thought to react with biological components in the immediate vicinity of the metal ion, leading to loss of associated cellular function. This mechanism has received widespread acceptance and is supported indirectly by considerable data from various model studies.[71,72,74]

Radical Reactions in Phagosomal Environments

Hydroxyl radical toxicity has been investigated using radiolysis techniques.[75] In general, the deleterious effects of ionizing radiation appear to originate from reactions occurring within the bacterial cell, as predicted by the site-specific mechanism.[76] This conclusion is based primarily upon observations that adding hydroxyl radical scavengers to the suspending medium failed to protect the bacteria. However, when the medium contained Cl$^-$ [69,77,78] or HCO$_3^-$ [69] ions at concentration levels mimicking the phagosomal environment, radiolytic damage by extracellularly generated OH was also observed.

The effect can be ascribed to formation of secondary radicals that are sufficiently long-lived to selectively oxidize vulnerable cellular targets. A persistent chlorinating agent with the reactivity of HOCl accumulated upon irradiation when Cl⁻ ion was present, and the carbonate radical was identified by electron spin resonance analyses of irradiated bicarbonate buffers containing the spin trap DMPO.[69] The bacteria could be protected by including low concentrations of reducing agents in the reaction medium; for example, 10 μM H_2O_2 was sufficient to completely block enhanced killing of bacteria suspended in bicarbonate-buffered physiological saline solutions. Since OH radical oxidation of Cl⁻ ion is relatively slow in neutral solutions,[79] the reductants undoubtedly prevent its formation by directly competing for the radiolytically generated OH radical. In contrast, reaction with HCO_3^- ion is relatively rapid,[80] so that at the concentration levels projected for the respiring neutrophil its reaction with OH radical will predominate regardless of other reactants in the medium. Protection by reducing agents in this case must arise primarily by their reaction with the CO_3^- radical that is formed. This reactivity pattern is summarized in Figure 8. Thus, although the relatively long-lived CO_3^- radical is preferentially formed from OH radical in these media and is toxic to bacteria, its lethal reactions are not expressed in dilute suspensions of the cells because it is effectively scavenged in competing reactions with the chemical reductants.

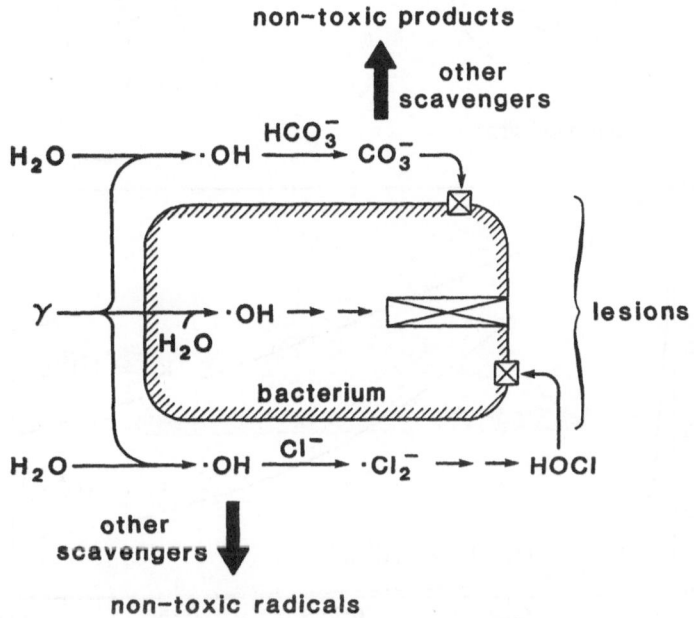

Figure 8. Hypothetical mechanisms for Cl⁻ and HCO_3^--enhanced radiolytic killing of a bacterium by extracellular pathways. Cl_2^- radical formation is suppressed by OH-reactive compounds in solution, but CO_3^- radical formation is not. In this case, protection by free radical scavengers must arise by direct competition for the CO_3^- radical. The reaction scheme also includes an intracellular pathway, although target sites are not specified.

If one considers the spatial dimensions of the phagosome, however, an entirely different reactivity pattern emerges. Specifically, the average diffusion distance before the CO_3^- radical encounters the bacterial envelope is markedly reduced in the confining space of the phagosome, so that competing reactions in the extracellular solution become less significant. A reaction model and typical results are given in Figure 9;

assuming reasonable kinetic parameters, the calculations show that reactions at the bacterial and phagosomal membrane surfaces will predominate, whereas radical quenching by solution components will be minor.[81] Thus, if conditions for CO_3^- radical generation exist within the phagosome, either by metal-catalyzed OH formation or direct reaction of the higher-valent metal ions with HCO_3^-, the toxicity of the CO_3^- radical should be expressed. This set of reactions constitutes a bactericidal mechanism that is fundamentally different from the site-specific mechanism; its expression is a unique consequence of the spatial constraints imposed by compartmentalization within the neutrophil, and is not generally applicable to other environments.

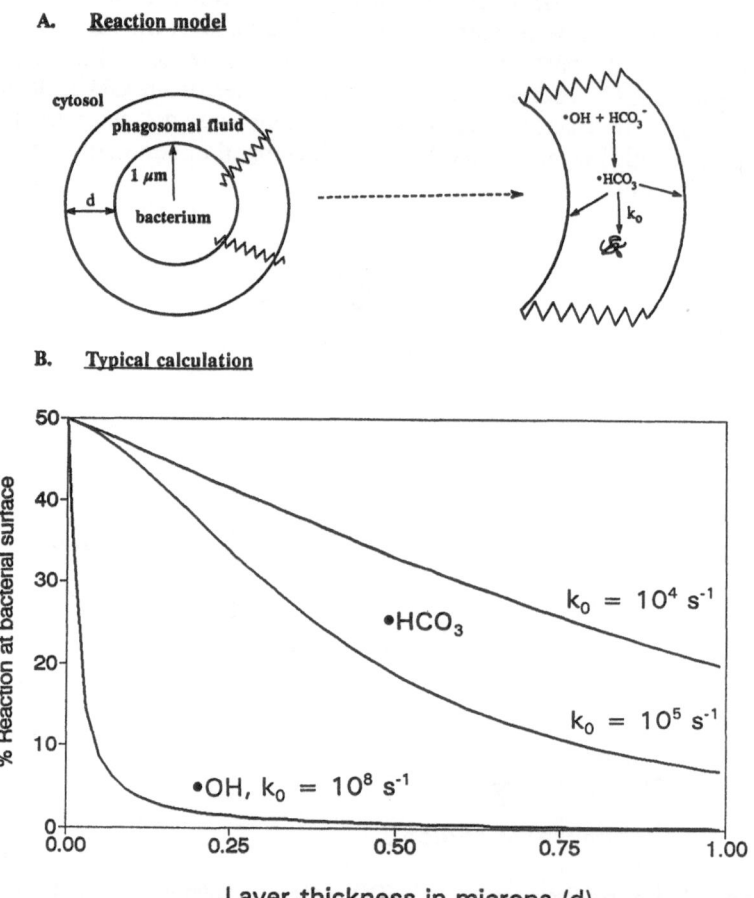

Figure 9. Dependence of reactions of postulated phagosome-generated free radicals with bacteria upon phagosomal dimensions. A. Kinetic model for CO_3^- reactions assuming spherical geometry. The radical is generated throughout the volume identified as "phagosomal fluid," and either reacts with solution components (k_0) or diffuses to the bacterial envelope or phagosomal membrane, where it reacts. The first-order rate constants are estimates based upon measured rate constants for typical biological compounds at 1 mM concentration levels. B. Typical results indicating the fraction of reaction occurring at the bacterial surface for varying thicknesses of the phagosomal fluid layer. In this calculation it was assumed that the bacterial and phagosomal surfaces were equally reactive. Even at fluid layer thickness approaching bacterial dimensions, a significant fraction of the CO_3^- radical reacts with the bacterium. However, the reactions of the more reactive OH radical are quenched at all distances beyond the immediate vicinity of the bacterial surface.

Bactericidal Mechanisms

Hydrogen peroxide-inflicted damage to eukaryotic cells appears to involve several intracellular loci, including cytosolic enzymes and nuclear DNA. These cellular reactions have been studied most extensively in macrophage-like tumor cells by Cochrane and associates.[82,83] They found that exposure to H_2O_2 caused inhibition of ATP synthesis by both glycolytic and respiratory pathways to levels which appeared to be sufficient to cause loss of homeostasis and subsequent cellular death. Inhibition of oxidative phosphorylation involved direct oxidative inactivation of the mitochondrial ATP synthase; inhibition of substrate-level phosphorylation involved both selective inactivation of a key glycolytic enzyme, glyceraldehyde-3-phosphate dehydrogenase, and depletion of its cofactor, NAD, through an indirect mechanism initiated by single-strand cleavage of nuclear DNA following oxidation of its ribose units. This nucleotide damage initiated activation of the repair enzyme, poly-ADP-ribose polymerase, whose enzymatic reactions involve NAD catabolism.

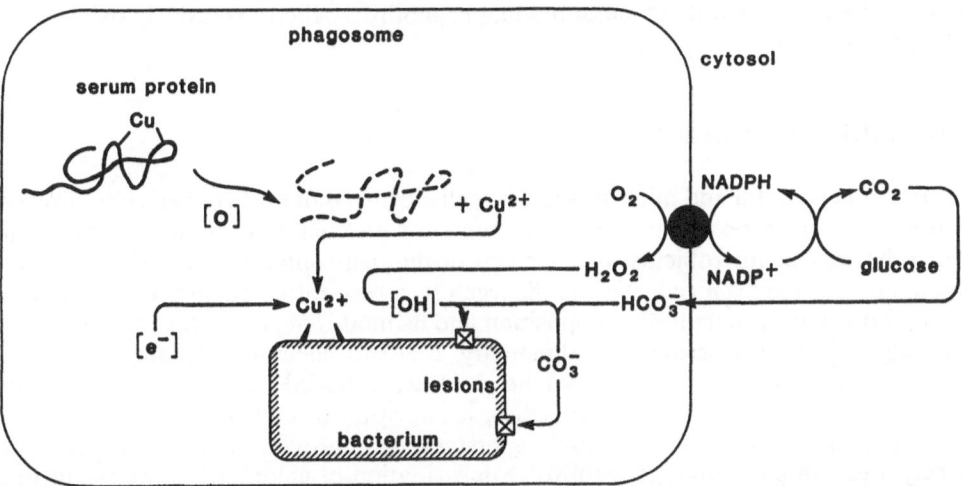

Figure 10. Speculative model for enhanced intraphagosomal toxicity. Serum Cu-binding protein undergoes oxidative [O] damage, releasing its Cu ion, which relocates on the bacterial envelope, where it either catalyzes formation of the lethal oxidant [Cu(III), OH] from H_2O_2 and endogenous reductants [e⁻], causing damage at that site, or catalyzes formation of CO_3^- radical, which selectively oxidizes vulnerable sites on the envelope.

Both single-strand and double-strand breaks were observed in the chromosomal DNA of the bacterium, *E. coli*, when exposed to aerobic media containing cupric ion and the reductant, ascorbic acid.[84] As recognized by the researchers,[85] the quantities of reagents used in this study were far in excess of the amounts usually required for killing. Subsequent research on paraquat (N,N'-dimethyl-4,4'-bipyridinium)-mediated toxicity established that the cellular content of ATP and potassium ions diminished roughly in proportion to cellular death and that active transport of leucine was inhibited in exposed cells, suggesting that the bacterial plasma membrane was the site of the lethal oxidative attack.[85] Localized damage to the bacterial envelope was also evident from electron micrographs of the bacteria. (Paraquat is thought to catalyze one-electron oxidation of O_2 by endogenous donors, forming O_2^- and, by subsequent disproportionation, H_2O_2, which then reacts with bacteria by the usual metal ion-mediated pathways.) Using a Cu(II)-ascorbate-H_2O_2 bactericidal assay system, we have

281

found that DNA from a plasmid-carrying *E. coli* underwent negligible chain cleavage at oxidant levels sufficient to kill the microbe.[69] Thus, unlike eukaryotic cells, the DNA of bacterial cells appears to be considerably less vulnerable to H_2O_2-induced oxidative degradation.

Although the results are preliminary, the cumulative evidence suggests that, as with HOCl, the bacterial toxicity of H_2O_2 might be attributed to reactions at the cellular envelope. If so, the activating metal ions (most likely copper) are almost certainly host-derived. The blood stream (and presumably interstitial fluids) contains several copper-binding proteins, which could be incorporated along with other extracellular components into the phagosome during its formation.[86] Oxidative degradation of these proteins might release the copper which, in the site-directed model would relocate to the bacterial envelope, rendering it vulnerable to oxidative damage or, otherwise, would catalyze formation of long-lived reactive species. Figure 10 presents a speculative model incorporating these ideas. A mechanism for "triggering" H_2O_2 toxicity by metal ion release has not yet been identified, however. Both the copper-binding serum protein, ceruloplasmin,[87] and the iron-binding lactoperoxidase[88] retain their metal-binding capabilities when extensively oxidized by HOCl, for example.

CONCLUDING REMARKS

In the neutrophil and other phagocytic cells involved in host resistance to disease, Nature has evolved remarkable cellular agents for microbial destruction. Prominent among the various microbicidal mechanisms of the neutrophil are oxidative reactions initiated by one-electron reduction of dioxygen by a respiratory chain that is unique in biology, both with respect to its composition and its modes of activation. Although our knowledge of these processes is rudimentary, it seems clear that dynamic control is maintained in a manner that cannot be duplicated outside the cell by the subtle interplay of inherent chemical selectivities, localization of endogenous catalysts, and compartmentation. As our conceptual understanding improves, we should be prepared for surprises. In particular, as with MPO peroxidation of chloride ion, we might find that other cellular components that are generally regarded as benign are, in fact, critical elements of the oxidative reaction sequences. For example, O_2^- could play an essential role in preventing MPO accumulation in its unreactive compound II oxidation state[53] (Figure 5) and in generation of the bactericidal agent in MPO-independent oxidative reactions (Figure 10), and the respiratory end-product and physiological buffer, bicarbonate ion, might potentiate radical-inflicted damage by forming carbonate radical, which is a relatively long-lived and selective oxidant (Figure 10). We have much to learn about these fascinating systems, but the recent pace of progress is encouraging.

ACKNOWLEDGEMENT

In his research, the author has benefited from consultation and collaboration with numerous individuals including, most prominently, Terrence Green, Seymour J. Klebanoff, and Henry Rosen, and in his own laboratory, J. Michael Albrich, William C. Barrette, Jr., and Diane Hannum. His work is supported by Public Health Service Grant AI-15834 from the National Institute of Allergy and Infectious Diseases.

REFERENCES

1. A.J. Jesaitis and E.A. Dratz. "The Molecular Basis of Oxidative Damage by Leukocytes," CRC Press, Boca Raton (1992).
2. J.K. Hurst and W.C. Barrette, Jr., Leukocytic oxygen activation and microbicidal oxidative toxins, *CRC Crit. Rev. Biochem. Mol. Biol.* 24:271 (1989).
3. S.J. Weiss, Tissue destruction by neutrophils, *New Engl. J. Med.* 320:365 (1989).
4. S.J. Klebanoff, Phagocytic cells: products of oxygen metabolism, *in:* "Inflammation: Basic Principles and Chemical Correlates," J.I. Gallin, I.M. Goldstein, and R. Snyderman, eds., Raven Press, New York (1988).
5. A.I. Tauber and B.M. Babior, Neutrophil oxygen reduction: the enzymes and the products, *Adv. Free-Radical Biol. Med.* 1:265 (1985).
6. S.J. Klebanoff and R.J. Clark. "The Neutrophil Function and Clinical Disorders," North-Holland, Amsterdam (1978).
7. See, e.g., S.J. Klebanoff, W.R. Henderson, Jr., E.C. Jong, A. Jörg, and R.M. Locksley, Role of peroxidase in eosinophil function, *in:* "Immunobiology of the Eosinophil," T. Yoshida and M. Torisu, eds., Elsevier, Amsterdam (1983), for leading references.
8. S.J. Klebanoff, Myeloperoxidase: occurrence and biological function, *in:* "Peroxidases in Chemistry and Biology," Vol. 1, J. Everse, K.E. Everse, and M.B. Grisham, eds., CRC Press, Boca Raton (1991).
9. K.R. Johnson and W.M. Nauseef, Molecular biology of MPO, *in:* "Peroxidases in Chemistry and Biology," Vol. 1, J. Everse, K.E. Everse, and M.B. Grisham, eds., CRC Press, Boca Raton (1991).
10. G.L. Mandell, Bactericidal activity of aerobic and anaerobic polymorphonuclear leukocytes, *Infect. Immun.* 9:337 (1974).
11. W.A.C. Vel, F. Namavar, M.J. Verweig, A.N.B. Pubben, and D.M. McLaren, Killing capacity of human polymorphonuclear leukocytes in aerobic and anaerobic conditions, *J. Med. Microbiol.* 18:173 (1984).
12. M.E. Selsted, Y.-Q. Tang, W.L. Morris, P.A. McGuire, M.J. Novotny, W. Smith, A.H. Henschen, and J.S. Cullor, Purification, primary structures, and antibacterial activities of β-defensins, a new family of antimicrobial peptides from bovine neutrophils, *J. Biol. Chem.* 268:6641 (1993).
13. T. Ganz, M.E. Selsted, D. Szklarek, S.S. Harwig, L. Daher, and R.I. Lehrer, Defensins: natural peptide antibiotics of human neutrophils, *J. Clin. Invest.* 76:1427 (1985).
14. P. Elsbach and J. Weiss, A reevaluation of the roles of O_2-dependent and O_2-independent microbicidal systems of phagocytes, *Rev. Infect. Dis.* 5:843 (1983).
15. F. Morel, J. Doussiere, and P.V. Vignais, The superoxide-generating oxidase of phagocytic cells: physiological, molecular and pathological aspects, *Eur. J. Biochem.* 201:523 (1991).
16. D. Rotrosen, C.L. Yeung, T.L. Leto, H.L. Malech, and C.H. Kwong, Cytochrome b_{558}: the flavin-binding component of the phagocyte NADPH oxidase, *Science* 256:1459 (1992).
17. A. Abo, A. Boyhan, I. West, A.J. Thrasher, and A.W. Segal, Reconstitution of neutrophil NADPH oxidase activity in the cell-free system by four components: p67-*phox*, p47-phox, p21*rac*1 and cytochrome b_{245}, *J. Biol. Chem.* 267:16767 (1992).
18. A.W. Segal, I. West, F. Wientjes, J.H.A. Nugent, A.J. Chavan, B. Haley, R.C. Garcia, H. Rosen, and G. Scrace, Cytochrome b_{245} is a flavocytochrome containing FAD and the NADPH-binding site of the microbicidal oxidase of phagocytes, *Biochem. J.* 284:781 (1992).
19. T. Miki, L.S. Yoshida, and K. Kakinuma, Reconstitution of superoxide-forming NADPH oxidase activity with cytochrome b_{558} purified from porcine neutrophils. Requirement of a membrane-bound flavin enzyme for reconstitution of activity, *J. Biol. Chem.* 267:18695 (1992).
20. K. Kakinuma, Y. Fukuhara, and M. Kaneda, The respiratory burst oxidase of neutrophils. Separation of an FAD enzyme and its characterization, *J. Biol. Chem.* 262:12316 (1987).
21. C.A. Parkos, R.A. Allen, C.G. Cochrane, and A.J. Jesaitis, Purified cytochrome b from human granulocyte plasma membrane is comprised of two polypeptides with relative molecular weights of 91,000 and 22,000, *J. Clin. Invest.* 80:732 (1987).
22. A.W. Segal, Absence of both cytochrome b_{245} subunits from neutrophils in X-linked chronic granulomatous disease, *Nature* 326:88 (1987).
23. M.T. Quinn, M.L. Mullen, and A.J. Jesaitis, Human neutrophil cytochrome b contains multiple hemes. Evidence for heme association with both subunits, *J. Biol. Chem.* 267:7303 (1992).
24. J.K. Hurst, T.M. Loehr, J.T. Curnutte, and H. Rosen, Resonance Raman and electron paramagnetic resonance structural investigations of neutrophil cytochrome b_{558}, *J. Biol. Chem.* 266:1627 (1991).
25. J.H.A. Nugent, W. Gratzer, and A.W. Segal, Identification of the haem-binding subunit of cytochrome b_{245}, *Biochem. J.* 264:921 (1989).

26. T. Yamaguchi, T. Hayakawa, M. Kaneda, K. Kakinuma, and A. Yoshikawa, Purification and some properties of the small submit of cytochrome b_{558} from human neutrophils, *J. Biol. Chem.* 264:112 (1989).

27. M.C. Dinauer, E.A. Pierce, G.A.P. Bruns, J.T. Curnutte, and S.H. Orkin, Human neutrophil cytochrome b light chain (p22-phox). Gene structure, chromosomal location, and mutations in cytochrome-negative autosomal recessive chronic granulomatous disease, *J. Clin. Invest.* 86:1729 (1990).

28. J. Butler, W.H. Koppenol, and E. Margoliash, Kinetics and mechanism of the reduction of ferricytochrome c by the superoxide ion, *J. Biol. Chem.* 257:10747 (1982).

29. J.K. Hurst, Myeloperoxidase: active site structure and catalytic mechanisms, *in:* "Peroxidases in Chemistry and Biology," Vol. 1, J. Everse, K.E. Everse, and M.B. Grisham, eds., CRC Press, Boca Raton (1991).

30. A. Hata-Tanaka, T. Chiba, and K. Kakinuma, ESR signals from stimulated and resting porcine blood and neutrophils, *FEBS Lett.* 214:279 (1987).

31. I. Ueno, S. Fujii, H. Ohya-Nishiguchi, T. Iizuki, and S. Kanegasaki, Characterization of neutrophil b-type cytochrome in situ by electron paramagnetic resonance spectroscopy, *FEBS Lett.* 281:130 (1991).

32. T. Miki, H. Fujii, and K. Kakinuma, EPR signals from cytochrome b_{558} purified from porcine neutrophils, *J. Biol. Chem.* 267:19673 (1992).

33. See, e.g., J.C. Salerno and J.S. Leigh, Crystal field of atypical low-spin ferriheme complexes, *J. Am. Chem. Soc.* 106:2156 (1984).

34. S.S. Sibbett and J.K. Hurst, Structural analysis of myeloperoxidase by resonance Raman spectroscopy, *Biochemistry* 23:3007 (1984).

35. G.T. Babcock, R.T. Ingle, W.A. Oertling, J.C. Davis, B.A. Averill, C.L. Hulse, D.J. Stufkens, B.G.M. Bolscher, and R. Wever, Raman characterization of human leukocyte myeloperoxidase and bovine spleen green haemoprotein. Insight into chromophore structure and evidence that the chromophores of myeloperoxidase are equivalent, *Biochim. Biophys. Acta* 828:58 (1985).

36. T. Kitagawa, S. Hashimoto, J. Teraoka, S. Nakamura, H. Yajima, and T. Hosoya, Distinct heme-substrate interactions of lactoperoxidase probed by resonance Raman spectroscopy: difference between animal and plant peroxidases, *Biochemistry* 22:2788 (1983).

37. S.S. Sibbett, S.J. Klebanoff, and J. K. Hurst, Resonance Raman characterization of the heme prosthetic group in eosinophil peroxidase, *FEBS Lett.* 189:271 (1985).

38. L.A. Andersson, T.M. Loehr, C.K. Chang, and A.G. Mauk, Resonance Raman spectroscopy of metallochlorins: models for green heme protein prosthetic groups, *J. Am. Chem. Soc.* 107:182 (1985).

39. L.A. Andersson, T.M. Loehr, C. Sotiriou, W. Wu, and C.K. Chang, Resonance Raman spectroscopy of metallochlorins. II. Properties of meso-substituted systems, *J. Am. Chem. Soc.* 108:2908 (1986).

40. A.W. Nichol, L.A. Angel, T. Moon, and P.S. Clezy, Lactoperoxidase haem, an iron-porphyrin thiol, *Biochem. J.* 247:147 (1987).

41. V. Thanabal and G.N. La Mar, A nuclear Overhauser effect investigation of the molecular and electronic structure of the heme crevice in lactoperoxidase, *Biochemistry* 28:7038 (1989).

42. M.J. Chatfield, G.N. La Mar, and R.J. Kauten, Proton NMR characterization of isomeric sulfmyoglobins: preparation, interconversion, reactivity patterns, and structural features, *Biochemistry* 26:6939 (1987), and references cited therein.

43. L.B. Dugad, G.N. LaMar, H.C. Lee, M. Ikeda-Saito, K.S. Booth, and W.S. Caughey, A nuclear Overhauser effect study of the active site of myeloperoxidase. Structural similarity of the prosthetic group to that on lactoperoxidase, *J. Biol. Chem.* 265:71173 (1990).

44. J. Zeng and R.E. Fenna, X-ray crystal structure of canine myeloperoxidase at 3 Å resolution, *J. Mol. Biol.* 226:185 (1992).

45. C. Sotiriou and C.K. Chang, Synthesis of the heme d prosthetic group of bacterial terminal oxidase, *J. Am. Chem. Soc.* 110:2264 (1988).

46. M. Sono, A.M. Bracete, A.M. Huff, M. Ikeda-Saito, and J.H. Dawson, Evidence that a formyl-substituted iron porphyrin is the prosthetic group of myeloperoxidase: magnetic circular dichroism similarity of the peroxidase to *Spirographis* heme-reconstituted myoglobin, *Proc. Natl. Acad. Sci. USA* 88:11148 (1991).

47. R. Wever, W.A. Oertling, H. Hoogland, B.G.J.M. Bolscher, Y. Kim, and G.T. Babcock, Denaturation and renaturation of myeloperoxidase. Consequences for the nature of the prosthetic group, *J. Biol. Chem.* 266:24308 (1991).

48. J. Eccles and B. Honig, Charged amino acids as spectroscopic determinants for chlorophyll *in vivo*, *Proc. Natl. Acad. Sci. USA* 80:4959 (1983).

49. B.G.J.M. Bolscher and R. Wever, A kinetic study of the reaction between human myeloperoxidase, hydroperoxides and cyanide. Inhibition by chloride and thiocyanate, *Biochim. Biophys. Acta* 788:1 (1984).

50. H. Iwamoto, T. Kobayashi, E. Hasegawa, and Y. Morita, Reaction of human myeloperoxidase with hydrogen peroxide and its true catalase activity, *J. Biochem.* 101:1407 (1987).

51. R.A. Cuperus, A.O. Muijsers, and R. Wever, The superoxide dismutase activity of myeloperoxidase: formation of compound III, *Biochim. Biophys. Acta* 871:78 (1986).

52. A.J. Kettle and C.C. Winterbourn, Superoxide modulates the activity of myeloperoxidase and optimizes the production of hypochlorous acid, *Biochem. J.* 252:529 (1988).

53. A.J. Kettle and C.C. Winterbourn, Influence of superoxide on myeloperoxidase kinetics measured with a hydrogen peroxide electrode, *Biochem. J.* 263:823 (1989).

54. J.E. Harrison, T. Araiso, M.M. Palic, and H.B. Dunford, Compound I of myeloperoxidase, *Biochem. Biophys. Res. Commun.* 94:34 (1980).

55. J.E. Harrison and J. Schultz, Studies on the chlorinating activity of myeloperoxidase, *J. Biol. Chem.* 251:1371 (1976).

56. S.J. Weiss, R. Klein, R. Slivka, and M. Wei, Chlorination of taurine by human neutrophils, *J. Clin. Invest.* 70:598 (1982).

57. C.S. Foote, T.E. Goyne, and R.I. Lehrer, Assessment of chlorination by human neutrophils, *Nature* 301:715 (1983).

58. J.K. Hurst, J.M. Albrich, T.R. Green, H. Rosen, and S.J. Klebanoff, Myeloperoxidase-dependent fluorescein chlorination by stimulated neutrophils, *J. Biol. Chem.* 259:4812 (1984).

59. J.M. Albrich, C.A. McCarthy, and J.K. Hurst, Biological reactivity of hypochlorous acid: implications for microbicidal mechanisms of leukocyte myeloperoxidase, *Proc. Natl. Acad. Sci. USA* 78:210 (1981).

60. J.C. Morris, Kinetics of reactions between aqueous chlorine and nitrogen compounds, *in:* "Principles and Applications of Water Chemistry," S.D. Faust and J.V. Hunter, eds., Wiley, New York (1967).

61. E.L. Thomas, Myeloperoxidase-hydrogen peroxide-chloride antimicrobial system: effect of exogenous amines on antibacterial action against *Escherichia coli*, *Infect. Immun.* 25:110 (1979).

62. E.L. Thomas, Myeloperoxidase, hydrogen peroxide, chloride antimicrobial system: nitrogen-chlorine derivatives of bacterial components in bactericidal action again *Escherichia coli*, *Infect. Immun.* 23:522 (1979).

63. J.K. Hurst, W.C. Barrette, Jr., B.R. Michel, and H. Rosen, Hypochlorous acid and myeloperoxidase-catalyzed oxidation of iron-sulfur clusters in bacterial respiratory dehydrogenases, *Eur. J. Biochem.* 202:1275 (1991).

64. W.C. Barrette, Jr., J.M. Albrich, and J.K. Hurst, Hypochlorous acid-promoted loss of metabolic energy in *Escherichia coli*, *Infect. Immun.* 55:2518 (1987).

65. J.M. Albrich, J.H. Gilbaugh III, K.B. Callahan, and J.K. Hurst, Effects of the putative neutrophil-generated toxin, hypochlorous acid, on membrane permeability and transport systems of *Escherichia coli*, *J. Clin. Invest.* 78:177 (1986).

66. W.C. Barrette, Jr., D.M. Hannum, W.D. Wheeler, and J.K. Hurst, General mechanism for the bacterial toxicity of hypochlorous acid: abolition of ATP production, *Biochemistry* 28:9172 (1989).

67. See, e.g., F.M. Harold. "The Vital Force. A Study of Bioenergetics," Freeman, New York (1986).

68. J.M. Albrich and J.K. Hurst, Oxidative inactivation of *Escherichia coli* by hypochlorous acid. Rates and differentiation of respiratory from other reaction sites, *FEBS Lett.* 144:157 (1982).

69. D.M. Hannum, H. Elzanowska, and R.C. Wolcott, unpublished observations.

70. H. Rosen and S.J. Klebanoff, Role of iron and ethylenediamine-tetraacetic acid in the bactericidal activity of a superoxide anion-generating system, *Arch. Biochem. Biophys.* 208:512 (1981).

71. M.G. Simic, K.A. Taylor, J. Ward, and C. von Sonntag. "Oxygen Radicals in Biology and Medicine," Plenum Press, New York (1988).

72. B. Halliwell and J.M. Gutteridge, Role of free radicals and catalytic metal ions in human disease: an overview, *Methods Enzymol.* 186:1 (1990).

73. H.C. Sutton and C.C. Winterbourn, On the participation of higher oxidation states of iron and copper in Fenton reactions, *Free Radical Biol. Med.* 6:53 (1989).

74. M. Chevion, A site-specific mechanism for free radical induced biological damage: the essential role of redox-active transition metals, *Free Radical Biol. Med.* 5:27 (1988).

75. C. von Sonntag. "Chemical Basis of Radiation Biology," Taylor and Francis, New York (1987).

76. A. Samuni and G. Czapski, Radiation-induced damage in *Escherichia coli B*: the effect of superoxide radicals and molecular oxygen, *Radiat. Res.* 76:624 (1978).

77. T. Brustad and E. Wold, Long-lived species in irradiated N_2O-flushed saline phosphate buffer, with toxic effect upon *E. coli* K-12, *Radiat. Res.* 66:215 (1976).

78. A. Matsuyama, M. Namiki, and Y. Okazawa, Alkali halides as agents enhancing the lethal effects of ionizing radiation on microorganisms, *Radiat. Res.* 30:687 (1967).

79. G.G. Jayson, B.J. Parsons, and A.J. Swallow, Some simple, highly reactive, inorganic chlorine derivatives in aqueous solution. Their formation using pulses of radiation and their role in the mechanism of the Fricke dosimeter, *J. Chem. Soc. Faraday Trans. I* 69:1597 (1973).

80. G.V. Buxton and A.J. Elliot, Rate constant for reaction of hydroxyl radicals with bicarbonate ions, *Radiat. Phys. Chem.* 27:241 (1986).

81. S.V. Lymar, unpublished results.

82. P.A. Hyslop, D.B. Hinshaw, W.A. Halsey, Jr., I.U. Schraufstätter, R.D. Sauerheber, R.G. Spragg, J.H. Jackson, and C.G. Cochrane, Mechanisms of oxidant-mediated cell injury. The glycolytic and mitochondrial pathways of ADP phosphorylation are major intracellular targets inactivated by hydrogen peroxide, *J. Biol. Chem.* 263:1665 (1988).

83. I.U. Schraufstätter, D.B. Hinshaw, P.A. Hyslop, R.G. Spragg, and C.G. Cochrane, Oxidant injury of cells. DNA strand-breaks activate polyadenosine diphosphate-ribose polymerase and lead to depletion of nicotinamide adenine dinucleotide, *J. Clin. Invest.* 77:1312 (1986).

84. R. Kohen, M. Szyf, and M. Chevion, Quantitation of single- and double-strand DNA breaks *in vitro* and *in vivo, Anal. Biochem.* 154:485 (1986).

85. R. Kohen and M. Chevion, Cytoplasmic membrane is the target organelle for transition metal mediated damage induced by paraquat in *Escherichia coli, Biochemistry* 27:2597 (1988).

86. M.C. Linder. "Biochemistry of Copper," Plenum Press, New York (1992).

87. B. Halliwell, O.I. Aruoma, M. Wasil, and J.M.C. Gutteridge, The resistance of transferrin, lactoferrin and caeruloplasmin to oxidative damage, *Biochem. J.* 256:311 (1988).

88. C.C. Winterbourn and A.L. Molloy, Susceptibilities of lactoferrin and transferrin to myeloperoxidase-dependent loss of iron-binding capacity, *Biochem. J.* 250:613 (1988).

METALLOPORPHYRINS AS BIOMIMETIC OXIDANTS AND SYNTHETIC LIGNINASES

David Dolphin

Department of Chemistry
University of British Columbia
2036 Main Mall, Vancouver, B.C.
Canada V6T 1Z1

Heme proteins serve many varied rules biologically even though all contain the same prosthetic group (heme, iron protoporphyrin IX, Figure 1). Those hemeproteins which are activated by hydrogen peroxide (catalase, peroxidase, ligninase) and those which activate oxygen, via a "peroxide" bound intermediate (cytochrome P-450), all function through the same, or similar intermediates.[1] Indeed when most iron porphyrins either protein bound or protein free react with hydrogen peroxide their reaction may be summarized as follows:

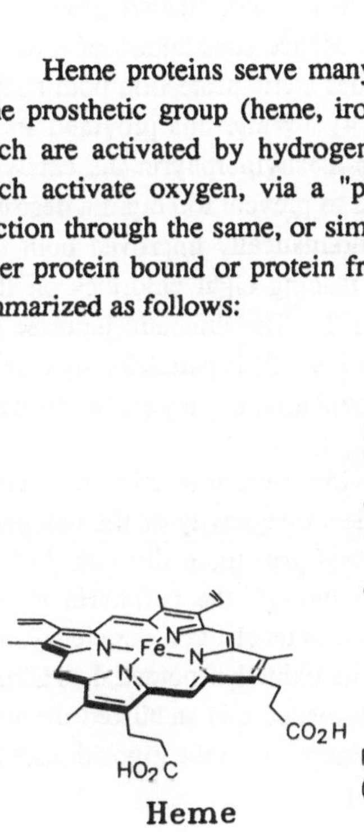

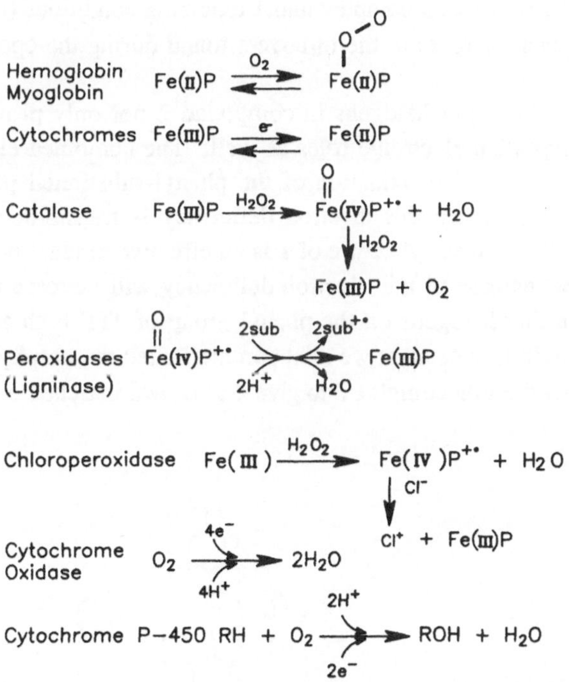

Figure 1: Major classes of hemeproteins using heme (iron protoporphyrin IX) as prosthetic group

The Activation of Dioxygen and Homogeneous Catalytic Oxidation,
Edited by D.H.R. Barton *et al.*, Plenum Press, New York, 1993

$$\text{Fe(III)P} \xrightarrow[\quad]{\quad H_2O_2 \quad H_2O \quad} \overset{\overset{\displaystyle O}{\|}}{\text{Fe(IV)P}^{+\bullet}} \qquad \textbf{1}$$

Complex **1**, an oxo-ferryl(FeIV) porphyrin π-cation radical has been well characterized for catalase, horseradish peroxidase, and ligninase,[2] and all of the chemistry of cytochrome P-450 points towards a similar high oxidation state intermediate.[3] In the enzymatic systems this intermediate is known as compound I.

In addition to the activation of cytochrome P-450 by dioxygen and two electrons the enzyme may be shunted by peroxides or other oxygen atom donors such as iodosyl benzene. Many simple ferric porphyrins behave in the same way when treated with peroxides, iodosyl benzenes, peracids, hypochlorite and even ozone. However, without the stabilization provided by the protein, simple iron porphyrins are rapidly destroyed when presented with these oxidizing agents due to their oxidative degradation. Groves[4] first showed that steric protection of the porphyrin periphery could stabilize these high oxidation states and thereby enhance the catalytic turnover. *meso*-Tetraporphyrins usually contain a flat aromatic porphyrin macrocycle where the four phenyl groups are constrained to be "perpendicular" to the porphyrin ring. Hence substitution of a bulky substituent on each of the eight o-phenyl positions provides steric protection both above and below the plane of the porphyrin ring (see Figure 4). While this provided some modest increase in turnover number in the case of tetra-mesitylporphyrin the catalytic epoxidation of norbornene still had to be carried at -80°C to prevent too rapid a destruction of the catalyst.[4] We and the Professors Traylor[5] dramatically improved both the stability and efficacy of iron tetraphenylporphyrins by placing eight chlorines on the phenyl rings, one at each of the ortho-positions to give **2**. The dramatic increase in stability of the iron complex under oxidizing conditions (Figure 2) is paralleled by a very significant increase in the turnovers found during the epoxidation of norbornene (Figure 3).

The eight halogens in compound **2** not only provide steric protection but serve two important electronic roles as well. The combined electronegativity of the halogens makes destructive oxidation of the phenyl-substituted porphyrin more difficult, but of equal importance this electron deficiency is transmitted through the porphyrin to the iron. The oxo-ferryl centre of **1** is an effective oxidant due to its electron deficiency and further increase in this electron deficiency will increase its oxidation potential. Having shown that halogens on the phenyl group of TPP both activated and stabilized the iron porphyrin as a catalyst we next perchlorinated the porphyrin rings at the β-positions,[6] by treating the iron complex **3** to give **4** as shown in Scheme 1.

    2

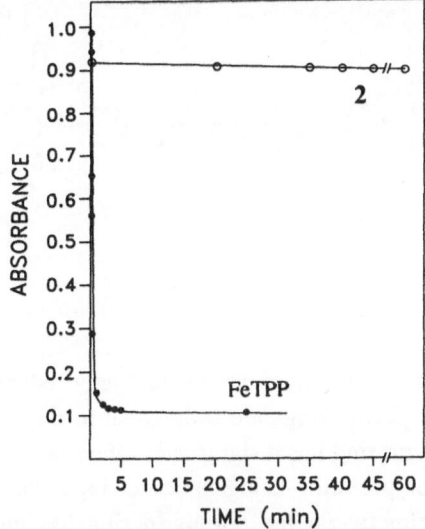

Figure 2: Bleaching (degradation) of iron *meso*-tetraphenylporphyrin and compound **2** by *m*-chloroperbenzoic acid at 1 mM. Porphyrin concentration $\simeq 10\mu M$. Absorbance measured at the Soret band

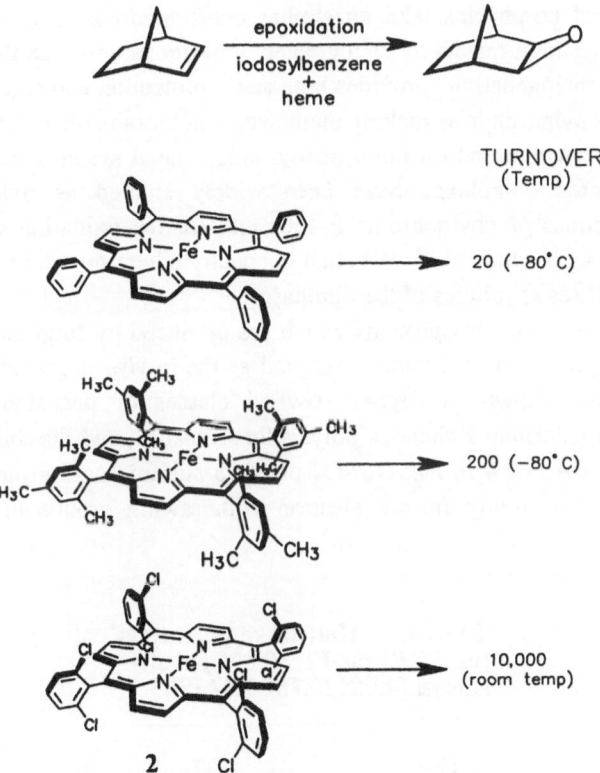

Figure 3: Turnover numbers of three iron *meso*-tetraphenylporphyrins catalyzing the epoxidation of norbornene

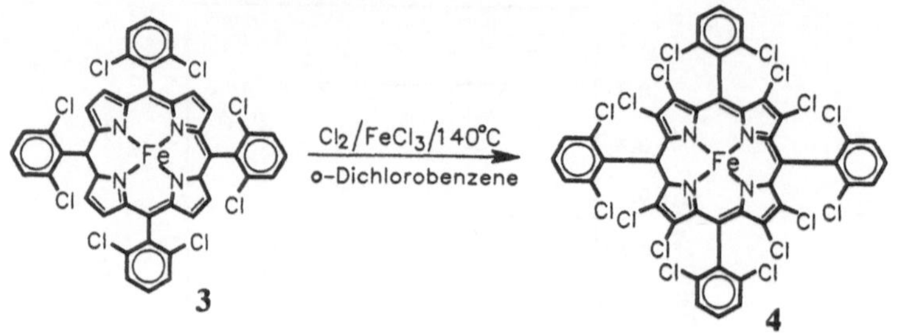

Scheme 1

While halogenation as the phenyl groups can only reduce electron density at the iron inductively, since the phenyl rings are orthogonal to the porphyrin ring, halogenation directly on the porphyrin ring has a significant effect on the electron density at iron as shown in Table 1. Thus per-ortho-halogenation changes the Fe^{2+}/Fe^{3+} couple by 60 mV while halogenation directly on the porphyrin ring has almost a ten fold greater effect.[7]

In addition to providing dramatic electronic activation the β-halogens also cause a significant change in the conformation of the porphyrin ring. Smith and his colleagues[8] have shown that bulky groups on the β-positions of *meso*-tetraphenylporphyrins can cause the normally flat aromatic porphyrin to take up a saddle shape where the four pyrrole rings alternately point up and down with respect to the mean porphyrin plane. The perhalogenated porphyrins take up similar conformations as shown in Figure 4. This saddle conformation results in even greater steric protection than the planar conformation. Thus "perhalogenation" provides both steric protection and electronic activation of metallo tetraphenylporphyrins making them excellent biomimetic catalysts.

Since we first reported on the o-phenyl halogenated systems they and the more recent β-halogenated complexes have been widely studied as oxidation catalysts particularly as mimics of chytochrome P-450[9] and for the oxidation of alkanes using dioxygen without a coreductant.[10] We shall concentrate here on the use of these highly chlorinated complexes as mimics of the ligninases.

The ligninases are hemeproteins which are produced by fingi and used in nature to assist in the degradation of lignin.[11] As well as the *in vivo* degradation of lignin the ligninases are also known to degrade several classes of persistant environmental pollutants such as chlorinated phenols, polycyclic aromatics, and dioxins.[12] Sulfonation of 2 and the free base of 4 in H_2SO_4/SO_3 places a sulfonic acid group in each of the phenyl rings thereby adding further electron withdrawing groups to give the water

TABLE 1. Half-wave potentials (vs. SCE) for Fe^{2+}/Fe^{3+} porphyrins in PhCN (0.1M $TBAPF_6$)

TPP	-0.37
2	0.04
perchloroTPP	0.27

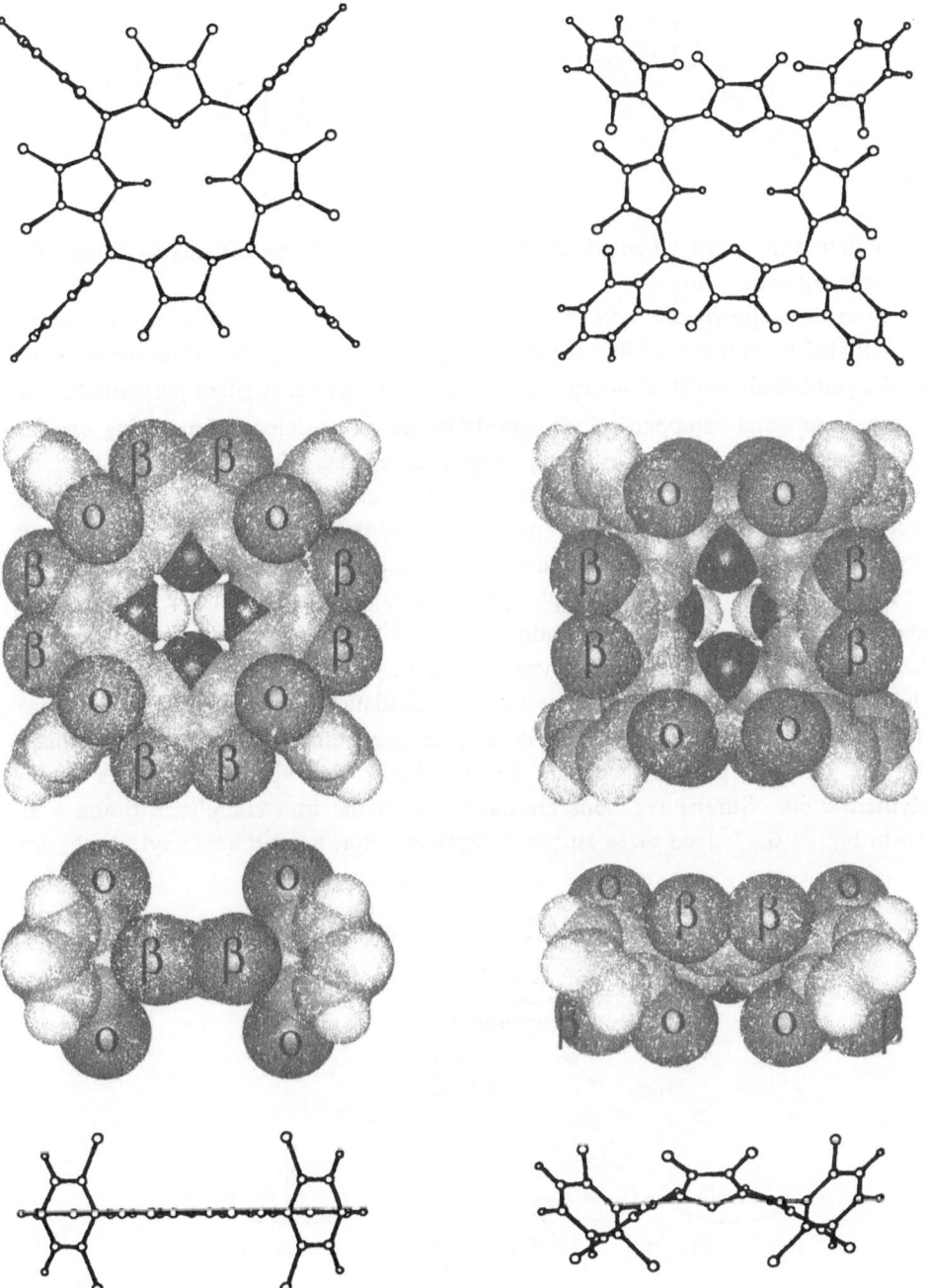

Figure 4 Space fitting and skeletal views of **4** Left column idealized flat porphyrin and orthogonal phenyl rings Right column energy minimized structures using Biosym InsightII Chlorine atoms are at the ortho phenyl (*o*) and porphyrin β positions

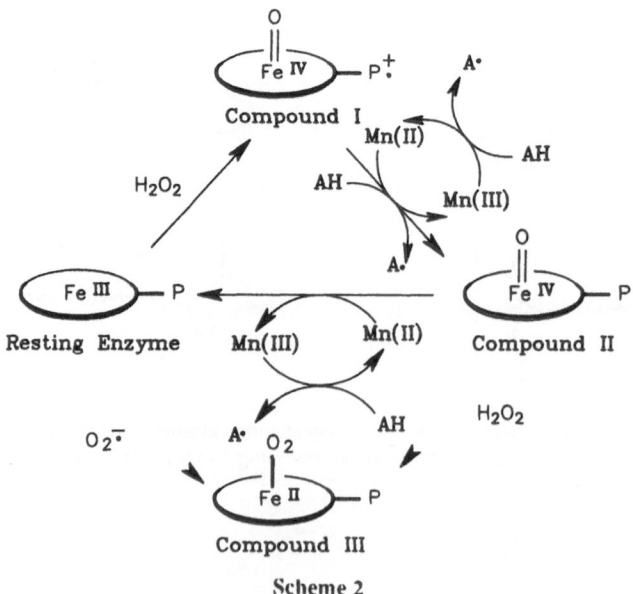

soluble complexes which when metallated to give **5** and **6** can effectively mimic the ligninases in aqueous solution.

Both the depolymerization of lignin and the degradation of aromatic pollutants are all initiated by single one-electron oxidations of the substrate.[13] Thus the catalytic cycle of ligninase shown in Scheme 2 is typical of that of many plant peroxidases and both compound I and compound II (O = Fe(IV)P) act as one electron oxidizing agents. Manganese dependent peroxidases bring about two separate $Mn^{2+} \rightarrow Mn^{3+}$ oxidations and the Mn^{3+} is thought to diffuse through the lignin and to bring about single one-electron oxidations.[14] Lignin is a complex three dimensional polymer containing a variety of functional groups and structural features all derived from the polymerization of the highly oxygenated monomeric phenyl propenoid unit as shown in Figure 5. The complexity and inhomo- geneity of lignin makes it difficult to study and so a variety of model systems for lignin have been developed and our understanding of lignin degradation has been derived from these simple models. Oxidation of the dimer **7** by ligninase and fragmentation of the same molecule in the mass spectrometer (Figure 6) shows that a single one electron process accounts for the bond cleavage that initiates lignin depolymerization. Similar reactions are catalyzed by the iron complexes **5** and **6** as shown in Figure 6. Indeed these simple halogenated iron *meso*-tetraphenyl porphyrins

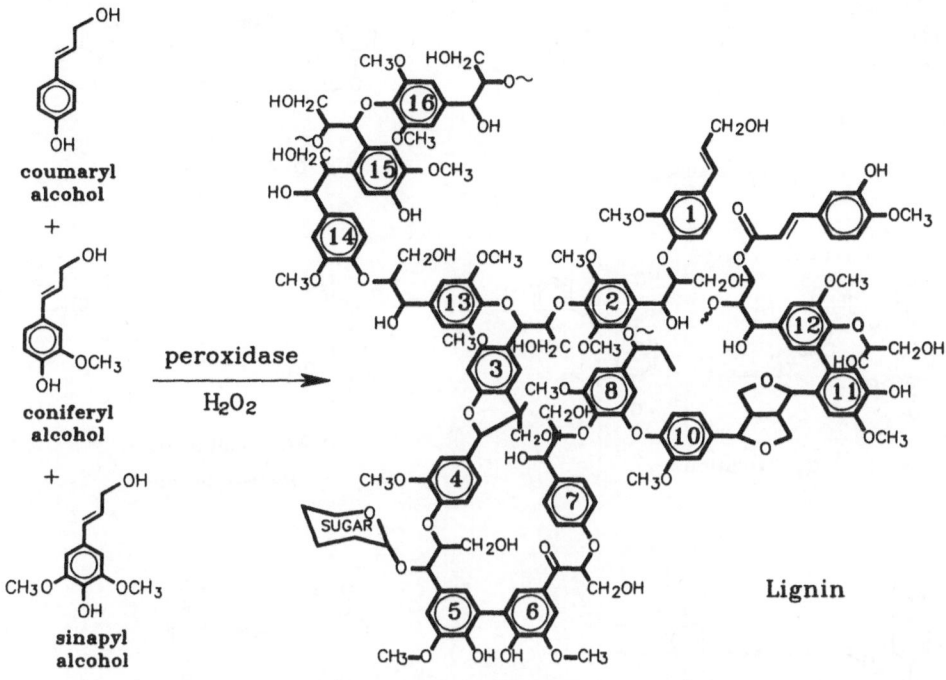

Figure 5: Partial structure of lignin showing the major substructural features

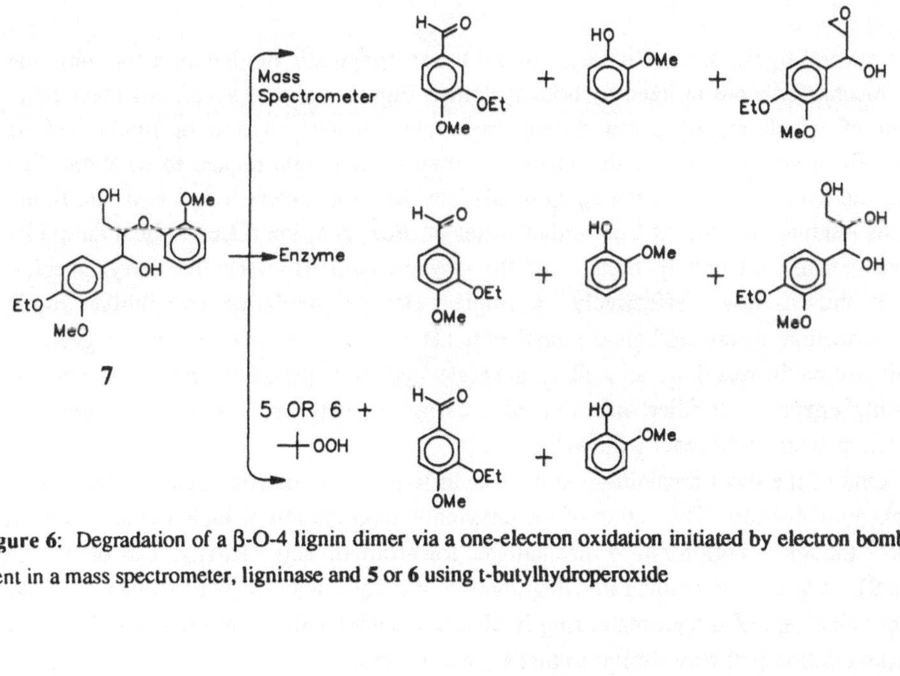

Figure 6: Degradation of a β-O-4 lignin dimer via a one-electron oxidation initiated by electron bombardment in a mass spectrometer, ligninase and 5 or 6 using t-butylhydroperoxide

293

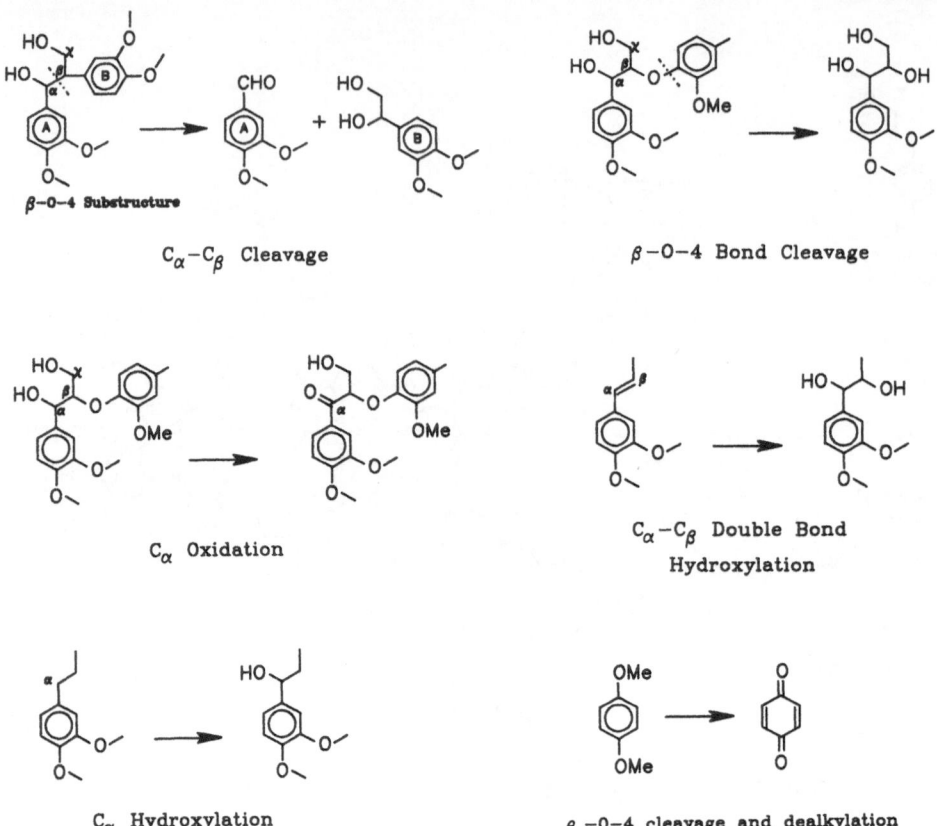

β-0-4 Substructure

C_α-C_β Cleavage

β-0-4 Bond Cleavage

C_α Oxidation

C_α-C_β Double Bond
Hydroxylation

C_α Hydroxylation

β-0-4 cleavage and dealkylation

Figure 7: Major cleavage reactions of lignin brought about by ligninase and iron *meso*-tetraphenylporphyrins

degrade several lignin dimers in ways similar to the enzymatic degradation showing that similar mechanisms are utilized by both systems. Figure 7 shows several of these reactions all of which are observed during the enzymatic degradation of lignin and its models. Even though some of this chemistry may at first sight appear to be P-450-like (such as the hydroxylation of the c_α-c_β bond) any additional oxygen inserted into lignin during its degradation comes from either water or dioxygen, via a free radical autoxidation mechanism, but not by transfer of the oxygen atom from an oxo-ferryl species. Figure 8 shows how effectively a single electron oxidation can initiate lignin depolymerization where carbon centered radicals can react with dioxygen to generate autoxidation chain reactions as well as generate hydrogen peroxide which can activate the resting enzyme. Further oxidation of a carbon centered radical can also generate carbocations which will react with water.

One of the most recalcitrant structures in lignin and environmental polluntants is the biphenyl substrate. The nature of the enzymatic degradation of biphenyl structures is unclear[15] though straightforward mechanisms for aromatic ring cleavage can be written (Figure 9). We have examined the degradation of dehydrovanillin[16] as shown in Figure 10 where cleavage of the aromatic ring is observed which can be rationalized via a one-electron oxidation pathway similar to that shown in Figure 9.

294

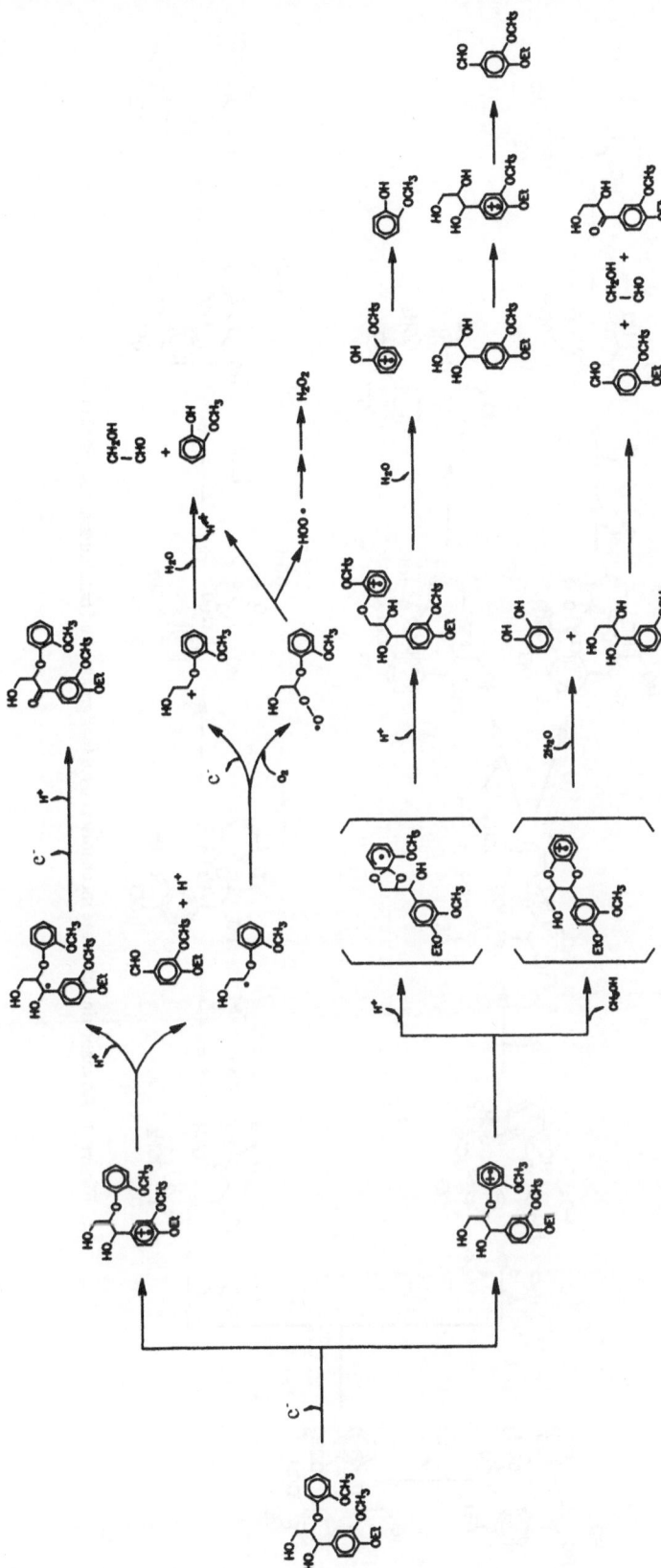

Figure 8: Mechanistic pathways for the one-electron initiated degradation of a lignin model. Note that any oxygen incorporated into degradation products arises from H_2O or O_2 and not from any O=Fe porphyrin species

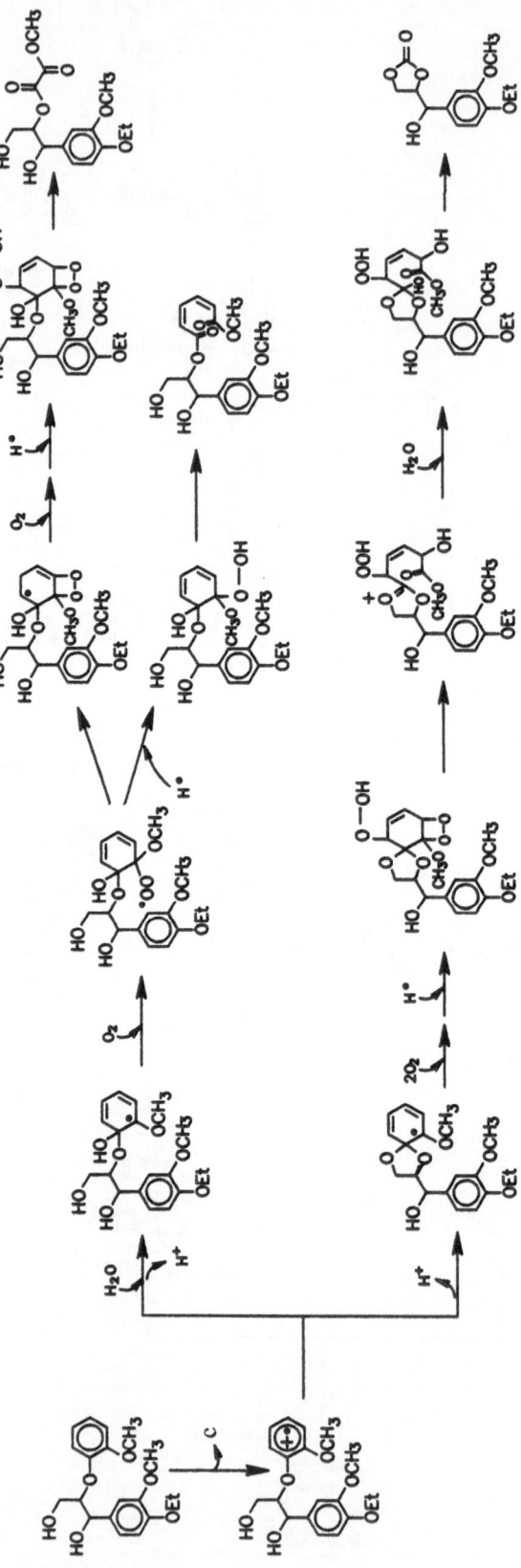

Figure 9: Aromatic ring cleavage mechanisms used by ligninase and biomimetic iron porphyrins

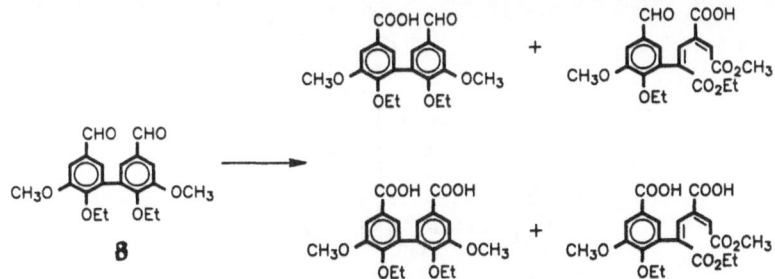

Figure 10: Ring cleavage products of dehydrovanillin (**8**) using **5** or **6**

As noted above lignin peroxidase catalyses the oxidation of polycyclic aromatic compounds. Pyrene (**9**) was oxidized to the 1,6- and 1,8-quinones (**10**, **11**) and benzo[a]pyrene gave the 1,6-, 3,6-, and 6,12-quinones. Pyrene was rapidly oxidized by **3** at pH 3 using *m*-chloroperbenzoic acid to give **10** and **11** (Figure 11). Further oxidation of these two quinones was slow using **3** but oxidation of **9** using **6** (as its Mn complex) which is more stable than **3** gave polymeric black tar insoluble in either water or organic solvents. Indeed the polymerization of polycyclic aromatics and simple phenols and aromatic amines using catalysts such as **5** and **6** suggests a means of removing such materials from industrial waste streams particularly using supported catalysts.

Figure 11 shows the pH rate profile for the oxidation of pyrene. In fact all substrates show complex pH rate profiles which also change depending upon the oxidant and metalloporphyrin. This is not surprising since the rate of oxidation of the substrate may be pH dependent (especially so with phenols), the rate of ligand exchange at the metal, so that oxidant may bind, will be pH dependent as will the binding of oxidant and finally cleavage of the O-O bond in peroxides or peracids will also be pH dependent.

We have examined the oxidation of a number of chlorinated phenols using **5** and **6** as their Fe and Mn complexes.[17] The rate of oxidation was dependent upon the oxidant with *m*-chloroperbenzoic acid > potassium monoperoxysulfate > hydrogen

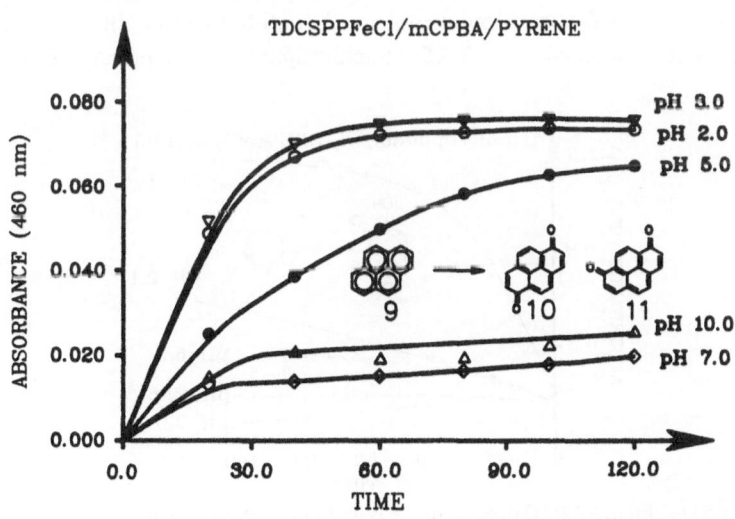

Figure 11: Oxidation of pyrene (**9**) using the iron complex **3** and *m*-chloroperbenzoic acid

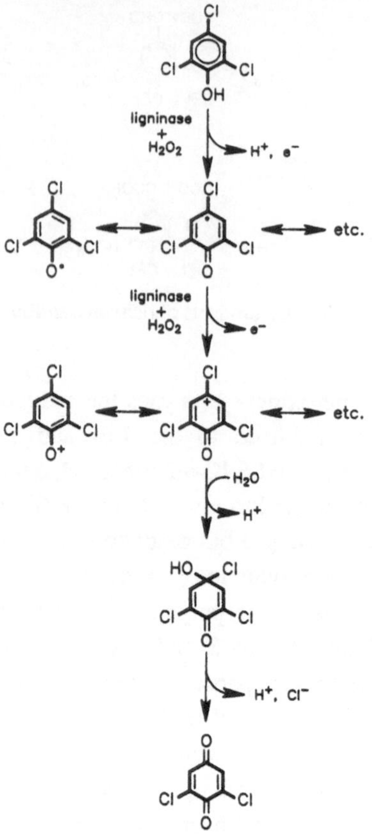

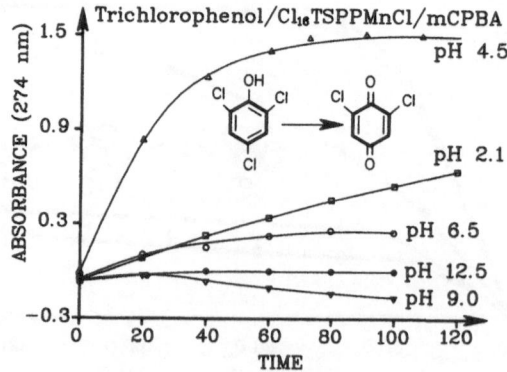

Scheme 3

peroxide > t-butylhydroperoxide, in general catalyst **5** (Fe) at pH <4 was the most effective catalyst. The oxidation rate was also strongly dependent upon the number and position of chlorine substitution on the phenol with 2,4-dichlorophenol > 2,4,5-trichylorophenol > 2-chlorophen ≈ 4-chlorophenol > 3,4-dichlorophenol ≈ 2,4,6-trichlorophenol > 2,3,4,6-tetrachlorophenol ≈ 2,3,5,6-tetrachlorophenol > 3-chlorophenol > 3,5-dichlorophenol ≈ 2,3,4,5-tetrachlorophenol.   Products derived from

Figure 12: Oxidation of 2,4,6-trichlorophenol using the manganese complex of **5** and *m*-chloroperbenzoic acid

phenolic radical coupling and quinones were the major products and once again suggests that the reactions are initiated by single one-electron oxidations. Scheme 3 shows a postulated mechanism for the reactions of 2,4,6-trichlorophenol and Figure 12 shows the pH rate profile. The major product from the oxidation of 2,4,6-trichlorophenol was the 2,6-dichloro-1,4-benzoquinone and with high substrate concentration the dimer **12** was observed.

12

The ability to degrade lignin and phenolics suggest the possible use of these metalloporphyrins as catalysts in the pulp and paper industry for both the bleaching of pulp and the treatment of waste effluents. The production of Kraft pulp involves the initial delignification followed by several bleaching sequences often using Cl_2 and ClO_2 which are aimed at removing the residual lignin since this colors the pulp and allows for subsequent decolorization of the paper. The extent of residual lignin in the pulp is measured by the Kappa number. Treatment of softwood Kraft pulp with either catalyst **5** or **6** (as their Fe complexes) and t-butylhydroperoxide gave a significant decrease in

TABLE 2. Bleaching of softwood Kraft pulp

Iron Porphyrin (% w/w)	t-butylhydroperoxide (% w/w)	Kappa number
0	0.0	17.9
0	0.5	17.6
6: 0.023	0.5	10.4
6: 0.045	0.5	9.5
5: 0.023	0.5	14.1
5: 0.045	0.5	14.1

Reaction Conditions: 18 hours, 60°C, 2.5% consistency, pH 4.8
Extraction Conditions: 90 minutes, 60°C, 2.5% consistency, 2.5% (w/w) sodium hydroxide

TABLE 3. Effluent Decolorization with Porphyrins

Temperature (°C)	Time to Decolorize 50% (minutes)
40	20
50	16
60	13

Reaction conditions: Porphyrin 6: 0.00028% (w/v)
 t-butyl hydroperoxide: 0.8% (w/v)
No decolorization occurred in the presence of peroxide without porphyrin present.

Kappa number (Table 2) which was accompanied by a decrease in viscosity showing that the cellulose as well as the lignin was oxidized.[18]

Highly colored effluents are produced in pulp mills which gives rise to environmental problems. We have seen very promising results in that undiluted effluent was decolorized by greater than 50% in less than thirteen minutes using low concentrations of catalyst and t-butylhydroperoxide (Table 3).[18]

Acknowledgements

This work was supported by the Institute for Chemical Science and Technology, the Canadian Natural Sciences and Engineering Research Council and the British Columbia Science Council. Tilak Wijesekera, Futong Cui, and Kelly Sveinson made major contributions to the work described here.

References

1. D. Dolphin, in "Inorganic and Organic Radicals: Their Biological and Clinical Relevance", H.A.O. Hill, ed., The Royal Society, London, 1986.

2. V. Renganathan and M.H. Gold, Biochem., 25:1626 (1986).

3. Cytochrome P-450, P.R. Ortiz de Montellano, ed., Plenum, New York (1986).

4. J.T. Groves and T.E. Nemo, J. Am. Chem. Soc., 105:5786 (1983).

5. P.S. Traylor, D. Dolphin, and T.G. Traylor, J. Chem. Soc. Chem. Commun., 279 (1984).

6. F. Cui, D. Dolphin, T. Wijesekera, R. Farrell, and P. Skerker, in "Applications of Biotechnology of Pulp and Paper Manufacture", T.K. Kirk and H.-M. Chang, eds., Butterworth-Heinemann, Boston, Massachusetts, U.S.A., 1990.

7. T. Wijesekera, A. Matsumoto, D. Lexa, and D. Dolphin, Angewandte Chemie I.E., 29:1028 (1990).

8. K.M. Barkigia, M.D. Berber, J. Fajer, C.J. Medforth, M.W. Renner, and K.M. Smith, J. Am. Chem. Soc., 112:8851 (1990).

9. B. Meunier, Chem. Rev., 92:1411 (1992).

10. P.E. Ellis Jr. and J.E. Lyons, Coord. Chem. Rev., 105:181 (1990).

11. M. Tien and T.K. Kirk, Proc. Natl. Acad. Sci. USA, 81:2280 (1984).

12. K.E. Hammel, B. Kalyanaraman, and T.K. Kirk, J. Biol. Chem., 261:16948 (1986).

13. H.E. Shoemaker, P.J. Harvey, R.M. Bowen, and J.M. Palmer, FEBS Letters, 183:7 (1985).

14. H. Wariishi, L. Aklieswaran, and M.H. Gold, Biochem., 27:5365 (1988).

15. Y. Kamana and T. Higuchi, Wood. Res., 70:25 (1984).

16. F. Cui, Ph.D. Thesis, University of British Columbia, (1990).

17. K.P. Sveinson, M.Sc. Thesis, University of British Columbia, (1992).

18. P.S. Skerker, R.L. Farrell, D. Dolphin, F. Cui, and T. Wijesekera, in "Applications of Biotechnology of Pulp and Paper Manufacture", T.K. Kirk and H.-M. Chang. eds., Butterworth-Heinemann, Boston, Massachusetts, U.S.A., (1990).

METHANE MONOOXYGENASE: MODELS AND MECHANISM

Katherine E. Liu, Andrew L. Feig, David P. Goldberg, Stephen P. Watton, and Stephen J. Lippard

Department of Chemistry
Massachusetts Institute of Technology
Cambridge, MA 02139

INTRODUCTION

There is much current interest in non-heme diiron carboxylate proteins. Included are the hydroxylase enzyme of methane monooxygenase (MMO), hemerythrin, ribonucleotide reductase, and purple acid phosphatase, all of which contain a dinuclear iron center at their active site. We ultimately desire an understanding of how these units are tuned in each protein to exhibit diverse functions ranging from the reversible binding of dioxygen in hemerythrin to activation of dioxygen for converting methane to methanol in MMO. In pursuit of this objective, we are investigating the proteins of the MMO system and exploring the fundamental chemistry of the hydroxylase diiron center. In the present article we review some of our recent progress in this area.

METHANE MONOOXYGENASE

The enzyme system MMO can be isolated from methanotrophic bacteria, which use methane as their only source of carbon and energy.[1] MMO catalyzes the first step in the metabolic pathway of the organism, eq. 1.[2] Research in our

$$CH_4 + O_2 + H^+ + NADH \longrightarrow CH_3OH + H_2O + NAD^+ \qquad (1)$$

laboratory has focused primarily on the soluble MMO isolated from *Methylococcus capsulatus* (Bath), although some work has also been carried out with MMO from

The Activation of Dioxygen and Homogeneous Catalytic Oxidation,
Edited by D.H.R. Barton *et al.*, Plenum Press, New York, 1993

Methylosinus trichosporium OB3b. These enzyme systems have three components: a hydroxylase, a coupling protein B, and a reductase.[3,4] The hydroxylase contains up to two non-heme, dinuclear iron centers, and is the site of substrate binding.[4] The reductase contains one FAD and one 2Fe-2S cluster. Its function is to accept electrons from NADH and pass them ultimately to the dinuclear center in the hydroxylase.[5-7] The coupling protein B regulates electron transfer,[7,8] but its role in catalysis is more complex.[9]

Here we summarize our structural studies of the hydroxylase dinuclear iron center, work that identifies interactions among the MMO component proteins, and probes into the mechanism of the hydroxylation reaction.

Structure of the Hydroxylase Dinuclear Iron Center

EXAFS spectroscopy has been used to study the hydroxylase in three oxidation states, the oxidized $Fe^{III}Fe^{III}$ (H_{ox}), mixed-valent $Fe^{II}Fe^{III}$ (H_{mv}), and fully reduced $Fe^{II}Fe^{II}$ (H_{red}) forms.[10,11] The Fe···Fe distance is 3.4 Å in both H_{ox} and H_{mv}, whereas no Fe···Fe backscattering was observed in H_{red}. The coordination sphere of the iron atoms was assigned to a mixture of O and N donor atoms, with an average Fe-O/N distance ranging from 2.04 Å in H_{ox} to 2.15 Å in H_{red}. No short Fe-O (1.80 Å) distance was observed in any oxidation state, implying the lack of an oxo bridge between the iron atoms. The absence of any optical features above 300 nm in H_{ox} further confirms the absence of an oxo bridge and rules out coordination by phenolate ion.[10]

EPR spectroscopy is a valuable diagnostic tool for characterizing dinuclear iron sites in proteins.[12] The H_{mv} protein exhibits a characteristic rhombic signal with $g_{av} = 1.83$[10,13] and weak antiferromagnetic exchange coupling (J = -32 cm^{-1}) as measured from EPR power saturation experiments.[10] This value is consistent with a hydroxo, alkoxo, or monodentate carboxylato bridge between the iron atoms.[10] Mössbauer spectral parameters, especially $\Delta E_Q = 1.05$ mm/s for H_{ox} and 3.014 mm/s for H_{red}, are close to those expected for an oxo-bridged species, but a hydroxo-bridged system with pentacoordinate iron atoms is also a reasonable interpretation.[10]

Recent Q-band proton ENDOR spectra of H_{mv} have provided new evidence concerning the nature of the bridging ligand.[14] Proton ENDOR results for both H_{mv} and the related semimet HrN3 revealed the existence of three different classes of protons. The first class, which exhibits resonances with A ≤ 4 MHz, are mostly nonexchangeable with D_2O and are assigned to protons that are > 3 Å from the iron center. A second class of protons was also identified. In semimet HrN3, these resonances are highly anisotropic and strongly coupled to the diiron center, with A ~ 12 - 28 MHz. They are assigned to the H atom of the hydroxo bridge[15-17] between the iron atoms. Proton ENDOR spectroscopy thus provides a direct method for identifying bridging hydroxide ligands in the mixed-valent diiron centers of proteins.[14] In H_{mv}, strikingly similar resonances appear in the ENDOR spectrum, with A ~ 14 - 30 MHz. Accordingly, these resonances were assigned to a bridging hydroxide ligand in H_{mv}, providing the first positive identification of the bridging unit for the diiron center in this protein. A third class of protons, observed in H_{mv} but not semimet HrN_3, has isotropic resonances with A ~ 8 MHz. These protons are assigned to terminal water or hydroxide ligands, which are not present in semimet HrN_3.[16] Both the second and third classes of protons are readily exchangeable with D_2O.[14] Taken together, these results provide strong evidence for the core structure shown in Figure 1.

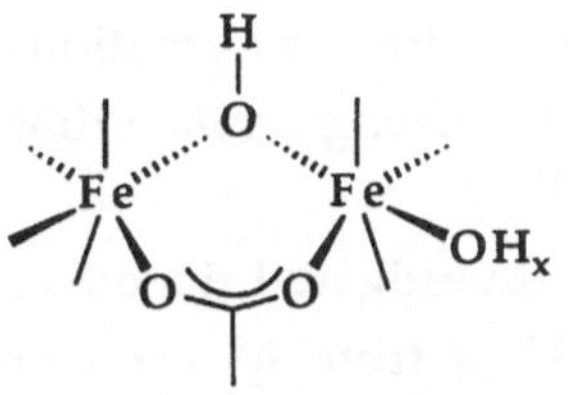

Figure 1. Postulated structure of the diiron center in the hydroxylase component of methane mono-oxygenase. The unlabelled ligands are a mixture of oxygen and nitrogen donors.

Component Interactions

Although considerable effort has been made to understand the structure and function of the hydroxylase, less is known about the interactions among the three component proteins of MMO. Previous work on this subject revealed the ability of the coupling protein B to regulate electron transfer between the hydroxylase and the reductase.[8,18] The hydroxylase is isolated as H_{ox}, but H_{mv} and H_{red} are accessible following chemical reduction with sodium dithionite in the presence of appropriate mediators. Through redox titrations monitoring the EPR signal of H_{mv}, the reduction potentials of the hydroxylase were determined to be +48 and -135 mV vs NHE, with a slight diminution to +40 and -160 mV following addition of the substrate propylene.[8] In the presence of a stoichiometric amount of the coupling protein and the reductase, however, no reduction of the hydroxylase was observed. In contrast, under the same conditions but with added substrate, a two electron reduction of the hydroxylase, from H_{ox} to H_{red}, was observed at potentials above 150 mV. No H_{mv} was detected, which suggests that this oxidation state is not physiologically significant. This result is consistent with kinetic work in which coupling protein B was demonstrated to inhibit NADH oxidation by the MMO system in the absence of substrate, a finding that was reversed by addition of substrate.[7]

Other workers have discovered that addition of coupling protein B to the hydroxylase in MMO from *M. trichosporium* OB3b decreases the midpoint of the reduction potential of the diiron center by -120 mV;[19] this result implies that interaction of B with the hydroxylase affects the diiron center. This interpretation is supported by studies of MMO from the same organism, which indicate that coupling protein B perturbs the EPR signal of H_{mv}[20] and dramatically alters the hydroxylation product distributions.[9] In the latter work, it was hypothesized that changes in the hydroxylase structure upon complexation with coupling protein B caused the substrate molecule to interact differently with the active site. Moreover, coupling protein B from *M. trichosporium* OB3b increases the initial velocity of the complete reconstituted system, but decreases initial velocities in reactions of H_{ox} with H_2O_2.[9] Direct evidence exists for the formation of complexes among the three protein components of the *M. trichosporium* OB3b MMO from studies using chemical cross-linking agents.[20]

Quite recently we have investigated the effect of coupling protein B on the single turnover reactivity of H_{red} from *M. capsulatus* (Bath) with nitrobenzene, previously demonstrated to be a suitable substrate with MMO from *M. trichosporium* OB3b.[21] In particular, the coupling protein increases the rate of formation of 4-nitrophenol, the pseudo-first order rate constant increasing from 0.012 s^{-1} without the protein to roughly 0.400 s^{-1} at a ratio of 1.5 equivalents of coupling protein to H_{red}.[22] These results are illustrated in Figure 2. Further addition of coupling protein above this ratio slightly diminishes the rate constant, which levels off at approximately 0.264 s^{-1}. The coupling protein B therefore induces a maximal increase of greater than 30-fold in the rate constant for the reaction. Addition of the reductase, in contrast, does not affect the rate significantly.

From these results it is apparent that there are at least three known functions of the *M. capsulatus* (Bath) coupling protein. It regulates electron transfer from the reductase to the hydroxylase so as to occur only in the presence of substrate.[7,8] It shifts the reduction potentials of the substrate/reductase/hydroxylase complex such that $E_2° > E_1° > +100$ mV,[8] and it increases k_{obs} in single turnover reactions of H_{red}.[22] A unifying explanation for these observed functions has yet to be determined. Future studies may identify additional roles of the coupling protein B from *M. capsulatus* (Bath), which will be interesting to compare with those of the analogous protein from *M. trichosporium* OB3b.

Mechanism of Hydroxylation

The catalytic mechanism of hydrocarbon oxidation by the MMO hydroxylase has been frequently compared to that of its heme analog, cytochrome P-450.[21,23-25] A widely accepted mechanism for P-450 hydroxylations invokes a high valent iron oxo, or ferryl, species which is then postulated to abstract a hydrogen atom from bound substrate.[26] The resulting substrate radical can then recombine with the iron-bound hydroxyl radical in a so-called "rebound" mechanism. Evidence for the presence of substrate radical intermediates in cytochrome P-450[27-29] includes the use of probes termed radical clocks.[30] Such substrates rearrange rapidly upon the

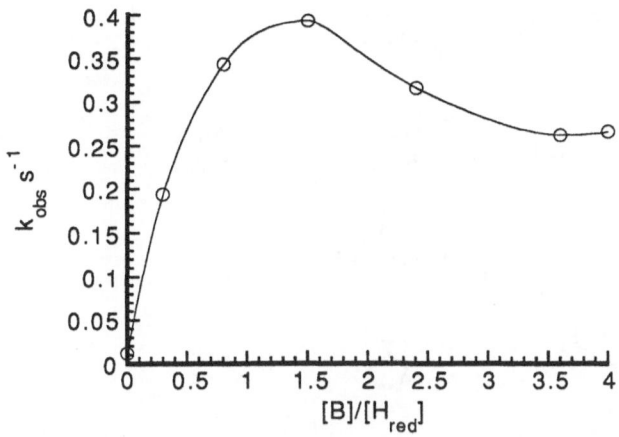

Figure 2. Effect of added coupling protein B on the rate of 4-nitrophenol formation from nitrobenzene in single turnover reactions of H_{red} and dioxygen.

abstraction of a hydrogen atom in the enzyme active site, affording skeletally modified hydroxylation products. The presence of such products is taken as evidence that a substrate radical intermediate participates in the mechanism. The substrate probe 1,1-dimethylcyclopropane has been employed previously to study MMO from *M. trichosporium* OB3b,[24] and the products recovered were consistent with both radical and cationic substrate intermediates. We have recently applied this methodology with MMO from *M. capsulatus* (Bath) and *M. trichosporium* OB3b using several radical clock substrates,[31] including two much faster probes than those reported previously for experiments with cytochrome P-450.

Results with MMO from *M. capsulatus* (Bath)[31] are listed in Figure 3, together with the known rearrangement rates[28,32,33] for the substrate-derived radical. No products consistent with the formation of a substrate radical were detected. This result implies that the mechanism of hydroxylation for the substrates employed in

Substrate	Unrearranged Product(s)	Rearranged Alcohol(s) Expected	Observed Ratio of Unrearranged to Rerranged Alcohol	Rate Constant For Radical Rearrangement, s-1 at 45 °C
 (structure: Ph-cyclopropane)	 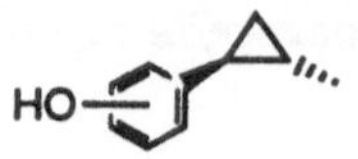 (structures)	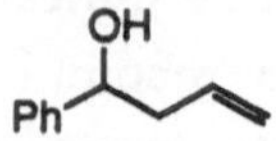 (structure: OH)	>100:1	4×10^{11}
 (structure: Ph, Ph cyclopropane)	 (structure)	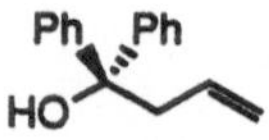 (structure: Ph, Ph HO)	>25:1	5×10^{11}
 (structure: methylcyclopropane)	 (structure: —OH)	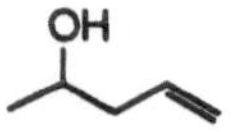 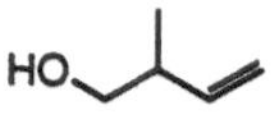 (structures)	both >100:1	$2.0 - 2.4 \times 10^8$
 (structure: bicyclic)	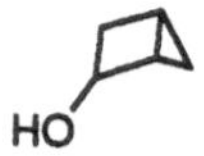 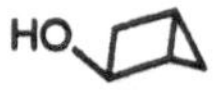 (structures)	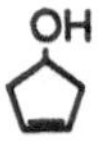 (structure: OH)	80:1[a]	3×10^9

[a]Rearranged product is accounted for through thermal rearrangement.

Figure 3. Expected products and observed ratios for hydroxylation reactions with MMO from *M. capsulatus* (Bath).

these studies may not involve the generation of substrate radical or carbocation intermediates. This conclusion contrasts with literature reports that radical intermediates can be trapped during methane hydroxylation by MMO from *M. capsulatus* (Bath),[34] although the amount of trapped radical was not quantitated in those experiments. Rearranged product with the probe *trans*-1,2-phenylmethylcyclopropane was observed in experiments with MMO from *M. trichosporium* OB3b, however.[31] From the ratio of unrearranged to rearranged products, a rebound rate constant for the substrate radical with an iron-bound hydroxyl radical was calculated to be 6 - 10 x 10^12 s-1. The formation of a substrate radical intermediate for *M. trichosporium* OB3b is consistent with literature reports.[24,25,35,36]

The radical clock experiments with MMO from *M. capsulatus* (Bath) and *M. trichosporium* OB3b carried out in our laboratory indicate that there may not be a single mechanism operative for these enzyme systems.[31] Instead, the mechanism may depend on factors such as the steric and energetic requirements of the substrate, as well as the temperature employed in the MMO hydroxylation reaction. Differences in the MMO systems from the two different organisms include their optimal hydroxylation temperatures as well as sequence variations in the coupling protein B from the two organisms.[37]

In addition to radical clock studies, kinetic isotope effects were measured by using deuterated *trans*-1,2-phenylmethylcyclopropane probes.[31] No significant intermolecular isotope effect was found in comparing reactions with probe deuterated at the methyl position versus experiments with undeuterated probe. Comparison of hydroxylation products for mono- and dideuterated derivatives of *trans*-1,2-phenylmethylcyclopropane, however, revealed an intramolecular isotope effect, k_H/k_D = 5.15 and 5.03, respectively. These results indicate that, whereas C-H bond breaking is not the rate-determining step in the overall enzymatic reaction, the hydroxylation step does involve a substantial C-H bond-breaking component. The magnitude of this intramolecular isotope effect for MMO from *M. capsulatus* (Bath) can be compared to values of 4.2 for MMO from *M. trichosporium* OB3b[36] and 7-14 for cytochrome P-450 hydroxylations,[26] although these literature values were determined for different substrates. In addition, dimethyldioxirane, which oxidizes hydrocarbons by an insertion process, has a k_H/k_D = 5.[38]

MODEL COMPLEXES

In a parallel approach to unraveling the chemistry of the non-heme diiron center in the MMO hydroxylase and related diiron carboxylate proteins, we have undertaken the synthesis, characterization, and exploration of biomimetic reaction chemistry of appropriately designed model compounds. Here we report progress in several areas. One is the preparation of kinetically stabilized carboxylate-bridged (μ-oxo)diiron(III) units to study terminal ligand substitution and other reactions. By using this approach, we have synthesized a model complex containing pendant phenoxyl radicals as a model for the R2 subunit in ribonucleotide reductase. Additionally, we have explored the basic reaction chemistry of bridged diiron(II) complexes with dioxygen. In this research the objectives are to determine the stoichiometry of the O_2 oxidation reaction and to identify intermediates and products through spectroscopic, kinetic, and structural investigations.

Dinuclear Complexes Stabilized by Molecular Cleft Ligands

We have been investigating the assembly of dinuclear iron model complexes using cleft-shaped dicarboxylate receptors designed by Rebek and coworkers for molecular recognition studies.[39,40] These rigid molecules are organized so that the two carboxylate groups are positioned toward the center of the cleft, resulting in kinetic and thermodynamic stabilization in their binding interactions. As an example, the ligand XDK (Figure 4), where XDK = xylenediamine bis(Kemp's triacid imide), a chelating dicarboxylate ligand obtained by condensing two Kemp triacid moieties with a xylyl spacer,[40,41] has been used to prepare a number of dinuclear ferric model complexes. This ligand directs the spontaneous and exclusive formation of a dinuclear μ-oxo complex in methanol solution, $[Fe_2O(XDK)(MeOH)_5(H_2O)](NO_3)_2$ (1), shown in Figure 5. Moreover, it does so in the absence of the 'capping' ligands normally required to prevent the formation of higher nuclearity aggregates.[42] In contrast to dinuclear complexes of simple carboxylates, this compound is sufficiently robust to permit stepwise substitution of

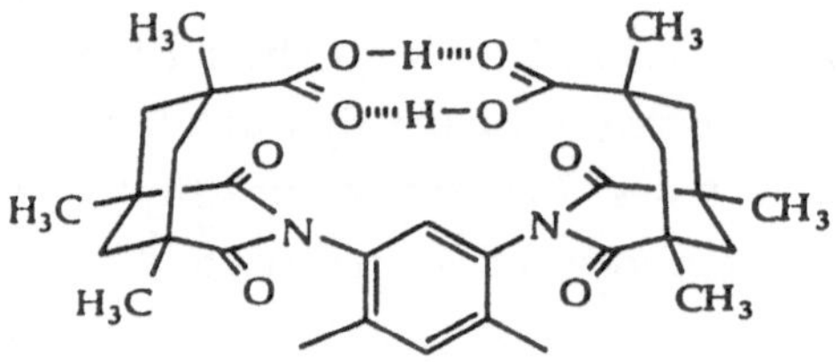

Figure 4. Structure of XDK ligand.

the terminal solvent ligands by monodentate and bidentate donors. Such substitution reactions can be monitored by changes in both the visible and resonance Raman spectra.[42] For example, titration of the complex with six equivalents of 1-methyl imidazole (1-MeIm) in methanol results in stepwise substitution at each of the terminal sites on the complex to afford the hexakis(1-MeIm) derivative. This approach should afford a variety of non-heme diiron units that have not been previously isolated.

R2 Model with a Stable Pendant Phenoxyl Radical

With the use of the unusually stable solvento complex 1 described above,[42] it was possible to construct a complex containing both an oxo/carboxylato-bridged

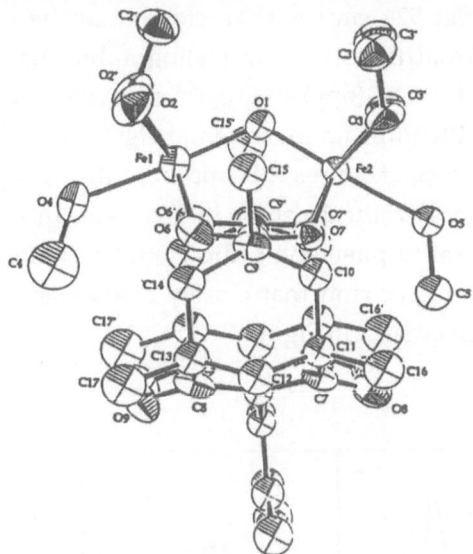

Figure 5. Molecular geometry of [Fe$_2$O(XDK)(MeOH)$_5$(H$_2$O)](NO$_3$)$_2$ as determined in a single crystal X-ray diffraction investigation.

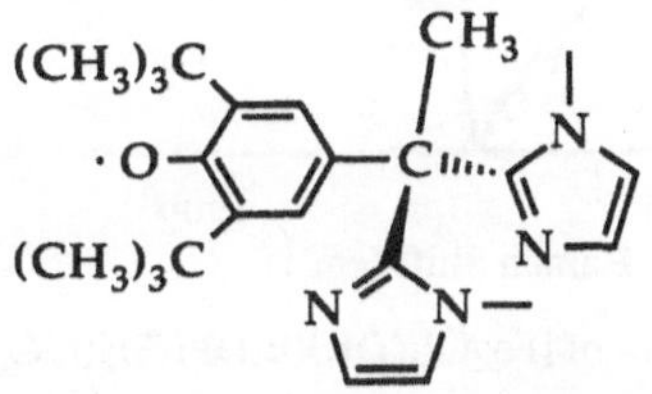

Figure 6. Structure of BIDPhE.

diiron(III) core and a proximate phenoxyl radical. The compound [Fe$_2$O(XDK)(BIDPhE)$_2$(NO$_3$)$_2$]·CH$_2$Cl$_2$ (**2**) was synthesized from **1** by addition of two equivalents of BIDPhE, a bidentate nitrogen donor ligand containing a pendant phenoxyl radical arm, illustrated in Figure 6. Complex **2** was isolated as a stable, brown-green crystalline solid from pentane/methylene chloride mixtures at -30 °C. Two bands are evident in the Raman spectrum of **2** dissolved in methylene chloride (Figure 7). The peak at 1503 cm^{-1} is assigned to the C-O stretch of the radical,[43] and an intense resonance at 524 cm^{-1} is characteristic of the symmetric Fe-O-Fe vibration in the (μ-oxo)diiron(III) core.[44] Crystallographic characterization of the analogous compound [Fe$_2$O(XDK)(bpy)$_2$(NO$_3$)$_2$][42] suggests that the terminal sites in **2** are filled by bidentate BIDPhE and monodentate NO$_3^-$ ligands.

From these spectroscopic features it is apparent that compound **2** is a good model for the diiron/tyrosyl radical center in R2. The incorporation of a (μ-oxo)diiron(III) core and nearby phenoxyl radical into a single, stable compound affords a resonance Raman spectrum that closely resembles that of the protein. This comparison can clearly be seen in Figure 7.

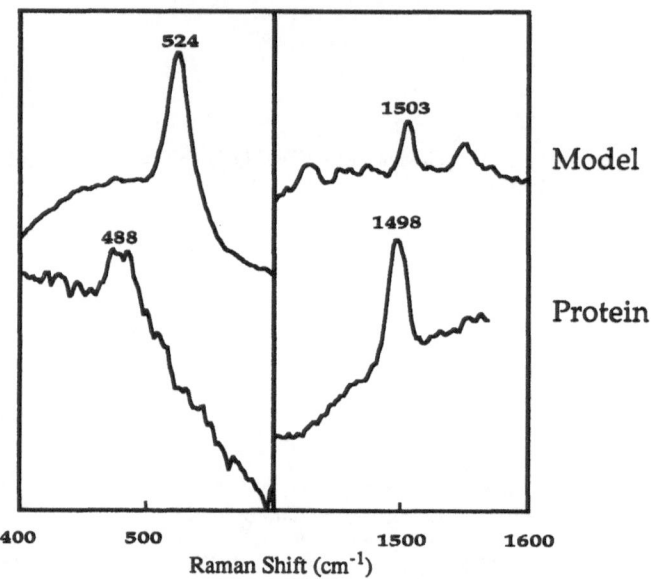

Figure 7. Resonance Raman spectra of [Fe$_2$O(XDK)(BIDPhE)$_2$(NO$_3$)$_2$] and the R2 subunit of ribonucleotide reductase.

Reaction Chemistry of Diiron(II) Complexes with Dioxygen

It is the reduced states of diiron carboxylate proteins that react with dioxygen. Accordingly, we, and others, have begun to explore the fundamental chemistry of diiron(II) model complexes with dioxygen as an approach to understanding the diverse reactivity of these proteins.[45,46] Examples of diiron(II) model compounds described in the literature are presented in Figure 8.[47-51] For kinetic studies of the reactivity of these complexes with dioxygen, a few guidelines have been followed. The oxidation reaction should afford a single product in high yield; the starting material and product should be structurally characterized; and the time scale of the oxidation must be able to be recorded by stopped flow

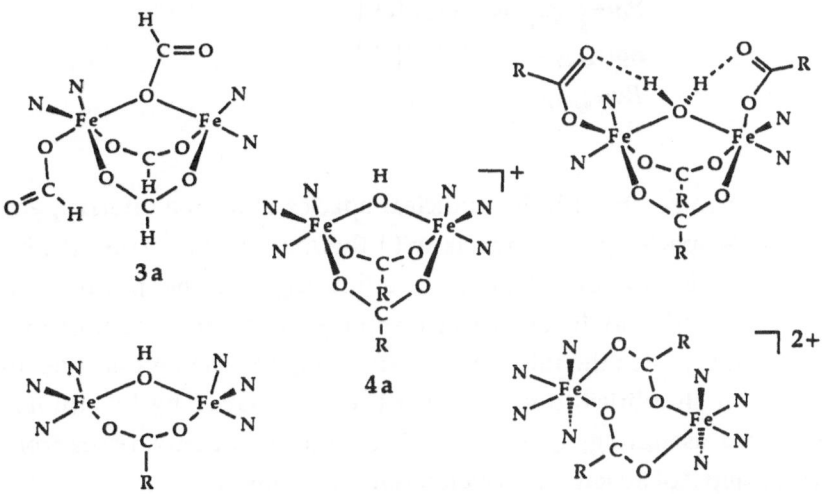

Figure 8. Examples of complexes prepared as models for the diiron(II) forms of carboxylate-bridged protein cores.

spectrophotometry. With these limitations in mind, [Fe$_2$(BIPhMe)$_2$(O$_2$CH)$_4$] (**3a**) (BIPhMe = 2,2'-bis(1-methylimidazolyl)phenylmethoxymethane) and [Fe$_2$(OAc)$_2$-(OH)(Me$_3$TACN)$_2$]$^+$ (**4a**) (Me$_3$TACN = 1,4,7-trimethyl-1,4,7-triazacyclononane) were chosen for initial studies. Compound **3a** presents an asymmetric core in which one iron atom is pentacoordinate, similar to the reduced form of Hr, and **4a** has a hydroxo-bridged structure that may be present in H$_{red}$ of MMO. Upon oxidation of **3a** and **4a**, (μ-oxo)bis(μ-carboxylato)diiron(III) species **3b** and **4b** are formed. The oxo bridge atom in both cases is derived from dioxygen, as revealed through ^{18}O labeling experiments.[46,47] Oxidations of both **3a** and **4a** consume 0.5 mole equivalents of dioxygen per dinuclear unit as measured by manometry. In addition, when oxidized in the presence of triphenylphosphine, no triphenylphosphine oxide is produced.

The reaction orders with respect to complexes were found to be 1.5 ± 0.1 and 1.9 ± 0.1 for the oxidations of **3a** and **4a**, respectively. In chloroform, the reactions of both **3a** and **4a** showed a first order dependence (0.9 ± 0.1) on dioxygen. In contrast, the reaction rate of **4a** in methanol was [O$_2$]-independent (zero order). The oxidation of **3a** could not be studied in methanol since a different product formed. The experimental rate laws for the two oxidations are given in equations 2-4. The observation of a fractional order with respect to **3a** (eq. 2) requires a dissociation equilibrium to occur prior to the rate-determining step. Several possible mechanisms could explain this result, but the favored one is shown in Scheme 1. In this

$$Rate_{3a:CHCl_3} = k[3a]^{\frac{1}{2}}[O_2] \tag{2}$$

$$Rate_{4a:CHCl_3} = k[4a]^2[O_2] \tag{3}$$

$$Rate_{4a:MeOH} = k[4a]^2 \tag{4}$$

mechanism, dioxygen reacts with the dinuclear species to form a diferric peroxide intermediate. This species in turn reacts with the mononuclear complex in two successive steps, the first of which is rate-limiting. If the predissociation equilibrium is fast relative to the other steps, and if steady-state is assumed for the intermediates I_1 and I_2, then the rate law corresponding to Scheme 1 is given by eq. 5. This expression simplifies to eq. 6 when [M] is replaced by its equilibrium concentration, and assuming $k_2[A] \ll k_{-1}$. This latter inequality arises from the assumption that step 2 of Scheme 1 is overall rate-determining.

$$Rate = 2\frac{k_1 k_2 [M][A][O_2]}{k_{-1}+k_2[A]} \tag{5}$$

$$Rate \cong 2\sqrt{K}\frac{k_1 k_2}{k_{-1}}[A]^{\frac{1}{2}}[O_2] \tag{6}$$

The postulated formation of the polynuclear peroxo intermediate has precedence in the literature of Fe(II)-porphyrin systems.[52,53] In addition, the only structurally characterized non-heme iron(III) peroxide complex, $[Fe_6(O_2)O_2(OBz)_{12}]$ (5), has a structural motif similar to that in the key intermediate, namely, a μ,μ-η^2,η^2-peroxide species in which four ferric ions are coordinated to the peroxide dianion (I_3 in Scheme 1).[54] Because 5 has all ferric ions, this peroxide-bridged complex is stable. The species in Scheme 1, on the other hand, is mixed-valent $Fe^{II}_2Fe^{III}_2$, and can therefore decompose by redox chemistry to form the stable (μ-oxo)diiron(III) product.

A mechanistic interpretation of the kinetic results on the oxidation of 4a is presented in Scheme 2 and the rate law derived from this mechanism is given in equation 7. The key feature to understanding this reaction is its order with respect to dioxygen. To account for the observation that in methanol the rate is independent of dioxygen concentration, there must be a reversible step occurring *prior* to the step involving dioxygen. With the appropriate assumptions, this step leads to simplification of the rate law such that the order in $[O_2]$ cancels (eq. 8). We postulate that this reversible step involves a shift of one of the bridging carboxylates to a monodentate binding mode accompanied by solvent binding to the newly opened site. Since chloroform is a non-coordinating solvent, the carboxylate-shifted intermediate is unstable (k_{-1} is large) and the dominant

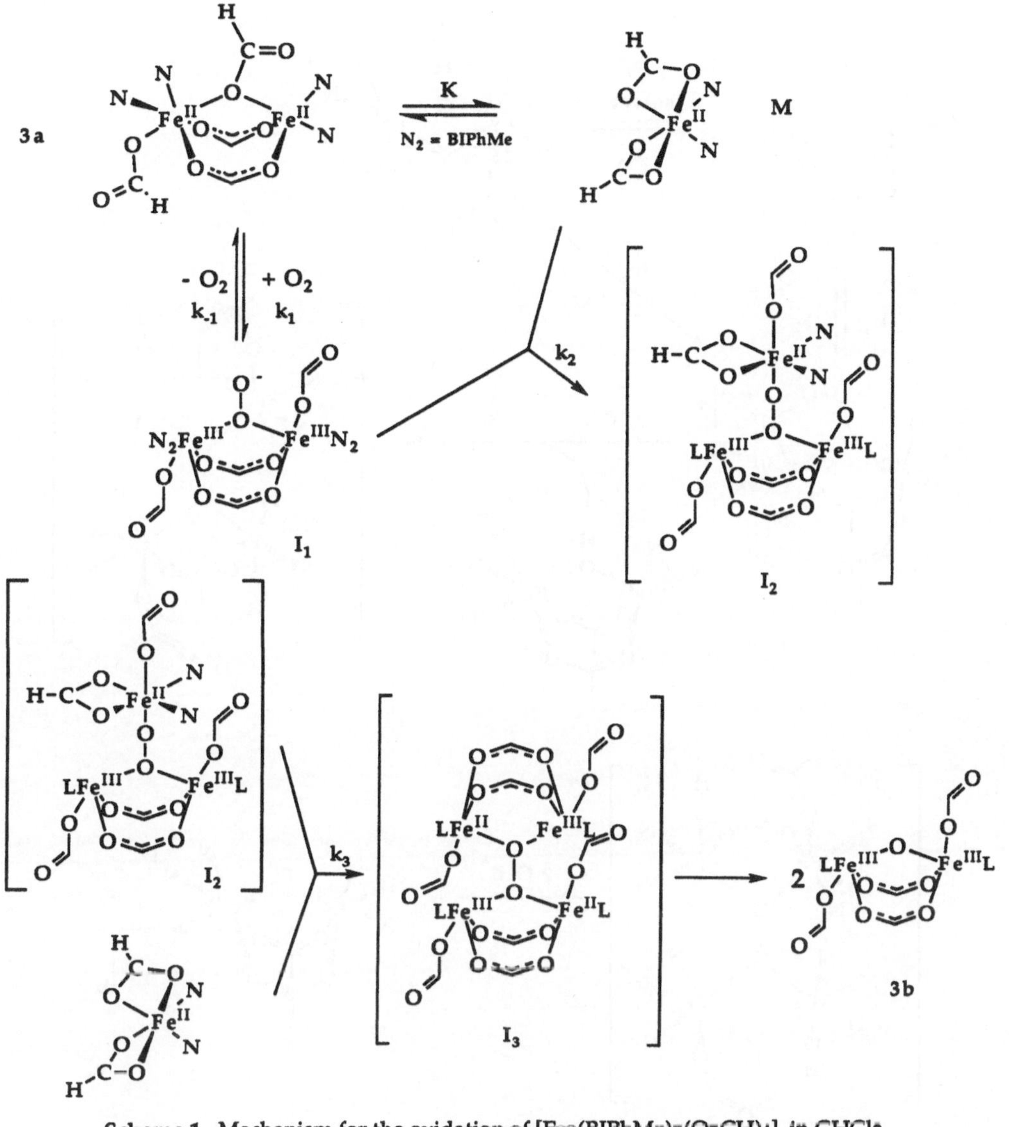

Scheme 1. Mechanism for the oxidation of [Fe₂(BIPhMe)₂(O₂CH)₄] in CHCl₃.

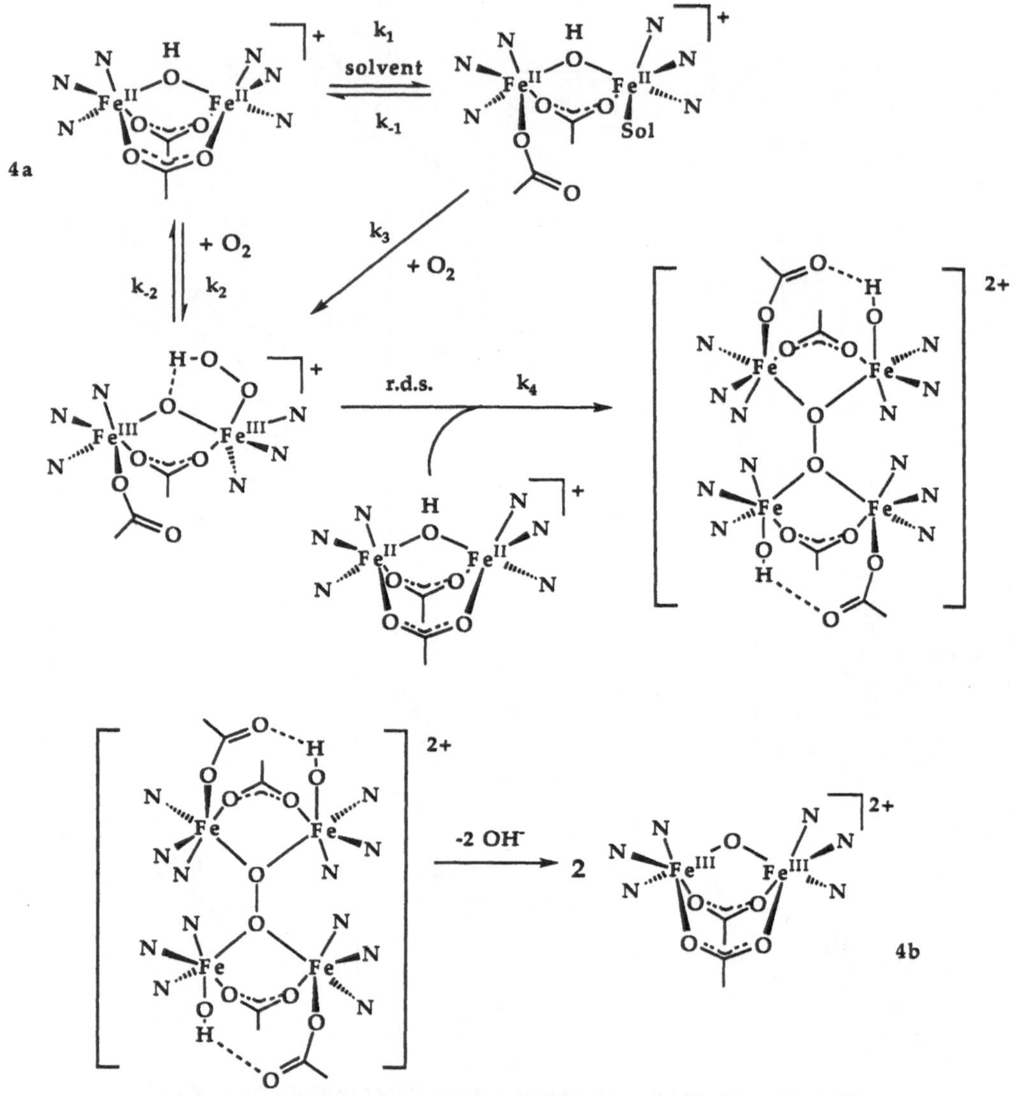

Scheme 2. Mechanism for the oxidation of [Fe$_2$(OAc)$_2$(OH)(Me$_3$TACN)$_2$]$^+$.

oxidation route occurs via reaction 2 in Scheme 2. The resulting rate law shows a first order dependence on dioxygen, as indicated in eq. 9. An intermediate in this reaction is observed spectroscopically under some conditions, but its physical properties have not yet been fully elucidated.

$$rate = \frac{2k_3k_4(k_{-1}+k_2[O_2])[A]^2[O_2]+2k_1k_2k_4[A]^2[O_2]}{(k_{-3}+k_4[A])(k_{-1}+k_2[O_2])}$$

(7)

in both solvents: $k_{-3} \gg k_4[A]$

in MeOH: $k_2[O_2] \gg k_{-1}$ and $k_1 \gg k_3[O_2]$

$$rate_{MeOH} \cong \frac{2k_4[A]^2(k_3[O_2]+k_1)}{k_{-3}} \approx \frac{2k_1k_4[A]^2}{k_{-3}}$$

(8)

in CHCl$_3$: $k_{-1} \gg k_2[O_2]$

$$rate_{CHCl_3} \cong \frac{2k_4[A]^2[O_2](k_{-1}k_3+k_1k_2)}{k_{-1}k_{-3}}$$

(9)

Although compounds **3a** and **4a** yield very similar products following oxidation, the kinetics of the two reactions highlight important differences. Compound **4a** is coordinatively saturated, but undergoes rapid carboxylate exchange at room temperature. The intermediate in the exchange process is the carboxylate-shifted form, which is probably stabilized by methanol coordination. Carboxylate shifts of this kind have been previously delineated,[41] and the present data require such a shift for meaningful interpretation. Monomer/dimer equilibria have also been observed in some diiron(II) compounds[49,51] and the consequences of such an equilibrium are manifest in the kinetics.

It is noteworthy that the kinetic data for these reactions can be interpreted without invoking high valent ferryl species. Neither **3a** nor **4a** exhibit any oxo-transfer activity when exposed to dioxygen in the presence of a large excess of triphenylphosphine. The formation of triphenyphosphine oxide from triphenylphosphine occurs rapidly in both non-heme dinuclear[55] and porphyrin[52] iron systems postulated to contain high-valent iron oxo units. It therefore appears that the reactions studied here do not involve such a species.

A characteristic feature of the present model systems is their dimerization to yield a tetrametallated peroxide intermediate or transition state. In the proteins the iron domains are sterically isolated from one another, thereby precluding such a dimerization reaction. Future efforts to model the chemical reactivity of these non-heme diiron protein cores must be directed toward the synthesis of complexes in which the metal centers are sufficiently hindered. In such systems, reversible binding of dioxygen, or oxidation of a substrate molecule by the diferric peroxide or a species derived therefrom, may be achieved.

ACKNOWLEDGMENTS

The work reviewed in this article was supported by a research grant from the National Institute of General Medical Sciences. We are very grateful to our many co-workers and collaborators, cited in the individual references, for their contributions.

REFERENCES

(1)   H. Dalton. "Oxidation of hydrocarbons by methane monooxygenases from a variety of microbes." In Adv. Appl. Microbiol., 26. Academic Press, 1980.

(2)   C. Anthony. "The Biochemistry of Methylotrophs." New York: Academic Press, 1982.

(3)   J. Colby and H. Dalton. Resolution of the methane monooxygenase of *Methylococcus capsulatus* (Bath) into three components. *Biochem. J.*, 171: 461 (1978).

(4)   B. G. Fox, W. A. Froland, J. E. Dege, and J. D. Lipscomb. Methane monooxygenase from *Methylosinus trichosporium* OB3b. *J. Biol. Chem.*, 264: 10023 (1989).

(5)   J. Lund and H. Dalton. Further characterization of the FAD and Fe_2S_2 redox centers of component C, the NADH: acceptor reductase of the soluble methane monooxygenase of *Methylococcus capsulatus* (Bath). *Eur. J. Biochem.*, 147: 291 (1985).

(6)   J. Lund, M. P. Woodland, and H. Dalton. Electron transfer reactions in the soluble methane monooxygenase of *Methylococcus capsulatus* (Bath). *Eur. J. Biochem.*, 147: 297 (1985).

(7)   J. Green and H. Dalton. A stopped-flow kinetic study of soluble methane monooxygenase from *Methylococcus capsulatus* (Bath). *Biochem. J.*, 259: 167 (1989).

(8)   K. E. Liu and S. J. Lippard. Redox properties of the hydroxylase component of methane monooxygenase from *Methylococcus capsulatus* (Bath). *J. Biol. Chem.*, 266: 12836 (1991).

(9)   W. A. Froland, K. K. Andersson, S.-K. Lee, Y. Liu, and J. D. Lipscomb. Methane monooxygenase component B and reductase alter the regioselectivity of the hydroxylase component-catalyzed reactions. *J. Biol. Chem.*, 267: 17588 (1992).

(10)   J. G. Dewitt, J. G. Bentsen, A. C. Rosenzweig, B. Hedman, J. Green, S. Pilkington, G. C. Papaefthymiou, H. Dalton, K. O. Hodgson, and S. J. Lippard. X-Ray absorption, Mössbauer and EPR studies of the dinuclear iron center in the hydroxylase component of MMO. *J. Am. Chem. Soc.*, 113: 9219 (1991).

(11)   A. Ericson, B. Hedman, K. O. Hodgson, J. Green, H. Dalton, J. G. Bentsen, R. H. Beer, and S. J. Lippard. Structural characterization by EXAFS spectroscopy of the binuclear iron center in protein A of methane monooxygenase from *Methylococcus capsulatus* (Bath). *J. Am. Chem. Soc.*, 110: 2330 (1988).

(12)   L. Que Jr. and A. E. True. Dinuclear iron- and manganese-oxo sites in biology. *Prog. Inorg. Chem.*, 38: 97 (1990).

(13)   M. P. Woodland, D. S. Patil, R. Cammack, and H. Dalton. ESR studies of protein A of the soluble methane monooxygenase from *Methylococcus capsulatus* (Bath). *Biochim. Biophys. Acta*, 873: 237 (1986).

(14)   V. J. DeRose, K. E. Liu, D. M. Kurtz Jr., B. M. Hoffman, and S. J. Lippard. Proton ENDOR identification of bridging hydroxide ligands in mixed-valent diiron centers of proteins, submitted for publication.

(15)   M. J. Maroney, D. M. Kurtz Jr., J. M. Nocek, L. L. Pearce, and L. Que Jr. ^{1}H NMR probes of the binuclear iron cluster in hemerythrin. *J. Am. Chem. Soc.*, 108: 6871 (1986).

(16)   L. L. Pearce, D. M. Kurtz Jr., Y.-M. Xia, and P. G. Debrunner. Reduction of the binuclear iron site in octameric methemerythrins. Characterization of intermediates and a unifying reaction scheme. *J. Am. Chem. Soc.*, 109: 7286 (1987).

(17)   J. M. McCormick, R. C. Reem, and E. I. Solomon. Chemical and spectroscopic studies of the mixed-valent derivatives of the non-heme protein hemerythrin. *J. Am. Chem. Soc.*, 113: 9066 (1991).

(18)   J. Green and H. Dalton. Steady-state kinetic analysis of soluble methane monooxygenase from *Methylococcus capsulatus* (Bath). *Biochem. J.*, 236: 155 (1986).

(19)   M. T. Stankovich, K. E. Paulsen, Y. Liu, and J. D. Lipscomb. Redox properties of methane monooxygenase hydroxylase and regulation by component B. *Abstracts of the 205th ACS Meeting*, INOR566, (1993).

(20)   B. G. Fox, Y. Liu, J. Dege, and J. D. Lipscomb. Complex formation between the protein components of methane monooxygenase from *Methylosinus trichosporium* OB3b. *J. Biol. Chem.*, 266: 540 (1991).

(21)   K. K. Andersson, W. A. Froland, S.-K. Lee, and J. D. Lipscomb. Dioxygen independent oxygenation of hydrocarbons by methane monooxygenase hydroxylase component. *New J. Chem.*, 15: 411 (1991).

(22) K. E. Liu, A. Masschelein, and S. J. Lippard, unpublished results.

(23) J. Green and H. Dalton. Substrate specificity of soluble methane monooxygenase. *J. Biol. Chem.*, 264: 17698 (1989).

(24) F. Ruzicka, D.-S. Huang, M. I. Donnelly, and P. A. Frey. Methane monooxygenase catalyzed oxygenation of 1,1-dimethylcyclopropane, evidence for radical and carbocationic intermediates. *Biochemistry*, 29: 1696 (1990).

(25) M. J. Rataj, J. E. Kauth, and M. I. Donnelly. Oxidation of deuterated compounds by high specific activity methane monooxygenase from *Methylosinus trichosporium*. *J. Biol. Chem.*, 266: 18684 (1991).

(26) P. R. Ortiz de Montellano. "Cytochrome P-450 structure, mechanism, and biochemistry." ed. P. R. Ortiz de Montellano. New York: Plenum Publishing Corp., 1986.

(27) P. R. Oritz de Montellano and R. A. Stearns. Timing of the radical recombination step in cytochrome P-450 catalysis with ring-strained probes. *J. Am. Chem. Soc.*, 109: 3415 (1987).

(28) V. W. Bowry, J. Lusztyk, and K. U. Ingold. Calibration of a new horologery of fast radical "clocks". Ring-opening rates for ring- and α-alkyl-substituted cyclopropylcarbinyl radicals and for the bicyclo[2.1.0]pent-2-yl radical. *J. Am. Chem. Soc.*, 113: 5687 (1991).

(29) V. W. Bowry and K. U. Ingold. A radical clock investigation of microsomal cytochrome P-450 hydroxylation of hydrocarbons. Rate of oxygen rebound. *J. Am. Chem. Soc.*, 113: 5699 (1991).

(30) D. Griller and K. U. Ingold. Free-radical clocks. *Acc. Chem. Res.*, 13: 317 (1980).

(31) K. E. Liu, C. C. Johnson, M. Newcomb, and S. J. Lippard. Radical clock substrate probes and kinetic isotope effect studies of the hydroxylation of hydrocarbons by methane monooxygenase. *J. Am. Chem. Soc.*, 115: 939 (1992).

(32) M. Newcomb, M. B. Manek, and A. G. Glenn. Ring-opening and hydrogen atom transfer trapping of the bicyclo[2.1.0]pent-2-yl radical. *J. Am. Chem. Soc.*, 113: 949 (1991).

(33) M. Newcomb, C. C. Johnson, M. B. Manek, and T. R. Varick. Picosecond radical kinetics. Ring openings of phenyl-substituted cyclopropylcarbinyl radicals. *J. Am. Chem. Soc.*, 114: 10915 (1992).

(34) P. C. Wilkins, H. Dalton, I. D. Podmore, N. Deighton, and M. C. R. Symons. Biological methane activation involves the intermediacy of carbon-centered radicals. *Eur. J. Biochem.*, 210: 67 (1992).

(35) B. G. Fox, J. G. Borneman, L. P. Wackett, and J. D. Lipscomb. Haloalkene oxidation by the soluble methane monooxygenase from *Methylosinus trichosporium* OB3b: mechanistic and environmental implications. *Biochemistry*, 29: 6419 (1990).

(36)   N. D. Priestley, H. G. Floss, W. A. Froland, J. D. Lipscomb, P. G. Williams, and H. Morimoto. Cryptic stereospecificity of methane monooxygenase. *J. Am. Chem. Soc.*, 114: 7561 (1992).

(37)   H.-C. Tsien and R. S. Hanson. Soluble methane monooxygenase component B gene probe for identification of methanotrophs that rapidly degrade trichloroethylene. *Appl. Environ. Microbiol.*, 58: 953 (1992).

(38)   W. Adam, R. Curci, and J. O. Edwards. Dioxiranes: a new class of powerful oxidants. *Acc. Chem. Res.*, 22: 205 (1989).

(39)   L. Marshall, K. Parris, J. Rebek Jr., S. V. Luis, and M. I. Burguete. A new class of chelating agents. *J. Am. Chem. Soc.*, 110: 5192 (1988).

(40)   J. Rebek Jr., L. Marshall, R. Wolak, K. Parris, M. Killoran, B. Askew, D. Nemeth, and N. Islam. Convergent functional groups: synthetic and structural studies. *J. Am. Chem. Soc.*, 107: 7476 (1985).

(41)   R. L. Rardin, W. B. Tolman, and S. J. Lippard. Monodentate carboxylate complexes and the carboxylate shift: implications for polymetalloprotein structure and function. *New J. Chem.*, 15: 417 (1991).

(42)   S. P. Watton, A. Masschelein, J. Rebek Jr., and S. J. Lippard. Synthesis and characterization of a bioinorganic chip for the carboxylate-bridged (μ-oxo)diiron(III) metalloprotein core, submitted for publication.

(43)   G. N. R. Tripathi and R. H. Schuler. The resonance Raman spectrum of phenoxyl radical. *J. Chem. Phys.*, 81: 113 (1984).

(44)   J. Sanders-Loehr, W. D. Wheeler, A. K. Shiemke, B. A. Averill, and T. M. Loehr. Electronic and Raman spectroscopic properties of oxo-bridged dinuclear iron centers in proteins and model compounds. *J. Am. Chem. Soc.*, 111: 8084 (1989).

(45)   A. Sauer-Masarwa, N. Herron, C. M. Fendrick, and D. H. Busch. Kinetics and intermediates in the autoxidation of synthetic, non-porphyrin iron(II) dioxygen carriers. *Inorg. Chem.*, 32: 1086 (1993).

(46)   A. Masschelein, A. L. Feig, and S. J. Lippard. Oxidation mechanisms of diiron(II) model compounds of non-heme iron enzymes by oxygen, submitted for publication.

(47)   W. B. Tolman, S. Liu, J. G. Bentsen, and S. J. Lippard. Models of the reduced forms of polyiron-oxo proteins: an asymmetric, triply carboxylate bridged diiron(II) complex and its reaction with dioxygen. *J. Am. Chem. Soc.*, 113: 152 (1991).

(48)   J. Hartman, R. L. Rardin, P. Chaudhuri, K. Pohl, K. Wieghardt, B. Nuber, J. Weiss, G. Papaefthymiou, R. Frankel, and S. Lippard. Synthesis and characterization of (μ-hydroxo)bis(μ-acetato)diiron(II) and (μ-oxo)bis(μ-acetato)diiron(III) 1,4,7-trimethyl-1,4,7-triazacyclononane complexes as models for binuclear iron centers in biology; properties of the mixed valence diiron(II,III) species. *J. Am. Chem. Soc.*, 109: 7387 (1987).

(49)   K. S. Hagen and R. Lachicotte. Diiron(II) μ–aquabis(μ-carboxylato) models of reduced dinuclear iron sites in proteins. *J. Am. Chem. Soc.*, 114: 8741 (1992).

(50)   N. Kitajima, N. Tamura, M. Tanaka, and Y. Moro-oka. Synthesis and molecular structures of diferrous complexes containing a bis(hydroxo) or a hydroxo carboxylato bridge. *Inorg. Chem.*, 31: 3342 (1992).

(51)   S. Menage, Y. Zang, M. P. Hendrich, and L. Que Jr. Structure and reactivity of a bis(μ-O,O'-acetato)diiron(II) complex, [Fe$_2$(O$_2$CCH$_3$)$_2$(TPA)$_2$](BPh$_4$)$_2$. A model for the diferrous core of ribonucleotide reductase. *J. Am. Chem. Soc.*, 114: 7786 (1992).

(52)   A. L. Balch. The reactivity of spectroscopically detected peroxy complexes of iron porphyrins. *Inorg. Chim. Acta*, 198-200: 297 (1992).

(53)   I. R. Paeng, H. Shiwaku, and K. Nakamoto. Detection of the Fe-O-O-Fe intermediate in the oxidation reaction of ferrous porphyrins by resonance Raman spectroscopy. *J. Am. Chem. Soc.*, 110: 1995 (1988).

(54)   W. Micklitz, S. G. Bott, J. G. Bentsen, and S. J. Lippard. Characterization of a novel μ_4-peroxide tetrairon unit of possible relevance to intermediates in metal-catalyzed oxidations of water to dioxygen. *J. Am. Chem. Soc.*, 111: 372 (1989).

(55)   R. Leising, B. Brennan, and L. Que Jr. Models for non-heme iron oxygenases: a high-valent iron-oxo intermediate. *J. Am. Chem. Soc.*, 113: 3988 (1991).

ALKANE FUNCTIONALIZATION AT NONHEME IRON CENTERS:

MECHANISTIC INSIGHTS

Randolph A. Leising, Takahiko Kojima, and Lawrence Que, Jr.*

Department of Chemistry, University of Minnesota
Minneapolis, Minnesota 55455

INTRODUCTION

Biological catalysts capable of functionalizing alkanes include cytochrome P450[1] (which contains a heme active site) and methane monooxygenase[2] (MMO, which has a nonheme diiron active site). For cytochrome P450, an oxoiron(IV)(porphyrin cation radical) complex derived from the heterolysis of the O-O bond in an intermediate ferric peroxide is strongly implicated as the active species in the cytochrome P-450 mechanism.[3] For MMO, an analogous mechanism (Figure 1) is proposed by substituting the heme center with a diiron active site.[2] As has been found for cytochrome P450, Lipscomb et al. recently demonstrated that diferric MMO can hydroxylate alkanes with H_2O_2 via a peroxide shunt pathway, thus implicating a diferric peroxide complex in its mechanism of alkane hydroxylation.[4] The peroxide intermediate is then proposed to decompose to a high valent iron-oxo species which is capable of abstracting hydrogen from alkanes. In our efforts to model the chemistry of MMO and other nonheme iron enzymes, we have explored the capabilities of a series of nonheme iron complexes to catalyze alkane functionalization chemistry and provide some mechanistic insight into this intriguing reaction.[5] Modeling efforts towards this goal have entailed the combination of an iron complex with peroxide or O_2/reductant.[6-8] Most prominent of these are the systems developed by Barton et al.[7] using iron salts collectively known as the "Gif" systems; an alkyl hydroperoxide has recently been identified to be the precursor of the alcohol and ketone products observed. This hydroperoxide is proposed to derive from O_2 trapping of an alkyliron(V) species produced by the interaction of the alkane

The Activation of Dioxygen and Homogeneous Catalytic Oxidation,
Edited by D.H.R. Barton *et al.*, Plenum Press, New York, 1993

with a high valent iron-oxo species. We have been interested in the nature of the oxidizing species that is involved in alkane functionalization reactions by nonheme iron centers. In order to define more clearly the coordination environment of the metal center in such reactions, we have used iron complexes of tetradentate tripodal ligands as catalysts for the *tert*-butyl hydroperoxide-dependent oxidation of cyclohexane. The tripodal ligands selected for this study offer the important advantage of systematic control over the ligand environment at a labile iron center.

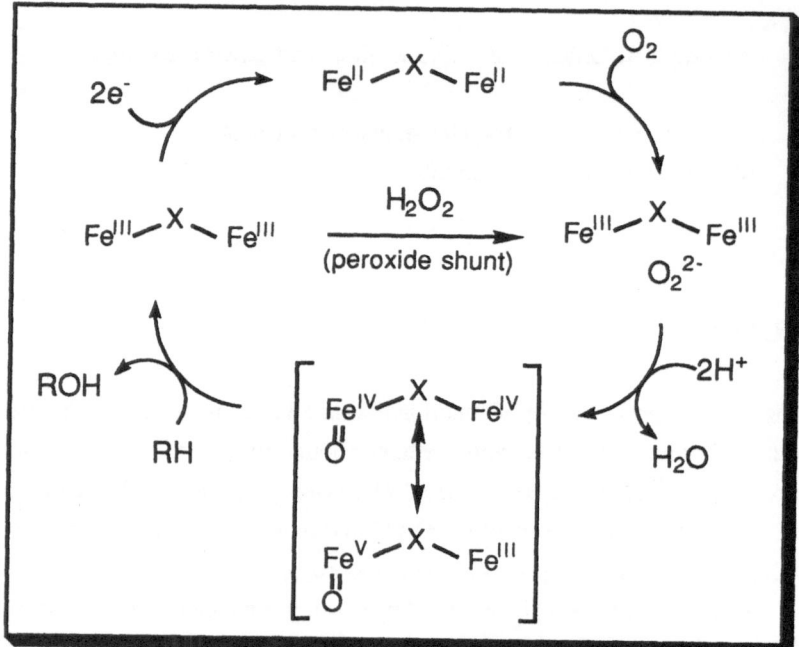

Figure 1. Proposed Mechanism for Methane Monooxygenase

CATALYTIC FUNCTIONALIZATION OF CYCLOHEXANE

We have found that both mononuclear and oxo-bridged dinuclear TPA (TPA = tris(2-pyridylmethyl)amine) complexes are quite effective in catalyzing cyclohexane oxidation[5,9] relative to other complexes reported in the literature (Table 1).

Table 1. Product Distributions for the Iron-Catalyzed TBHP Oxidation of Cyclohexane in Acetonitrile Under 1 atm of Argon at 25 °C.

Catalyst	Rxn Time (h)	Products[a] OH	O	OO^{t}Bu	(A+K)/P	TN/h[b]
effect of nuclearity						
[FeTPACl$_2$]ClO$_4$[c]	2	15	12	8	3.4	28
[Fe$_2$O(TPA)$_2$OAc](ClO$_4$)$_3$	**0.25**	**9**	**11**	**16**	**1.3**	**252**
effect of the terminal ligand						
[Fe$_2$BPA$_2$O(OBz)$_2$](ClO$_4$)$_2$	16	15	13	10	2.8	3.8
[Fe$_2$NTB$_2$O(OAc)](ClO$_4$)$_3$	16	8	7	0.7	20	1.5
[Fe$_2${HB(pz)$_3$}$_2$O(OAc)$_2$]	16	0	0	0	—	0
effect of the bridging anion						
[Fe$_2$TPA$_2$O(O$_2$CC$_6$H$_4$-4-OCH$_3$)](ClO$_4$)$_3$	1	11	13	18	1.3	73
[Fe$_2$TPA$_2$O(CO$_3$)](ClO$_4$)$_2$	4	8	8	6	2.7	9
[Fe$_2$TPA$_2$O(phthalate)](ClO$_4$)$_2$	20	15	14	5	5.8	2.6

[a] moles of product per moles of catalyst; A = cyclohexanol, K = cyclohexanone, P = ROOR'.
[b] turnover expressed in terms of moles TBHP required to produce observed products
 (1 for alcohol, 2 for ketone, and 2 for ROOR')
[c] one mole of chlorocyclohexane also produced per mole catalyst

Representative complexes for our studies are [Fe(TPA)Cl$_2$]ClO$_4$ (**1**) and [Fe$_2$O(OAc)(TPA)$_2$](ClO$_4$)$_3$ (**2**), whose structures are shown in Figure 2. **1** is a mononuclear six-coordinate complex,[10] while **2** is a (μ-oxo)(μ-acetato)diferric complex with distinct iron sites because of the different orientations of the TPA ligands relative to the bridging groups.[11] A typical reaction consists of a 0.77 M

TPA

solution of cyclohexane and 0.10 M t-butyl hydroperoxide (TBHP) in acetonitrile reacted in the presence of 0.70 mM catalyst at 25 °C under argon. For **2**, TBHP was consumed in 15 minutes. Cyclohexanol, cyclohexanone, and t-butylperoxycyclohexane were formed (Table 1), and complete mass balance of substrate was established. The dinuclear catalyst showed remarkable stability; after 15 minutes, its characteristic visible and NMR spectra remained

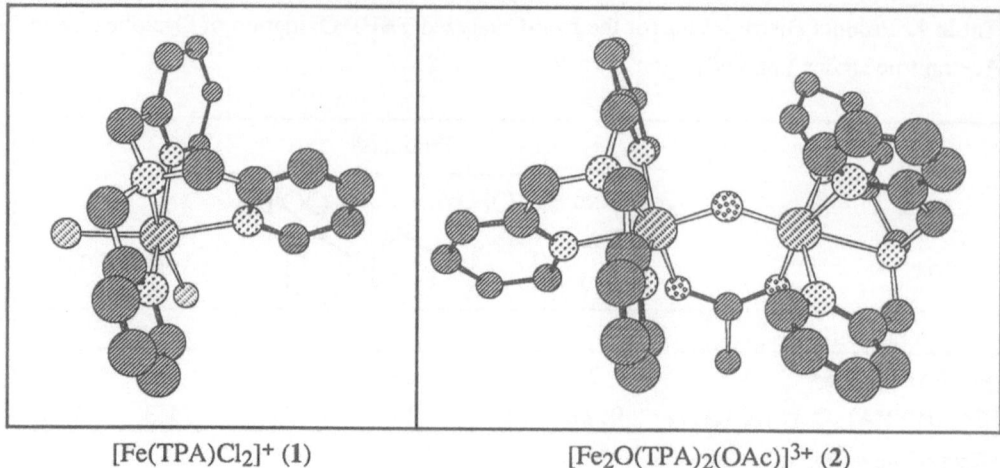

$[Fe(TPA)Cl_2]^+$ (**1**) $[Fe_2O(TPA)_2(OAc)]^{3+}$ (**2**)

Figure 2. Chem3D® representations of the x-ray structures of complexes **1** and **2**.

unchanged. Furthermore the addition of another aliquot of TBHP resulted in its consumption within the same time period with the same conversion efficiency into products.

For **1**, the same products were observed (Table 1). But TBHP was completely consumed after 2 hours, and the catalyst became inactive. NMR monitoring of the reaction indicated that the mononuclear catalyst was converted to $[Fe_2O(TPA)_2Cl_2]^{2+}$, which could be independently synthesized and found to be incapable of catalyzing these reactions. In addition, an equivalent of chlorocyclohexane was observed as product.

Variation of the tripodal ligand and coordinated anion resulted in differences in the length of time required for complete consumption of TBHP (Table 1). These results suggested that the metal center played a central role in activating the peroxide for alkane functionalization. The data presented in Table 2 show that the TPA complexes were the most efficient in the activation of TBHP and that the use of more electron donating ligands served to slow down the reactions. These observations led us to focus on the TPA catalysts for mechanistic studies.

BPA   **NTB**   **HB(pz)$_3$**

The nature of the C-H bond cleavage step was probed for these reactions. Significant kinetic isotope effects (6-12) were observed in a competitive oxidation of cyclohexane and cyclohexane-d_{12}. When the oxidation of cyclohexane was conducted in 90% CH_3CN/10% CH_2Br_2, significant amounts of bromocyclohexane were formed, demonstrating the involvement of radicals in this reaction.[12] These radicals appeared to be trapped immediately since no dicyclohexyl was observed.[13] The alkyl radical may be generated by the reaction of alkane with species derived from TBHP heterolysis, i. e. an iron(V)-oxo moiety, or TBHP homolysis, i. e. an alkoxy radical.

The participation of the two mechanisms is indicated by the effect of added dimethyl sulfide. The presence of dimethyl sulfide suppressed formation of cyclohexanol, cyclohexanone and halocyclohexane but did not affect the production of t-butylperoxy-cyclohexane. In place of the former products, dimethyl sulfoxide was found. Since sulfides can act as a trap for a two-electron oxidant such as an iron-peroxo or an iron-oxo species,[14] the formation of alcohol, ketone and haloalkane must involve heterolysis of the alkyl hydroperoxide, i.e.

$$Fe^{III}TPA + ROOH \rightarrow [(TPA)Fe^V=O] + ROH$$
$$[(TPA)Fe^V=O] + R'-H \rightarrow Fe^{III}TPA + R'=OH$$
$$[(TPA)Fe^V=O] + R'-OH \rightarrow Fe^{III}TPA + R'=O$$
$$[(TPA)Fe^V=O] + (CH_3)_2S \rightarrow Fe^{III}TPA + (CH_3)_2SO$$

On the other hand, the persistence of the dialkyl peroxide indicates that it must not derive from a mechanism involving such oxidants. Indeed such dialkyl peroxides are readily formed by metal-catalyzed decomposition of alkyl hydroperoxides and involve alkoxy and alkylperoxy radicals.[15] The mechanism for dialkyl peroxide formation shown below is adapted for FeTPA from previously proposed schemes:

initiation
$$Fe^{III}TPA + ROOH \rightarrow Fe^{II}TPA + ROO \cdot$$
propagation
$$Fe^{II}TPA + ROOH \rightarrow Fe^{III}TPA + RO \cdot + OH^-$$
$$RO \cdot + R'H \rightarrow ROH + R' \cdot$$
termination
$$R' \cdot + ROO \cdot \rightarrow R'OOR$$
or
$$R' \cdot + ROOH \rightarrow [R' \cdot (ROOH)]$$
$$[R' \cdot (ROOH)] + Fe^{III}TPA \rightarrow R'OOR + Fe^{II}TPA + H^+$$

Support for this hypothesis comes from the fact that the product distributions observed depended on the nature of the catalyst (Table 1). Indeed the proportion of dialkyl peroxide formed diminished as the ligands on the iron center became more electron donating. The increased electron density at the metal center should shift the redox potential of the metal center to more negative values and disfavor the reduction of the metal catalyst in either the initiation or termination steps of the dialkyl peroxide formation mechanism.

STOICHIOMETRIC ALKANE HALOGENATION

The formation of haloalkane in the reactions catalyzed by the mononuclear [Fe(TPA)X$_2$]$^+$ complexes provides important mechanistic insights. Only one equivalent of haloalkane is produced per molecule of catalyst.[5] This stoichiometry is maintained even in the presence of excess halide (10-750 equiv), indicating that free halide is not incorporated into product. When [Fe(BPA)Br$_3$] (BPA = bis(2-pyridylmethyl)amine) is used as catalyst, two equivalents of bromocyclohexane were formed.[5b] Taken together, these observations implicate the metal center directly in the formation of the haloalkane products.

Table 2. Comparison of the Selectivities of Various Mononuclear Catalysts for the Halogenation of Alkanes.

Complex	k_H/k_D for cyclohexane	3°/2° (TBHP)[a] adamantane	3°/2° (CHP)[a] adamantane
[Fe(TPA)Br$_2$]$^+$	7	5.4	6.1
[Fe(TPA)Cl$_2$]$^+$	8	9.7	9.1
[Fe(TPA)(N$_3$)$_2$]$^+$	9	10	11
[Fe(N-Et-NTB)Br$_2$]$^+$	10	7.6	—
[Fe(N-Et-NTB)Cl$_2$]$^+$	—	12.8	—

[a]TBHP = *tert*-butyl hydroperoxide; CHP = cumyl hydroperoxide.

When the [Fe(TPA)X$_2$]$^+$ catalysts are reacted with stoichiometric amounts of TBHP in the presence of alkanes, only the alkyl halide is formed and in good yield. 70-80 % of the TBHP oxidizing equivalents is converted into alkyl halide product. The absence of additional products greatly facilitates mechanistic interpretation of the results. For example, the band at 500 nm in [Fe(TPA)Br$_2$]$^+$ attributed to bromide-to-iron(III) charge transfer decays upon addition of TBHP concomitant with the appearance of the alkyl halide product. NMR studies of the reaction mixture show that the starting complex is converted to a (μ-oxo)diferric species during the course of this reaction. Addition of dimethyl sulfide inhibits the formation of product, while a radical scavenger such as 4-methyl-2,6-di-*tert*-butylphenol does not.

Table 2 summarizes isotope effect data on cyclohexane halogenation and selectivity data on adamantane halogenation as a function of mononuclear catalyst. The data demonstrate that the nature of oxidant depends on the tripodal ligand and the bound halide, but not on the alkyl hydroperoxide. These results lead us to conclude that the oxidant in the halogenating reaction is an iron(V)-oxo species with bound halide (Figure 3). The alkyl hydroperoxide displaces one of the bound halides in the initial interaction and undergoes O-O bond heterolysis to form the iron-oxo intermediate. The iron-oxo moiety then abstracts a hydrogen from substrate generating an iron(IV) species and an alkyl radical in a solvent

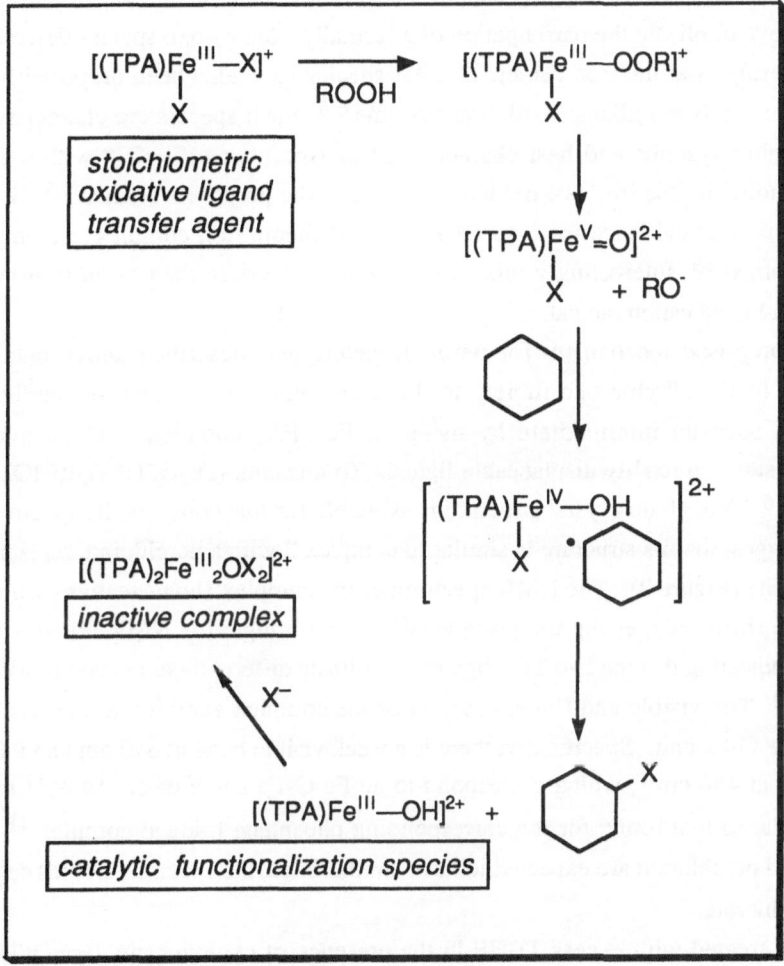

Figure 3. Proposed Mechanism for the Oxidative Ligand Transfer Reaction
Catalyzed by [Fe(TPA)X$_2$]$^+$ Complexes

cage, by analogy to the mechanism ascribed to models for cytochrome P450. At this point, the radical can either combine with the bound hydroxide to form alcohol (oxygen rebound) or with the bound halide to form alkyl halide (ligand transfer). The absence of any alcohol in the stoichiometric oxidations indicate that halide transfer is favored over oxygen rebound, which is consistent with the lower oxidation potentials of halide vs hydroxide. The transfer of halide leaves behind $[Fe(TPA)OH]^{2+}$, which probably converts to a (μ-oxo)diferric species that is responsible for the alcohol, ketone, and dialkyl peroxide products observed under catalytic conditions. In the absence of added ROOH, the catalyst picks up halide to form the catalytically inactive $[Fe_2O(TPA)_2X_2]^{2+}$ complex.

A HIGH VALENT NONHEME IRON-OXO INTERMEDIATE

The dimethyl sulfide trapping experiments in the alkane functionalization reactions described above implicate the participation of a formally iron(V)-oxo species derived from peroxide heterolysis at the iron center. Similar species have also been proposed in other nonheme iron-catalyzed alkane oxidation systems.[6-8] Such species are characterized in heme-containing systems and best characterized as $[(porphyrin)Fe=O]^+$ with oxidizing equivalents stored on the iron (+4 oxidation state) and the porphyrin (radical).[3] However direct spectroscopic evidence for a corresponding nonheme iron complex has only been recently obtained.[16] Interestingly this species is also best described as an iron(IV)-oxo species with a ligand cation radical.

The proposed mechanism for peroxide heterolysis described above entails the coordination of the alkylperoxo moiety to the iron center. We sought to facilitate the formation of such an intermediate by using an Fe(TPA) complex with an available coordination site or a readily displaceable ligand. To this end, $[Fe_2O(TPA)_2](ClO_4)_4$ was synthesized.[16] Though no crystal structure is available for this complex, its spectroscopic properties suggest that its structure is similar to complex 2 with a perchlorate replacing the bridging acetate (Figure 2). The NMR spectrum of the complex shows features with small paramagnetic shifts, suggesting the presence of an oxo bridge; the NMR features have a multiplicity indicating that the two TPA ligands coordinate differently to the two iron centers, just as in 2.[11] The visible and Raman spectra of the complex exhibit features associated with a bent Fe-O-Fe unit. Specifically, there is a weak visible band at 640 nm and a Raman ν_s (Fe-O-Fe) at 456 cm^{-1}, which correspond to an Fe-O-Fe angle of ca. 140°;[17] such an angle is similar to that found for the corresponding phosphate bridged complex.[11] Since phosphate and perchlorate are expected to have similar bites, the spectral features suggest a bridging perchlorate.

When treated with excess TBHP in the presence of cyclohexane, $[Fe_2O(TPA)_2]$-$(ClO_4)_4$ indeed consumed TBHP faster than 2 and gives rise to cyclohexanol, cyclohexanone, and t-butyldioxycyclohexane. Treatment with H_2O_2 instead resulted in the appearance of a flash of brilliant green color. When cooled to -40 °C, this transient intermediate could be stabilized for two hours, which allowed its spectroscopic characterization.[16]   Treatment of this intermediate with Ph₃P causes the immediate decomposition of the intermediate, forming Ph₃PO and regenerating the starting complex.

The intermediate is presently best formulated as $[(TPA)Fe=O]^{3+}$. It exhibits an intense visible absorption band at 614 nm (ε_M 3500) associated with its brilliant green color. Excitation into this band elicits a resonance enhanced Raman vibration at 676 cm⁻¹ which shifts to 648 cm⁻¹ when the intermediate is treated with $H_2^{18}O$. The pronounced effect of $H_2^{18}O$ excludes the assignment of this feature to a ν(O-O) of a bound peroxide, since a

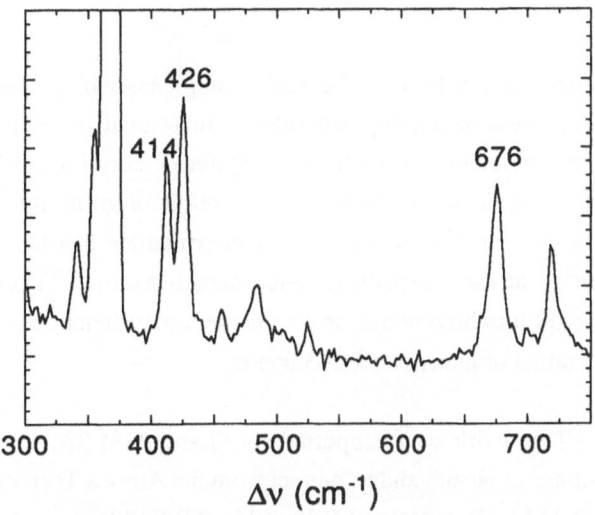

Figure 4. Resonance Raman spectrum of $[(TPA)Fe=O]^{3+}$ in CD₃CN

peroxide moiety would not be expected to exchange with solvent. Indeed the observed isotopic shift is almost precisely what is expected for an Fe-O vibration ($\Delta\nu_{calcd.}$= -29 cm⁻¹). The value for the iron-oxo stretch is low when compared to those observed for synthetic iron-oxo porphyrins (800-850 cm⁻¹);[18] however it flanks the values observed for the Compounds I of Fe horseradish peroxidase (737 cm⁻¹)[19] and Mn horseradish peroxidase (622 cm⁻¹).[20] The metal-oxo vibration in porphyrin complexes has been shown to be sensitive to the electronic properties of the other ligands, particularly the basicity of the axial ligand.[21] A cogent explanation for the Raman properties of the nonheme derivative will require a better understanding of its structural and electronic properties.

$[(TPA)Fe=O]^{3+}$ exhibits an EPR spectrum with g values at 4.5, 3.95, and 2.0, corresponding to an $S = 3/2$ center. The presence of three unpaired electrons is consistent with an Fe(V) oxidation state, but Mössbauer studies unequivocally identify the iron center to be in +4 oxidation state, having an isomer shift ($\delta = 0.11$ mm/s) similar to those found for Compounds I and II of horseradish peroxidase and related model compounds.[22] In order to reconcile the EPR and Mössbauer data, the remaining oxidizing equivalent must be assigned to the TPA ligand as a cation radical,[16] a description analogous to that applied to corresponding porphyrin species.[3] As in the heme species, the $S = 3/2$ spin state is proposed to result from the ferromagnetic coupling of an $S = 1$ Fe(IV) state with an $S = 1/2$ radical state on the ligand. The precise description for this ligand cation radical is presently being pursued using NMR spectroscopy.

SUMMARY

We have explored the reactivity of Fe(TPA) complexes with peroxides and found that such centers are capable of activating peroxides to functionalize alkanes in a catalytic fashion. Because of the efficient stoichiometric transfer of coordinated halide onto the alkane substrate, we have been able to deduce the participation of an $[(TPA)Fe(X)=O]^{2+}$ species in these reactions. Finally the high valent intermediate can be stabilized under appropriate conditions to allow its spectroscopic characterization. These experiments provide significant insight into how nonheme iron centers may function in enzyme active sites for the functionalization of unactivated C-H bonds.

Acknowledgments. This work was supported by Grants GM-33162 and GM-38767 from the National Institutes of Health and a contract from the Amoco Technology Company. R.A.L. acknowledges a N.I.H. Postdoctoral Fellowship (GM-13343).

REFERENCES

1. Ortiz de Montellano, P. R., Ed. *Cytochrome P-450: Structure, Mechanism, and Biochemistry*; Plenum: New York, 1985.
2. (a) Fox, B. G.; Froland, W. A.; Dege, J. E.; Lipscomb, J. D. *J. Biol. Chem.* **1989**, *264*, 10023.
   (b) Green, J.; Dalton, H. *J. Biol. Chem.* **1989**, *264*, 17698.
3. Groves, J. T. *J. Chem. Ed.* **1985**, *62*, 928.
4. Andersson, K. K., Froland, W. A., Lee, S.-K., Lipscomb, J. D. *New J. Chem.* **1991**, *15*, 411.
5. (a) Leising, R. A., Norman, R. E., Que, L., Jr. *Inorg. Chem.* **1990**, *29*, 2553.
   (b) Leising, R. A., Zang, Y., Que, L., Jr. *J. Am. Chem. Soc.* **1991**, *113*, 8555.
6. (a) Groves, J. T.; Van Der Puy, M. *J. Am. Chem. Soc.* **1976**, *98*, 5290.
   (b) Sugimoto, H.; Sawyer, D. T. *J. Org. Chem.* **1985**, *50*, 1784.

7. Barton, D. H. R.; Doller, D. *Acc. Chem. Res.* **1992**, *25*, 504.

8. Fish, R. H.; Konings, M. S.; Oberhausen, K. J.; Fong, R. H.; Yu, W. M.; Christou, G.; Vincent, J. B.; Coggin, D. K.; Buchanan, R. M. *Inorg. Chem.* **1991**, *30*, 3002.

9. Que, L., Jr. in *Bioinorganic Catalysis*; Reedijk, J., Ed.; Marcel Dekker: New York, 1993; p. 347.

10. Yan, S.; Que, L., Jr., results to be published.

11. Norman, R. E.; Yan, S.; Que, L., Jr.; Backes, G.; Ling, J.; Sander-Loehr, J.; Zhang, J. H.; O'Connor, C. J. *J. Am. Chem. Soc.* **1990**, *112,* 1554.

12. Groves, J. T.; Kruper, W. J., Jr.; Haushalter, R. C. *J. Am. Chem. Soc.* **1980**, *102*, 6377.

13. Farmer, E. H.; Moore, C. G. *J. Chem. Soc.,* **1951**, 131.

14. Labeque, R.; Marnett, L. J. *J. Am. Chem. Soc.* **1989**, *111*, 6621.

15. (a) Srinivasan, K.; Perrier, S.; Kochi, J. K. *J. Mol. Catal.,* **1986**, *36*, 297.
    (b) Kharasch, M. S.; Fono, A. *J. Org. Chem.* **1959**, *24,* 72.

16. Leising, R. A.; Brennan, B. A.; Que, L., Jr.; Fox, B. G.; Münck, E. *J. Am. Chem. Soc.* **1991**, *113*, 3988.

17. Norman, R. E.; Holz, R. C.; Ménage, S.; Que, L., Jr.; Zhang, J. H.; O'Connor, C. J. *Inorg. Chem.* **1990**, *29,* 4629.

18. (a) Paeng, I. R.; Shiwaku, H.; Nakamoto, K. *J. Am. Chem. Soc.* **1988**, *110*, 1995.
    (b) Kincaid, J. R.; Schneider, A. J.; Paeng, K.-J. *J. Am. Chem. Soc.* **1989**, *111,* 735.

19. Paeng, K.-J.; Kincaid, J. R. *J. Am. Chem. Soc.* **1988**, *110,* 7913.

20. Nick, R. J.; Ray, G. B.; Fish, K. M.; Spiro, T. G.; Groves, J. T. *J. Am. Chem. Soc.* **1991**, *113,* 1838.

21. (a) Su, Y. O.; Czernuszewicz, R.; Miller, L. A.; Spiro, T. G. *J. Am. Chem. Soc.* **1988**, *110,* 4150.
    (b) Czernuszewicz, R.; Su, Y. O.; Stern, M. K.; Macor, K. A.; Kim, D.; Groves, J. T.; Spiro, T. G. *J. Am. Chem. Soc.* **1988**, *110,* 4158.

22. Schulz, C. E.; Rutter, R.; Sage, J. T.; Debrunner, P. G.; Hager, L. P. *Biochemistry* **1984**, *23,* 4743.

CHARACTERIZATION OF PAGE BANDS FROM 3'-LABELED SHORT DNA FRAGMENTS RESULTING FROM OXIDATIVE CLEAVAGE BY "Mn-TMPyP/KHSO₅". DRASTIC MODIFICATIONS OF BAND MIGRATIONS BY 5'-END SUGAR RESIDUES

Marguerite Pitié, Geneviève Pratviel, Jean Bernadou, and
Bernard Meunier

Laboratoire de Chimie de Coordination du CNRS
205 route de Narbonne
31077 Toulouse cedex, France

SUMMARY

Recent experiments with the chemical nuclease Mn-TMPyP/KHSO$_5$ showed highly specific cleavage on both 3'-sides of A.T base pair triplets of double stranded oligonucleotides.[1] Hydroxylation at the 5'-carbon of a deoxyribose represents the initial damage and leaves a 3'-phosphate and a 5'-aldehyde at the ends. Careful examination of PAGE bands resulting from oxidative cleavage of two selected double-stranded oligodeoxyribonucleotides which have been labeled either at 5' or 3' ends indicated that 5'-end-labeled fragments with a 3'-phosphate terminus were unsensitive to different chemical or thermal treatments and migrated according to the corresponding fragments obtained by Maxam-Gilbert sequencing. Whereas 3'-end-labeled fragments with a 5'-aldehyde terminus (-CHO) migrated differently after reduction (-CH$_2$OH), oxidation (-COOH), heating (5'-P terminus and release of free base and furfural), heating and alkaline phosphatase treatment (5'-OH terminus and loss of free base, furfural and inorganic phosphate). The precise knowledge of the chemical lesion at the oxidized site of DNA allowed us to localize the exact position of the break and attribute the structures of the different fragments observed on PAGE analysis after the manganese porphyrin-mediated cleavage of two ODNs containing A.T rich sequences.

INTRODUCTION

Over the past several years, an increasing interest has been devoted to porphyrins and metalloporphyrins, natural or synthetic, able to cause chemical

The Activation of Dioxygen and Homogeneous Catalytic Oxidation,
Edited by D.H.R. Barton *et al.*, Plenum Press, New York, 1993

or light-mediated DNA and cellular damages. Following chemical activation (ascorbate, iodosylbenzene, monopersulfate, ...), some metalloporphyrins, especially those containing manganese, promote efficient DNA cleavage. Mn-TMPyP, *meso*-tetrakis(4-N-methylpyridiuniumyl)porphyrinatomanganese(III) pentaacetate, associated to potassium monopersulfate, $KHSO_5$, degrades double- or single-stranded DNA and RNA.[2-6] Previous analyses on the binding mode to DNA and degradation of DNA showed its preference for AT sequences.[3, 7-10] Recently we reported the high specific cleavage on both 3'-sides of A.T base-pair (bp) triplets (for short, three contiguous A.T bp will be named as A.T triplet) of double-stranded oligonucleotides ODNs;[1] we have shown that immediate DNA strand scission occurred through oxidative C-H bond activation of DNA sugar at C-5' leading to 3'-phosphate and 5'-aldehyde termini at the sites of breaks. Therefore, interpretations of polyacrylamide gel electrophoresis (PAGE) data are very easy in the case of 5'-labeling since the 3'-terminus of fragments is a phosphate: they migrate exactly as the corresponding fragments obtained by the Maxam-Gilbert methods.[11] However this is not the case when studying the migration of short 3'-labeled oligonucleotides resulting from the cleavage of double-stranded 12- and 22-mers. In the absence of a thermal step, which eliminates the oxidatively damaged sugar residue at the 5'-end, band migration is highly dependent on the 5'-modified sugar end. Then differences of several bases can be found with the Maxam-Gilbert scale and misinterpretation of PAGE bands is possible in the absence of a precise knowledge of the chemistry going on at the 5'-end of these short oligonucleotides. The relationship between the sugar damages and the oxidative cleavage conditions allowed us to discuss the difference between metalloporphyrin-mediated DNA cleavage involving non-diffusable metal-oxo species or diffusible oxygenated radicals (metal-oxo *versus* HO°).

EXPERIMENTAL PROCEDURES

Materials

Potassium monopersulfate (in fact a triple salt $2KHSO_5.KHSO_4.K_2SO_4$, Curox®) was a gift of Interox. Mn-TMPyP was synthesized as previously described.[2] Alkaline phosphatase (EC 3.1.3.1, from *Escherichia coli*, type III) was from BRL. T4 polynucleotide kinase (EC 2.7.1.78), terminal deoxynucleotidyl transferase (EC 2.7.7.31), [γ-^{32}P]ATP and [α-^{32}P]ddATP were from Boehringer. Maxam-Gilbert kit was from NEN (Dupont). Other chemicals were from Sigma or Aldrich.

Oligonucleotides

ODNs I, II, III (sequences are indicated in Figure 1) and reference fragments (ACGA and GACGA) were synthesized on a Cyclone Plus DNA synthesizer from Milligen/Biosearch and purified by PAGE. Concentrations of ODNs and their fragments were determined according to their absorption coefficient.[12]

Gel electrophoresis analyses of cleavage pattern of the two ^{32}P-end-labeled 12- and 22-mer ODNs (general protocole)

The three unlabeled ODNs I, II and III (see Figure 1 for sequences) were separately treated with T4 polynucleotide kinase and [γ-^{32}P]ATP or with terminal deoxynucleotidyl transferase and [α-^{32}P]ddATP. Before reaction, 5'- or 3'-labeled ODNs were annealed with their complementary strand (I and II, III and III) by heating at 90°C for 1 min and subsequent slow cooling at room temperature. All following reactions were done on ice-bath.

KHSO$_5$ as oxygen atom donor. The reaction mixtures (total volume = 11 μL) contained 1 μM 12-mer or 22-mer duplex, 5 x 10^3 cpm of ^{32}P end labeled ODN and 0.5 μM Mn-TMPyP in 100 mM NaCl and 40 mM Tris/HCl pH 8 buffer at + 4°C. Following 15 min pre-incubation, cleavage of duplexes was initiated by the addition of a freshly prepared solution of 40 μM KHSO$_5$ (final concentration). After a 15 min-incubation for cleavage at 4°C, 2 μL of 0.5 M Hepes buffer (pH 8) were added as stopping reagent. Some experiments were heated 30 min at 90°C or incubated 30 min at ambient temperature with NaBH$_4$ or with NaOI. The samples (11 μL) were then diluted with 1 μL of 1 mg/mL salmon testes DNA and 250 μL of 0.3 M sodium acetate pH 5.2 buffer containing 0.1 mg/mL yeast tRNA, precipitated with ethanol, washed with ethanol/water (3/1, v/v), dried as pellets, redissolved in loading buffer and run on 15 % (22-mer) or 20 % (12-mer) denaturing polyacrylamide gels (3 hours at 1,700 V).

H$_2$O$_2$ as oxygen atom donor. Alternatively, 5 mM H$_2$O$_2$ instead of 40 μM KHSO$_5$ was used. In these experiments the Mn-TMPyP concentration was 5 μM, the incubation time for cleavage 1 hour at + 4°C, the reaction was quenched with 1 μL catalase (200 units), the precipitation was obtained with 5 μL of 1 mg/mL DNA. Before PAGE analysis, some samples were redissolved in 100 μL of 1 M piperidine and heated at 90 °C for 30 min.

Heating in alkaline conditions. In some experiments, analysis of strand I was performed after incubating the 22-bp 5'-^{32}P-end-labeled duplex I/II (1 μM) either (i) with 0.5 μM Mn-TMPyP in 100 mM NaCl and 40 mM Tris/HCl (pH 8) buffer in the presence or not of 40 μM KHSO$_5$ (lanes 8 and 6, respectively, Figure 2) or (ii) with 2 μM Mn-TMPyP, without KHSO$_5$, in 5 mM phosphate (pH 7.2) buffer (lanes 12 and 14, Figure 2). After incubation for 15 min at 4°C, 100 (i) or 5 (ii) mM Hepes (pH 8) buffer was added and the medium was heated for 30 min at 90°C after addition of 1.2 μL 1 M NaOH. Then PAGE analysis was performed according to the general protocole indicated above.

Phosphate hydrolysis with alkaline phosphatase

Some reaction samples were incubated for 30 min at 37°C, with or without a preliminary heating step (30 min at 90°C), in the presence of 150 units of alkaline phosphatase.

NaBH$_4$ reduction of 5'-aldehyde derivatives

Some reaction samples were incubated for 30 min at ambient temperature in the presence of 0.1 M NaBH$_4$ (final concentration). Reaction was stopped by further addition of 1 μL acetone (10 equivalents with respect to NaBH$_4$). Then the general protocole was applied.

NaOI oxidation of 5'-aldehyde derivatives

Some reaction samples were incubated for 30 min at ambient temperature in the presence of 0.1 M Tris pH 9 buffer, 1 M I_2 and 1 M NaI. Then the general protocole was followed excepted samples were washed twice with ethanol/water (3/1, v/v) before gel preparation.

Gel electrophoresis analyses of references fragments

The 3'-labeling of ACGA and GACGA was performed with terminal deoxynucleotidyl transferase and $[\alpha\text{-}^{32}P]$ddATP. After synthesis and purification on a C_{18} Sep-Pack cartridge, these ODNs were phosphorylated at the 5'-end with ATP and T4 polynucleotide kinase. After dissolution in loading buffer, they were run on 15 % polyacrylamide gel, 3 h at 1,700 V.

RESULTS

Cleavage of a ^{32}P-end-labeled 22-mer ODN containing two independent A.T base pairs triplets under various conditions

To confirm the sequence selectivity observed on short ODNs for A.T base pairs triplets, we studied the oxidative cleavage of a 22-mer duplex containing two A.T triplets (Figure 1, duplex I/II) with Mn-TMPyP activated by $KHSO_5$.

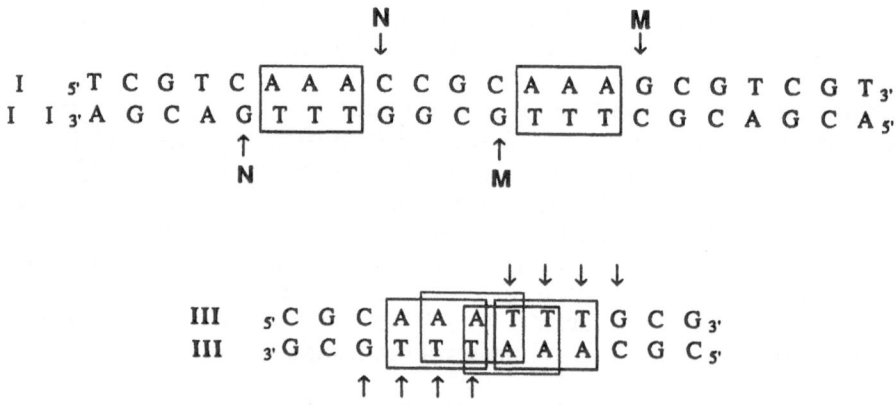

Figure 1. The selectivity of cleavage of short oligonucleotide duplexes by Mn-TMPyP/$KHSO_5$ depends on the presence and the location of an A.T triplet. Arrows indicate the sites of selective cleavage on the 3'-side of the A.T triplets. Duplex I/II which contains two independent triplets gives two independent cleavage sites on each strand (M and N sites), duplex III/III which contains four overlapping triplets gives four consecutive cleavage sites.

In a preliminary report[1], using the 5'-[^{32}P]-end-labeled ODNs I and II, we have shown that a very selective breakage occured at the two expected sites M and N on each strand, always located on the 3'-side of the two A.T triplets (see Figures 1 and 2, lane 7). Migration of the different fragments were unsensitive to heat or $NaBH_4$ treatment (data not shown) which is in agreement with the presence of 3'-phosphate termini. Therefore they may be compared directly with the chemical sequencing lanes of the Maxam-Gilbert method.

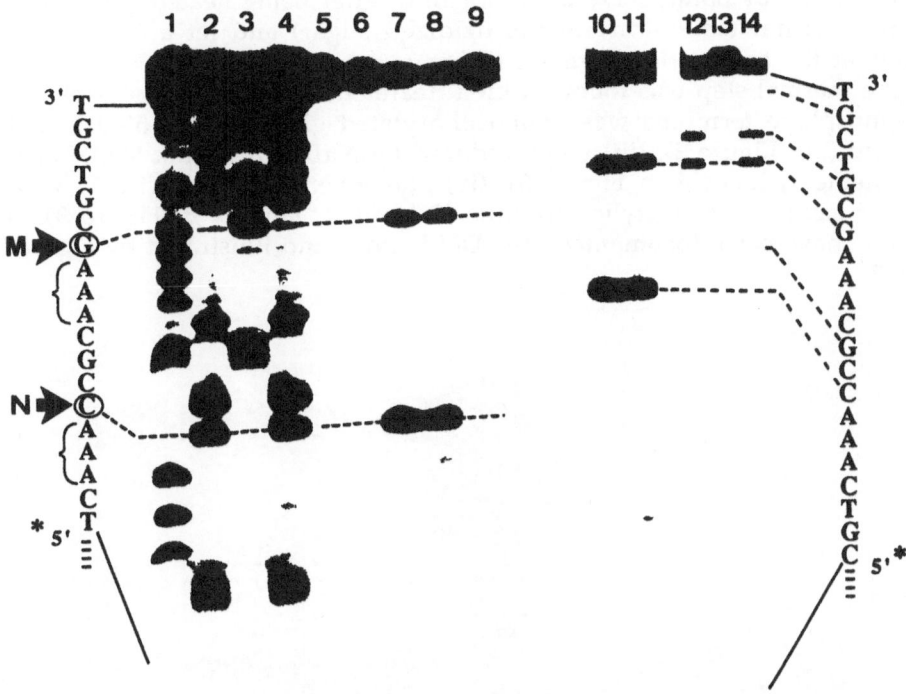

Figure 2. Polyacrylamide gel electrophoretic analysis of strand I after treatment of the 22-bp 5'-^{32}P end labeled duplex I/II (or ODN I, lane 14) with activated Mn-TMPyP. Reaction mixtures contained 1 µM of ODN. Lanes 1, 2, 3 and 4, Maxam-Gilbert sequencing reactions (A + G, G + C, G and C respectively). Lane 5 to 11: 40 mM Tris/HCl pH 8 buffer and 100 mM NaCl; lane 5, duplex control; lane 6, 0.5 µM Mn-TMPyP followed by alkali + heating treatment; lane 7, 0.5 µM Mn-TMPyP + 40 µM KHSO$_5$; lane 8, 0.5 µM Mn-TMPyP + 40 µM KHSO$_5$ followed by alkali + heating treatment; lane 9, control duplex heating; lane 10, 5 µM Mn-TMPyP + 5 mM H$_2$O$_2$; lane 11, 5 µM Mn-TMPyP + 5 mM H$_2$O$_2$ followed by piperidine + heating treatment. Lanes 12 to 14: 5 mM phosphate pH 7.2 buffer; lane 12, double-stranded ODN I/II incubated with 2 µM Mn-TMPyP followed by alkali + heating treatment; lane 13, control double-stranded ODN I/II incubated with Mn-TMPyP; lane 14, single-stranded ODN I incubated with 2 µM Mn-TMPyP followed by alkali + heating treatment.

In the case of 3'-labeling, we want to point out the necessity of a careful examination of PAGE bands before localizing the cleavage site(s). There is no straightforward correlations between the Maxam-Gilbert scale and the fragment bands corresponding to short ODNs having a 5'-modified sugar residue. Figure 3 displays the cleavage patterns of 3'-[^{32}P]-end-labeled ODNs I and II. Several products created by activated Mn-TMPyP, at least five and four for I and II respectively, were seen in lanes 3. Only two of them in each case (one at each site, N or M) can be related to the 5'-phosphate terminus fragments generated by breaks at the expected sites, since they co-migrated with the corresponding DNA sequencing bands. Various chemical treatments of the cleavage reaction samples allowed a clear identification of the other bands: the detailed interpretation is given in Scheme 1 and Figure 3. The initial damage was due to hydroxylation at C5' on one strand giving a 3'-phosphate terminus on the 5'-side of the break and a 5'-aldehyde terminus on the 3'-side. Concerning this last fragment, four points have to be noted: (i) after being heated it lost a free base and furfural (as residue of the oxidized sugar) and let a 5'-phosphate terminus at the end of the fragment (Scheme 1, both lanes 4 in Figure 3), (ii) when the thermal step was followed by a treatment with alkaline phosphatase, the 5'-phosphate terminus was dephosphorylated giving a free 5'-OH at the end (lanes 5 in Figure 3), (iii) when reduced by $NaBH_4$ it gave a 5'-OH at the end (Scheme 1, lanes 6 in Figure 3), (iv) and when oxidized with $I_2 + OH^-$ (NaOI) it gave a 5'-carboxylic terminus (Scheme 1, lanes 7 in Figure 3). All these data have been documented on ODN I and II and illustrated by Figure 3, lanes 4-7.

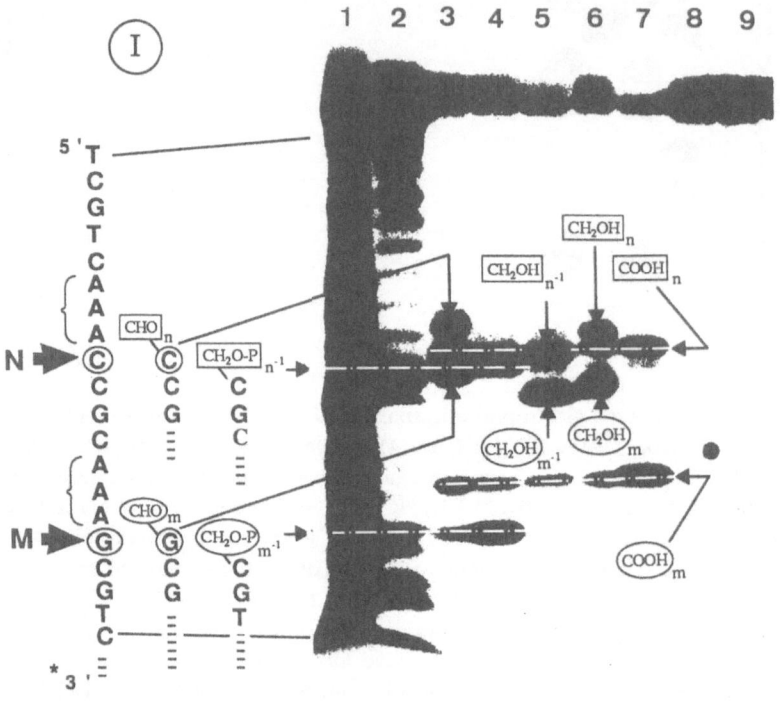

Figure 3 I. Polyacrylamide gel electrophoretic analysis of the 22-bp 3'-^{32}P-end-labeled duplex I/II following treatment with Mn-TMPyP + KHSO$_5$. Reaction mixtures contained 1µM of duplex, 40mMTris/HCl pH8 buffer and 100 mM NaCl. Experiments with 3'-end labeled strand I. Lanes 1 and 2, Maxam-Gilbert sequencing reactions (A + G and C + T, respectively); lane 3, 0.5 µM Mn-TMPyP + 40 µM KHSO$_5$; lane 4, 0.5 µM Mn-TMPyP + 40 µM KHSO$_5$ followed by heating treatment; lane 5, 0.5 µM Mn-TMPyP + 40 µM KHSO$_5$ followed by heating and alkaline phosphatase treatment; lane 6, 0.5 µM Mn-TMPyP + 40 µM KHSO$_5$ followed by NaBH$_4$ treatment; lane 7, 0.5 µM Mn-TMPyP + 40 µM KHSO$_5$ followed by I$_2$ + OH- treatment; lane 8, 40 µM KHSO$_5$; lane 9, duplex control. Other controls (0.5 µM Mn-TMPyP; NaBH$_4$ treatment; I$_2$ + NaOH treatment) were identical to that of lanes 8 and 9 (data not shown). Brackets indicate the selective affinity sites of the metalloporphyrin and arrows its selective cleavage sites. Letters m and n correspond to the nucleotide related to the cleavage sites M and N.

If we consider PAGE analysis of cleavage reaction without any co-treatment (lanes 3 in Figure 3) the pattern is now more clear. For example lane 3-I showed five bands which can be interpreted as follows: two bands corresponding to fragments with a 5'-aldehyde terminus which resulted directly from the initial hydroxylation at the two expected sites of cleavage; two fragments with a 5'-carboxylic terminus resulting from further oxidation of the aldehyde in the reaction conditions; two 5'-phosphate end fragments resulting from a partial evolution of the 5'-aldehyde fragments in the medium conditions with lost of base and furfural. In fact two bands on lane 3-I are superposed (so lane 3-I displays 5 bands, not 6): the shorter ODN fragment with a -CHO terminus comigrated with the larger one with a 5'-phosphate end. Interpretation of PAGE analysis profile for ODN II is similar as for ODN I excepted that no amount of $[COOH]_n$ fragment was detected and, consequently, lane 3-II displays only four bands.

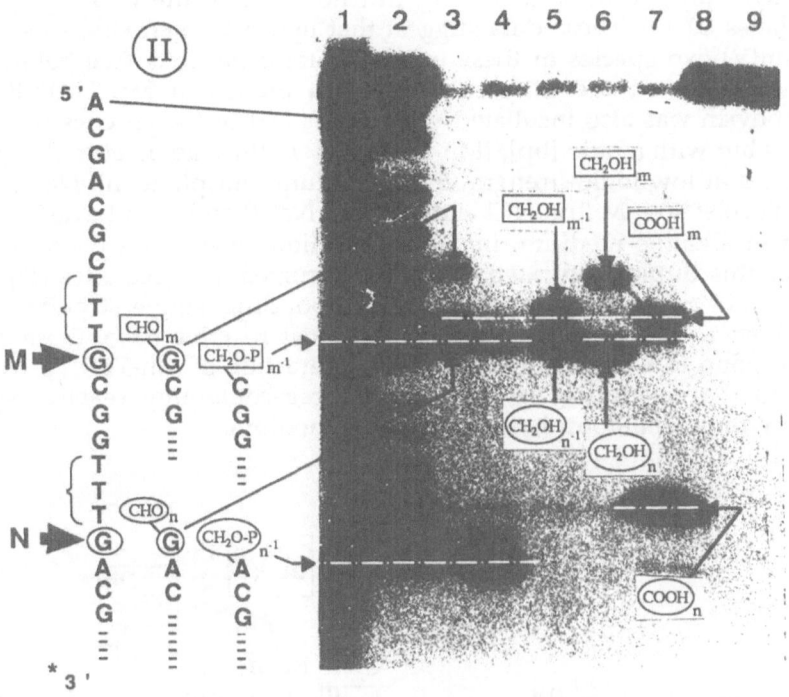

Figure 3 II. Polyacrylamide gel electrophoretic analysis of the 22-bp 3'-^{32}P-end-labeled duplex I/II following treatment with Mn-TMPyP + $KHSO_5$ as indicated in Figure 3I caption Experiments with 3'-end-labeled strand II See Figure 3I for lanes assignement

In Figure 3-II, lanes 5-6, for PAGE analysis of ODN II cleavage, we have to point out the puzzling behavior of the fragments ACGA (noted $[CH_2OH]_{n-1}$) and GACGA (noted $[CH_2OH]_n$): the longer fragment migrated faster than the shorter. In order to ensure our interpretation, we checked that synthetic reference fragments ACGA and GACGA effectively presented this surprising electrophoretic behavior (data not shown). In fact increasing the gel polyacrylamide content from 15 to 20 % allowed to get back the usual order of migration. Moreover with the 5'-phosphorylated corresponding oligonucleotides, the electrophoretic mobilities were as expected, *i.e.* the smaller oligomer migrated faster.

All the results presented here on the PAGE analyses of 3'-[^{32}P]-end-labeled duplexes cleaved by KHSO$_5$/Mn-TMPyP are in agreement with our preliminary study,[1] which was mainly carried out by HPLC on several short double-stranded ODNs. The cleavage occurs very selectively on both 3'-sides of A.T triplets.

In order to get an insight into reactive high-valent porphyrinatomanganese species involved in the cleavage reaction, some additional experiments were undertaken. The metalloporphyrin was activated with H$_2$O$_2$ as oxidant (Figure 2 lanes 10-11). The breaks were as specific as for KHSO$_5$ but with an intensity highly dependent on the nature of the buffer. Tris-HCl buffer allowed a much more efficient cleavage than phosphate buffer (not shown); treatment with piperidine and heating did not modify the cleavage pattern (Figure 2, lanes 10-11). These data suggest that hydrogen peroxide was able to generate Mn(V)-oxo species in these experimental conditions (see below for a general discussion on the different activation modes of Mn-TMPyP). The metalloporphyrin was also incubated with ODNs I/II in the absence of oxygen atom donors but with a ratio [bp]/[Mn-TMPyP] = 10 (instead of 44 in the general protocole) and at low ionic strength (5 mM sodium phosphate, no NaCl; in the general protocole: 40 mM Tris HCl and 100 mM NaCl) and then heated at 90° C for 30 min in alkaline medium; in these conditions inspection of the sites of cleavage for this duplex indicated that most occurred at G residues (Figure 2, lane 12); results were similar for the corresponding single-stranded ODN (Figure 2, lane 14); as guanine is by far the most reactive base, these results might correspond to a non sequence-specific interaction of Mn-TMPyP with the ODN and to the generation of diffusible oxygen-containing reactive species produced by the metalloporphyrin in alkaline medium.

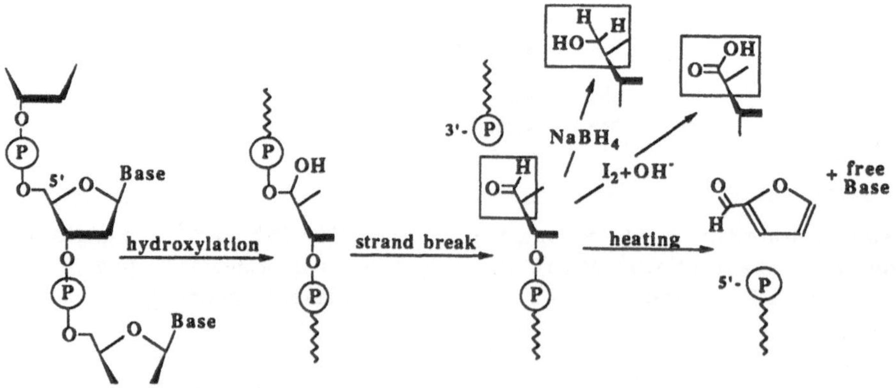

Scheme 1. Chemical structures involved in DNA breaks generated by the chemical nuclease Mn-TMPyP/KHSO$_5$.

Cleavage of a ^{32}P end labeled 12-mer ODN containing four overlapping A.T base pairs triplets

Sequence selectivity and scission mechanism described above are not limited to A.T triplets but could be extended to larger A.T rich regions. As example we present in Figure 4 the PAGE analysis of the 12-bp self-complementary 3'- and 5'-[^{32}P]-end-labeled duplex III/III which contains six consecutive A.T base-pairs. These six A.T bp actually represent four overlapping triplets (see duplex III/III in Figure 1). Cleavage treatment by activated Mn-TMPyP gave the four expected sites of cleavage on the 3'-sides of the four A.T triplets as illustrated by lanes 11-14 in Figure 4 for the 5'-[^{32}P]-end-labeled duplex analysis. In the case of 3'-[^{32}P]-end-labeled duplex (Figure 4, lanes 1-10), direct analysis of the cleavage reaction was more complicated (lane 1) but an additional thermal treatment simplified the gel and showed the four expected bands (lane 2) which co-migrated with the corresponding DNA sequencing bands. Modifications introduced by further alkaline phosphatase hydrolysis (lane 3) or reductive (lane 4) or oxidative (lane 5) treatments were in agreement with observations and the 5'-sugar residue/migration behavior relationships described above for the 22-mer. Present results unambiguously demonstrate that the activated metalloporphyrin, even in the case of overlapping A.T triplets, is able to cleave each of them specifically on the 3'-side.

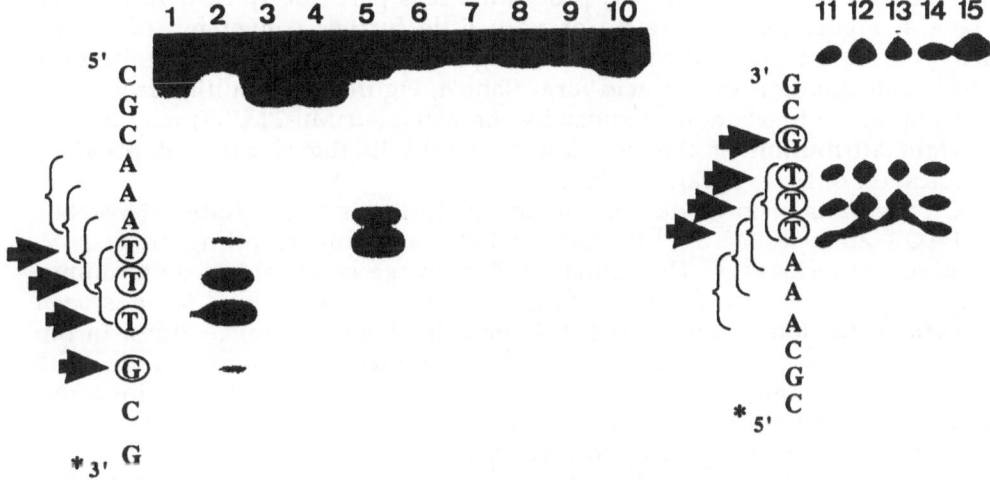

Figure 4. Polyacrylamide gel electrophoretic analysis of the 12 bp 3'- or 5'-^{32}P end labeled duplex III/III following treatment with Mn-TMPyP + KHSO$_5$. Reaction mixtures contained 1 μM of duplex, 40 mM Tris/HCl pH 8 buffer and 100 mM NaCl. Lanes 1-10, 3'-end labeling; l; lane 1, 0.5 μM Mn-TMPyP + 40 μM KHSO$_5$; lane 2, 0.5 μM Mn-TMPyP + 40 μM KHSO$_5$ followed by heating treatment; lane 3, 0.5 μM Mn-TMPyP + 40 μM KHSO$_5$ followed by heating and alkaline phosphatase treatment; lane 4, 0.5 μM Mn-TMPyP + 40 μM KHSO$_5$ followed by NaBH$_4$ treatment; lane 5, 0.5 μM Mn-TMPyP + 40 μM KHSO$_5$ followed by I$_2$ + NaOH treatment; lane 6, duplex control; lane 7, 0.5 μM Mn-TMPyP; lane 8, 40 μM KHSO$_5$; lane 9, control NaBH$_4$ treatment; lane 10, control I$_2$ + NaOH treatment. Lanes 11-15, 5'-end labeling; lanes 11-14, same conditions as lanes 1-4; lane 15, duplex control. Brackets indicate the selective affinity sites of the metalloporphyrin and arrows its selective cleavage sites.

DISCUSSION

Chemical cleavage of the double-stranded polynucleotides I/II and III/III

Both oligomers, when treated with the chemical nuclease Mn-TMPyP/KHSO$_5$ at + 4° C, pH 8 and [bp]/porphyrin above 20, showed important cleavage, ocurring specifically on both 3'-sides of A.T bp triplets. No or very little cleavage is observed at other bases. In the case of 5'-[^{32}P] labeling, the cleaved fragments migrated as those obtained in the Maxam-Gilbert procedure. Therefore they carry 3'-phosphate terminal groups. In the case of 3'-[^{32}P] labeling, direct interpretation is not obvious. As exemplified in lane 3 of Figure 3 (particularly in gel I), the determination of the number and the precise position of cleavage sites are not evident due to the presence of multiple bands. A precise knowledge of the chemical nature of the oxidative lesion is required to facilitate band attributions. As previously demonstrated scission occurs through hydroxylation of 5'-carbons which leads to the DNA break and leaves the 3'-terminus with an aldehyde function (Scheme 1). So, further oxidation by alkaline iodine of this aldehyde clearly indicates only two scission sites of this strand (Figure 3, lane 7); reduction with NaBH$_4$ gives the same result (Figure 3, lane 5), but it may be less evident if partial oxidation of the 5'-aldehyde terminus occurred during the cleavage reaction itself. At last, an heating step, which allows release of furfural and free base through two β-eliminations, leads to 5'-phosphate termini like in the case of usual DNA sequencing bands obtained with the Maxam-Gilbert procedure. It is particularly clear in the case of lane 4, Figure 3-II, where no or very little further conversion of the 5'-aldehyde in carboxylic acid occurred during the cleavage reaction. The presence of noticeable amounts of this acid form (lane 4, Figure 3-I), resulting from a re-oxidation of the 5'-position of sugar by the activated Mn-TMPyP, may lead to uncertain attribution of the position of breaks in the absence of previous identification of these 5'-carboxylic fragments.

On the basis of these data the main conclusion is to carefully interpret the PAGE analyses of 3'-[^{32}P]-end-labeled fragments resulting from a 5'-oxidative DNA cleavage. The initial DNA cleavage band (aldehyde terminus) always migrated slowly; *for short fragments the difference may be of several nucleotides*, for longer ones[13] the difference is about two nucleotides. In our previous work,[4] as in reference 3, the presence of several bands with migrations different of that of Maxam-Gilbert ones induced incorrect attribution of the cleavage positions (the exact position of breaks has to be shifted by about two nucleotides towards the 3'-end).

Again a precise knowledge of the molecular aspects of chemical DNA cleavages is necessary to ensure the exact localization of breaks when they are determined by PAGE analysis. Depending on the nature of the cleaving reagent and the targeted C-H bond, sugar residues (including kinetically unstable ones), may remain attached to either the 5'- or the 3'-phosphate and, consequently, may give or not additional PAGE bands in the case of respectively 3'- or 5'-labeling. Table 1 summarizes the main results in this field.

What about the nature of the activated Mn-TMPyP ?

Some informations on the oxidative species involved in the cleavage reaction should be obtained from the present experiments. In the general protocole the high specificity of the chemical cleavage can be deduced from the quasi absence of additional weaker bands close to the main fragments in PAGE analysis and it strongly supports a non-diffusible active species, namely a high-valent manganese-oxo porphyrin complex (Scheme 2) similar to the ferryl intermediate involved in cytochrome P-450 chemistry. In the case of a Fenton chemistry (free hydroxyl radicals) we should expect oxidative damages extending over several nucleotides and also lesions on nucleobases giving additional breaks after heating in alkaline conditions. In Figure 2 it is clear that in these last conditions (compare lanes 7 and 8 for activation of the metalloporphyrin with $KHSO_5$ and lanes 10-11 for activation with H_2O_2) no other additional cleavage sites were revealed; only very weak cleavage bands, also present in the control (lane 6), were added in alkaline and heating conditions.

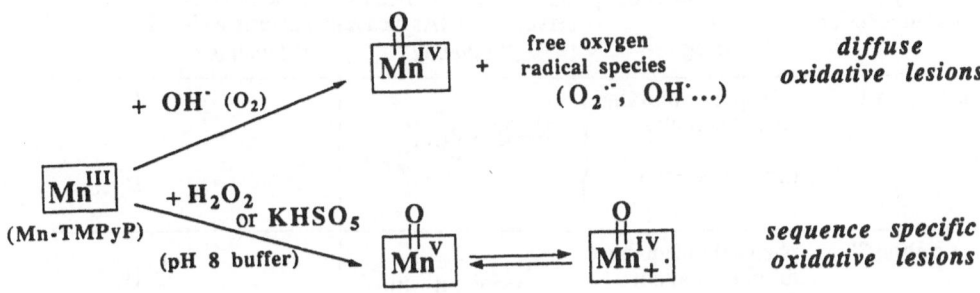

Scheme 2. Reactive species generated from Mn-TMPyP in various conditions. The porphyrin ligand is schematized by a square.

Nethertheless it could be possible to produce lesions on nucleobases and to induce limited amounts of cleavage when heating in alkaline conditions by changing the activation mode of Mn-TMPyP and modifying the experimental conditions (ionic strength, pH, bp/complex ratio). We have shown previously [4] that Mn-TMPyP is able to cleave restriction fragments at several sites in alkaline conditions. These data were obtained at low ionic strength with a low ratio [bp]/[Mn-TMPyP]. In these conditions predominant lesions occurred primarily at guanosine residues. Experiments on ODNs I/II in similar conditions of buffer and ionic strength gave the results presented in Figure 2 lanes 12-14 . The gel pattern is very different compared to that obtained when Mn-TMPyP was activated by an oxygen atom donor. The main bands are effectively localized at G sites but many additional secondary bands and smears are present suggesting a different mechanism of oxidative cleavage which might involve diffusible species reacting at bases or sugars without sequence specificity. The HO° radicals resulting from the homolytic cleavage of hydrogen peroxide by Mn-TMPyP are reasonable candidates, as depicted in Scheme 2 (see reference 14 for a recent review article on the activation of metalloporphyrins by various oxygen atom donors). At basic pH, Mn(III)-OH porphyrin complexes are oxidized to Mn(IV)=O species with the concomittant formation of superoxide anion which dismutes and generates hydrogen peroxide in the reaction medium. These Mn(IV)-oxo complexes are less reactive than a Mn(V)-oxo complex produced by potassium monopersulfate with respect to the hydroxylation of hydrocarbon C-H bonds (see reference 15 for a recent article on C-H bond hydroxylation and reference 16 for a discussion on manganese-oxo species). But Mn(IV)-oxo entities are still able to abstract electrons from the oxidizable sites of DNA. Then non-hydroxylating Mn(IV)-oxo species with free oxygen radical species should oxidize nucleobases or sugar residues without sequence specificity and produce DNA breaks without sequence specificity. This hypothesis was confirmed by the fact that the observed cleavage pattern was similar on single-stranded ODN compared with double-stranded ones.

Table I. Structures involved in additional PAGE bands depending on the end labeling and the site of oxidation.

| deoxyribose oxidative lesion | chemical nuclease or drug | postulated or identified metastable product observed on DNAs cleavage patterns with | | Ref. |
		5'-labeling	*3'-labeling*	
at carbon 1'	1,10-phenanthroline copper complex			17
	neocarzinostatin			18
at carbon 3'	phenanthrenequinone diimine rhodium complexes	$5' \sim\sim (P)\text{-}CH_2\text{-}CHO$		19
at carbon 4'	methidiumpropyl-EDTA-Iron(II)			20
	1,10-phenanthroline copper complex	$5' \sim\sim (P)\text{-}CH_2\text{-}COO^-$		17
	bleomycin			21 22 23
	neocarzinostatin	$5' \sim\sim (P)\text{-}CH_2\text{-}CO\text{-}CH=CH\text{-}CHO$		23
at carbon 5'	neocarzinostatin		OHC, O, Base	24
	Mn-TMPyP/KHSO$_5$		$3' \sim\sim (P)$	this work

CONCLUSION

In the field of chemical nucleases, Mn-TMPyP/KHSO$_5$ represents an original system able to selectively activate the $C_{5'}$-H bond at the 3' side to its highest affinity sites (namely an A.T triplet). Hydroxylation at the 5'-carbon of a deoxyribose represents the initial damage and leaves a 3'-phosphate and a 5'-aldehyde at the ends. Careful examination of PAGE bands resulting from oxidative cleavage of two selected double-stranded oligodeoxyribonucleotides which have been labeled either at 5' or 3' ends indicated that (i) 5'-end-labeled fragments with a 3'-phosphate terminus were unsensitive to different chemical or thermal treatments and migrated according to the corresponding fragments obtained by Maxam-Gilbert sequencing, (ii) whereas 3'-end-labeled fragments had an electrophoretic mobility highly dependent on the 5'-terminus; migrations were different with a 5'-aldehyde terminus (-CHO), after reduction (-CH$_2$OH), oxidation (-COOH), heating (5'-P terminus and release of free base and furfural) or heating and alkaline phosphatase treatment (5'-OH terminus and loss of free base, furfural and inorganic phosphate). The precise knowledge of the chemical lesion at the oxidized site of DNA provided the possibility to localize the exact position of breaks and attribute the structures of the different fragments observed on PAGE analysis after the manganese porphyrin-mediated cleavage of two ODNs containing A.T rich sequences. Additional experiments with hydrogen peroxide and alkali-activation of Mn-TMPyP suggest that sequence specific DNA cleavage by MnTMPyP/KHSO$_5$ is due to non-diffusible manganese(V)-oxo entities.

ACKNOWLEDGMENTS

This work was partially supported by grants from the French agency for AIDS research (ANRS), the Région Midi-Pyrénées and the Association pour la Recherche contre le Cancer (ARC, Villejuif). The authors are indebted to Dr. Martine Defais-Villani and Marie-Jeanne Pillaire for making possible gel electrophoresis analyses and for helpful discussions during this work.

REFERENCES

1. M. Pitié, G. Pratviel, J. Bernadou, and B. Meunier, Preferential hydroxylation by the chemical nuclease *meso*-tetrakis (4-N-methylpyridiniumyl)porphyrinatomanganese[III] pentaacetate/KHSO$_5$ at the 5' carbon of deoxyriboses on both 3' sides of three contiguous A·T base pairs in short double-stranded oligonucleotides, *Proc. Natl. Acad. Sci. U.S.A.*. 89:3967 (1992).

2. J. Bernadou, G. Pratviel, F. Bennis, M. Girardet, and B. Meunier, Potassium monopersulfate and a water-soluble manganese porphyrin complex, [Mn(TMPyP)](OAc)$_5$, as an efficient reagent for the oxidative cleavage of DNA, *Biochemistry* 28:7268 (1989).

3. J.C. Dabrowiak, B. Ward, and J. Goodisman, Quantitative footprinting analysis using a DNA-cleaving metalloporphyrin complex, *Biochemistry* 28:3314 (1989).

4. R.B. Van Atta, J. Bernadou, B. Meunier, and S.M. Hecht, On the chemical nature of DNA and RNA modification by a hemin model system, *Biochemistry* 29:4783 (1990).

5. G. Pratviel, M. Pitié, J. Bernadou, and B. Meunier, Furfural as a marker of DNA cleavage by hydroxylation at the 5' carbon of deoxyribose, *Angew. Chem. Int. Ed.* 30:702 (1991).

6. G. Pratviel, M. Pitié, J. Bernadou, and B. Meunier, Mechanism of DNA cleavage by cationic manganese porphyrins: hydroxylations at the 1'-carbon and 5'-carbon atoms of deoxyriboses as initial damages, *Nucleic Acids Res.* 19:6283 (1991).

7. M.J. Carvlin, E. Mark, R. Fiel, and J.C. Howard, Intercalative and nonintercalative binding of large cationic porphyrin ligands to polynucleotides, *Nucleic Acids Res.* 11:6141 (1983).

8. R.F. Pasternack, E.J. Gibbs, and J.J. Villafranca, Interactions of porphyrins with nucleic acids, *Biochemistry* 22:5409 (1983).

9. B. Ward, A. Skorobogaty, and J.C. Dabrowiak, DNA cleavage specificity of a group of cationic metalloporphyrins, *Biochemistry* 25:6875 (1986).

10. L. Ding, J. Bernadou, and B. Meunier, Oxidative degradation of cationic metalloporphyrins in the presence of nucleic acids: a way to binding constants? *Bioconjugate Chem.* 2:201 (1991).

11. A.M. Maxam, and W. Gilbert, Sequencing end-labeled DNA with base-specific chemical cleavages, *Methods in Enzymology* 65:499 (1980).

12. G.D. Fasman, Nucleic Acids, in "Handbook of Biochemistry and Molecular Biology" 3rd Edition, CRC Press Vol. I, p. 589 (1975).

13. I.H. Goldberg, Free radical mechanisms in neocarzinostatin-induced DNA damage, *Free Radical Biology & Medicine* 3:41 (1987).

14. B. Meunier, Metalloporphyrins as versatile catalysts for oxidation reactions and oxidative DNA cleavage, *Chem. Rev.* 92:1411 (1992).

15. P. Hoffmann, A. Robert, and B. Meunier, Preparation and catalytic activities of the manganese and iron derivatives of Br_8TMP and $Cl_{12}TMP$, two robust porphyrin ligands obtained by halogenation of tetramesityl-porphyrin, *Bull. Soc. Chim. Fr.* 129:85 (1992).

16. J.T. Groves, and M.K. Stern, Synthesis, characterization, and reactivity of oxomanganese(IV) porphyrin complexes, *J. Am. Chem. Soc.* 110:8628 (1988).

17. M. Kuwabara, C. Yoon, T. Goyne, T. Thederahn, and D.S. Sigman, Nuclease activity of 1,10-phenanthroline-copper ion: reaction with CGCGAATTC GCG and its complexes with netropsin and *EcoRI*, *Biochemistry* 25:7401 (1986).

18. L.S. Kappen, and I.H. Goldberg, Neocarzinostatin acts as a sensitive probe of DNA microheterogeneity: switching of chemistry from C-1' to C-4' by a G.T mismatch 5' to the site of DNA damage, *Proc. Natl. Acad. Sci. U.S.A.* 89: 6706 (1992).

19. A. Sitlani, E.C. Long, A.M. Pyle, and J.K. Barton, DNA Photocleavage by phenanthrenequinone diimine complexes of rhodium (III): shape-selective recognition and reaction, *J. Am. Chem. Soc.* 114:2303 (1992).

20. R.P. Hertzberg, and P.B. Dervan, Cleavage of DNA with methidiumpropyl-EDTA-iron (II): reaction conditions and product analyses, *Biochemistry* 23:3934 (1984).

21. Y. Hashimoto, H. Iijima, Y. Nozaki, and K. Shudo, Functional analogues of bleomycin: DNA cleavage by bleomycin and hemin-intercalators, *Biochemistry* 25:5103 (1986).

22. L.F. Povirk, Y.H. Han, and R.J. Steighner, Structure of bleomycin-induced DNA double-strand breaks: predominance of blunt ends and single-base 5' extensions, *Biochemistry* 28:5808 (1989).

23. B.L. Frank, L.Worth,Jr., D.F. Christner, J.W. Kozarich, J. Stubbe, L.S. Kappen and I.H. Goldberg, Isotope effects on the sequence-specific cleavage of DNA by neocarzinostatin: kinetic partitioning between 4'- and 5'-hydrogen abstraction at unique thymidine sites, *J. Am. Chem. Soc.* 113:2271 (1991).

24. P.C. Dedon, Z.W. Jiang, and I.H. Goldberg, Neocarzinostatin-mediated DNA damage in a model AGT.ACT site: mechanistic studies of thiol-sensitive partitioning of C4' DNA damage products, *Biochemistry* 31:1917 (1992).

NEW MODEL SYSTEMS FOR OXYGENASES

Daniel Mansuy

Université René Descartes, URA 400,
45 rue des Saints-Pères, 75270 Paris Cedex 06, France

1 - INTRODUCTION

The finding of new efficient and selective oxidation catalysts is an important challenge in fine chemistry and in bulk chemicals manufacture. In fact, the synthesis of many of the main chemicals of industrial chemistry, like ethylene and propylene oxides, phenol or terephthalic acid, involve oxidation catalysts. There is an always increasing demand of more efficient and selective catalysts for the transfer of oxygen atoms from readily available and cheap oxygen atom donors like O_2, H_2O_2 or alkylhydroperoxides (eq.1). In that respect, the hydroxylation of inert C-H bonds of alkanes under mild conditions, the enantioselective epoxidation of alkenes and the regioselective hydroxylation of aromatic rings by those oxygen atom donors remain very interesting challenges in organic chemistry.

$$RH + AO \xrightarrow{\text{Catalyst}} ROH + A \quad \text{(eq.1)}$$

A possible strategy to find new selective oxidation catalysts is to consider and mimic the enzymatic systems which have been developed by living organisms during life evolution. Two classes of enzymes are able to catalyze the transfer of oxygen atoms from dioxygen to substrates[1]. The first ones, called monooxygenases catalyze the reductive activation of dioxygen with consumption of two electrons and two protons and the transfer of only one oxygen atom from O_2 to substrates while the second oxygen atom is reduced to water (eq.2).

$$RH + O_2 + 2e^- + 2H^+ \xrightarrow{\text{Monooxygenase}} ROH + H_2O \quad \text{(eq.2)}$$

Most monooxygenases that are ubiquitous in living organisms contain an hemeprotein called cytochrome P450. The active oxygen species involved in the catalytic cycle of these enzymes seems to be an oxyferryl complex of the Fe(V)=O

The Activation of Dioxygen and Homogeneous Catalytic Oxidation,
Edited by D.H.R. Barton *et al.*, Plenum Press, New York, 1993

type. This very reactive, highly electrophilic species is able to transfer its oxygen atom to almost any organic substrate and to perform the hydroxylation of alkanes and arenes, the epoxidation of alkenes, the N-oxidation of amines and the S-oxidation of thioethers. At least in vitro, these monooxygenases also catalyze the transfer of an oxygen atom from H_2O_2 or alkylhydroperoxides to substrates (eq.3,4). Chemical model systems of such monooxygenases should exhibit several advantages because of the selectivity of the oxidations performed, even on alkanes, and the diversity of substrates that can be oxidized. However, their disadvantage is to transfer only one oxygen atom from O_2 and to consume one mol of reducing agent per mol of substrate hydroxylated.

$$RH + R'OOH \xrightarrow{\text{Monooxygenase}} ROH + R'OH \qquad (eq.3)$$

$$Fe^{III} + R'OOH \xrightarrow{\text{-R'OH}} [Fe^V=O] \xrightarrow{\text{+RH}} ROH + Fe^{III} \quad (eq.4)$$

The second class of oxygenases, the dioxygenases, do not present such disadvantages as they catalyze the transfer of the two oxygen atoms of O_2 into substrates without any consumption of a reducing agent (eq.5). From a chemical point of view, chemical catalysts able to catalyze selective dioxygenations of C-H bonds of hydrocarbons with the intermediate formation of an alkylhydroperoxide and the eventual formation of a ketone for instance would be highly interesting. Unfortunately, dioxygenases known so far only catalyze the oxidation of a quite limited number of substrates which are relatively reactive by themselves like catechols or unsaturated fatty acids.

$$RH + O_2 \xrightarrow{\text{Dioxygenase}} ROOH \longrightarrow \text{products} \quad (eq.5)$$

Many chemical model systems based on metalloporphyrin catalysts and mimicking cytochrome P450-dependent monooxygenases have been described during these last decade. Several review articles have been devoted to these systems [2-10]. In that context, very recent results about the preparation and catalytic properties of new homogeneous and supported catalysts will be described in a first chapter. In the second chapter, some preliminary results showing that the oxidation of alkanes by a dioxygenase-like mechanism could occur in the presence of iron porphyrin catalysts activated either photochemically or thermally, will be reported.

2 - BIOMIMETIC SYSTEMS FOR CYTOCHROME P450-DEPENDENT MONOOXYGENASES

Model systems for cytochromes P450 must involve a metalloporphyrin able to receive an oxygen atom from oxygen atom donors like PhIO, H_2O_2, or O_2 in the presence of a reducing agent, with intermediate formation of a high-valent metal-oxo complex which will transfer its oxygen atom to substrates. Two strategies have been used to build up such systems. In the first strategy, a metalloporphyrin is used as an homogeneous catalyst. In the second one, the metalloporphyrin is included in a polymer matrix leading to a supported catalyst which should be more easily separated from the reaction mixture and recycled. Moreover, such supported catalysts could lead to substrate shape selectivity and to a regioselectivity of the oxidations catalyzed because of specific interactions of the substrates with the polymer matrix.

2-1 : Development of a third generation of metalloporphyrins very efficient for the hydroxylation of alkanes and the oxidation of aromatic compounds.

An intense effort of research has been done during these last twelve years in order to find chemical systems mimicking either the short catalytic cycle of cytochrome P450 by associating various Fe(III) or Mn(III) porphyrins with oxygen atom donors like PhIO, ClO$^-$, ROOH, H_2O_2, $KHSO_5$, ClO_2^- , or the long catalytic cycle of this hemeprotein by using those metalloporphyrins and O_2 in the presence of a reducing agent such as BH_4^-, ascorbate, H_2+Pt or Zn. The purpose of this presentation is not to give a review of the corresponding results because this has been done in several review articles [2-10] but more to illustrate very recent tendencies in the field.

The first system reported by Groves et al.[11] used iodosylbenzene, PhIO, in the presence of a simple catalyst, Fe(TPP)Cl (Fig.1). It reproduced quite well most cytochrome P450 reactions at least from a qualitative point of view. However, such first generation metalloporphyrin catalysts like Fe(TPP)Cl and Mn(TPP)Cl undergo a fast oxidative degradation under the strong oxidizing conditions used. Much better results were obtained with a second generation of Fe(III) and Mn(III) meso-tetra-arylporphyrins bearing several electron-withdrawing substituents at each meso-aryl groups. This is the case of Mn(TDCPP)Cl which bears eight Cl

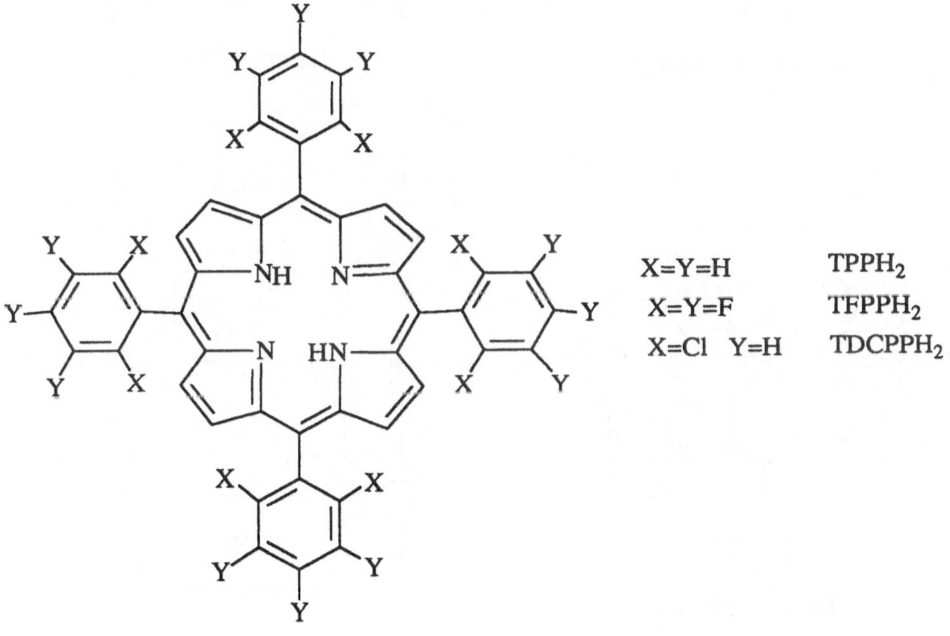

X=Y=H	$TPPH_2$
X=Y=F	$TFPPH_2$
X=Cl Y=H	$TDCPPH_2$

Figure 1. Formula of representative porphyrins used in biomimetic catalysts of the first and second generation.

substituents and of Fe(TFPP)Cl whose meso-aryl substituents are C6F5 groups (Fig.1). By using one of this second generation catalysts, Mn(TDCPP)Cl, an efficient system for hydrocarbon oxidation by diluted H_2O_2 was built. This system, which also used imidazole as a cocatalyst, involves a Mn(V)=O active species and performs very efficient stereospecific epoxidations of alkenes and the hydroxylation of alkanes and aromatic compounds with satisfactory yields [12]. However, the yields obtained for the hydroxylation of poorly reactive substrates like linear alkanes by using those second generation catalysts remained modest and there was an important oxidative degradation of the catalyst.

More recently, even more robust and more reactive polyhalogenated metalloporphyrins bearing electron-withdrawing substituents on the β-pyrrole positions (Fig.2) have been prepared and used as efficient oxidation catalysts [6-10]. The three following examples illustrate the better catalytic activities of these third generation metalloporphyrins.

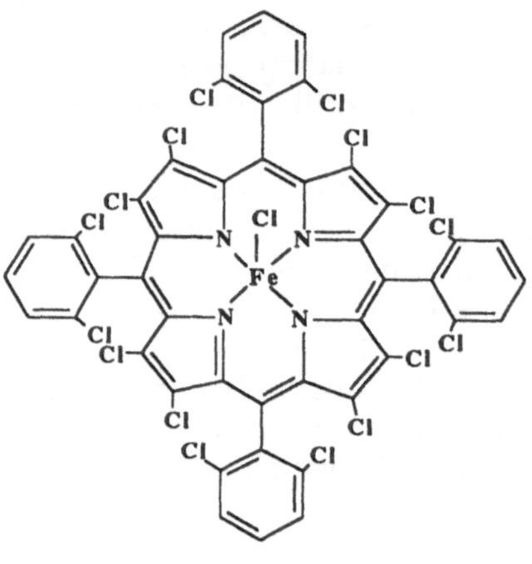

Fe(TDCPCl$_8$P)Cl

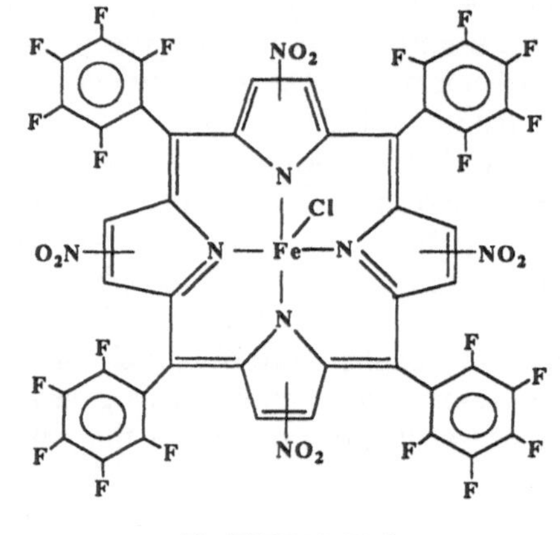

Fe(TFPN$_4$P)Cl

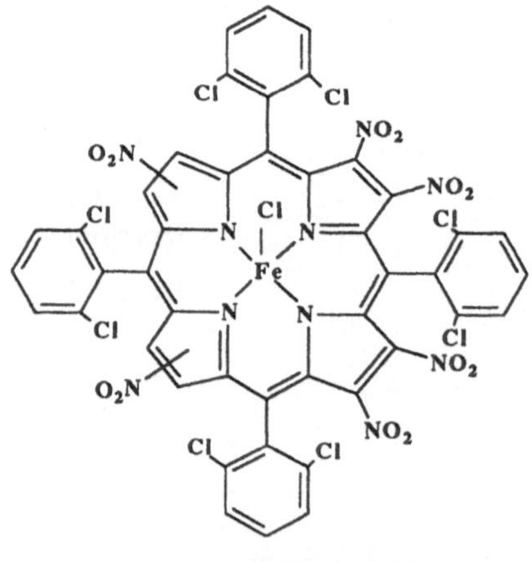

Fe(TDCPN$_6$P)Cl

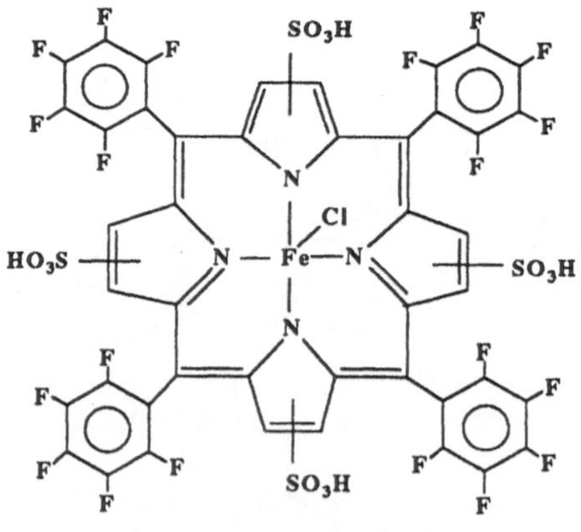

Fe(TFPS$_4$P)Cl

Figure 2. Formula of some third generation iron porphyrins used as oxidation catalysts.

Fe(TPP)Cl and Fe(TDCPP)Cl catalyze the oxidation of cyclododecane by PhIO with formation of the corresponding alcohol and ketone in moderate yields (total yields of 25 and 45% and minor formation of ketone), whereas the third generation catalysts Fe(TDCPCl8P) and Fe(TDCPBr8P) give almost quantitative yields based on PhIO after 1h at 20°C. Similar results have been observed for the oxidation of adamantane by PhIO, with an increase of the yield from 16% for Fe(TPP)Cl to 100% for Fe(TDCPCl8P)Cl (57% adamantan-1-ol, 38% adamantan-2-ol and 5% adamantan-2-one) (Fig.3). Even heptane, a poorly reactive alkane, is oxidized by PhIO in much higher yields (≈ 80%) in the presence of Fe(TDCPCl8P)Cl than in the presence of Fe(TDCPP)Cl (38%) [13]. Moreover, Fe(TDCPCl8P)Cl and Fe(TDCPBr8P)Cl exhibit a chemoselectivity toward the hydroxylation of a mixture of cyclooctene and heptane (1:100) different from that found for Fe(TDCPP)Cl, the cyclooctene oxide : heptanols ratio being 8.2 with Fe(TDCPP)Cl and only 2.9 for Fe(TDCPCl8P)Cl [13].

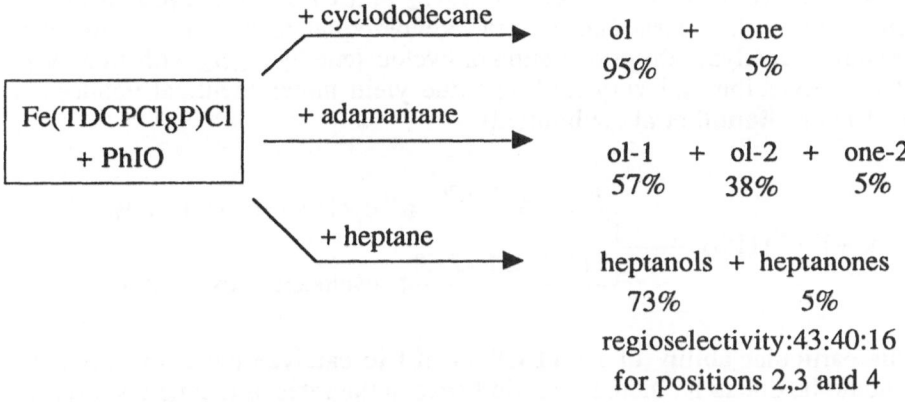

Figure 3. Hydroxylation of alkanes by PhIO catalyzed by Fe(TDCPCl8P)Cl (ref. 13 and D. Bouy, P. Battioni and D. Mansuy, in preparation).

The main differences observed between the catalytic properties of the second and third generation iron porphyrins should be due at least in part to an increase of the electrophilic character and reactivity of the high-valent iron-oxo species involved as intermediates in those oxidations. In fact, a recent X-ray structure determination of Ni(TF5PBr8P) [14] and recent molecular mechanics and semi-empirical quantum calculations [15] made on TDCPCl8PH2 and TDCPBr8PH2 have shown that these third generation porphyrins adopt a saddle-shape structure for their tetrapyrrole ring. Thus, while TDCPPH2, as most other porphyrins of the first and second generation, exhibit a planar porphyrin ring and four meso aryl groups almost perpendicular to this ring, TDCPBr8PH2 and TF5PBr8PH2 show a saddle-shape structure for the porphyrin and meso-aryl groups very much tilted in order to minimize the strong steric interactions between the β-halogen and meso-aryl groups. These structure differences between Fe(TDCPP)Cl and Fe(TDCPBr8P)Cl should lead to a different accessibility of the iron-oxo intermediates formed in the presence of PhIO. The different electrophilicity and accessibility of these iron-oxo intermediates should be at the origin of the different chemo- and regio-selectivities observed in the oxidation of a cyclooctene-heptane mixture catalyzed by these two iron porphyrins.

β-polynitroporphyrins have been prepared very recently by reaction of TDCPPH$_2$ and TFPPH$_2$ with concentrated HNO$_3$ (J.F. Bartoli, P. Battioni and D. Mansuy, submitted for publication). The major compound, TDCPN$_6$PH$_2$, prepared from TDCPPH$_2$ (50% yield) bears six nitro substituents at the β-pyrrole positions (from ^{1}H NMR and mass spectroscopy; mixture of regioisomers), while the one obtained by nitration of TF$_5$PPH$_2$ only bears four β-nitro groups (one per each pyrrole ring) (Fig. 2). Their Fe(III) complexes, Fe(TDCPN$_6$P)Cl and Fe(TF$_5$PN$_4$P)Cl, are also good catalysts for the hydroxylation of alkanes by PhIO, with yields only slightly lower than those obtained with Fe(TDCPCl$_8$P)Cl. However, the most interesting property of Fe(TDCPN$_6$P)Cl is its particular ability to catalyze the epoxidation of alkenes by H$_2$O$_2$ in the absence of any cocatalyst. In a general manner, iron porphyrins have been found so far as bad catalysts for the oxidation of alkenes and alkanes by H$_2$O$_2$ because of their great propensity to dismutate H$_2$O$_2$ [12,2-10]. In fact, although Fe(TDCPP)Cl and even the third generation catalysts Fe(TDCPCl$_8$P)Cl and Fe(TF$_5$PBr$_8$P)Cl give very low yields of conversion of cyclooctene by 3 eq. of H$_2$O$_2$ in a CH$_2$Cl$_2$:CH$_3$OH mixture (conversions lower than 10% and epoxide yields equal to or lower than 5%), Fe(TDCPN$_6$P)Cl catalyzes the epoxidation of cyclooctene by H$_2$O$_2$ with an almost quantitative conversion and very high epoxide yield under identical conditions (95%) (eq.6) (J.F. Bartoli et al., submitted).

$$\text{cyclooctene} + \text{H}_2\text{O}_2 \ (1{:}3) \longrightarrow \begin{array}{l} \xrightarrow{\text{Fe(TDCPP)Cl}} \text{cyclooctene oxide (5%)} \\ \xrightarrow{\text{Fe(TDCPN}_6\text{P)Cl}} \text{cyclooctene oxide (95%)} \end{array} \qquad \text{(eq.6)}$$

This particular ability of Fe(TDCPN$_6$P)Cl to catalyze the epoxidation of alkenes should be due to an increased ratio between the rates of alkene epoxidation and H$_2$O$_2$ dismutation. Applications of this Fe(TDCPN$_6$P)Cl-H$_2$O$_2$ system, which does not require the use of a cocatalyst contrary to the previously described Mn(TDCPP)Cl-H$_2$O$_2$-imidazole system, to other substrates is under study.

Another third generation catalyst, obtained by substitution of four β-hydrogens of Fe(TFPP) by SO$_3$H groups (one per pyrrole ring) was found particularly active for the oxidation of aromatic rings [16-18]. For instance, in the presence of this β-tetrasulfonated iron porphyrin, Fe(TFPS$_4$P) (Fig.2), several methoxyarenes are oxidized to the corresponding quinones in one step by the water soluble oxygen atom donor Mg-monoperoxyphthalate (MMP) in a CH$_3$CN-tartrate buffer pH3, with yields between 50 and 95% [16] (eq.7). This reaction was applied successfully to the preparation of the natural product methoxatin (or PQQ) by oxidation of a methoxyarene precursor [16]. Formation of quinones by the Fe(TFPS$_4$P)-MMP system also occurs on dimethoxyarenes bearing an electron-donating substituent. Interestingly, dimethoxyarenes bearing an electron-withdrawing substituent are oxidized by this system with the selective formation of muconic dimethylesters derived from the cleavage of the C-C bond bearing the two methoxy substituents [17] (eq.8). This reaction provides a new access to functional muconic diesters with yields around 40%. Sulfonated iron porphyrins of the first and second generation do not lead to the formation of muconic diesters under identical conditions. The formation of quinones or muconic diesters from dimethoxyarenes involve several steps and the consumption of two moles of oxidant per mole of substrate oxidized. The most likely first step of these reactions is a one-electron transfer from the aromatic substrate to the iron-oxo active species [16-18]. The good selectivity observed for the formation of quinones or muconic diesters in the presence of Fe(TFPS4P) should be due to a particularly good control of substrate-derived free radical intermediates by the iron during the catalytic cycle [16-18].

$$\text{(eq.7)}$$

$$\text{(eq.8)}$$

2.2. Biomimetic oxidation catalysts supported on inorganic matrices.

Several strategies have been used to obtain supported catalysts involving immobilized metalloporphyrins [10]. Mineral supports appear particularly attractive because of their inertness in strongly oxidizing media [19]. Immobilized metalloporphyrins were recently prepared by taking advantage of the selective substitution of the para-fluorine substituents of TFPPH2 by various nucleophiles [20] (Fig.4). This reaction was applied to the preparation of insoluble polymers of Fe or Mn(TFPP) [21] upon reaction of Fe (or Mn) (TFPP) with S^{2-} , and to the covalent binding of these metalloporphyrins to silica or montmorillonite bearing a (CH2)3 NH2 function [22]. A suspension of these supported Fe(III) or Mn(III) porphyrins in various solvents efficiently catalyze the epoxidation of alkenes and the hydroxylation of alkanes by PhIO [21,22].

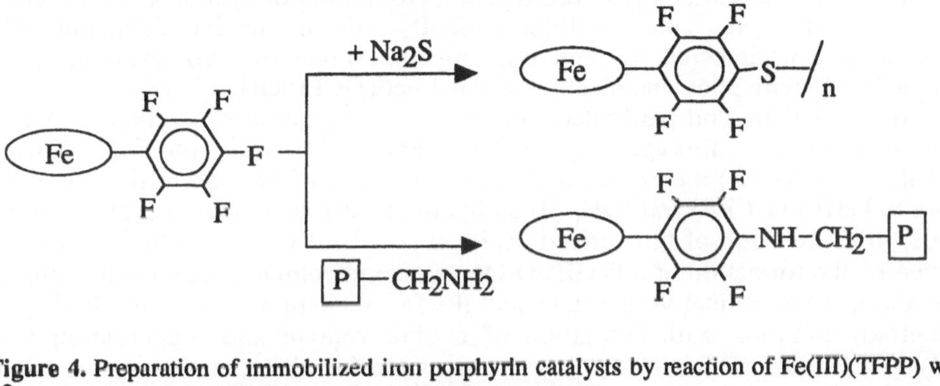

Figure 4. Preparation of immobilized iron porphyrin catalysts by reaction of Fe(III)(TFPP) with S^{2-} or functionalized supports (P = silica or montmorillonite).

Another approach was to intercalate metalloporphyrins into layered minerals such as clays [19]. For instance, the tetracationic Mn(TMPyP = meso-tetra-N-methyl-pyridiniumyl porphyrin) catalyst was intercalated into the interlayer space of montmorillonite by simple ion exchange with already present cations [23]. The corresponding supported catalyst was found to be much more efficient for hydroxylation of heptane by PhIO than the same Mn-porphyrin either simply adsorbed on silica or used as such as an homogeneous catalyst (60% yield instead of 40% and 3% respectively). Moreover, Mn(TMPyP) intercalated in montmorillonite was found to select substrates of different shapes. This was shown in oxidations of a pentane : adamantane (2:1) mixture by PhIO in the presence of Mn(TMPyP) intercalated into montmorillonite or Mn(TMPyP) adsorbed on silica. The latter catalyst led to an almost exclusive hydroxylation of the more reactive adamantane, whereas the former one led to an adamantanols-pentanols ratio much more in favor of pentane [23].

3 - IRON PORPHYRIN-CATALYZED OXIDATION OF ALKANES BY DIOXYGEN WITHOUT CONSUMPTION OF A REDUCING AGENT.

Iron porphyrins of the first and second generation, like Fe(TPP)Cl or Fe(TDCPP)Cl, are not able to catalyze the oxidation of cyclohexane by O_2 at room temperature in the absence of a reducing agent. However, we recently found that irradiation by a UV-visible light (350-450 nm), of a solution of Fe(TDCPP)OH in O_2-saturated cyclohexane led to a progressive formation of cyclohexanone (eq.9) [24].

$$\text{cyclohexane} + O_2 \xrightarrow[\text{h}\nu]{\text{(TDCPP)Fe}^{\text{III}}\text{-OH}} \text{cyclohexanone} \quad (30 \text{ turnovers h}^{-1}) \text{ (eq.9)}$$

This formation was linear as a function of time and cyclohexanol was only detected as a very minor product. Cyclooctane was similarly oxidized with the formation of cyclooctanone as a major product. Other iron(III)porphyrins involving different axial ligands, like Fe(TDCPP)Cl and Fe(TPP)Cl, were much less efficient and less selective. Irradiation of Fe(TDCPP)OH in the presence of O_2 and cis-stilbene failed to give any formation of stilbene epoxide, and cyclohexene was mainly oxidized on its allylic position with formation of cyclohex-2-enol and cyclohex-2-enone under such conditions. Finally, the intermediate formation of $\cdot$OH radicals was detected by spin trapping experiments during cyclohexane oxidation by O_2 with photochemically-activated Fe(TDCPP)OH.

All these data, and particularly the selective formation of cyclohexanone, are in favor of the "dioxygenase-like" mechanism [24] shown on Fig.5. This mechanism involves (i) a photodissociation of the Fe(III)-OH bond of the catalyst leading to Fe(II)(TDCPP) and $\cdot$OH, (ii) an hydrogen abstraction from cyclohexane by $\cdot$OH, (iii) a reaction of the derived cyclohexyl radical with O_2 which may be followed by the formation of a Fe(III)-O-O-cyclohexyl complex upon combination of the alkylperoxo radical with Fe(II), and the fast decomposition of the Fe(III)-alkylperoxo complex with formation of cyclohexanone and regeneration of Fe(III)(TDCPP)OH. The active species responsible for alkane activation in that system is $\cdot$OH and not an oxyferryl complex as in cytochromes P450. This explains why the Fe(TDCPP)OH-O_2-hν system oxidizes alkanes but does not epoxidize alkenes. Its mechanism (Fig.5) resembles that of dioxygenases as it involves the incorporation of the two oxygen atoms of O_2 into the substrate and a control of the intermediate substrate-derived peroxy radical by the iron catalyst which should be at the origin of the selective formation of cyclohexanone.

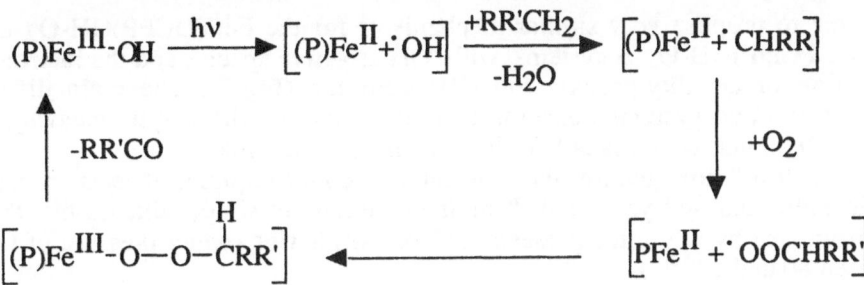

Figure 5. "Dioxygenase-like" mechanism for the oxidation of alkanes to ketones by O_2 catalyzed by photoactivated Fe(TDCPP)OH.

The key step for the oxidation of alkanes by O_2 shown in Figure 5 is the photoactivation of the Fe(III)-OH bond of the catalyst leading to the ˙OH active species. Such an homolytic cleavage of the (TDCPP)Fe(III)-OH bond does not occur by simple heating of a solution of Fe(TDCPP)OH in O_2-saturated cyclohexane at 80°C as shown by the lack of any formation of cyclohexanone and cyclohexanol after 20h under these conditions. Presumably, the strength of the (TDCPP)Fe(III)-OH bond is too high to be homolytically cleaved by simple heating at moderate temperature. However, we thought that the homolytic cleavage of a (porphyrin) Fe(III)-OH bond could be easier in the case of the third generation iron porphyrins because of the greater stability of their Fe(II) state. In agreement with this assumption, heating an O_2-saturated solution of Fe(TDCPCl8P)OH in cyclooctane at 80°C led to the formation of cyclooctanone and cyclooctanol (375 and 75 turnovers per h respectively) (J.F. Bartoli, P. Battioni and D. Mansuy, to be published and ref. 25). Reactions performed under identical conditions but in the absence of iron porphyrin or in the presence of Fe(TDCPP)OH did not give these oxidation products. Most characteristics of theoxidations of substrates by O_2 catalyzed by thermally activated Fe(TDCPCl8P)OH resemble those catalyzed by photochemically-activated Fe(TDCPP)OH.

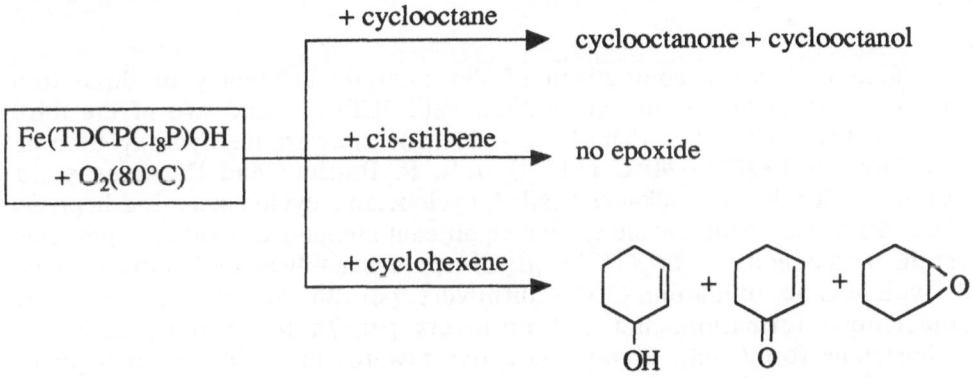

Figure 6. Oxidation of hydrocarbons by O_2 catalyzed by Fe(TDCPCl8P)OH (80°C, O_2 saturated hydrocarbon in benzene).

Cis-stilbene is not epoxidized and alkenes containing allylic C-H bonds like cyclohexene are mainly oxidized at allylic positions (Fig.6). Therefore, it is tempting to propose very similar mechanisms for the Fe(TDCPP)OH-O₂-hv and Fe(TDCPCl8P)OH-O₂-Δ systems, with ˙OH as active species and the intermediate formation of an alkylperoxo-iron (III) complex (Fig.5). The main difference between the two systems concerns their first step, the homolytic cleavage of an Fe(III)-OH bond, which is either photochemical or thermal [25].

Such a "dioxygenase-like" mechanism could explain, at least in part, the results published by Lyons et al. [26] on the oxidation of simple alkanes like propane and isobutane by O₂ in the presence of iron porphyrins (under pressure of O₂ and between 80 and 120°C).

Table 1. Oxidation of alkanes by O₂ in the presence of third generation iron porphyrins[1].

Alkane	Products	Fe(TDCPP)Cl	Fe(TDCPCl8P)Cl	Fe(TDCPN6P)Cl
cyclooctane	ol + one	0	1550	2125
	ol/one	-	0.3	0.25
cyclohexane	ol + one	0	317	640
	ol/one	-	1.26	1.3
heptane	ols+ones	0	24	190
	ols/ones	-	0.33	0.6
	-2/-3/-4	-	38/37/25	38/38/24

[1] Conditions : iron porphyrin (1mM) in a 1:1 alkane:benzene mixture; 90°C, PO₂:10 bars. ol and one are used for the alcohol and ketone derived from the starting alkane. Yields are given in mol of product per mol of catalyst per 2h. For heptane, heptan-2,-3, and 4-ol and heptan-2,-3, and 4-one are formed.

Table 1 shows a comparison of the catalytic efficiency of three iron porphyrins, one of the second generation, Fe(TDCPP)Cl, and two of the third generation, Fe(TDCPCl8P)Cl and Fe(TDCPN6P)Cl, toward the hydroxylation of alkanes by O₂ itself at 90°C (J.F. Bartoli, P. Battioni and D. Mansuy, in preparation). For the three alkanes studied, cyclooctane, cyclohexane and heptane, Fe(TDCPP)Cl was found unable to give significant amounts of oxidation products under the used conditions. Fe(TDCPCl8P)Cl was reasonably active for cyclooctane and cyclohexane oxidation (1550 turnovers per 2h for cyclooctanol and cyclooctanone formation, and 317 turnovers per 2h for cyclohexanol and cyclohexanone formation). It was less active towards the oxidation of heptane, with only 24 mol of heptanols and heptanones formed per mol of catalyst per 2h. Interestingly, Fe(TDCPN6P)Cl was much more active than Fe(TDCPCl8P)Cl for the three alkanes studies, the best increase of activity (8-fold) being observed in the case of heptane. This is another illustration of the catalytic potency of these third generation iron β-polynitroporphyrins, which have already been reported (in this paper) to act as particularly good catalysts for oxidation of hydrocarbons by H₂O₂.

4 - CONCLUSION

Two kinds of systems based on Fe(III) or Mn(III) porphyrins are available now for the oxidation of hydrocarbons. The first ones involve such a metalloporphyrin catalyst and an oxygen atom donor like PhIO, H_2O_2 or O_2 and a reducing agent [2-10]. They reproduce quite well the reactions catalyzed by cytochrome P450-dependent monooxygenases and involve a high-valent metal-oxo active species which is able to epoxidize alkenes, hydroxylate alkanes and aromatic compounds and perform N- or S- oxidations.

Third generation iron porphyrins, which bear several electron-withdrawing β-substituents, very often are much better oxidation catalysts than iron porphyrins of the first and second generation. For instance, Fe(TDCPCl8P)Cl is particularly active for the hydroxylation of alkanes by PhIO (80% yield for heptane), Fe(TDCPN6P)Cl is very active for the epoxidation of alkenes by H_2O_2 without cocatalyst, and Fe(TFPS4P) for the oxidation of dimethoxyarenes to quinones or muconic diesters. Such homogeneous or supported metalloporphyrin models for monooxygenases should be very good tools for oxidations in fine chemistry and for the study and prediction of the oxidative metabolism of drugs.

The second class of systems use iron porphyrin catalysts and O_2 in the absence of a reducing agent. After a photochemical or thermal activation of their Fe(III)-axial ligand bond, some iron porphyrins catalyze the oxidation of alkanes by O_2 to the corresponding ketones (and alcohols in some cases). Although more work is needed to completely determine their mechanisms, these systems seem to involve a free radical derived from their iron axial ligand as active species. This explains why they perform the oxidation of alkanes but not the epoxidation of alkenes contrary to monooxygenases. Such systems are attractive as they oxidize alkanes by O_2 without consumption of any reducing agent and seem to involve a "dioxygenase-like" mechanism though there is no dioxygenase so far reported to oxidize alkanes in living organisms. Third generation iron porphyrins, and particularly Fe(TDCPN6P)Cl, appear as more active catalysts also in these systems.

REFERENCES

1. D. Mansuy and P. Battioni, "Dioxygen Activation at Heme Centers in Enzymes and Synthetic Analogs", in : "Bioinorganic Catalysis", J. Reedjik, ed., Marcel Dekkker, Inc., New-York, (1993).
2. T.J. McMurry, J.T. Groves, "Cytochrome P-450 : structure, mechanism and biochemistry" (P.R. Ortiz de Montellano, ed.) p.1, Plenum Press, New York. (1986).
3. B. Meunier, *Bull. Soc. Chim., Fr.*, 4:578(1986).
4. T.C. Bruice, *Ann. N.Y. Acad. Sci.*, 471:83 (1987).
5. D. Mansuy, *Pure Appl. Chem.*, 59:759 (1987).
6. D. Mansuy, P. Battioni, J.P. Battioni, *Eur. J. Biochem.*, 184:267 (1989).
7. D. Mansuy, *Pure Appl. Chem.*, 62:741 (1991).
8. M.J. Gunter and P. Turner, *Coord. Chem. Rev.*, 108:115 (1991).
9. T.G. Traylor, *Pure Appl. Chem.*, 63:265 (1991).
10. B. Meunier, *Chem. Rev.*, 92:1411 (1992).
11. J.T. Groves, T.E. Nemo, R.S. Meyers, *J. Am. Chem. Soc.*, 101:1032 (1979).
12. P. Battioni, J.P. Renaud, J.F. Bartoli, M. Reina-Artiles, M. Fort, D. Mansuy, *J. Am. Chem. Soc.*, 110:8462 (1988) .
13. J.F. Bartoli O. Brigaud, P. Battioni and D. Mansuy, *J. C. S. Chem. Comm.*, 440-442 (1991).

14. D. Mandon, J. Fischer, R. Weiss, K. Jayaraj, R.N. Austin, A. Gold, P.S. White, O. Brigaud, P. Battioni and D. Mansuy, *Inorg. Chem.*, 31:2044 (1992).

15. O. Brigaud, P. Battioni, D. Mansuy and C. Giessner-Pretre, *New Journal of Chemistry*, 16:1031 (1992).

16 I. Artaud, K. Ben Aziza, C. Chopard and D. Mansuy,*J. Chem. Soc. Chem. Comm.*, 31-33 (1991)

17. I. Artaud, H. Grennberg and D. Mansuy *J.C.S.Chem. Comm.*, 15:1036 (1992).

18. I. Artaud, K. Benaziza, D. Mansuy, *J. Org. Chem.*, in press (1993).

19. L.Barloy, J.P. Lallier, P. Battioni, D. Mansuy, Y. Piffard, M. Tournoux, J.B. Valim and W. Jones, *New Journal of Chemistry*, 16:71 (1992).

20. P. Battioni, O. Brigaud, H. Desvaux, D. Mansuy and T.G. Traylor, *Tetrahedron Lett.*, 32:2893 (1991).

21. T.G. Traylor, Y. S. Byun, P.S. Traylor, P. Battioni and D. Mansuy, *J.Am. Chem.Soc.*, 113:7821(1991).

22. P. Battioni, J.F. Bartoli, D. Mansuy, Y.S. Byun and T.G. Traylor, *J.C.S. Chem. Comm.*, 15:1051 (1992).

23. L. Barloy, P. Battioni and D. Mansuy.*J. Chem. Soc. Chem. Comm.*, 19:1365 (1990) .

24. A. Maldotti, C. Bartocci, R. Amadelli, E. Polo, P. Battioni and D.Mansuy, *J.C. S. Chem. Comm.*,1487 (1991).

25. D. Mansuy, *Coord. Chem. Rev.*, (1993) in press.

26.P.E. Ellis and J.E. Lyons, *Coord. Chem. Rev.*, 105:181 (1990).

OXYGEN ACTIVATION BY TRANSITION METAL COMPLEXES OF MACROBICYCLIC CYCLIDENE LIGANDS

Bradley K. Coltrain, Norman Herron, and Daryle H. Busch

The Ohio State University
Columbus, Ohio 43210

ABSTRACT

The cyclidene ligand produces effective dioxygen binding in the metal ions cobalt(II) and iron(II). This and the facility with which the structure may be varied or expanded has led to the design of species with large cavities (vaulted cyclidenes) that bind both dioxygen and organic molecules. This provides *look-alike* models for the ternary complex that precedes the reaction cycle of cyctochromes P450. The synthesis and characterization of the iron, manganese and chromium derivatives of the vaulted cyclidenes has opened the way for their evaluation as functional models. The O_2 complex of the vaulted iron(II) cyclidene has been prepared in solution; the reactions of the bound O_2 have led to formation of peroxo complexes, analogous to those proposed for P450, and the promoted oxidations of organic substrates with O_2 have been explored. With surrogate oxidants, the vaulted cyclidenes display reactivities reminiscent of the iron porphyrins, and like the early versions of those oxidation catalysts, the cyclidenes are destroyed during the promoted reactions. Success in promoting reaction between activated metal centers and substrate that is clearly within the hydrophobic receptor has been limited. At this early stage it can be concluded that the cyclidene complexes display the same capabilities as oxidation catalysts that have previously been demonstrated for porphyrin complexes.

INTRODUCTION

The lacunar cyclidene complexes of cobalt(II) are extremely robust and display high dioxygen affinities under a wide range of conditions,[1-6] and the iron(II) complexes with these ligands enjoy the distinction of being the only non-porphrin iron(II) derivatives that behave as stable dioxygen carriers for extended periods of time under ambient conditions.[2,4,8-13] This motivates work designed to incorporate these species in model systems that proceed from the capability to bind O_2. The much studied family of enzymes, cytochromes P450, dominate research of this sort and many of the related investigations

The Activation of Dioxygen and Homogeneous Catalytic Oxidation,
Edited by D.H.R. Barton *et al.*, Plenum Press, New York, 1993

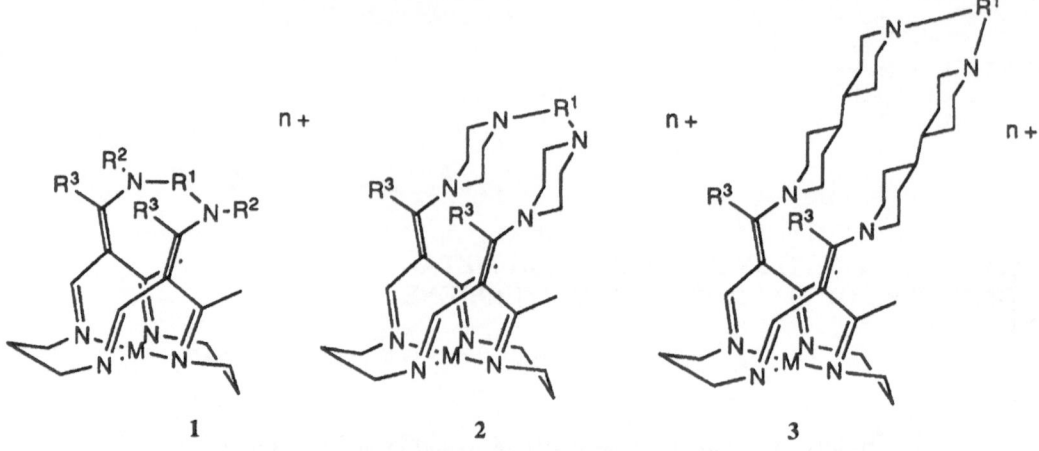

Figure 1. Structures of cyclidene complexes: 1) lacunar cyclidene complex; 2) vaulted cyclidene complex; 3) supervaulted cyclidene complex.

have made use of various porphyrin derivatives,[14-19] although other complexes, and even simpler species,[20-25] have been studied.

From among synthetic systems, the cyclidenes offer two persuasive features when one considers them as possible biomimics for the P450 enzymes: the iron(II)/O_2 complexes are well characterized and many elaborations of the cyclidene structure are readily accomplished.[2-4,7,12,26-34] **Figure 1** presents the *lacunar* cyclidene structure (**structure 1**); the m-xylylene = R^1 bridged species have provided the best iron(II) dioxygen carriers. **Structures 2 and 3** show the *vaulted* and *supervaulted* cyclidene ligands, derivatives with greatly expanded cavities. Extensive solution studies,[4,35-41] mostly using NMR relaxation methods, have shown that organic molecules form guest-host complexes with the vaulted and supervaulted cyclidene complexes in aqueous solutions. Further, substantial regio- and substrate selectivity has been observed. In a particularly intriguing experiment using a cobalt(II) vaulted complex and its O_2 adduct, it was shown that the ternary cobalt cyclidene : dioxygen : organic substrate complex is indeed formed in these systems.[42] Thus a cyclidene system provided the first still-life model of the ternary complex of cytochrome P450.

Ultimate goals of the research producing all of the results described in this brief summary are to imitate the function of the ternary complex of cytochrome P450. To that end, we report here the synthesis and characterization of the iron, manganese, and chromium complexes of vaulted cyclidene ligands. The ability of the vaulted iron(II) cyclidene complex to bind O_2 has been established in this work, and reactions of the dioxygen adduct with reducing agents have been studied. Attempts have been made to oxidize organic substrates, beginning with the dioxyen adduct of the iron complex, and the first investigations on the oxidations of organic substrates using surrogate oxidants are reported.

RESULTS AND DISCUSSION

Iron Vaulted Cyclidene Complexes

The ligands were synthesized by template reactions using nickel(II) as described earlier.[32,33,43] The ligands were liberated with HBr gas in methanol and the free ligand salts served as starting materials for the synthesis of the new complexes. The structures in **Figure 1** provide the basis for the abbreviated nomenclature used here to identify any given cyclidene ligand. Our ligand abbreviations are of the form "LR^2R^3(R^1)", where L stands for

Table 1. Temperature dependence of magnetic moment for [Fe{VMe(3,6-durene)}Cl]PF$_6$ in 311APW.

Temperature, K	μ_{eff}
288	5.03
283	5.00
278	4.99
273	4.94
268	4.90
263	4.85
258	4.78
253	4.65
248	4.52
243	4.39
238	4.28
233	4.22

lacunar, the bridge R^1 is specifically identified as a polymethylene chain or, perhaps, a m-xylylene group, and the substituents R^2 and R^3 are specified; "VR^3(R^1)", where V means *vaulted*, and the bridge R^1 and R^3 are specified (the riser is always piperazine); "UR^3(riser)", where U means *unbridged*, and the piperazine, piperidine (or other) riser and R^3 are specified.

The vaulted iron(II) cyclidene complexes were easily isolated as orange to red crystalline bis(hexafluorophosphate) salts or chloro hexafluorophosphate salts. The extreme air sensitivity and proclivity toward capturing varying numbers of molecules of solvent led to difficulties in obtaining simply definable and reliable elemental analyses. However, careful work led to pure samples and elemental analysis was an important part of the characterization of these compounds (see Experimental Section). Magnetic moments were determined by the Evans method at 25°C in acetonitrile solution: [Fe{VMe(3,6-durene)}Cl]PF$_6$, 5.0BM; [Fe{VC$_5$H$_{11}$(3,6-durene)}](PF$_6$)$_2$, 4.87BM; [Fe{VPh(9,10-anthracene)}](PF$_6$)$_2$, 5.41; and [Fe{UPh(piperidine)}](PF$_6$)$_2$, 4.68. The low values in the set may indicate a mixture of spin states as described by **equation 1**; such a process has been

$$Fe^{II}(V)py \ + \ S \ \Longleftrightarrow \ Fe^{II}(V)(py)S \tag{1}$$

studied in detail for one system and is described below. In the solid state [Fe{VMe(3,6-durene)}Cl]PF$_6$ exhibits Mossbauer parameters ($\delta = 0.93$ mm s^{-1}, $\Delta E_Q = 2.62$ mm s^{-1}) that confirm its high spin ferrous state, but in frozen 3:1:1 acetone/pyridine/water (henceforth 311 APW), the parameters ($\delta = 0.48$ mm s^{-1}, $\Delta E_Q = 0.61$ mm s^{-1}) indicate a low spin structure. This is consistent with studies on lacunar iron(II) complexes; as cavity size increases, the six-coordinate species becomes increasing important. The change in spin state has been quantitated over the temperature range 233-288K by magnetic susceptibility measurements on 311APW solutions (**Table 1**). The equilibrium of **equation 1** is characterized by $\Delta H = -4.3 \pm 0.2$ kcal/mole and $\Delta S = -19.4 \pm .6$ eu. Calculations assumed the low spin magnetic moment to be 0.60 Bohr magnetons and the high spin moment to be 5.31 BM. These observations confirm the binding of sixth ligands within the cavities of the vaulted iron(II) complex.

Electrochemical data for one unbridged and four vaulted cyclidene complexes of iron(II) appear in **Table 2**. The most obvious feature is the dependence of redox potential on the axial ligand. For mixed chloride/bromide derivatives, the potential is much more negative than that for complexes having neutral acetonitrile as the axial ligand. The potentials are generally consistent with values reported for lacunar iron(II) cyclidene

Table 2. Electrochemical data[a] for the first oxidation reaction of vaulted and unbridged iron(II) cyclidene complexes.

R^1	R3	Axial Ligand	E1/2(V)	ΔEp(V)
3,6-durene	Me	Cl Br	$\approx$-0.35[b]	0.18
9,10-anthracene	Me	Cl Br	$\approx$-0.35[b]	0.20
3,6-durene	n-C_5H_{11}	acetonitrile	-0.09	0.10
9,10-anthracene	C_6H_5	acetonitrile	0.04	0.08
-	C_6H_5	acetonitrile	-0.18	0.07

[a]0.1M TBAH/CH_3CN vs. Ag/$AgNO_3$ in CH_3CN at 100 mV.s^{-1}; working electrode is a Pt bead; 3-compartment cell employed; under N_2.
[b]Broad.

complexes.[10,31,45] For example, replacing R^3 = alkyl with phenyl shifts the potential anodically. Reversibility varied with ΔE_p values ranging from 70 to 200mV.

The iron(III) complex [Fe{VMe(3,6-durene)}](PF_6)·3HPF$_6$ was prepared by oxidation of the corresponding iron(II) complex with ceric ammonium nitrate in methanol. The purple crystalline product is clearly high spin under most conditions and the ESR parameters at 77K in various solvents are given in **Table 3**. In the presence of excess N-methylimidazole (NMI), or in neat pyridine, low spin signals are also observed, indicating the influence of strong axial ligands on spin state.

Chromium Vaulted Cyclidene Complexes

The complexes [Cr{VMe(3,6-durene)}(py)Cl]PF$_6$ and [Cr{UMe(piperidine)}(py)Cl]PF$_6$ were synthesized from Cr(Py)$_2$Cl$_2$, under N_2, using triethylamine to liberate the ligand from its salt. Even under carefully protected conditions, the chromium(III) complex was isolated. The product was recrystallized from pyridine/ethanol, yielding the six-coordinate complexes containing chloride and pyridine as axial ligands. Magnetic susceptibility measurements gave magnetic moments of 3.74 for the unbridged complex and 3.64 for the vaulted

Table 3. ESR spectral parameters for [Fe{VMe(3,6-durene)}(H$_2$O)$_2$](PF$_6$)$_3$·HPF$_6$·2H$_2$O in frozen glasses at 77K.

Solvent	g-Values
311 APW	6.32, 5.52, 2.00
DMF	6.33, 5.53, 2.00
Acetone	6.29, 5.48, 2.00
DMSO	6.38, 5.57, 2.02
KHP buffer, pH 4.0	6.31, 5.46, 2.00
20% NMI/MeOH	6.13, 5.37, 2.00 2.31, 2.19, 1.90
20% NMI/DMF	2.17, 1.96
Pyridine	5.60, 2.00 2.16, 1.95

complex. ESR spectroscopy is consistent with high spin d^3 in pseudo octahedral complexes, showing broad signals in the region around g = 3.48. Electrochemical studies on the unbridged complex showed two reversible waves at $E_{1/2}$ = 0.63V (ΔE_p, 90mV) and $E_{1/2}$ = 0.90V (ΔE_p = 100 mV). The lower potential process is assumed to involve the Cr^{4+}/Cr^{3+} couple and the higher potential is ascribed to a ligand oxidation. In acidic media the vaulted complex also gave two waves with $E_{1/2}$ = 0.83 V and 1.06V (ΔE_p = 100 mV in each case).

Manganese Vaulted Cyclidene Complexes

The method reported by Herron for the synthesis of lacunar cyclidene complexes of manganese[44] was used to prepare unbridged cyclidene complexes [Mn{UPh(piperidine)}Br](PF$_6$)$_2$ and [Mn{UMe(piperidine)}Br](PF$_6$)$_2$ and the lacunar complex [Mn{LMe(CH$_2$)$_4$}Cl](PF$_6$)$_2$, but attempts to make pure vaulted complexes of manganese failed. The starting material was tris(acetylacetonato)manganese(III). At 25°C, the magnetic moments in acetonitrile solution are 4.66BM, 4.73BM, and 4.82BM, respectively. Electrochemical measurements in acetonitrile revealed clean, reversible Mn^{4+}/Mn^{3+} redox couples: for [Mn{UMe(piperidine)}Br](PF$_6$)$_2$ $E_{1/2}$ = -0.47V (ΔE_p = 90mV); for [Mn{UPh(piperidine)}Br](PF$_6$)$_2$ $E_{1/2}$ = -0.34V (ΔE_p = 80mV); and for the lacunar complex $E_{1/2}$ = -0.55V (ΔE_p = 110mV). These values are slightly more negative than those reported for the iron complexes but consistent with literature values for the manganese derivatives;[44] however, with other metal ions, the most negative potential was not observed with lacunar species. As observed for iron and nickel, the phenyl substituent lowers the redox potential. In keeping with earlier reports on the lacunar complexes,[44] a distinct ESR spectrum was observed for these species in frozen solutions at 77K; the very broad, isotropic signal centered at g $\simeq$ 2.01 is split into a very poorly resolved sextet (A_{Mn} $\simeq$ 110G) by hyperfine coupling to the manganese nucleus (I = 5/2) in frozen acetonitrile.

Dioxygen Adducts of Vaulted Iron(II) Cyclidenes

In keeping with expectations generated by earlier work with lacunar cyclidene complexes, the vaulted iron(II) cyclidene complexes are extremely air sensitive.[10,12,46-48] The large cavity in the vaulted complex neither limits the dioxygen affinity nor protects the bound O$_2$ from other reagents. Stable dioxygen adduct formation could be observed with the vaulted iron(II) complexes in 311APW solutions at, or below, -30°C. Above this temperature rapid decomposition occurred. The solvent is uniquely convenient for these systems because the acetone provides good solubility for the PF$_6^-$ salts and the water facilitates replacement of axially bound chloride ligands by pyridine, and nitrogen axial bases strongly promote stable dioxygen adduct formation. Exposure of a solution of [Fe{VMe(3,6-durene)}Br]PF$_6$ to only 0.11 torr of O$_2$ results in total conversion to the dioxygen adduct as signaled by the appearance of a strong absorption band at 560 nm and a shoulder at 660 nm in the visible spectrum (**Figure 2**). This spectral change closely resembles those that occur upon formation of the well documented dioxygen adducts of the lacunar iron(II) cyclidenes.[8-10,12] The O$_2$ adduct was further identified from Mossbauer spectra measured on frozen 311APW solutions (120K) that had been exposed to 760 torr of O$_2$. The Mossbauer spectral parameters compared closely to those of a lacunar complex and of the picket fence porphyrin[49]: [Fe{LMeMe(m-xylene)}(py)O$_2$]$^{2+}$ δ = 0.23 mm s^{-1}, ΔE_Q = 2.39 mm s^{-1};   [Fe{VMe(3,6-durene)}(py)O$_2$] δ = 0.24 mm s^{-1}, ΔE_Q = 2.44 mm s^{-1}. Increasing the pressure by a factor of 5 has no detectable effect on the spectrum, indicating that the equilibrium is saturated; i.e., P_{50} << 0.1 torr. At the higher temperature of -28°C, exposure of a similar solution to 0.30 torr of O$_2$ is sufficient to saturate the adduct formation equilibrium. Further, prolonged purging with N$_2$ led to no appreciable decrease in the absorption band arising from the presence of the dioxygen adduct; this suggests that P_{50} is indeed much smaller that 0.1 torr. Our apparatus is inadequate for the determination

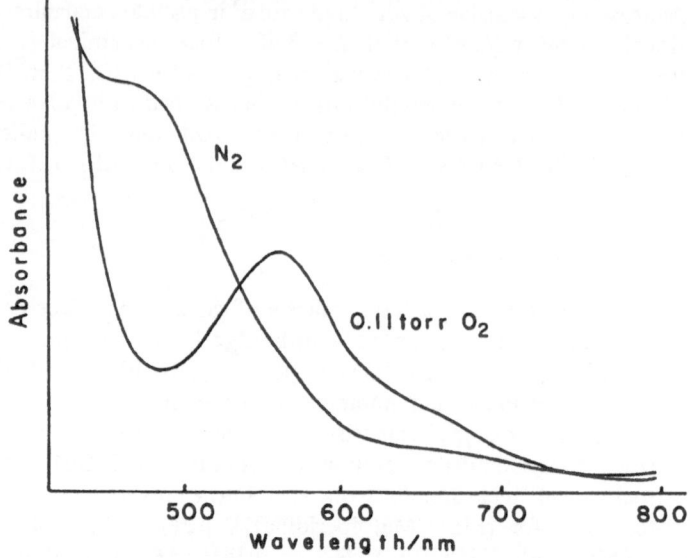

Figure 2. Changes in visible spectrum upon exposure of a solution of [Fe{VMe(3,6-durene)}Br]BF$_4$ to dioxygen at -40°C in 311APW.

of such large dioxygen binding constants (P$_{50}$ values so small). These observations confirm the expectation that the very large vaulted cavity would be accompanied by a very large dioxygen affinity. In contrast, the dioxygen affinity for the lacunar iron(II) cyclidene dioxygen carrier [Fe{LMe(PhCH$_2$)(m-xylene)}Cl]PF$_6$ is P$_{50}$ = 330 torr (K$_{O2}$ = .003 torr^{-1}) at -20°C.[10] Dioxygen adduct formation was also detected for the vaulted iron(II) cyclidene complexes in the solvents pyridine, DMF, and 20% N-methylimidazole/pyridine, but decomposition was more evident even at low temperatures. We conclude at this point that the vaulted iron(II) complex has been shown capable of accomplishing the second requirement for formation of the ternary complex that biomimics that of cytochrome P450 - - it forms a dioxygen complex. The first requirement is the ability to hydrophobically bind an organic substrate in the large vaulted cavity. This was demonstrated in earlier studies with corresponding complexes of divalent nickel, copper and cobalt.[35-42]

Reactions of the Bound Dioxygen

Many previous studies pave the way to seek the activation of the dioxygen that is bound to iron in the vaulted cyclidene complexes.[26-29] The results of detailed studies on the lacunar cyclidene complexes are particularly germane and helpful.[30,46-48,50] Mechanistic studies revealed that the destruction by dioxygen (autoxidation) of the lacunar complexes occurs by an outer sphere electron transfer mechanism and that the dioxygen adducts of the iron cyclidenes, and by inference other iron dioxygen carriers, are stable to this primary mechanism. A second important result is that the iron(III) product of this autoxidation may capture peroxide ion produced in the process and form a low spin peroxo adduct. A number of investigators have provided positive identification of end-on low spin peroxo complexes of iron(III) in simple porphyrins derivatives[51-53] and in natural products.[54-59] The outstanding characteristic is a rhombic, or more rarely axial, low spin ESR spectrum with distinctly small anisotropic splittings. It is particularly significant with respect to the thesis of this report that, in addition to autoxidation, the same peroxo complexes can be produced by a variety of independent routes (**equations 2-5**).[47] The fact that the product can be formed

$$[Fe(L)B(O_2)]^{2+} + e^- + H^+ \longrightarrow [Fe^{III}(L)B(HO_2^-)]^+ \qquad (2)$$

$$[Fe^{III}(L)B)]^{3+} + H_2O_2 \longrightarrow [Fe^{III}(L)B(HO_2^-)]^+ + H^+ \qquad (3)$$

$$[Fe^{II}(L)B)]^{2+} + HO_2 \longrightarrow [Fe^{III}(L)B(HO_2^-)]^+ \qquad (4)$$

$$[Fe^{II}(L)B)]^{2+} + PhI=O + OH^- \longrightarrow [Fe^{III}(L)B(HO_2^-)]^+ + PhI \qquad (5)$$

by all of these routes confirms the spectroscopic assignment of the structure as a peroxo complex of iron(III). The reaction of **Equation 5** is particularly significant because it establishes the reverse of the bond cleavage reaction that is critical to the activation of bound dioxygen.[47]

The autoxidation of $[Fe^{II}\{VMe(9,10\text{-anthracene})\}py]^{2+}$ in 311APW, at 0°C, in the presence of 38 torr of O_2, is accompanied by electronic spectral changes that parallel those for the related lacunar complexes (**Figure 3**). The band at 560 nm that is characteristic of the dioxygen adduct decreases in intensity and a new band at 660 nm grows in intensity, changes that are also observed in the autoxidation of the corresponding lacunar complexes. The ESR spectrum of the product solution at low temperatures and relatively short times is also closely related to those observed for the lacunar species. A high spin component (g = 6.50, 5.16) appears to be associated with the same product that is formed by ceric ion oxidation of the original iron(II) complex. However, a low spin species forms that is both familar and a bit unusual. It has the distinctly small splitting that identifies the low spin peroxo complexes, but it is an axial spectrum (g = 2.18, 1.96), rather than the rhombic spectrum that is most commonly observed for the lacunar species. The axial spectrum was observed for the lacunar complexes only in very dry acetonitrile solutions. An obvious conjecture is that the large cavity is indeed anhydrous, as assumed to be necessary for hydrophobically driven guest-host complexation.[35,37-39,42] Since evidence exists that the piperazine amines may be protonated under certain conditions, it might better be assumed that the axial signal reflects a different level of protonation of the coordinated peroxide, as

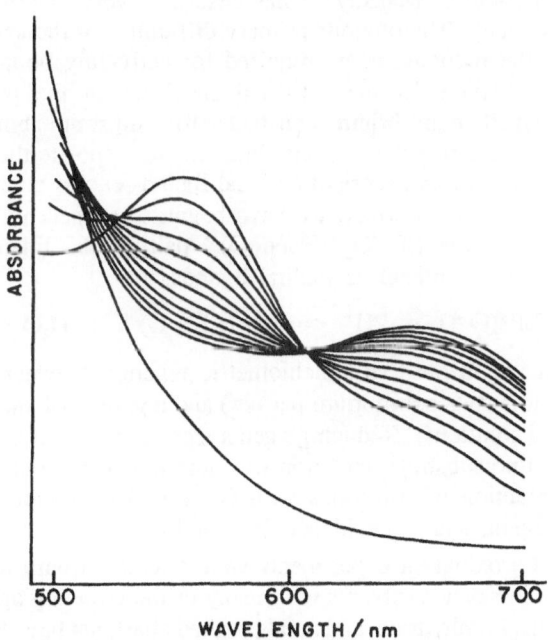

Figure 3. Changes in visible spectra during autoxidation of $[Fe\{VMe(9,10\text{-anthracene})\}Cl]PF_6$ in 311APW at 0°C and 38 torr of dioxygen.

might be expected in changing the polarity of the intra-cavity environment. At higher temperatures, the iron compounds become ESR silent, but Mossbauer data indicates that only high spin ferric species are present (δ = 0.41 mm s^{-1}, ΔE_Q = 0.68 mm s^{-1}). These long term products are almost certainly oxo or hydroxo bridged dinuclear species, as characterized for the lacunar iron(II) cyclidene complexes.[60]

The addition of small amounts ($\sim$1 equivalent) of ascorbic acid or dithionite to the dioxygen adduct of the vaulted iron cyclidene produced the axial signal attributed to the end-on peroxo complex (g = 2.18, 1.96). With dithionite, that signal is accompanied by a broader rhombic pattern (g = 2.37, 2.18, 1.91) that becomes the only spectral pattern when increasingly large amounts of dithionite are added. Similarly, excess ascorbic acid leads to a second dominant product, but in that case it is high spin (g = 6.94, 4.82, 2.00) and different from any previously reported high spin iron(III) cyclidene derivatives. It is suggested that the reducing agents or their oxidation products bind to the iron(III) in these cases. High spin species similar to that produced by adding an excess of ascorbic acid to the dioxygen adduct of the vaulted iron(II) complex were also observed in a variety of other related systems, all of which involved autoxidation of the complex. Ascorbic acid was replaced by hydroquinone, sodium acetate, sodium benzoate, or acetic acid. Further, pyridine was not necessary in order to observe these changes. Additional studies are needed, but it is suggested that these high spin species involve different oxygen donor ligands occupying the axial site, but with a peroxo group coordinated within the cavity; i.e., they may be the high spin peroxo complexes of the cyclidenes, a species that has been sought but not observed previously.[61]

Attempted Substrate Autoxidations

Developments described above comprise both proof of the formation of the ternary complex between the vaulted cyclidene, dioxygen, and substrate and demonstration that reducing agents cause the formation of the peroxo complex that is the stoichiometric equivalent of the activated species in the cytochrome P450 reaction cycle.[14-19] The stage is set for selective autoxidation of substrates by the vaulted cyclidene complexes and this *grail* has indeed been sought. Mansuy[14,15] has described very clearly the complexity of achieving this ultimate goal. The obvious primary difficulty for the systems at hand is the unavoidable fact that the reducing agent required for activating bound O_2 may compete effectively with the substrate for the activated species. In the present case, this is aggravated by the small equilibrium constants for substrate binding. Additional complications arise from the fact that the coordination site opposite the in-cavity site must be blocked by the presence of an excess of an axial ligand--excess axial ligand might also act as a competitive substrate. Further, we have begun to suspect that the equilibium of **equation 6** may vary for the LFe(O_2^{2-})/(LFe=O)$^{3+}$ pair as the ligand L changes from porphyrins to various other synthetic or natural ligands.[62]

$$LFe(O_2^{2-}) + 2H^+ \Longleftrightarrow (LFe=O)^{3+} + H_2O \qquad (6)$$

Attempts were made to observe stoichiometric reaction between the [Fe{VMe(3,6-durene)}(HO$_2^-$)]$^{2+}$ (or rather its equilibrium partner) and a variety of substrates in 311APW, acetonitrile, or acetone solutions. Reducing agents were also varied and included ascorbic acid and dithionite, but no substrate oxidation was detected. The possibility of vaulted iron cyclidene promoted reaction of substrates with O_2 was also explored in the absence of reducing agents but, again, with no evidence for reaction.

More traditional O_2 oxidation experiments were attempted with the vaulted iron and manganese complexes in several solvents with many of the reducing agents (ascorbic acid, sodium ascorbate, thiols, borohydride, and amalgamated zinc) that have been used elsewhere in work with porphyrins and their derivatives.[14-19] The more promising systems are mentioned here. Various thiols, such as thioglycol ethyl ester and β-mercaptoethanol, were added to solutions of iron(III) lacunar, vaulted, and unbridged complexes in acetonitrile. Substrates were added and the solutions were exposed to air. In the case of styrene, small, but detectable, amounts of benzaldehyde were formed during a period of one hour. Also for [Fe{VMe(3,6-durene)}]$^{3+}$ in acetonitrile with cyclohexene as substrate, the presence of β-mercaptoethanol resulted in detectable amounts of cyclohexene oxide which were formed in the course of 3 hours. Unfortunately, even the reactions that gave some product were poorly reproducible--in some repetitions no oxidation was observed.

In pyridine/water solution, the system making use of [Mn{VMe(piperidine)}Br]$^{2+}$, sodium borohydride, and O_2 produced a very small amount of cyclohexene oxide from cyclohexene. Similar systems involving different solvents, substrates and reducing agents failed to give evidence for autoxidation. The addition of acyl halides and acetic anhydride to these solutions in an effort to facilitate O-O bond cleavage by acylation also failed to produce detectable concentrations of products.[63]

Despite a very substantial accumulation of negative results, these few positive observations place the cyclidene chemistry in a situation that parallels early stages of developments with porphyrin autoxidation catalysts. Indeed the literature clearly shows that many attempts preceded substantial success and that early claims of success were not easily reproduced.[64] In fact, even today after enormous efforts by many investigators, the porphyrin promoted air oxidation of organic substrates has not been brought under adequate control. The main reason is characteristic of the difference between *in vivo* chemistry and laboratory chemistry; the various components in the multiple component system cannot be restricted to their intended roles in the reactions. In synthetic systems, substrates, activating reducing agents, and axial ligands compete interchangeably for all intended roles.

Oxidations with Iodosobenzene in Acetonitrile Solution

Surrogate oxidizing agents function effectively in the presence of iron, manganese, and chromium cyclidene complexes, as they are well know to function with porphyrins, other metal chelates and even metal salts.[14-25, 64d] Iodosobenzene has been chosen for these initial studies with cyclidenes, a choice in keeping with the relative bulk of literature in this area of research. Solubility considerations led us to use acetonitrile as the solvent for the first investigations even though we recognized that this would greatly lower the probability that oxygenation would occur inside the cavity of a vaulted complex since hydrophobic forces are responsible for guest-host complexes in these systems.[35-42] Iodosobenzene is sufficiently soluble to facilitate reaction in acetonitrile, but not in acetone or water. Lacunar, unbridged, and vaulted cyclidenes have all been used in order to discern those factors that are important in the promoted oxidations. Reactions were conducted at least in duplicate and in an inert atmosphere glove box to eliminate any complications O_2 might cause. Reaction time was 2 hours and temperature was ambient, ~25°. The yields were reproducible within 10-15% and blanks from which the catalyst was omitted yielded virtually no oxidation products.

Most generally, **Tables 4-7** report product yields and total turnovers for the substrates styrene, cyclohexene, cyclohexane, and methylcyclohexane. These data show that the reactions typically promoted by cytochrome P450 and its porphyrin-based models are also promoted by cyclidene complexes. Highly selective epoxidations are readily achieved and even saturated hydrocarbons undergo hydroxylation. As for porphyrins, the manganese cyclidenes generally give the best performance; however, the lacunar iron(III) cyclidene having a very short bridge [Fe{LMeMe(CH$_2$)$_4$Cl]$^+$ provides an intriguing exception with all substrates. The results are generally explainable on the assumption that reactivity of the activitated metal site varies in the expected order Fe > Mn > Cr. Chromium generally shows lesser capability because of an inherent low reactivity, but yields observed in most iron systems are lowered because the very reactive activated iron center destroys itself more rapidly that does the corresponding manganese center. This leads to the supposition that the high yields found for the short-bridged lacunar iron complex reflect an exceptionally long

Table 4. Product yields for oxidation of styrene by PhIO, catalyzed by cyclidene complexes in acetonitrile solution at ~25°C.

Catalyst Fe^{3+}			% yield[a]
LMeMe(CH$_2$)$_4$, Cl	97	3	4100
LMeMe(CH$_2$)$_6$, Cl	85	15	300
LMeMe(m-xylene), Cl	81	19	300
VMe(3,6-durene)	72	28	250
Mn^{3+}			
LMeMe(CH$_2$)$_4$, Cl	92	8	4000
LMeMe(m-xylene), Cl	79	21	2900
UMe(piperidine), Br	93	7	4700
UPh(piperidine), Br	93	7	3200
Cr^{3+}			
VMe(3,6-durene), Cl, py	58	42	400
UMe(piperidine), Cl, py	63	37	350

[a]Yields are based on catalyst concentrations.

Table 5. Product yields for oxidation of cyclohexene by PhIO, catalyzed by cyclidene complexes in acetonitrile solution at ~25°C.10

Catalyst Fe^{2+}				% yield[a]
LMeMe(CH$_2$)$_4$, Cl	67	27	6	1800
LMeMe(CH$_2$)$_6$, Cl	86	4	10	450
LMeMe(m-xylene), Cl	60	18	22	900
VMe(3,6-durene)	84	13	3	500
Mn^{3+}				
LMeMe(CH$_2$)$_4$, Cl	91	3.5	3.5	1700
LMeMe(m-xylene), Cl	92	8	0	1200
UMe(piperidine),Br	89	5.5	5.5	1800
UPh(piperidine),Br	85	10	5	2500
Cr^{3+}				
VMe(3,6-durene), Cl, py	61	25	14	400
UMe(piperidine), Cl, py	70	21	9	450

[a]Yields are based on catalyst concentrations.

Table 6. Product yields for oxidation of cyclohexane by PhIO, catalyzed by cyclidene complexes in acetonitrile solution at ~25°C.

Catalyst	OH	O	X [a]	% yield[b]
Fe²⁺				
LMeMe(CH₂)₄, Cl	75	18	7	40
LMeMe(CH₂)₆, Cl	60	20	20	10
LMeMe(m-xylene), Cl	53	13	33	30
VMe(3,6-durene)	0	100	0	2
Mn³⁺				
LMeMe(CH₂)₄, Cl	40	40	20	107
LMeMe(m-xylene), Cl	47	34	19	90
UMe(piperidine),Br	0	42	58	70
UPh(piperidine),Br	0	28	72	85
Cr³⁺				
VMe(3,6-durene), Cl, py)	32	41	27	35
UMe(piperidine), Cl, py	23	31	46	40

[a]Yields are based on catalyst concentration
[b]X = Cl or Br.

Table 7. Product yields for oxidation of methylcyclohexane by PhIO, catalyzed by cyclidene complexes in acetonitrile solution at ~25°C.

Catalyst	CH₂OH	CH₃ OH	CH₃ OH	CH₃ OH	% yield[a]
Fe²⁺					
LMeMe(CH₂)₄, Cl	0	91	9	Tr	160
LMeMe(CH₂)₆, Cl	0	100	Tr	Tr	10
LMeMe(m-xylene), Cl	0	70	30	Tr	30
VMe(3,6-durene)	0	100	Tr	Tr	15
Mn³⁺					
LMeMe(CH₂)₄, Cl	0	100	Tr	Tr	60
LMeMe(m-xylene), Cl	0	100	Tr	Tr	60
UMe(piperidine), Br	8	83	9	Tr	90
UPh(piperidine), Br	19	74	7	Tr	70
Cr³⁺					
VMe(3,6-durene), Cl, py	0	67	33	Tr	35
UMe(piperidine), Cl, py	0	62	38	Tr	20

[a]Yields are based on catalyst concentrations.

life for that particular iron catalyst. The most obvious consequence of the short bridge is the improbability of having either substrate or a fifth ligating group inside the cavity. This suggests that the good results come about because the activated species is a 5-coordinated $(LFe=O)^{3+}$, whereas the other iron complexes may form 6-coordinated activated forms. Analogously, one perspective on the role of the presence of mercaptide as the 5th donor in cytochrome P450 is that under highly oxidative conditions this ligand might be oxidized to the radical leaving a 5-coordinated activated species. For a given level of oxidation, the lower coordination number would enhance the electrophilicity of the Fe=O group. In keeping with the assertion that yields are determined by catalyst degradation, careful GC-MS studies on the reactions solutions have yielded fragments of ligand destruction. When the unbridged piperidine complex was employed as catalyst, significant amounts of acetylpiperidine were detected (structure 4), indicating cleavage of the vinyl group of the ligand. Detailed mechanistic studies on the autoxidation of lacunar cyclidenes have confirmed this mechanism and traced it to the acidity of the R^3 methyl groups, which participate in a conjugate base-ligand oxidation mechanism.[65,66]

4

Olefin epoxidations are both productive, with up to 47 catalytic turnovers, and moderately selective, with manganese showing the greatest selectivity. The product yields are greatly diminished on changing the substrate from olefin to alkane, a commonly observed result that is in keeping with the reactivities of the substrates, especially in systems where the catalyst is being destroyed. The most interesting products are the cyclohexyl chloride and bromide (the halide depends on the axial ligand of the catalyst); similar products were observed long ago in porphyrin catalyzed oxidations, and these results are consistent with the accepted mechanism of oxidation of alkanes under the catalysis of porphyrins.[16,67,68]

Yields remained low with methylcyclohexane, but the observed selectivity is interesting. Despite steric demands, the tertiary hydrogen was abstracted in the vast majority of cases, with most of the remaining reaction at the secondary hydrogens. Only two catalysts led to attack on the primary methyl hydrogens and these were the least sterically hindered manganese catalysts. In contrast to the suggestion for iron at the end of the preceding paragraph, one must conclude that the most accessible manganese atom forms the most reactive (least selective) activated form. Alternatively, the lack of selectivity may indicate a different mechanism traceable to the bromide axial ligand of these unbridged catalysts. In support of this view, it should be noted that these catalysts produce about 3 times the % yield of cyclohexyl halide as do the corresponding manganese catalysts that have chloride as axial ligands. A more interesting but not necessarily better justified point of view proceeds from the assumption that the halide remains bound to the side opposite the natural cleft of the unbridged complex. The rather bulky piperidine groups might make it difficult for the cyclohexane ring to approach the metal center, thereby favoring attack at the smaller methyl group. Finally to obtain an additional calibration point for the reactivities of these catalysts, catalytic oxidation of styrene by iodosobenzene was attempted using the simple compound, ferric chloride. Both styrene oxide and benzaldehyde were formed but in a combined yield of only 27%, about 200 times less than that due to our best cyclidene catalysts.

Oxidations with Hydrogen Peroxide in Water

The vaulted cyclidenes were designed to facilitate selective oxidation of organic substrates through mimicking the action of the ternary complex of cytochrome P450. Because

Table 8. Oxygenation of toluene by hydrogen peroxide in aqueous solution at ~25°C, catalyzed by iron cyclidenes of various cavity sizes.

Catalyst	A. Cresols(%)	B. Benzaldehyde(%)	Ratio(A/B)
[Fe{LMeMe(CH$_2$)$_4$}Cl]$^+$	11.9	6.0	2.0
[Fe{LMeMe(m-xylene)}Cl]$^+$	15.9	7.4	2.1
[Fe{LMeMe(CH$_2$)$_6$}Cl]$^+$	25.0	6.6	3.8
[Fe{VMe(3,6-durene)}]$^{2+}$	30.9	4.6	6.7
*[Fe$_2${DHMe(M-xylene)py]$^{4+}$	39.9	3.6	11.1

*The symbol D here indicates the dinuclear ligand having structure 5.

of the great oxidizing power of the activated center in cytochrome P450, the selectivity is believed to be derived from selective substrate binding, and that aspect of enzyme is central to our vaulted cyclidenes. Our first oxidation studies in acetonitrile provided little liklihood of taking advantage of the selective binding capabilities of the vaulted cyclidene system because the driving force for guest-host complexation with these species is hydrophobic. A series of iron cyclidenes have been used to promote the oxidation of toluene by hydrogen peroxide in aqueous solution (**Table 8**), and the results suggest reaction within the cavity of the vaulted cyclidenes. Low yields of the expected products, cresols and benzaldehyde, were observed because of the rapid destruction of the catalyst under the reaction conditions. The most significant observation is that the ratio of cresols to benzaldehyde depends on the cavity size of the macrobicycle, and the overall yield increases in the same direction. The cavity size increases from lacunar to vaulted to dimeric cyclidene. The latter complex has **structure 5** with a very large cavity between the two metal binding sites. Referring back to **Tables 4-7** and the results of reactions of the iron complexes in acetonitrile, no similar correlation was evident, and that is totally consistent with the conclusion that the reactions occur outside the cavity in nonaqueous solvents. Returning to the hydroxylation of toluene

5

in aqueous media, presumably the very restrictive lacunar complexes with m-xylylene and tetramethylene bridges would permit no oxidations within their small cavities (they barely accept such small ligands as O_2 and HO_2^-). Thus the ratio of cresols to benzaldehyde for these complexes(~2) probably represents the limiting value for sterically unhindered oxidations that occur on the side opposite the bridge. On the other hand, the other complexes have cavities large enough to permit some reactivity inside the voids and it is just these instances where steric considerations could influence the site of oxidative attack. The data strongly suggest that at least some of the oxidation events are occuring inside the cavities of the complexes with the more commodious voids and that the prefereed orientation of toluene within those cavities appears to favor attack on the aromatic moiety. These results provide the first confirmation among the cyclidenes of the design principle for selective catalysis which has been put forth in these studies.

CONCLUSIONS

The iron and chromium complexes of macrobicyclic vaulted cyclidene ligands and manganese complexes of macrocyclic unbridged cyclidene ligands have been synthesized and characterized. The binding to the vaulted iron(II) complexes by O_2 and the reactions of bound O_2 have been investigated. The dioxygen affinities of vaulted iron(II) cyclidenes are very great with P_{50} probably much smaller than 0.1 torr. Reduction of the bound O_2 yields a species identified from its ESR spectrum as a low spin complex of peroxide; this is a presumed intermediate along the activation path for cytochrome P450 and its mimics. Intense efforts to observe stoichiometric oxidations of substrates with solutions of these peroxo complexes failed, but several tantalizing but poor examples of autoxidation of substrates were found under catalytic conditions. In contrast to the difficulties associated with the promotion of oxidations with O_2, iodosobenzene oxidations of substrates in acetonitrile solutions were catalytic for the iron, manganese and chromium cyclidene complexes and both saturated and unsaturated hydrocarbons were attacked. Largely because of the solvent, the iodosobenzene reactions occurred with the substrate outside the cavities of the vaulted cyclidene catalysts. However, the reactivities and selectivities were reminiscent of porphyrin behavior and showed that cyclidene catalysts are capable of performing many of the cytochrome P450 reactions, including hydroxylation of primary, secondary and tertiary saturated hydrocarbon atoms. The dependence of the yields of reaction products on the metal ions also paralleled those of the porphyrins, and the behavior generally implies the same mechanism or mechanisms. At this early stage, and again like the porphyrins, the product yields are limited by competing destruction of the cyclidene catalyst, and the general nature of that reaction has been found. In aqueous solution, using hydrogen peroxide as the surrogate oxidant and toluene as the catalyst, selective attack at the aromatic nucleus increases as the cavity size of the catalyst increases, a result implying that a major fraction of the oxidation events occur within the catalyst cavities.

EXPERIMENTAL

Reagents and Solvents

Solvents used in chromium(III), manganese(III), and iron(II) reactions were dried by conventional means and degassed with nitrogen. All other reagents, solvents, and chemicals were reagent grade and were used without further purification.

Physical Measurements

Elemental analyses were performed by Galbraith Laboratories, Inc., Knoxville, Tennessee. Infrared spectra were obtained from 4000 to 200 cm^{-1} using a Perkin-Elmer

Model 283B Infrared Spectrophotometer. Samples were Nujol mulls held between potassium bromide plates, or as KBr pellets. Electronic spectra were obtained using a Cary 17D Recording Spectrophotometer or a Varian 2300 Spectrophotometer. Exposure of solutions of iron(II) complexes to dioxygen at various partial pressures was achieved with a series of calibrated rotameters (Matheson, E. Rutherford, N.J.) to accurately mix dioxygen and dinitrogen. The gas mixture was first saturated with solvent vapor in a simple bubble cell followed by passage through the sample solution in 1cm quartz optical bubbling cells (Precision Cells, Inc., Hicksville, N.Y.) equipped with a gas inlet and bubbling tube. The cell was thermostated (±0.3°C) by using a Neslab or FTS constant-temperature circulation system with methanol coolant flowing through a Varian double-jacketed cell holder within the nitrogen-flushed cell compartment of the spectrophotometer. The cell temperature was monitored with a calibrated copper-constantan thermocouple attached to the cell holder.

Carbon-13 NMR spectra were recorded on a Bruker WP80 fourier transform spectrometer (30MHz). Deuterated solvents were used for all NMR experiments. Magnetic susceptibilities were measured on a Bruker WM300 or Bruker WH90 NMR spectrometer by the Evans method[69] using a Wilmad coaxial capillary tube fitted with a ground glass joint.

Electrochemical data were obtained by Dr. Ken Goldsby using a Princeton Applied Research Corp. Potentiostat/Galvanostat Model 173 equipped with a Model 175 linear programmer and a Model 179 digital coulometer. Current vs. potential curves were measured on a Houston Instruments Model 200XY recorder. The working electrode for voltammetric curves was a platinum disc with potentials measured vs. an Ag°/Ag^{+} (0.1M) reference. Peak potentials were measured from cyclic voltammograms at a 100 MV.s^{-1} scan rate. Half-wave potentials were taken as the average of the anodic and cathodic peak potentials.

EPR spectra were obtained using a Varian E-112 Spectrometer in the X band at 9.3 GHz with g values quoted relative to external dpph (g = 2.0036). Samples were run as frozen glasses at 77K in quartz tubes.

Mossbauer studies were done at the University of Illinois with the assistance of Mr. Dan English. The Mossbauer transmission spectra were collected by using a vertically positioned Ranger drive, a Reuter-Stokes Ar-CO_2 proportional counter, and a Canberra Series 30 analyzer operating in multiscaler mode. Synchronization was achieved by using the twice-integrated square wave signal (MSB) from the multichannel analyzer to generate the drive signal. The source consisted of approximately 45 m Ci of ^{57}Co diffused in a rhodium matrix. Calibrations were made by using the known splittings of the metallic iron spectrum;[70] all isomer shifts are reported relative to room temperature iron metal. In these calibrations spectral line widths of approximately 0.27 mm s^{-1} were normally observed. The temperature was monitored with either a GaAs or Si diode. The spectral parameters (isomer shifts, quadrupole splittings, line widths, and relative areas) were obtained by a least-squares fitting to a sum of Lorentzian components. The fits were performed by using a modified Gauss-Newton iteration procedure. Enriched $^{57}Fe_2O_3$ (86.06%) was obtained from Oak Ridge National Laboratory and converted to 20 to 33% ^{57}Fe enriched $FeCl_2$ by the method of Chang, DiNello, and Dolphin.[71] The $^{57}FeCl_2$ was then converted to $^{57}Fe(py)_2Cl_2$[73] which was used to insert the iron into the macrocyclic ligands as described below. The Mossbauer spectra were obtained on solids or on frozen solutions at 120K.

Gas chromatography was performed on a Carlo Erba 4160 series GC equipped with a flame ionization detector, a Carlo Erba Model 200 cryogenic unit, and a Houston Instruments B-5000 strip chart recorder. A J and W 30M x 0.32 mm, 1.0 μm film thickness, DB-5 capillary column was employed using a splitless injection at 20°C and programming 20 to 180°C at 10°/min. Standard curves were obtained for the expected oxidation products and were used for quantitation experiments.

GC-MS data was obtained at the University of Colorado, Boulder, with the assistance of Prof. Robert Sievers' research group on an HP-5710A GC coupled with an HP 5980A mass spectrometer, an HP 5933A data system, and an HP 2100S computer. The samples were run under the same conditions listed above for the GC experiments.

Oxidations Using Fe(III) Complexes and H_2O_2

Into a 25 ml Erlenmeyer flask was added 20mg of the Fe(III) catalyst, 0.1 g of the substrate, and 1 ml of CH_3CN. Five ml of distilled water and 3 ml of an 0.975 M stock solution of H_2O_2 were then added with vigorous stirring. The reaction was allowed to continue stirring until no additional color change was detected (usually 45 minutes to 1.5 hours). The mixture was extracted 3 times with 2 ml of CH_2Cl_2. The extracts were combined, dried over $MgSO_4$, filtered, and diluted to 10 ml for analysis. An 0.5 μL injection was used for the GC, and a 2.0 μL injection was used for the GC-MS.

Oxidation with PhIO

Ten mg of the appropriate catalyst was weighed into a 25 ml Erlenmeyer flask and taken into a glove box. The solid was dissolved in 3 ml of CH_3CN and 10 drops of substrate ($\approx$ 0.1 g) was added. Solid PhIO (0.200 g) was added to the solution which was stoppered and allowed to stir for 2 hours. The solution was removed from the glove box, and 10 ml of H_2O was added. This solution was extracted 2 times with 2 ml of CH_2Cl_2, acidified with 5 drops of concentrated H_2SO_4, and extracted 3 times with 2 ml of CH_2Cl_2. The extracts were combined, dried over $MgSO_4$, filtered, and diluted to an appropriate volume in a volumetric flask. An 0.5 μL sample was used for the GC injection with quantitative results being obtained by comparison with a standard curve. The standard curve was prepared by obtaining GC peak heights from a series of concentrations of the expected product. Product identification was made by comparison with authentic samples (GC) or by comparison with mass spectra of the expected products (GC-MS).

Synthesis of Iron(II) Complexes

[Chloro(2,9,10,17,19,25,33,34-octamethyl-3,6,13,16,20,24,27,31-octaazapentacyclo[16.7.7.2^{8,11}.2^{3,6}.2^{13,16}]-octatriaconta-1,8.10,17,19,24,26,31,33-nonaene-κ^4N)iron(II)] Hexafluorophosphate, [Fe{VMe(3,6-durene)Cl}PF$_6$. This compound was synthesized by the method previously reported by Takeuchi.[73]

[Bromo)2,21,23,29-tetramethyl-3,6,17,20,24,28,31,35-octaazaheptacyclo[20.7.7.6.^{8,15}2.^{3,6}2^{17,20}.0^{9,14}.0^{37,42}]-hexatetraconta-1,8(37),9,11,13,15(42),21,23,28,30,35,38,40-tridecaene-κ^4N)iron(II) Hexafluorophosphate, [Fe{VMe(9,10-anthracene)}Br]PF$_6$. To a slurry of 0.3 g (0.30 mmol) of [H$_4${VMe(9,10-anthracene)}](Br)$_4$ and 0.1 g (0.5 mmol) of [(dichloro)(diacetonitrile)iron(II)] in 15 ml of ethanol was added 15 dops of triethylamine. This addition causes all of the solid to dissolve and yielded a deep red solution. The solution was warmed for fifteen minutes and allowed to cool to room temperature before filtering. A solution of 0.15 g (0.92 mmol) of ammonium hexafluorophosphate dissolved in 10 ml of methanol was added dropwise to induce the precipitation of a golden-brown microcrystalline product. Yield: 0.25 g (86%). Anal. Calc for Fe $C_{42}H_{54}N_8BrPF_6.H_2O$: C, 52.02; H, 5.82; N, 11.56; Fe, 5.76. Found: C, 52.28; H, 6.12; N, 11.48; Fe, 5.73.

[(23,29-Dimethyl-2,21-diphenyl-3,6,17,20,24,28,31,25-octaazaheptacyclo[20.7.7.6^{8,15}.2^{3,6}.2^{17,20}.0^{9,14}.0^{37,42}]-hexatetraconta-1,8(37),9,11,13,15(42),21,23,28,30,35,38,40-tridecaene-κ^{4}N)iron(II)] Hexafluorophosphate, [Fe{VPh(9,10-anthracene)}](PF$_6$)$_2$. To a solution of 0.55 g (0.44 mmol) of [H$_4${VPh(9,10-anthracene)}]Br(PF$_6$)$_3$ in 20 ml of acetonitrile was added 0.18 g (0.9 mmol) of [(dichloro)(diacetonitrile)iron(II)] and 25 drops of triethylamine. The resulting deep red solution was warmed for 15 minutes and then filtered. The solution was then evaporated to dryness, and the residue was dissolved in 10 ml of ethanol. A solution of 0.3 g (1.8 mmol) of ammonium hexafluorophosphate in 10 ml of ethanol was added. The volume of the resulting solution was slowly reduced to effect precipitation of an orange product. The product was recrystallized from an acetonitrile/ethanol mixture. Yield: 0.15 g (30%). Anal. *Calc. for FeC$_{25}$H$_{58}$N$_8$P$_2$F$_{12}$.CH$_3$CN.2H$_2$O: C, 53.26; H, 5.38; N, 10.35; Fe 4.59. Found: C, 53.03; H, 5.34; N, 10.17; Fe 4.23.

[(9,10,19,25,33,34-Hexamethyl-2,17-di-n-pentyl-3,6,13,16,20,24,27,31-octaazapentacyclo[16.7.7.2^{8,11}.2^{13,16}]-octatriaconta-1,8,10,17,19,24,26,31,33-nonaene-κ^{4}N)iron(II)] Hexafluorophosphate, [Fe{V(C$_5$H$_{11}$)(3,6-durene)}](PF$_6$)$_2$. This reaction was performed as described for the preceding reaction using 0.69 g (0.55 mmol) of [H$_4${V(C$_5$H$_{11}$)(3,6-durene)}]Br(PF$_6$)$_3$, 0.21 g (1.1 mmol) of [(dichloro)(diacetonitrile)iron(II)], and 35 drops of triethylamine in 20 ml of acetonitrile. The resulting powder was recrystallized from an acetonitrile/ethanol mixture yielding large red crystals. Yield 0.25 g (42%). Anal. Calc. for FeC$_{46}$H$_{74}$N$_8$P$_2$F$_{12}$.CH$_3$CN.H$_2$O: C, 50.40; H, 6.96; N, 11.02; Fe, 4.88. Found: C, 50.30; H, 6.62; N, 11.30; Fe 4.52.

[(2,12-Dimethyl-3,   11-bis[1-(piperidino)benzylidene]-1,5,9,13-tetraazacyclohexadeca-1,4,9,12-tetraene-κ^{4}N)iron(II)]   Hexafluorophosphate, [Fe{UPh(piperidine)}](PF$_6$)$_2$. This reaction was performed as described for the preceding reaction using 0.43 g (0.35 mmol) of [H$_4${Ph(piperidine)}]Cl(PF$_6$)$_3$, 0.14 g (0.73 mmol) of [(dichloro)(diacetonitrile)iron(II)], and 20 drops of triethylamine in 20 ml of acetonitrile. The resulting brown microcrystalline product was recrystallized from acetonitrile/ethanol to yield a red-brown crystalline product which was collected, washed with ethanol, and dried in vacuo. Yield: 0.31 g (86%). Anal. Calc. for FeC$_{38}$H$_{50}$N$_6$P$_2$F$_{12}$.2CH$_3$CH$_2$OH: C, 49.03; H, 6.07; N, 8.17; Fe, 5.43. Found: C, 48.73; H, 6.10; N, 8.22; Fe, 5.21.

Synthesis of ^{57}Fe Substituted Complexes

The synthesis of 20% to 33% ^{57}Fe enriched iron(II) chloride was performed according to the published procedure.[72] This latter comound was used as the iron source for metal insertion into the ligand salts as described below.

Chloro(2,3,11,12,14,20-hexamethyl-3,11,15,19,22,26-hexaazatricyclo-[11.7.7.1^{5,9}]-octacosa-1,5,7,9(28),12,14,19,21,26-nonaene-κ^{4}N)iron(II)*] Hexafluorophosphate, [Fe*{LMeMe(m-xylene)}Cl](PF$_6$). This compound was synthesized according to the published procedure[10] by substituting ^{57}Fe enriched [(dichloro)(dipyridine)iron(II)] for [(dichloro)(diacetonitrile)-iron(II)]. No analytical data was obtained.

[Chloro(2,9,10,17,19,25,33,34-octamethyl-3,6,13,16,20,24,27,31-octaazapentacyclo[16.7.7.2^{8,11}.2^{3,6}.2^{13,16}]-octatriaconta-1,8,10,17,19,24,26,31,33-nonaene-κ^{4}N)iron(II)*] Hexafluorophosphate, [Fe*{VMe(3,6-durene)}Cl](PF$_6$). To a solution of 0.26 g (0.23 mmol) of [H$_4${VMe(3,6-durene)}]Br(PF$_6$)$_3$ and 0.070 g (0.25 mmol) of 33% ^{57}Fe enriched [(dichloro)(dipyridine)iron(II)] in 15 ml of acetonitrile was added 17 drops of triethylamine. The resulting deep red solution was warmed for 15 minutes, cooled to room temperature, and filtered. This solution was evaporated to near dryness and 15 ml of ethanol was added to induce the precipitation of a golden-yellow microcrystalline solid. This solid was filtered, washed with ethanol, and dried. Yield: 0.15 g (76%). No analytical data was obtained.

Synthesis of Iron(III) Complex

[(2,9,10,17,19,25,33,34-Octamethyl-3,6,13,20,24,27,31-octaazapentacyclo[16.7.7.2^{8,11}.2^{3,6}.2^{13,16}]-octatriaconta-1,8,10,17,19,24,26,31.33-nonaene-κ^4N)iron(III)] Hexafluorophosphate, [Fe{VMe(3,6-durene)}H$^+$](PF$_6$)$_4$. To a slurry of 0.35 g (0.40 mmol) of [Fe{VMe(3,6-durene)}Cl](PF$_6$) in 10 ml of methanol was added 0.21 g (0.40 mmol) of ceric ammonium nitrate in 2 ml of methanol. The resulting solution immediately turned blue and a blue precipitate was formed. The solution was allowed to stir for 2 hours to insure the completion of the reaction. The blue precipitate was collected, washed with a small amount of methanol, dried, and removed from the glove box. The blue solid (0.25 g) was dissolved in 10 ml of water, and 3 drops of concentrated nitric acid were added to prevent the formation of μ-oxo dimer. The solution was heated (*caution:* too much heating will demetallate the complex), and a solution of 0.36 g (2.2 mmol) of ammonium hexafluorophosphate in 5 ml of water was added dropwise to induce the precipitation of a purple crystalline product. The product was collected, washed with water, and air dried. This complex should be stored in a glove box since it is somewhat moisture-sensitive. Yield: 0.17 g (35%). Anal. Calc. for FeC$_{38}$H$_{59}$N$_8$P$_4$F$_{24}$·4H$_2$O: C, 34.17; H, 5.06; N, 8.39; Fe, 4.18; P, 9.27. Found: C, 34.17; H, 4.89; N, 8.49; Fe, 4.09; P, 8.99.

Synthesis of Manganese(III) Complexes

[Bromo(2,12-dimethyl-3,11-bis[1-(piperidino)ethylidene]-1,5,9,13-tetraazacyclohexadeca-1,4,9,12-tetraene-κ^4N)manganese(III)]Hexafluorophosphate, [Mn{UMe(piperidine)}Br](PF$_6$)$_2$. A 1.0 g (1.0 mmol) sample of [H$_4${UMe(piperidine)}]Br(PF$_6$)$_3$ was dissolved in 50 ml of acetonitrile, and to this was added tris(2,4-pentanedionato)manganese(III) (0.53 g, 1.5 mmol). The solution was refluxed for 45 minutes and then pumped to dryness. The residue was slurried with toluene and then slurried with ethanol to remove any unreacted Mn(acac)$_3$. The sandy brown powder which remained was recrystallized from acetonitrile/ethanol to yield dark red-brown crystals. Yield: 0.49 g (55%). Anal. Calc. for MnC$_{28}$H$_{46}$N$_6$BrP$_2$F$_{12}$: C, 37.72; H, 5.20; N, 9.43; Mn, 6.16; Br, 8.96. Found: C, 37.88; H, 5.49; N, 9.45; Mn, 6.01; Br, 9.01.

[Bromo(2,12-dimethyl-3,11-bis[1-(piperidino)benzylidene]-1,5,9,13-tetraazacyclohexadeca-1,4,9,12-tetraene-κ^4N)manganese(III)]Hexafluorophosphate, [Mn{UPh(piperidine)}Br](PF$_6$)$_2$. This reaction was performed in the same manner as the preceding reaction using 0.65 g (0.59 mmol) of [H$_4${UPh(piperidine)}]Br(PF$_6$)$_3$ and 0.33 g (0.94 mmol) of tris(2,4-pentanedionato)manganese(III) in 30 ml of acetonitrile. Recrystallization from acetonitrile/ethanol yielded a deep brown crystalline product. It should be noted that the recrystallization was done on the bench-top rather than in a glove box. Yield: 0.31 g (50%). Anal. Calc. for MnC$_{38}$H$_{50}$N$_6$P$_2$F$_{12}$Br·2H$_2$O: C, 43.40; H, 5.18; N, 7.99; Mn, 5.22; Br, 7.60. Found: C, 43.40; H, 4.82; N, 7.97; Mn, 5.20; Br, 7.59.

[Chloro(2,3,8,9,11,17-hexamethyl-3,8,12,16,19,23-hexaazabicyclo-[8.7.7]-tetracosa-1,9,11,16,18,23-hexaene-κ^4N)manganese(III)] Hexafluorophosphate, [Mn{LMeMe(CH$_2$)$_4$}Cl](PF$_6$)$_2$. A 0.85 g (0.96 mmol) sample of [H$_4${LMeMe(CH$_2$)$_4$}]Cl(PF$_6$)$_3$ was slurried in 10 ml of acetonitrile, and to this was added 0.47 g (1.3 mmol) of tris(2,4-pentanedionato)manganese(III) dissolved in 5 ml of acetonitrile. The deep red-brown solution was stirred and warmed for 45 minutes and then pumped to dryness. The residue was slurried with toluene to removed any unreacted Mn(acac)$_3$. The remaining brown solid was recrystallized three times from acetonitrile/ethanol. The repeated recrystallizations were apparently necessary in order to remove some unreacted ligand salt. A brown crystalline solid was obtained. Yield: 0.45 g (59%). <u>Anal</u>. Calc. for MnC$_{24}$H$_{40}$N$_6$ClP$_2$F$_{12}$: C, 36.35; H, 5.08; N, 10.60; Mn, 6.93; Cl, 4.47. Found: C, 36.42; H, 5.06; N, 10.73; Mn, 6.09; Cl, 4.47.

Synthesis of Chromium(III) Complexes

trans-[(Chloro)(pyridine)(2,12-dimethyl-3,11-bis[1-(piperidino)-ethylidene]-1,5,9,13-tetraazacyclohexadeca-1,4,9,12-tetra ene-κ^4N)chromium(III)] Hexafluorophoshate, [Cr{UMe(piperidine)}Cl(py)](PF$_6$)$_2$. A 1.32 g (1.3 mmol) sample of [H$_4${UMe(piperidine)}]Cl(PF$_6$)$_3$ was dissolved in 20 ml of acetonitrile, and to this was added 0.45 g (1.6 mmol) of [(dichloro)(dipyridine)chromium(II)][76] and 40 drops of triethylamine. The deep red solution was stirred and warmed for 20 minutes and then reduced in volume to 5 ml. Ethanol was then added dropise to induce the precipitation of a yellow powder which was collected and washed with ethanol. Recrystallization from pyridine/ethanol yielded a yellow microcrystalline product. Yield: 0.50 g (42%). <u>Anal</u>. Calc. for CrC$_{33}$H$_{51}$N$_7$ClP$_2$F$_{12}$: C, 42.93; H, 5.57; N, 10.62; Cr, 5.63. Found: C, 43.30; H, 5.45; N, 10.61; Cr, 5.34.

trans-[(Chloro)(pyridine)(2,9,10,17,19,25,33,34-octamethyl-3,6,13,16,20,24,27,31-octaazapentacyclo[16.7.7.2^{8,11}.2^{3,6}.2^{13,16}]-octatriaconta-1,8,10,17,19,24,26,31,33-nonaene-κ^4N)chromium(III)] Hexafluorophosphate, [Cr{VMe(3,6-durene)}Cl(py)](PF$_6$)$_2$. This reaction was performed as was the preceding reaction using 1.07 g (0.88 mmol) of [H$_4${VMe(3,6-durene)}]Br(PF$_6$)$_3$, 0.4 g (1.2 mmol) of [(dichloro)(dipyridine)chromium(II)], 70 drops of triethylamine, and 25 ml of acetonitrile. Recrystallization from pyridine/ethanol yielded a greenish-yellow microcrystalline product. Yield: 0.37 g (39%). <u>Anal</u>. Calc. for CrC$_{43}$H$_{63}$N$_9$ClP$_2$F$_{12}$: C, 47.67; N, 5.86; N, 11.64; Cr, 4.80. Found: C, 47.61; H, 5.78; N, 11.50; Cr, 4.52.

ACKNOWLEDGEMENT

The financial support of the National Institutes of Health is gratefully acknowledged. We thank Dr. Ken Goldsby for his help with the electrochemical experiments, Dr. Tom Meade for his contribution to a variety of NMR experiments, Dr. Dan English and Professor David Hendrickson for the use of their laboratory and for assistance with the Mossbauer measurements, Professor Robert Sievers for providing his facilities and guidance for the GC-MS determinations, and to Professor Kenneth Takeuchi for contributing analytical data on one compound.

REFERENCES

1. H. E. Tweedy, N. W. Alcock, N. Matsumoto, P. A. Padolik, N. A. Stephenson, and Daryle H. Busch, *Inorg. Chem.* 29:616 (1990).
2. R. Thomas, C. M. Fendrick, W.-K. Lin, M. W. Glogowski, M. Y. Chavan, N. W. Alcock, and D. H.Busch, *Inorg. Chem.* 27:2292 (1988).
3. D. H. Busch, Synthetic Dioxygen Carriers for Dioxygen Transport, *in*: Oxygen Complexes and Oxygen Activation by Transition Metals, A. E. Martell and D. T. Sawyer, Ed., Plenum Publishing Corp., N.Y., (1988).

4. J. H. Cameron, M. Kojima, B. Korybut-Daszkiewicz, B. K. Coltrain, T. J. Meade, N. W. Alcock, and D. H. Busch, *Inorg. Chem.*, 26:427 (1987).
5. J. C. Stevens, P. J. Jackson, W. P. Schammel, G. G. Christoph, and D. H. Busch, *J. Am. Chem. Soc.*, 102:3283 (1980).
6. J. C. Stevens and D. H. Busch, *J. Am. Chem. Soc.*, 102:3285 (1980).
7. W.-K. Lin, N. W. Alcock, and D. H. Busch, *J. Am. Chem. Soc.*, 113:7603 (1991).
8. N. Herron and D. H. Busch, *J. Am. Chem. Soc.*, 103:1236 (1981).
9. N. Herron, J. H. Cameron, G. L. Neer, and D. H. Busch, *J. Am. Chem. Soc.*, 105:298 (1983).
10. N. Herron, L. L. Zimmer, J. J. Grzybowski, D. J. Olszanski, S. C. Jackels, R. W. Callahan, J. H. Cameron, G. G. Christoph, and D. H. Busch, *J. Am. Chem. Soc.*, 105:6585 (1983).
11. K. A. Goldsby, B. D. Beato, and D. H. Busch, *Inorg. Chem.*, 25:2342 (1986).
12. D. H. Busch, *La Transfusione del Sangue*, 33:57 (1988).
13. P. A. Padolik, A. J. Jircitano, N. W. Alcock, and D. H. Busch, *Inorg. Chem.*, 30:2713 (1991).
14. D. Mansuy, *Pure & Applied Chem.*, 62:4:741 (1990).
15. D. Mansuy, P. Battioni, and J. P. Battioni, *Eur. J. Biochem.*, 184:267 (1989).
16. D. Ostovic and T. C. Bruice, *Accounts Chem. Res.*, 24:244 (1991); T. C. Bruice, ibid., 25:314 (1992).
17. C. L. Hill, Ed., "Activation and Functionalization of Alkanes," Wiley-Interscience, New York (1989).
18. P.R. Ortiz de Montellano, "Cytochrome P450: Structure, Mechanism, and Biochemistry," Plenum Purblishing Corp., New York (1986).
19. M. J. Gunter, and P. Turner, *Coord. Chem. Rev.*, 108:115 (1991).
20. Y. Yang, F. Diederich, and J. S. Valentine, 112:7826 (1990).
21. D. Mota de Freitas, L. J. Ming, N. Ramasamy, W. Nam, and J. S. Valentine, *J. Am. Chem. Soc.*, 112:4877 (1990).
22. J. S. Valentine, R. B. Van Atta, L. D. Margerum, and Y. Yang. *Stud. Org. Chem.* (Amsterdam), 33:175 (1988).
23. W. Nam, R. Ho, and J. S. Valentine, *J. Am. Chem. Soc.*, 113:7052 (1991).
24. Y.-D. Wu, K. N. Houk, and J. S. Valentine, *Inorg. Chem.*, 31:718 (1992).
25. Y. Yang, F. Diederich, and J. S. Valentine, *J. Am. Chem. Soc.*, 113:7195 (1991).
26. D. H. Busch and N. A. Stephenson, *in* "Inclusion Compounds Volume 5: Inorganic and Physical Aspects of Inclusion," J. Atwood, E. Davies, and D. MacNicol, ed. Oxford U. Press, Oxford, 276 (1991).
27. N. W. Alcock, P. A. Padolik, G. A. Pike, M. Kojima, C. J. Cairns, and D. H. Busch, *Inorg. Chem.*, 29:2599 (1990).
28. D. H. Busch and N. A. Stephenson, *J. Inclusion Phen.*, 7:137 (1989).
29. D. H. Busch, and C. Cairns, *in* "Prog. Macrocycle Chem.", R. M. Izatt and J. J. Christensen, ed., page 1, Wiley-Interscience, New York (1987).
30. N. Herron, M. Y. Chavan, and D. H. Busch, *J. Chem. Soc., Dalton Trans.*, 1491 (1984).
31. B. Korybut-Daskiewicz, M. Kojima, J. H. Cameron, N. Herron, M. Y. Chavan, A. J. Jircitano, B. K. Coltrain, G. L. Neer, N. W. Alcock and D. H. Busch, *Inorg. Chem.* 23:903 (1984).
32. D. H. Busch, S. C. Jackels, R. Callahan, J. J. Grzybowski, L. L. Zimmer, M. Kojima, D. J. Olszanski, W. P. Schammel, J. C. Stevens, K. A. Holter, and J. Mocak, *Inorg. Chem.*, 20:2834 (1981).
33. D. H. Busch, G. G. Christoph, L. L. Zimmer, S. C. Jackels, J. J. Grzybowski, R. Callahan, M. Kojima, K. A. Holter, J. Mocak, N. Herron, M. Chavan, and W. P. Schammel, *J. Am. Chem. Soc.*, 103:5107 (1981).

34. D. H. Busch, Pure & Appl. Chem., 52:2477 (1980).

35. T. J. Meade, N. W. Alcock, and D. H. Busch, *Inorg. Chem.*, 29:3766 (1990).

36. N. Hoshino, A. Jircitano, and D. H. Busch, *Inorg. Chem.*, 27:2292 (1988).

37. T. J. Meade, W.-L. Kwik, N. Herron, N. W. Alcock, and D. H. Busch, *J. Am. Chem. Soc.*, 108:1954 (1986).

38. W.-L. Kwik, N. Herron, K. Takeuchi, and D. H. Busch, *J. Chem. Soc., Chem. Commun.* 409 (1983).

39. K. J. Takeuchi and D. H. Busch, *J. Am. Chem. Soc.*, 105:6812 (1983).

40. K. J. Takeuchi, D. H. Busch, and N. W. Alcock, *J. Am. Chem. Soc.*, 105:4261 (1983).

41. K. J. Takeuchi, D. H. Busch, and N. W. Alcock, *J. Am. Chem. Soc.*, 103:2421 (1981).

42. T. J. Meade, K. J. Takeuchi, and D. H. Busch, *J. Am. Chem. Soc.*, 109:725 (1987).

43. C. J. Cairns and D. H. Busch, *Inorg. Synth.*, 27:261 (1990).

44. N. Herron and D. H. Busch, *Inorg. Chem.*, 22:3470 (1983).

45. D. H. Busch, D. J. Olszanski, J. C. Stevens, W. P. Schammel, M. Kojima, N. Herron, L. L. Zimmer, K. A. Holter, J. Mocak, *J. Am. Chem. Soc.*, 103:1472 (1981).

46. L. D. Dickerson, A. Sauer-Masarwa, N. Herron, C. M. Fendrick and D. H. Busch, *J. Am. Chem. Soc.*, in press.

47. A. Sauer-Masarwa, N. Herron, C. M. Fendrick, and D. H. Busch, *Inorg. Chem.*, in press.

48. A. Sauer-Masarwa, L. D. Dickerson, N. Herron, and D. H. Busch, *Coord. Chem. Revs.*, in press.

49. J. P. Collman, R. R. Gagne, C. A. Reed, T. R. Halbert, G. Lang, W. T. Robinson, *J. Am. Chem. Soc.*, 97:1427 (1975).

50. N. Herron, L. Dickerson, and D. H. Busch, *J. Chem. Soc., Chem. Commun.*, 884 (1983).

51. K. Tajima, *Kagaku to Kogyo* (Tokyo), 44:248 (1991).

52. K. Tajima, J. Jinno, K. Ishizu, H. Sakurai, H. Ohya-Nishiguchi, *Inorg. Chem.*, 28:709 (1989).

53. K. Tajima, M. Shigematsu, J.Jinno, K. Ishizu, H. Ohya-Nishiguchi, *J. Chem. Soc., Chem. Commun.*, 144 (1990).

54. J. Stubbe and J. W. Kozarich, *Chem. Revs.*, 87:1107 (1987).

55. M. C. R. Symons and R. L. Peterson, *Biochim. Biophys. Acta*, :5, 241 (1978).

56. N. Bartlett and M. C. R. Symons, *Biochim. Biophys. Acta*, 744:110 (1982).

57. Z. Gasyna, F.E.B.S. Letts., 106:213 (1979).

58. R. Kappl, M. Hohn-Berlage, J. Huetterman, N. Bartlett, M. C. R. Symons, *Biochim. Biophys. Acta*, 827:327 (1985).

59. R. Davydov, R. Kappl, J. Huetterman, J. A. Peterson, *F.E.B.S. Letts.*, 295:113 (1991).

60. P. Padolik, Ph.D. thesis, The Ohio State University, Columbus, Ohio, (1989).

61. High spin peroxo complexes have long been known in other systems, for example: a) E. McCandlish, A.R. Miksztal, M. Nappa, A. Q. Sprenger, J. S. Valentine, J.D. Stong, and T. G. Spiro, *J. Am. Chem. Soc.*, 102:4268 (1980); b) C. H. Welborn, D. Dolphin and B. R. James, *J. Am. Chem. Soc.*, 103:2869 (1981); c) A. Shirazi and H. M. Goff, *J. Am. Chem. Soc.*, 104:6318 (1982).

62. Despite detailed and repeated measurements by various techniques, almost no evidence can be found for the formation of $(LFe=O)^{3+}$ in the cyclidene systems, despite the fact that peroxo complexes form readily and by one route that constitutes the reverse of the reaction of **equation 6**. The results reported here also stand in contrast to recent studies based on non-heme oxygenase model systems that indicate the possibility of multiple pathways, implying that (hydroperoxo)iron(III) species might have catalytic activity; e.g., a) Y.-D. Wu, K. N. Houk, J. S. Valentine, and W. Nam, *Inorg. Chem.*, 31:718 (1992); b) W. Nam, R. Ho, and J. S. Valentine, *J. Am.*

Chem. Soc., 113:7052 (1991); c) Y. Yang, F. Diederich, and J. S. Valentine, *J. Am. Chem. Soc.*, 113:7195 (1991); d) J. S. Valentine, J. N. Burstyn, and L. D. Margerum, in "Oxygen Complexes and Oxygen Activation by Transiton Metals," A.E. Martell and D. T. Sawyer, ed., Plenum Publishing Corp., 175, New York (1988).

63. J. T. Groves, Y. Watanabe, and T. J. McMurry, *J. Am. Chem. Soc.*, 105:4489 (1983).

64. C. C. Franklin, R. B. Van Atta, A. F. Tai, and J. S. Valentine, *J. Am. Chem. Soc.*, 106:814 (1984).

65. M. Masarwa, P. R. G. Warburton, W. E. Evans and D. H. Busch, submitted for publication.

66. K. A. Goldsby, A. J. Jircitano, D. L. Nosco, J. C. Stevens and D. H. Busch, *Inorg. Chem.*, 29:2523 (1990).

67. T. J. McMurry and J. T. Groves, in "Cytochrome P450: Structure, Mechanism and Biochemistry," P. R. Ortiz de Montellano, ed., Plenum Publishing Corp., 1, New York (1986).

68. C. L. Hill, J. A. Smegal, T. J. Henly, J. Org. Chem., 48:3277 (1983).

69. D. Evans, *J. Chem. Soc.*, 2005 (1959).

70. R. S. Preston, S. S. Hanna, and J. J. Heberle, *Phys. Rev.*, 128:2207 (1962).

71. C. K. Chang, R. K. DiNello, and D. Dolphin, *Inorg. Synth.*, 20:147 (1980).

72. G. J. Long, D. L. Whitney, and J. E. Kennedy, *Inorg. Chem.*, 10:1406 (1971).

73. K. J. Takeuchi, Ph.D. thesis, The Ohio State University, Columbus, Ohio, 1981.

74. R. C. Edwards, Ph.D. thesis, The Ohio State University, Columbus, Ohio, 1976.

DIOXYGEN REACTIVITY MODELS FOR CYTOCHROME C OXIDASE: SYNTHESIS AND CHARACTERIZATION OF OXO AND HYDROXO-BRIDGED PORPHYRIN-IRON/COPPER DINUCLEAR COMPLEXES

Alaganandan Nanthakumar,[1] Stephen Fox,[1] Sarwar M. Nasir[1]
Natarajan Ravi,[2] Boi H. Huynh,[2] Robert D. Orosz,[2] Edmund P.
Day,[2] Karl S. Hagen[3] and Kenneth D. Karlin[1] *

[1]Department of Chemistry, The Johns Hopkins University
  Baltimore, MD 21218, USA
[2]Department of Physics, Emory University, Atlanta, Georgia 30322
[3]Department of Chemistry, Emory University, Atlanta, Georgia 30322

INTRODUCTION

The synthesis of appropriate transition-metal complexes to model the structural, spectroscopic, and magnetic properties of a metalloprotein active-site provides an opportunity to consider the function and associated mechanism of that metalloprotein at the molecular level. One nice example is the dinuclear cuprous amine-*bis*-pyridyl complex, which effects arene hydroxylation (albeit of the ligand *m*-xylyl spacer) using molecular oxygen (O_2).[1] This extraordinary reaction involves cleavage of the O-O bond and subsequent insertion of an oxygen atom into an arene C-H bond under essentially ambient conditions, to model the function of copper monooxygenases such as tyrosinase. Another excellent example is the generation of dicupric *trans*-μ-1,2-peroxo complexes from cuprous precursors and O_2, reversibly,[2-4] to model the oxygen-transport property of the protein hemocyanin, which subsequently was discovered to bind O_2 in $\eta^2{:}\eta^2$ fashion, as shown in Figure 1.[4] The metalloprotein cytochrome c oxidase,[5] however, due to its combination of diverse and unusual active-site metal centers, has eluded a convincing model description. As for its function, it probably binds O_2 at a dinuclear site comprising heme-iron and histidyl-copper coordination; it then cleaves the O-O bond, *via* reduction, (*vide infra*).[5] The structural changes associated with this dinuclear site during turnover, and the intermediates produced therefrom, are by no means clearly understood. In the resting state, the dinuclear site exhibits strong antiferromagnetic coupling (-J =200 cm^{-1}) suggesting the involvement of a bridging ligand, often postulated as μ-sulfido, μ-chloro, or μ-hydroxo. Thus, we have endeavored to synthesize model complexes of this enigmatic dinuclear site.

---

* Author to whom correspondence should be addressed.

The Activation of Dioxygen and Homogeneous Catalytic Oxidation,
Edited by D.H.R. Barton *et al.*, Plenum Press, New York, 1993

381

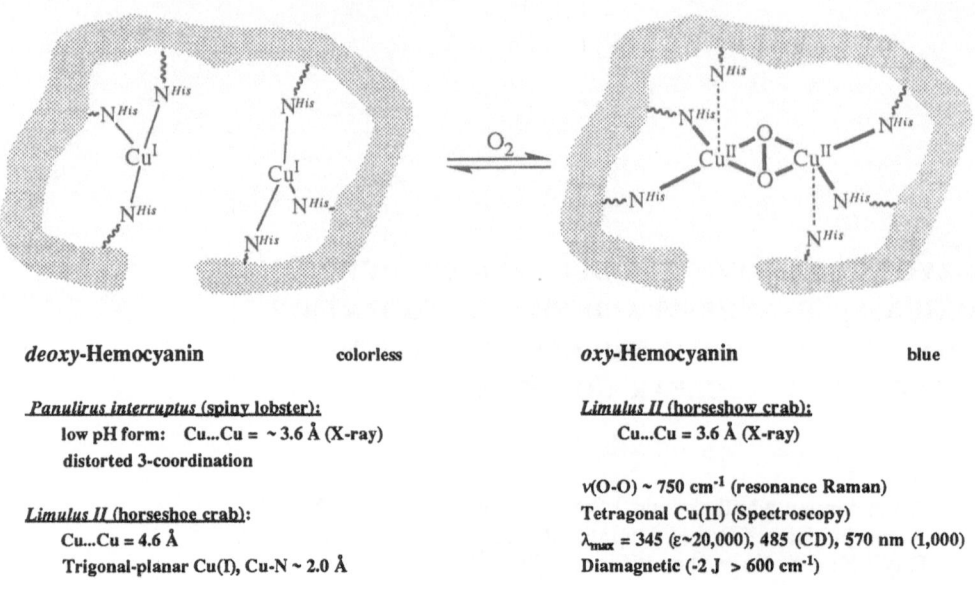

deoxy-Hemocyanin   colorless   **oxy-Hemocyanin**   blue

Panulirus interruptus (spiny lobster):
   low pH form: Cu...Cu = ~3.6 Å (X-ray)
   distorted 3-coordination

Limulus II (horseshoe crab):
   Cu...Cu = 4.6 Å
   Trigonal-planar Cu(I), Cu-N ~ 2.0 Å

Limulus II (horseshow crab):
   Cu...Cu = 3.6 Å (X-ray)

ν(O-O) ~ 750 cm^{-1} (resonance Raman)
Tetragonal Cu(II) (Spectroscopy)
λ_{max} = 345 (ε~20,000), 485 (CD), 570 nm (1,000)
Diamagnetic (-2 J > 600 cm^{-1})

Figure 1: Structural, physical and magnetic properties of *oxy* and *deoxy-* Hemocyanin.

Dioxygen Reactivity of the Tris[2-pyridylmethyl]amine]Cu(I) Complex: Generation and Characterization of a *Trans-μ-*1,2-Peroxo Dicopper(II) Complex.

We have shown in previous studies that the mononuclear Cu(I) complex [(TMPA)Cu(RCN)]$^+$ (1) forms the structurally characterized *trans-μ-*1,2-(O$_2$$^{2-}$)-dicopper(II) complex (2) when reacted with O$_2$ at -80 °C (Figure 2).[2,3] Dioxygen binds strongly at low temperature in dichoromethane or propionitrile solvents to form an intense purple solution (Figure 3; λ_{max} = 440 (ε = 2000 M^{-1}cm^{-1}), 525 (11500), 590 (sh, 7600) and a *d-d* band at 1035 (180) nm). Although the binding of O$_2$ is strong at low temperature, it is reversible as demonstrated by vacuum cycling (application of vacuum while subjecting the solution to mild warming, Figures 2 and 3). Dioxygen can also be displaced from 2 by reaction with CO or PPh$_3$ to give the adducts [(TMPA)Cu(CO)]$^+$ and [(TMPA)Cu(PPh$_3$)]$^+$, respectively.

Figure 2: Reversible binding of O$_2$ to [(TMPA)Cu(RCN)]$^+$ (1) to give a structurally characterized *trans-μ-*1,2-peroxo dicopper(II) complex [{(TMPA)Cu}$_2$(O$_2$)]$^{2+}$ (2).

X-ray data obtained for crystals of [{(TMPA)Cu}$_2$(O$_2$)](PF$_6$)$_2$·5Et$_2$O at -90 °C revealed a *trans*-μ-1,2-O$_2{}^{2-}$ group bridging the two Cu(II) ions. The Cu atoms are pentacoordinate with distorted trigonal bipyramidal geometry and the peroxo oxygen atoms occupy the axial sites. The Cu-Cu' distance is 4.359 Å and the O-O' bond length is 1.432 Å. Resonance Raman studies showed an intraperoxide stretch (832 cm^{-1}) and a Cu-O stretch (561 cm^{-1}).[3]

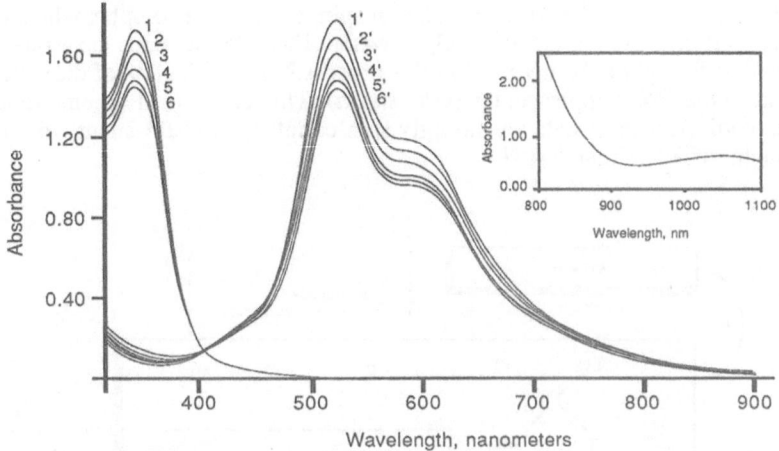

Figure 3. UV-vis spectra demonstrating the reversible O$_2$-binding behavior of [(TMPA)Cu(RCN)]$^+$ (1) in propionitrile to give the dioxygen adduct (2) (*vacuum cycling*). Reaction of 1 (λ_{max} = 345 nm) with O$_2$ at -80 °C produces an intensely violet solution of 2, spectrum 1', λ_{max} = 525 nm. The inset shows the d-d band in the near-IR region.

A detailed kinetic study has been carried out for the formation of 2 from 1 and O$_2$. Figure 4 shows the kinetic scheme deduced from a low-temperature stopped-flow kinetic/spectroscopic study.[6] The 1:1 intermediate [LCu(O$_2$)]$^+$ species was detected at

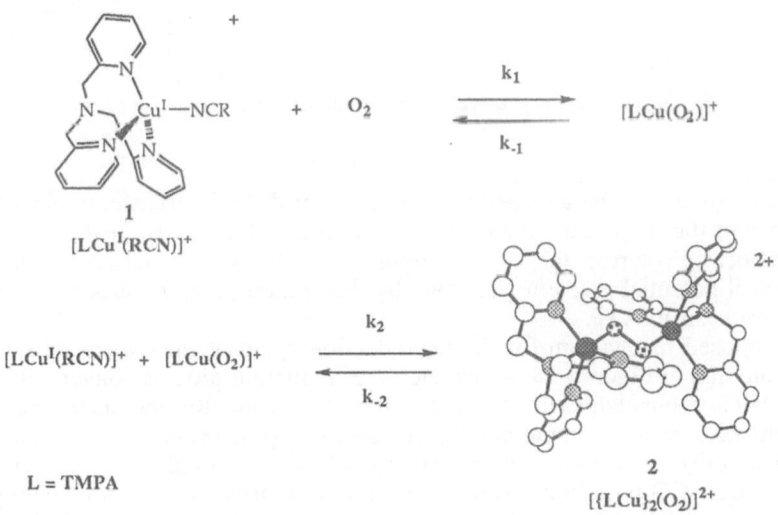

Figure 4: Kinetic scheme for the formation of [{(TMPA)Cu}$_2$(O$_2$)]$^{2+}$.

temperatures below -75° C as a transient species (λ_{max} = 410 nm, ε = 4000 $M^{-1}cm^{-1}$). The exact formulation of this species as a Cu(II)-superoxide or Cu(I)-O_2 adduct is as yet undetermined. The rate of formation of this Cu-O_2 adduct is much faster than that of most Co(II) ligand systems and is comparable to the dioxygen reactivity of Fe(II) porphyrin model complexes.

Dioxygen Reactivity in the Cytochrome *c* Oxidase Enzyme: Role of Copper

Cytochrome *c* oxidase (CcO) is a terminal respiratory protein complex which catalyzes the four-electron four-proton reduction of O_2 to water. Figure 5 diagrams the arrangement of Fe and Cu centers thought to be involved in this process.[5] The reduction of dioxygen occurs at a dinuclear heme-iron/copper center (*vide supra*). This consists of a heme a_3 and Cu_B which in the oxidized resting state are strongly spin-coupled, and EPR silent, with a Cu...Fe distance thought to be less than 5 Å.[5f]

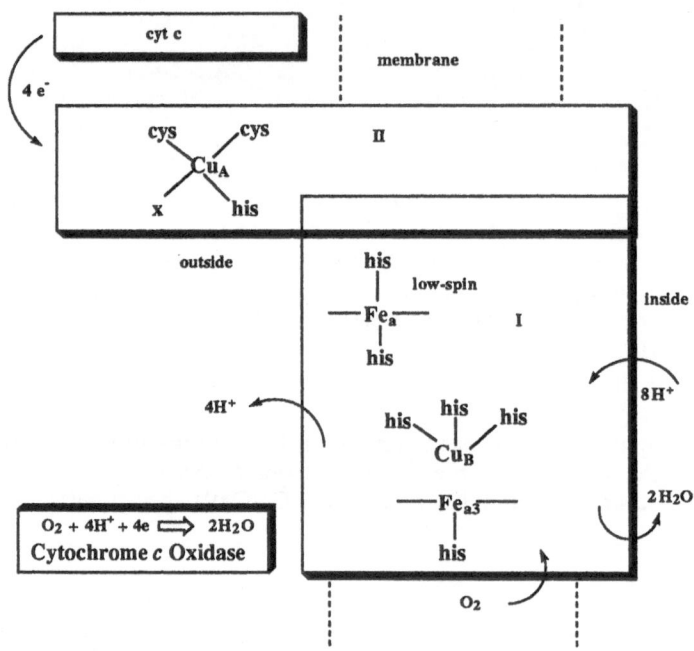

Figure 5: The cytochrome *c* oxidase enzyme and its metal-containing active sites.

A second heme (heme *a*) and a second copper (Cu_A) mediate the transfer of electrons from cytochrome *c* to the dinuclear center. The electron transfer process is directly linked to proton translocation across the cell membrane and the O_2-reduction reaction. The electrochemical potential gradient generated by this proton-pumping process is ultimately used in the synthesis of ATP.

The proposed mechanism for the O_2 reduction cycle of the cytochrome *c* oxidase enzyme is outlined in Figure 6.[5g] A two-step 2-e⁻ reduction process converts the inactive oxidized Fe^{3+}-Cu^{2+} bimetallic site to an active Fe^{2+}-Cu^+ state. Recent kinetic/spectroscopic studies implicate Cu_B as the initial binding site for dioxygen,[7,8] thus affirming (for the first time) that Cu(I)/O_2 interactions are important in CcO. This evidence comes from flash-photolysis of Cu_B-CO protein derivatives[7,8a] as well as from observations (stopped-flow kinetics/spectroscopy) regarding the direct reaction of O_2 with reduced CcO.[8b] The intermediate $Fe-O_2$ species was proposed based on optical and time resolved resonance Raman spectroscopic measurements.[5b,c] The electron transfer associated with the formation of the subsequent peroxo-iron(III) species is expected to be rapid in order to trap the bound dioxygen. After protonation and the third electron transfer, the O-O bond is cleaved to generate a ferryl Fe^{IV}-oxo species.[5b,c] The resting oxidized state is regenerated after the fourth electron transfer and protonation. The third and fourth electron transfer are coupled to the proton translocation process and Cu_B provides the entry point for these electrons.

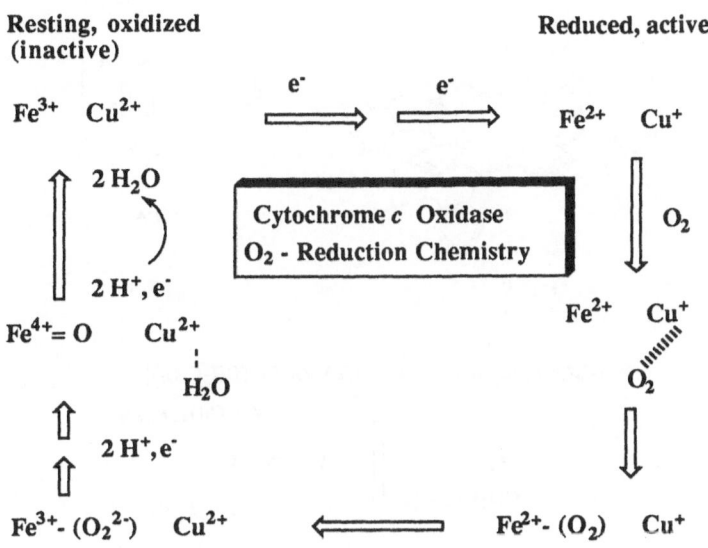

Figure 6: Simplified illustration of the O_2 reduction catalytic cycle for the cytochrome c oxidase enzyme.

Inorganic modeling of this active site dinuclear complex has the potential to be very helpful in elucidating aspects of structure, associated spectroscopy, and mechanism of O_2 reduction. Most of the previous activity associated with such chemistry has been directed at the attempted synthesis and modeling of the magnetic properties of the resting state enzyme, *via* the generation of mixed-metal iron(III)-Cu(II) complexes having oxo, imidazolato, sulfur containing or other bridging ligands.[9,10] Although the reactivity of dioxygen with both Fe(II) porphyrins and Cu(I) complexes have separately been studied in detail, few attempts have been made to characterize the products associated with the reaction of O_2 in the presence of both.[11] Such studies would illuminate structural and mechanistic aspects of the O_2 reduction pathway of the cytochrome c oxidase enzyme. With our considerable previous experience with $Cu(I)/O_2$ reaction chemistry,[2,12,13] we recently decided to also explore porphyrin-Fe with Cu-ligand/O_2 reactivity.

SYNTHESIS OF AN OXO-BRIDGED IRON-COPPER DINUCLEAR COMPLEX

Synthesis from Iron(II) Porphyrin, [Cu(I)(TMPA)]⁺, and Dioxygen

Our initial approach was to use $[(TMPA)Cu(RCN)]^+$ (TMPA = tris[2-pyridylmethyl]-amine) (**1**), and the reduced iron(II) porphyrins {(F$_8$-TPP)Fe(II)-B$_2$, B = pyridine, piperidine, F$_8$-TPP = tetrakis(2,6-difluorophenyl)porphyrin} in the presence of molecular oxygen, since ortho-halogenated iron tetraphenylporphyrins are known to be resistive to the formation of μ-oxo dimers. TMPA was found to be an appropriate ligand for such studies since a recent report indicates a four N-ligand coordination for Cu_B in cytochrome ba_3 from *Thermus thermophilus*.[14a] However, other studies on bacterial oxidases implicate three His ligands for Cu_B.[14b,c] (TPP)Fe(II)-B (TPP = tetraphenylporphyrin) is known to bind dioxygen reversibly at low temperatures (-80° C) in methylene chloride to form an iron porphyrin dioxygen complex.[15,16]

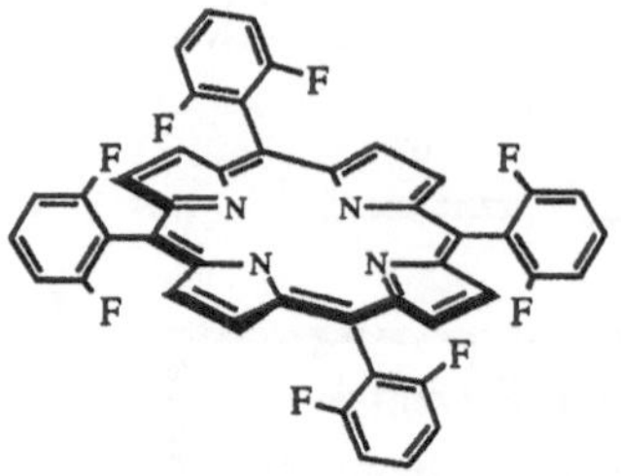

F_8-TPP

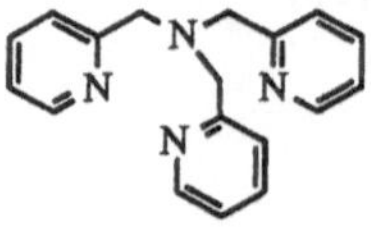

TMPA

$(F_8\text{-TPP})Fe^{II}(pip)_2$ + $[(TMPA)Cu^I(CH_3CN)]Y$

$Y = ClO_4, PF_6$

O_2

CH_2Cl_2

-80° C to 0° C

$[(F_8\text{-TPP})Fe\text{-}(O^{2-})\text{-}Cu(TMPA)]Y$
(3)

Figure 7: Synthetic route for the generation of **3** from dioxygen.

When an equimolar mixture of $(F_8\text{-TPP})Fe(II)pip_2$ (pip = piperidine) and $[(TMPA)Cu]^+$ (**1**) were allowed to react at -80° C in CH_2Cl_2 in the presence of dioxygen and warmed to 0° C, a purple-red solid could be isolated by precipitation with heptane (yield > 80%, Figure 7). A microcrystalline solid formulated as $[(F_8\text{-TPP})Fe\text{-}(O^{2-})\text{-}Cu(TMPA)]^+$ (**3**) was isolated by dissolution of the crude product in acetonitrile and reprecipitation by slow addition of diethylether (Figure 7) in an overall isolated yield of 50 %. Low temperature UV-visible spectroscopic monitoring of this reaction did not indicate the presence of any intermediate species. The above reaction also occurs with $(TPP)Fe(II)pip_2$ and $(OEP)Fe(II)Py_2$ (OEP = octaethylporphyrin; Py = pyridine) but the analogous products could not be isolated as completely pure solids, presumably due to their greater reactivity, i.e., instability.[17]

Elemental analysis for both ClO_4^- and PF_6^- salts of **3** are satisfactory. If **3** is decomposed by addition of NH_4OH(aq) and the CH_2Cl_2 extract is analyzed, a ^{1}H NMR spectrum shows the presence of the $(F_8\text{-TPP})Fe\text{-}OH$ and TMPA ligand signals in a 1:1 mole ratio. Conductivity measurements indicate **3** is a 1:1 electrolyte in CH_3CN, consistent with the formulation. Compound **3** is soluble in CH_2Cl_2 and CH_3CN but is sparingly soluble in toluene and benzene, similar to other known cationic Cu complexes. Immediate decomposition is observed in the presence of protic solvents such as methanol.

Synthesis from Hydroxo-Iron(III) Porphyrin, [Cu(II)(TMPA)]$^{2+}$, and Base

Interestingly, the reaction of equimolar amounts of (F$_8$-TPP)Fe-OH,[18] [Cu(TMPA)(CH$_3$CN)][ClO$_4$]$_2$, and NEt$_3$ in CH$_3$CN under argon (Figure 8) generated a red micro-crystalline product, in 80% yield, identified as **3** on the basis of its identical UV-vis and ^{1}H NMR spectra. Subsequent recrystallization from CH$_2$Cl$_2$/heptane yielded single crystals which gave comparable elemental analysis to **3** generated from dioxygen.

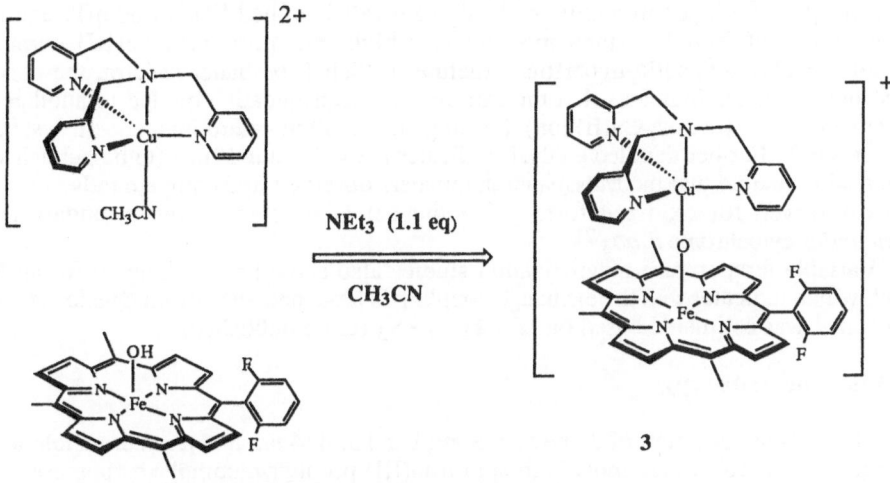

Figure 8: Generation of **3** from deprotonation of hydroxoiron(III) porphyrin.

An X-ray structure determination of **3** established the Fe-O-Cu angle to be 172(2)° with Fe⋯Cu = 3.56 Å, Fe-O = 1.74 (2) Å, and Cu-O = 1.84 (2) Å (Figure 9). These short, virtually co-linear, metal-oxygen bonds implicate considerable double bond character.[19] Complex **3** reacts immediately in CH$_2$Cl$_2$ with SO$_2$, CO$_2$, and equimolar HPF$_6$ to yield distinct products, which we are currently characterizing.

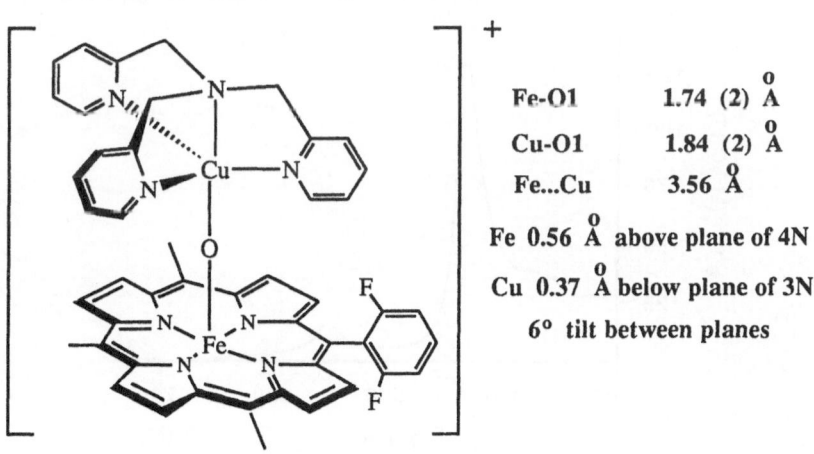

Fe-O1	1.74 (2) Å
Cu-O1	1.84 (2) Å
Fe...Cu	3.56 Å

Fe 0.56 Å above plane of 4N

Cu 0.37 Å below plane of 3N

6° tilt between planes

Figure 9: Structural parameters of oxo-bridged complex **3**.

PHYSICAL METHODS OF CHARACTERIZATION

Mössbauer Spectroscopy and Magnetic Measurements

The Mössbauer spectroscopic properties of solid $[Fe(III)-(O^{2-})-Cu(II)](ClO_4)$ (3-ClO_4) are particularly revealing. The 4.2 K zero-field spectrum is a very sharp quadrupole doublet, with fitted parameters $\delta = 0.47 \pm 0.01$ mm/s and $\Delta E_Q = 1.26 \pm 0.02$ mm/s, typical for high-spin Fe(III)-heme (S = 5/2). However, the quadrupolar splitting value is unusually high relative to other high-spin iron(III) porphyrins but is in the region of that observed for many *non-heme* μ-oxo bridged iron dimers.[20] With a parallel applied field of 60 mT, significant broadening is observed. Since mononuclear high-spin porphyrin Fe(III) complexes generally display magnetic hyperfine structure in their Mössbauer spectra, the observed Mössbauer behavior indicates that the iron in 3 is electronically coupled to another half-integer spin center (i.e. the Cu(II) ion), forming an overall non-zero integer spin system, i.e. S = 2 or S = 3. Further detailed analysis indicates a S = 2 ground state (to be published) It is interesting to note that the Mössbauer parameters observed in 3 compare rather closely to values observed for oxidized forms of both beef heart cytochrome *c* oxidase and *T. thermophilus* cytochrome $c_1 aa_3$.[21]

Variable temperature magnetization studies also show an excellent fit for an S = 2 model, which indicates that the exchange coupling is large and antiferromagnetic. A J value of -89 cm^{-1} was estimated based on $H = 2J\ S_1 \cdot S_2$ (to be published).

UV-Vis Spectroscopy

The UV-vis spectrum of 3 shows a Soret band at 434 nm and another visible band at 554 nm. (Figure 10). Since most high-spin iron(III) porphyrin complexes possess a Soret band in the 400-415 nm region, the observed position at 434 nm for 3 is unusually low in energy. However, this compares well with that seen for the high-spin peroxo Fe(III) complex (with side-on η^2:η^2-peroxo coordination), $[(TPP)Fe(O_2^{2-})]$ characterized by Valentine and co-workers;[22] the red-shift of the Soret band (relative to typical high-spin (P)Fe(III) complexes) has been attributed to the dinegative charge of the peroxo ligand.[22b] The red shift of the Soret band for 3 is surprising since μ-oxo bridged iron porphyrin dimers do not exhibit this trend indicating that the copper-TMPA moiety dramatically alters the electronic properties of the iron-oxo moiety. A similar shift of the Soret band is observed for the (TPP) and (OEP) analogs of 3.[17]

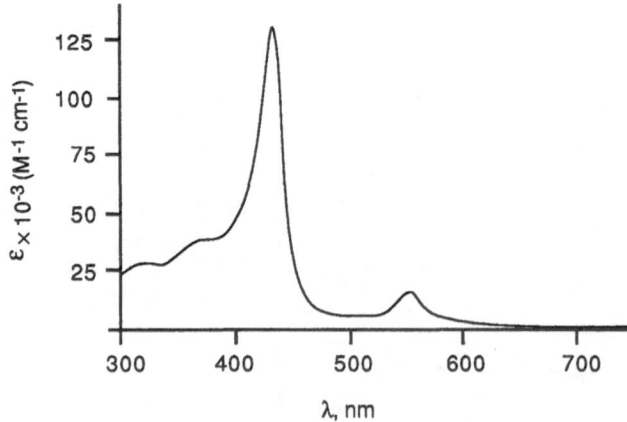

Figure 10: UV-vis spectrum of 3 in CH_3CN at room temperature.

Another unusual feature in the visible spectrum of $[(TPP)Fe(O_2^{2-})]^-$ was the appearance of shoulders in the α, β bands.[22b] $[(F_8\text{-}TPP)Fe(O_2^{2-})]^-$, generated by addition of 2 equivalents of superoxide ion to $[(F_8\text{-}TPP)Fe\text{-}Cl)]$ in dimethylsulfoxide also showed a shoulder in the visible band observed at 560 nm. The apparent splitting of these bands was attributed to the rhombic nature of the structure due to the side-on bound peroxo ligand which eclipses two of the pyrrole nitrogen atoms and pulls the iron above the plane of the porphyrin ligand. However, the visible band at 554 nm for **3** (Figure 10) and the α, β bands of its (TPP) and (OEP) analogs are symmetric in shape, thus indicating an axially symmetric structure consistent with the observed linear Fe-O-Cu bond.

NMR Spectroscopy

The proton NMR spectrum of **3** in CD_3CN and indicates a pyrrole H-resonance at 65 ppm, a split meta-phenyl signal at 9.6 and 9.2 ppm and a para-phenyl signal at 7.8 ppm. A ^2H NMR spectrum of the pyrrole deuteriated analog of **3** confirmed the pyrrole peak assignment and indicated the presence of only one porphyrin species. The observed porphyrin chemical shifts are comparable with other high-spin iron(III) five coordinate complexes.[23] However, the pyrrole signal is shifted upfield relative to a typical $S = 5/2$ complex such as (TPP)Fe-Cl, which shows a pyrrole signal at 80 ppm. The upfield shift may be due to the antiferromagnetic coupling between the Fe ($S = 5/2$) and Cu ($S = 1/2$) which would result in a $S = 2$ electronic ground state. High-spin Fe(II) ($S = 2$) porphyrin complexes generally exhibit pyrrole shifts over a wide range between 30 and 61 ppm.[24] Although spin-admixed ($S = 3/2$, $S = 5/2$) systems may also cause an upfield shift in the pyrrole proton signal[25,26] the observed Mössbauer parameters and the visible spectrum do not support such an assignment. A plot of chemical shift vs reciprocal temperature ($1/T$: K^{-1}) indicated a linear relationship and hence a Curie law behavior in the temperature range 298-210 K. However, the plot does not extrapolate at infinite temperature to a chemical shift expected for a diamagnetic compound. This non-ideal behavior could be a result of the antiferromagnetic coupling between the iron and copper centers. The room temperature magnetic moment of **3** is 5.2 ± 0.2 B.M. (Evans Method in CD_3CN), a value consistent with the suggestion that **3** is a coupled $S = 5/2$ (i.e., heme) and $S = 1/2$ (i.e., Cu(II)) system. The complex **3** is EPR silent at 77 K. However, the close proximity of the Fe and Cu centers could induce rapid electronic relaxation to cause broadening of the EPR signals and hence an apparent EPR silent behavior at 77 K.[27]

Additional signals at 4.5, -6.8, -21.5 and -100.5 ppm are also observed in the NMR spectrum of **3** and are tentatively assigned to the TMPA ligand. The signals at 4.5, -6.8 and -21.5 ppm are of approximately equal integration and roughly correspond to 3 protons (if integrated with respect to the para or meta-phenyl signals); the furthest upfield shifted signal at -100.5 ppm integrates to a higher proton count. Hence this signal could be tentatively assigned to the methylene ($-CH_2$) protons of TMPA. Specific labelling of the methylene protons is needed to confirm this assignment. Precise integration was not possible due to the broad nature of the peaks. All four signals showed an upfield movement upon lowering the temperature.

IR Spectroscopy

The IR spectrum of **3** (solid, nujol) showed a new band at 856 cm^{-1}, which is in the region where the Fe-O-Fe antisymmetric stretch has been reported for most oxo-bridged iron complexes.[20] Use of $^{18}O_2$ gas in the synthesis resulted in lowering of the intensity of this band with corresponding changes in the 780-790 cm^{-1} region. However a firm assignment cannot be made at this time, due to the overlap of strong porphyrin and TMPA ligand absorptions in this region. The antisymmetric vibration of $[(F_8\text{-}TPP)Fe\text{-}O\text{-}Fe(F_8\text{-}TPP)]$ occurs at 867 cm^{-1}.

SYNTHESIS AND CHARACTERIZATION OF A HYDROXO-BRIDGED Fe-Cu COMPLEX

Reaction of an equimolar mixture of [(F$_8$-TPP)Fe-OH)] and [(TMPA)Cu(CH$_3$CN)](ClO$_4$)$_2$ in a non-polar solvent such as dichloromethane or toluene results in the formation of a new species which is tentatively formulated as the hydroxo bridged [(F$_8$-TPP)Fe-(OH)-Cu(TMPA)]$^{2+}$ complex **4** (Figure 11). This species exhibits a pyrrole proton NMR signal at 68.5 ppm in CD$_2$Cl$_2$ which is shifted upfield from the parent [(F$_8$-TPP)Fe-OH] complex, presumably due to a bridging interaction of the hydroxo ligand between the Fe and Cu centers. A bridging hydroxo group may provide a weaker ligand field strength at the axial position of the iron porphyrin so as to cause mixing of an intermediate spin (S = 3/2) state and result in an upfield shift of the pyrrole proton signal. Antiferromagnetic coupling between the Fe and Cu centers could also cause this behavior, similar to the μ-oxo bridged complex **3**.

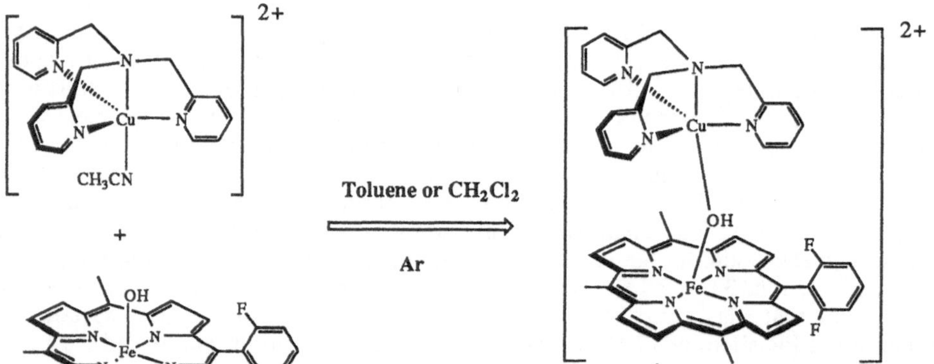

Elemental: C; 54.01 (53.91); 3.03 (2.84); 7.33 (8.11).
^{1}H NMR (CD$_2$Cl$_2$): 68.5 ppm (pyrrole)
λ_{max} = 410 (Soret), 570 nm

Figure 11: Synthesis of the putative hydroxo-bridged complex [(F$_8$-TPP)Fe-(OH)-Cu(TMPA)]$^{2+}$ (**4**).

The formulation of **4** is supported by the stoichiometric conversion of **4** to **3** upon titration with 1 equivalent of triethylamine as observed by NMR spectroscopy. A ^{2}H NMR spectrum of the pyrrole deuterated [(F$_8$-TPP)Fe-OH] in CH$_2$Cl$_2$ shows a pyrrole signal at 80 ppm as expected for axially symmetric high-spin iron(III) complexes. Upon addition of an equivalent of [(TMPA)Cu(CH$_3$CN)](ClO$_4$)$_2$ the pyrrole signal moves to 67.5 ppm which is comparable to the proton NMR signal of the isolated complex **4**. Addition of an equivalent of triethylamine converts this peak to 64 ppm which is comparable to that observed for the oxo-bridged complex **3**. The above experiment points to an acid-base relationship between the oxo-bridged (**3**) and hydroxo (**4**) complexes as indicated in Figure 12.

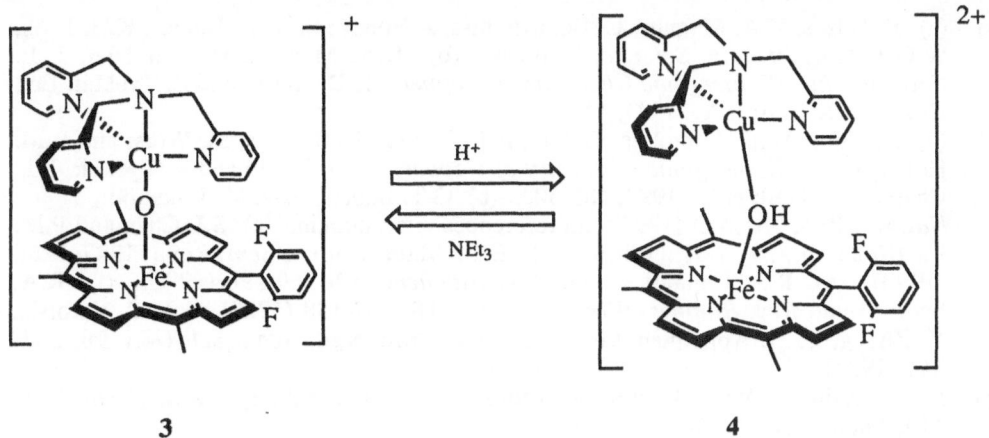

Figure 12: Proposed acid-base relationship of hydroxo and oxo-bridged iron-copper complexes.

CONCLUSION

The reaction of O_2 with porphyrin-Fe(II) and Cu(I) complexes leads to the dinuclear oxo-bridged $[Fe(III)-(O^{2-})-Cu(II)]^+$ (**3**) species, which indicates the occurrence of O-O bond cleavage at some point. Experiments are currently underway to elucidate the exact role of dioxygen in the reaction scheme. The reaction of $(F_8\text{-}TPP)Fe\text{-}OH$ and $[Cu(TMPA)(CH_3CN)][ClO_4]_2$ with equimolar NEt_3 also generated **3** in excellent yield; the same reaction without NEt_3, in toluene, gave the putative μ-hydroxo complex **4**. We are presently attempting the protonation of **3** to generate **4**.

Spectroscopically detected intermediates involving heme a_3 and O_2 or reduced derivatives (e.g., peroxo or ferryl) have been implicated in CcO action;[5] bridged Fe/Cu[5a,b,30] or discrete copper-dioxygen species[7,8] also may be involved. Such studies could also extend to other Cu-dioxygen complexes characterized by our group.[2,12,13] The results described here represent a conspicuous step towards developing systems which may aid in understanding O_2-reduction mechanism(s), and structures and protonation steps involving both porphyrin iron and copper ion.

ACKNOWLEDGMENT

We are grateful for the support of the National Institute of Health (GM 28962, K.D.K; GM 32394, E.P.D) and the National Science Foundation (DMB9001530, B. H. H.).

REFERENCES

(1) M. S. Nasir, B.I. Cohen, K.D. Karlin, *J. Am. Chem. Soc.,* **114**, 2482-2494 (1992).
(2) (a) Z. Tyeklár, R. R. Jacobson, N. Wei, N. Narasimha Murthy, J. Zubieta, K. D. Karlin, *J. Am. Chem. Soc.,* **115**, 2677-2689 (1993). (b) R.R. Jacobson, Z. Tyeklár, A. Farooq, K.D. Karlin, S. Liu, J. Zubieta, *J. Am. Chem. Soc.,* **110**, 3690-3692 (1988).
(3) M.J. Baldwin, P.K. Ross, J.E. Pate, Z. Tyeklár, K.D. Karlin, E.I. Solomon, *J. Am. Chem. Soc.,* **113**, 8671-8679 (1991).

(4)   (a) B. Hazes, K.A. Magnus, C. Bonaventura, J. Bonaventura, Z. Dauter, K.H. Kalk, W.G.J. Hol, *Protein Science*, in press. (b) K.A. Magnus, H. Ton-That, J. E. Carpenter *in "Bioinorganic Chemistry of Copper,"* K.D. Karlin and Z. Tyeklár, Ed., Chapman & Hall: N.Y., 1993, 143-150.

(5)   (a) J.A. Fee, W.E. Antholine, C. Fan, R.J. Gurbiel, K. Surerus, M. Werst and B.M. Hoffman *in "Bioinorganic Chemistry of Copper,"* K.D. Karlin and Z. Tyeklár, Ed., Chapman & Hall: N.Y., 1993, 485-500. (b) G.T. Babcock and M. Wikström, *Nature* ., **356**, 301-309 (1992) and references cited therein. (c) S.I. Chan and P.M. Li, *Biochem.*, **29**, 1-12 (1990). (d) B.G. Malmström, *Chem. Rev.*, 1247-1260 (1990). (e) R.A. Capaldi, *Annu. Rev. Biochem.*, **59**, 569-596 (1990). (f) R.A. Scott, *Annu. Rev. Biophys. Biophys. Chem.*, **18**, 137-158 (1989). (g) C. Varotsis, Y. Zhang, E. H. Appelman, G. T. Babcock, *Proc. Natl. Acad. Sci. USA*, **90**, 237-241 (1993).

(6)   K.D. Karlin, N. Wei, B. Jung, S. Kaderli, A.D. Zuberbühler, *J. Am. Chem. Soc.*, **113**, 5868-5870 (1991).

(7)   W.H. Woodruff, O. Einarsdóttir, R.B. Dyer, K.A. Bagley, G. Palmer, S.J. Atherton, R.A. Boldbeck, T.D. Dawes, D.S. Kliger, *Proc. Natl. Acad. Sci. USA.*, **88**, 2588-2592 (1991).

(8)   (a) M. Oliveberg and B.G. Malmström, *Biochemistry.*, **31**, 3560-3563 (1992). (b) R.S. Blackmore, C. Greenwood, Q.H. Gibson, *J. Biol. Chem.*, **266**, 19245-19249 (1991).

(9)   Model compounds containing porphyrin-Fe(III) include: (a) T. Prosperi and A.A.G. Tomlinson, *J. C. S. Chem. Comm.*, 196-197 (1979). (b) M.J. Gunter, L.N. Mander, G.M. Mclaughlin, K.S. Murray, K.J. Berry, P.E. Clark, D.A. Buckingham, *J. Am. Chem. Soc.*, **102**, 1470-1473 (1980). (c) K.J. Berry, P.E. Clark, M.J. Gunter, K.S. Murray, *Nouv. J. Chim.*, **4**, 581-585 (1980). (d) M.J. Gunter, L.N. Mander, K.S. Murray, *J. C. S. Chem. Comm.*, 799-801 (1981). (e) M.J. Gunter, L.N. Mander, K.S. Murray, P.E. Clark, *J. Am. Chem. Soc.*, **103**, 6784-6787 (1981). (f) E.A. Deardorff, P.A.G. Carr, J.K. Hurst, *J. Am. Chem. Soc.*, **103**, 6611-6616 (1981). (g) B. Lukas, J.R. Miller, J. Silver, M.T. Wilson, *J. C. S. Dalton Trans.*, 1035-1040 (1982). (h) C.M. Elliott and K. Akabori, *J. Am. Chem. Soc.*, **104**, 2671-2674. (1982). (i) C.K. Chang, M.S. Koo, B. Ward, *J. C. S. Chem. Comm.*, 716-719 (1982). (j) S.E. Dessens, C.L. Merrill, R.J. Saxton, R.L. Ilaria, Jr., J.W. Lindsey, L.J. Wilson, *J. Am. Chem. Soc.*, **104**, 4357-4361 (1982). (k) R.J. Saxton, L.W. Olson, L.J. Wilson, *J. C. S. Chem. Comm.*, 984-986 (1982). (l) C.K. Schauer, K. Akabori, M. Elliott, O.P. Anderson, *J. Am. Chem. Soc.*, **106**, 1127-1128 (1984). (m) M.J. Gunter, K.J. Berry, K.S. Murray, *J. Am. Chem. Soc.*, **106**, 4227-4235 (1984). (n) R.J. Saxton and L.J. Wilson, *J. C. S. Chem. Comm.*, 359-361 (1984). (o) V. Chunplang and L.J. Wilson, *J. C. S. Chem. Comm.*, 1761-1763 (1985). (p) C.M. Elliott, N.C. Jain, B.K. Cranmer, A.W. Hamburg, *Inorg. Chem.*, **26**, 3655-3659 (1987). (q) C.T. Brewer, G.A. Brewer, *Inorg. Chem.*, **26**, 3420-3422 (1987). (r) B.R. Serr, C.E.L. Headford, C.M. Elliott, O.P. Anderson, *J. C. S. Chem. Comm.*, 92-94 (1988). (s) An imidazolate-bridged Mn-porphyrin-Cu complex is, C.A. Koch, B. Wang, G. Brewer, C.A. Reed, *J. C. S. Chem. Comm.*, 1754-1755 (1989). (t) C.A. Koch, C.A. Reed, G.A. Brewer, N.P. Rath, W.R. Scheidt, G. Gupta, G. Lang, *J. Am. Chem. Soc.*, **111**, 7645-7648 (1989). (u) B.R. Serr, C.E.L. Headford, O.P. Anderson, C.M. Elliott, C.K. Schauer, K. Akabori, K. Spartalian, W.E. Hatfield, B.R. Rohrs, *Inorg. Chem.*, **29**, 2663-2671 (1990). (v) G.P. Gupta, G. Lang, C.A. Koch, B. Wang, W.R. Scheidt, C.A. Reed, *Inorg. Chem.*, **29**, 4234-4239 (1990). (w) L. Salmon, J.-B. Verlhac, C. Bied-Charreton, C. Verchére-Beaur, A. Gaudemer, R.F. Pasternack, *Tet. Lett.*, **31**, 519-522 (1990). (x) V. Bulach, D. Mandon, R. Weiss, *Angew. Chem. Int. Ed. Engl.*, **30**, 572-575 (1991). (y) B.R. Serr, C.E.L. Headford, O.P. Anderson, C.M. Elliott, K. Spartalian, V.E. Fainzilberg, W.E. Hatfield, B.R. Rohrs, S.E. Eaton, G.E. Eaton, *Inorg Chem.*, **31**, 5450-5465 (1992).

(10) Species containing non-porphyrin Fe moities: (a) R.H. Petty and L.J. Wilson, *J. C. S. Chem. Comm.*, 483-485 (1978). (b) R.H. Petty, B.R. Welch, L.J. Wilson, L.A. Bottomley, K.M. Kadish, *J. Am. Chem. Soc.*, **102**, 611-620 (1980). (c) W. Kanda, H. Okawa, S. Kida, *Bull. Chem. Soc. Jpn.*, **57**, 1159-1160 (1984). (d) G.A. Brewer and E. Sinn, *Inorg. Chem.*, **23**, 2532-2537 (1984). (e) G.A. Brewer and E. Sinn, *Inorg. Chem.*, **26**, 1529-1535 (1987). (f) P. Chaudhuri, M. Winter, P. Fleischhauer, W. Haase, U. Florke, H. Haupt, *J. C. S. Chem. Comm.*, 1728-1730 (1990). (g) I. Morgenstern-Badarau, D. Laroque, E. Bill, H. Winkler, A.X. Trautwein, F. Robert, Y. Jeannin, *Inorg. Chem.*, **30**, 3180-3188 (1991). (h) P.A. Chetcuti, A. Liegard, G. Rihs, G. Rist, *Helvetica Chimica Acta.*, **74**, 1591-1599 (1991).

(11) For one such study, see K. J. Berry, M. J. Gunter and K. S. Murray in "Oxygen and Life" Second BOC Priestley Conference, Birmingham, U.K., Sept. 1980, pp. 170-179, Special Publication No. 39. The Royal Society of Chemistry, London, U.K.

(12) (a) Z. Tyeklár and K.D. Karlin, *Acc. Chem. Res.* **22**, 241-248 (1989). (b) K.D. Karlin, Z. Tyeklár, A.D. Zuberbühler, *in "Bioinorganic Catalysis*, Reedijk," J., Ed., Marcel Dekker: N.Y., 1992, Chapter 9, 261-315. (c) K.D. Karlin and Z. Tyeklár, *Adv. Inorg. Biochem.*, **9**, 123-172 (1993).

(13) (a) I. Sanyal, R.R. Strange, N.J. Blackburn, K.D. Karlin, *J. Am. Chem. Soc.*, **113**, 4692-4693 (1991). (b) M.S. Nasir, B.I. Cohen, K.D. Karlin, *J. Am. Chem. Soc.*, **114**, 2482-2494 (1992). (c) K.D. Karlin, Z. Tyekár, A. Farooq, M.S. Haka, P. Ghosh, R.W. Cruse, Y. Gultneh, J.C. Hayes, J. Zubieta, *Inorg. Chem.*, **31**, 1436-1451 (1992).

(14) (a) K.K. Surerus, W.A. Oertling, C. Fan, R.J. Gurbiel, O. Einarsdottir, W.E. Antholine, R.B. Dyer, B.M. Hoffman, W.H. Woodruff, J.A. Fee, *Proc. Natl. Acad. Sci. U.S.A.*, **89**, 3195-3199 (1992). (b) J. Minagawa, T. Mogi, R.B. Gennis, Y.J. Anraku, *J. Biol. Chem.*, **267**, 2096-2104 (1992). (c) J.P. Shapleigh, J.P. Hosler, M.M.J. Tecklenburg, Y. Kim, G.T. Babcock, R.B. Gennis, S. Ferguson-Miller, *Proc. Nat. Acad. Sci., U.S.A.*, **89**, 4786-4790 (1992).

(15) D.L. Anderson, C.J. Weschler, F. Basolo, *J. Am. Chem. Soc.*, **96**, 5599 (1974).

(16) C.J. Weschler, D.L. Anderson, F. Basolo, *J. Am. Chem. Soc.*, **97**, 6707-6713 (1975).

(17) For the complex presumed to be [(TPP)Fe-(O^{2-})-Cu(TMPA)](ClO$_4$) and [(OEP)Fe-(O^{2-})-Cu(TMPA)](ClO$_4$) λ_{max} = 439 (Soret), 558 and 598 (α,β bands) and λ_{max} = 423 (Soret), 541 and 573 (α,β bands) respectively in CH_3CN or CH_2Cl_2 solvent.

(18) This compound was made by reacting [(F$_8$-TPP)Fe-Cl] with NaOH(aq), see. A. Nanthakumar, H.M. Goff, *Inorg. Chem.*, **30**, 4460-4464 (1991).

(19) (a) J.D. Dunitz, L.E. Orgel, *J. Chem. Soc.*, 2594-2596 (1954). (b) B.O. West, *Polyhedron*, **8**, 219-274 (1989).

(20) D.M. Kurtz, *Chem. Rev.*, **90**, 585-606 (1990).

(21) (a) T.A. Kent, L.J. Young, G. Palmer, J.A. Fee, E. Münck, *J. Biol. Chem.*, **258**, 8543-8546 (1983). (b) T.A. Kent, E. Münck, W.R. Dunham, W.F. Filter, K.L. Findling, T. Yoshida, J.A. Fee, *J. Biol. Chem.*, **257**, 12489-12492 (1982). (c) F.M. Rusnak, E. Münck, C.I. Nitsche, B.H. Zimmerman, J.A. Fee, *J. Biol. Chem.*, **262**, 16328-16332 (1987).

(22) (a) E. McCandlish, A.R. Miksztal, M. Nappa, A.Q. Sprenger, J.S. Valentine, J.D. Stong, T.G. Spiro, *J. Am. Chem. Soc.*, **102**, 4268-4271 (1980). (b) J.N. Burstyn, J.A. Roe, A.R. Miksztal, B.A. Shaefitz, G. Lang, J.S. Valentine, *J. Am. Chem. Soc.*, **110**, 1382-1388 (1988).

(23) M.A. Phillippi and H.M. Goff, *J. Am. Chem. Soc.*, **104**, 6026-6034 (1982).

(24) A. Shirazi and H.M. Goff, *J. Am. Chem. Soc.*, **104**, 6318-6322 (1982).

(25) H.M. Goff and E. Shimomura, *J. Am. Chem. Soc.*, **102**, 31-37 (1980).

(26) C.A. Reed, T. Masiko, S.P. Bently, M.E. Kastner, W.R. Scheidt, K. Spartalian, G. Lang, *Inorg. Chem.*, **101**, 2948-2958 (1979).

(27) M.A. Phillippi, N. Baenziger, H.M. Goff, *Inorg. Chem.*, **20**, 3904-3911 (1981).

(28) C.H. Welborn, D. Dolphin, B.R. James, *J. Am. Chem. Soc.*, **103**, 2869-2871 (1981).

(29) N. Kitajima, T. Koda, Y. Iwata, Y. Moro-oka, *J. Am. Chem. Soc.*, **112**, 8833-8839 (1990).

(30) R.W. Larsen, L.-P. Pan, S.M. Musser, Z. Li, S.I. Chan, *Proc. Natl. Acad. Sci. USA* , **89**, 723-727 (1992).

MECHANISTIC ASPECTS OF O$_2$-ACTIVATION ON NICKEL(II) TETRAHYDROSALEN COMPLEXES

Arnd Böttcher, Horst Elias,[*] Andreas Huber, and Lutz Müller

Eduard-Zintl-Institut für Anorganische Chemie
Technische Hochschule Darmstadt
Hochschulstrasse 10
64289 Darmstadt, FRG

INTRODUCTION

The activation of dioxygen in the coordination sphere of transition-metal centers is equally important for chemistry and biology and has been known for a long time for Fe-, Co- and Cu-containing active sites and the analogous model complexes.[1] There is only very limited information on nickel(II) complexes interacting with dioxygen. Kimura et al. and others[2] described nickel(II) complexes with macrocyclic N5 ligands, which are capable of both O$_2$ addition and activation. H. Kanatomi[3] reported that specifically substituted derivatives of tetrahydrosalen[4] form nickel(II) complexes which, in aerated organic solution and in the presence of added hydroxide ions or pyridine, are subject to oxidative dehydrogenation to finally form the corresponding Schiff base complexes after prolonged heating. In an attempt to model the co-enzyme F$_{430}$, Berkessel et al.[5] prepared a nickel(II) complex of the Ni(ONNOS) type, in which the nickel is bound to a dihydrosalen ligand with a pendant thioether group. This complex obviously is subject to intramolecular oxidation by dioxygen and subsequent degradation. Very recently, we reported several O$_2$ active nickel(II) complexes with the open-chain N$_2$O$_2$ ligand tetrahydrosalen,[4] carrying specific substitution patterns on the benzene rings of the ligand.[6] The activation of dioxygen by these latter complexes takes place under mild conditions (ambient temperature) and does not necessitate the addition of bases. The reaction with O$_2$ is strongly solvent dependent, leads to oxidative dehydrogenation of one C-N bond under the

---

[*] Presenter

The Activation of Dioxygen and Homogeneous Catalytic Oxidation,
Edited by D.H.R. Barton *et al.*, Plenum Press, New York, 1993

given conditions (formation of the corresponding dihydrosalen nickel(II) complex) and produces O=PPh₃ in the presence of PPh₃.[6] Darensbourg et al.[7] found that certain nickel thiolato chelate complexes of the Ni(SNNS) type are able to add O_2 to produce different oxygenates in which the thiolato sulfur is oxidized to sulfinato sulfur.

In comparison to the extensive literature on transition metal complexes with the tetradentate Schiff base ligand salen[4] and, in particular, on the dioxygen affinity of Co(salen) and its derivatives,[8] rather little is known about tetrahydrosalen complexes.[9-20] Comparing the ligand properties of salen and tetrahydrosalen, one expects increased N-basicity and greater flexibility as a consequence of C=N bond hydrogenation. Tetrahydrosalen ($=H_2[H_4]$salen[4]) should thus coordinate more easily in a folded fashion as well, which is indeed found.

Borer et al.[15] prepared the dinuclear complex $[Fe([H_4]$salen$)(OH)]_2 \cdot 2H_2O \cdot 2py$, which contains two edge-sharing octahedral iron(III) units, bridged by two hydroxyl ions and capped by the folded anion $[H_4]$salen^{2-}. We reported recently[21] on the dinuclear complex $[Fe\{[H_4](Me)L^1\}(OMe)]_2,$[4] which is another example for this folded mode of coordination.

In mononuclear complexes such as Ni($[H_4]$salen) and Zn($[H_4]$salen) the ligand $H_2[H_4]$salen was suggested[9] to be coordinated in its planar, unfolded form. The paramagnetism of Ni($[H_4]$salen) ($\mu_{exp} = 2.94$ BM) was taken as evidence however for additional intermolecular Ni-O interaction, making the nickel six-coordinate.[9] We presented recently the results of the first single crystal X-ray structure analysis of a mononuclear nickel(II) tetrahydrosalen complex.[6] The nickel in the cherry-red, practically diamagnetic complex Ni$\{[H_4](H)L^1\}$[4] has indeed coordinated the tetradentate ligand in square-planar fashion, the NiN_2O_2-plane being slightly distorted though. The similarity of the corresponding vis spectra suggests[6] that the complexes Ni$\{[H_4](Me)L^1\}$[4] and Ni$\{[H_4](Cl)L^1\}$[4] are also square-planar.

The most remarkable property of the planar nickel(II) complexes Ni$\{[H_4](X^5)L^1\}$[4] is their interaction with dioxygen.[6] In organic solution, the complexes are subject to oxidative dehydrogenation in the sense that, in the presence of dioxygen, one of the two C-N bonds is dehydrogenated to form a C=N imine bond. Further dehydrogenation of this half-salen species to the salen complex with two C=N bonds is not observed under the given experimental conditions.

The observed O_2-activating properties of complexes Ni$\{[H_4](X^5)L^1\}$ led us to prepare a variety of tetrahydrosalen ligands (see Chart I) and study the corresponding nickel(II) complexes in their interaction with dioxygen. Some copper(II) and cobalt(II) complexes were also investigated. The permethylated ethylene bridge in complexes M$\{[H_4](X^5)L^2\}$ (X^3=t-Bu; X^5=Me,Cl)[4] was deliberately chosen to increase, in addition to the effect of the t-butyl group, steric crowding around the metal center and thus hinder M-O-M and/or M-O-O-M bridging. Ligands $H_2[H_4]L^3$ and $H_2[H_2]L^3$ were chosen to introduce asymmetry and observe its effect on O_2-induced C=N bond formation. The methods used were UV/vis spectrophotometry, X-ray structure analysis, spectrophotometric titration, and kinetic control based on spectrophotometry and HPLC techniques.

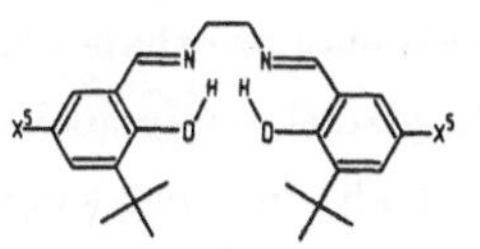

$H_2(X^5)L^1$ 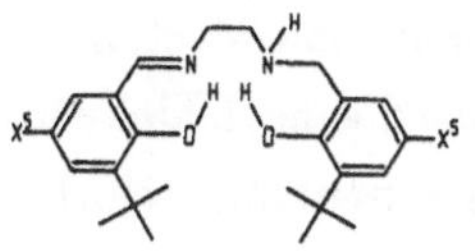 $H_2[H_2](X^5)L^1$ 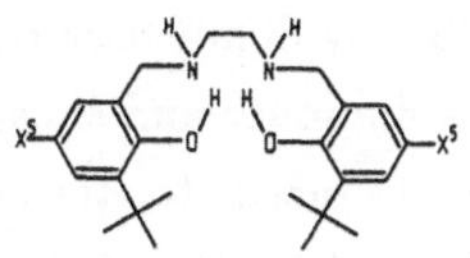 $H_2[H_4](X^5)L^1$

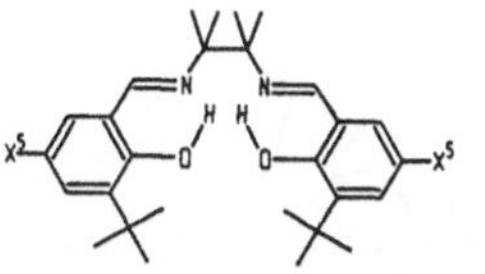

$H_2(X^5)L^2$ 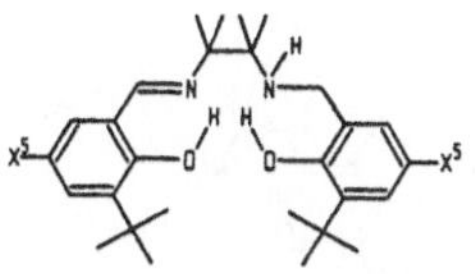 $H_2[H_2](X^5)L^2$ 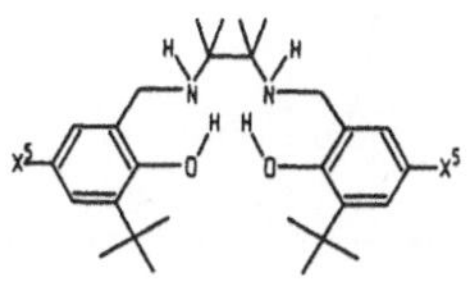 $H_2[H_4](X^5)L^2$

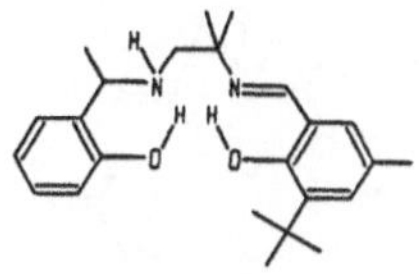

$H_2[H_2]L^3$ 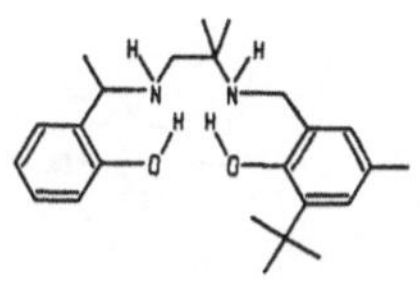 $H_2[H_4]L^3$

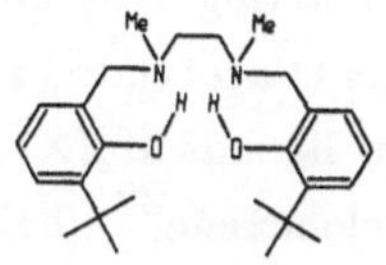

H_2L^4 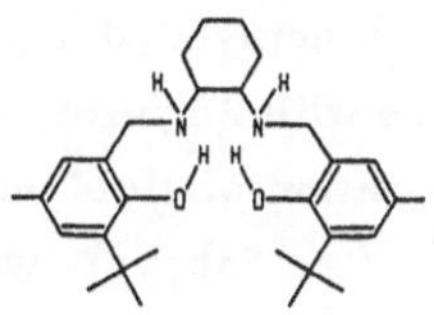 $(+)$-trans-H_2L^5

Chart I. Structural Formulae of the Ligands and Abbreviations.

The present work contributes to the more general aspects of reaction (1) with the pattern of substitution on the ligand, the solvent, and the metal M being the

$$[M([H_4]salen)] \xrightarrow[-2H]{O_2} [M([H_2]salen)] \xrightarrow[-2H]{O_2} [M(salen)] \tag{1}$$

variables. The relevance of reaction (1) for the understanding of the role of metal ions in biological systems and for catalysis is obvious.

RESULTS AND DISCUSSION

Preparative Aspects

The symmetrical salen derivatives $H_2(X^5)L^1$ and $H_2(X^5)L^2$, yellow solids, are easily accessible by Schiff base condensation reactions in methanol or ethanol from the corresponding diamines and substituted salicylaldehydes. In glacial acetic acid solution they can be reduced to the corresponding colorless tetrahydrosalen compounds $H_2[H_4](X^5)L^1$ and $H_2[H_4](X^5)L^2$ by stepwise addition of $Na[BH_3(CN)]$.

The synthesis of the asymmetrical mono Schiff base $H_2[H_2]L^3$ and of the dihydro analogue $H_2[H_4]L^3$ was carried out in a series of steps:

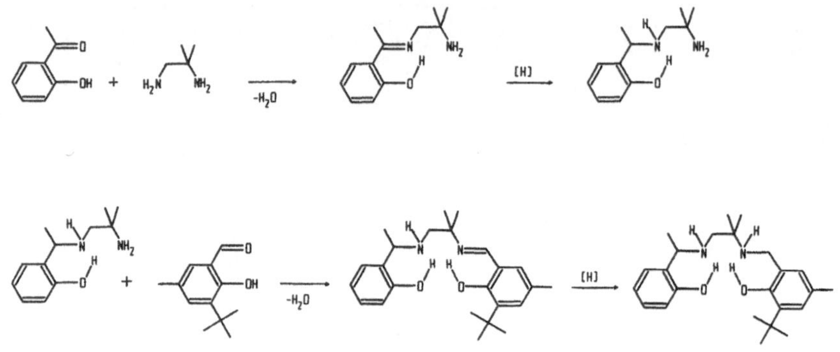

The N,N'-dimethylated ligand H_2L^4 was obtained by reacting N,N'-dimethyl-ethylenediamine with 2-hydroxy-3-t-butyl-benzylbromide (in its O-acetylated form).

The preparation of H_2L^5 and $H_2[H_4]L^5$ followed that of ligands $H_2(X^5)L^1$ and $H_2[H_4](X^5)L^1$ (see above) with (+)-trans-1,2-diaminocyclohexane[22] being the diamine.

To obtain the complexes $Ni\{(X^5)L\}$ and $Ni\{[H_4](X^5)L\}$, the ligands were reacted with nickel acetate in O_2-stripped ethanol. After recrystallization the complexes were characterized by elemental analysis and IR spectroscopy.

Visible Absorption of the Complexes in Acetone

The vis absorption data of the various nickel complexes in acetone are summarized in Table I. The tetrahydrosalen complexes are characterized by a d-d band at about 510 nm ($\epsilon = 970 - 1400 \text{ M}^{-1}\text{cm}^{-1}$) and by a CT band at about 360 nm ($\epsilon = 1900 - 2700 \text{ M}^{-1}\text{cm}^{-1}$). The spectra of the corresponding salen complexes exhibit a strong CT band at about 420 nm ($\epsilon = 6500 - 8500 \text{ M}^{-1}\text{cm}^{-1}$) and two shoulders at approximately 450 and 390 nm, respectively. The effect of substituent X^5 (= H, Me, t-Bu, OMe, Cl) on the absorption properties is minor, as is the effect of the solvent.

It follows from the data for the half-salen complexes Ni[H$_2$]L^3 and Ni{[H$_2$](Me)L^1} that their absorption properties are somewhere in between those of the tetrahydrosalen and salen analogues, with bands at 360 - 370 nm ($\in \approx$ 4600 M^{-1}cm^{-1}) and 420-430 nm ($\in \approx$ 3000-4000 M^{-1}cm^{-1}) and a shoulder at about 460 nm.

Table I. Visible Absorption of the Nickel Complexes in Acetone.

complex	λ_{max}/nm($\in_{max}$/M^{-1}cm^{-1})
Ni{[H$_4$](Cl)L^1}	352(2450); 512(1200)
Ni{[H$_4$](H)L^1}	352(1950); 512(1000)
Ni{[H$_4$](Me)L^1}	362(2150); 514(1130)
Ni{[H$_4$](tBu)L^1}	360(1930); 513(990)
Ni{[H$_4$](OMe)L^1}	372(2300); 519(1235)
Ni{[H$_2$](Me)L^1}	375(4300); 424 (3240)
Ni{(Cl)L^1}	420(7360)
Ni{(H)L^1}	414(6850)
Ni{(Me)L^1}	422(7500)
Ni{(tBu)L^1}	422(8590)
Ni{(OMe)L^1}	438(7060)
Ni{[H$_4$](Me)L^2}	366(2740); 514(1400)
Ni{[H$_4$](Cl)L^2}	360(2400); 510(1050)
Ni{(Me)L^2}	420(6500)
Ni{(Cl)L^2}	420(6900)
Ni{[H$_4$]L^3}	356(2100); 510(980)
Ni{[H$_2$]L^3}	364(4625); 424(3975)
NiL^4	366(1220); 526(850)
Ni{(+)-trans-[H$_4$]L^5}	364(1890); 518(970)
Ni{(+)-trans-L^5}	424(7830)

X-ray Structures and Magnetic Behavior

The single crystal X-ray structure analysis of Ni{[H$_4$](H)L^1} was the first of a nickel(II) tetrahydrosalen complex.[6] It proved the expectation that the N$_2$O$_2$ coordination around the nickel is practically square-planar. Figure 1 gives a view of the coordination geometry for the complex Ni{(+)-trans-[H$_4$]L^5}[23] which confirms the square-planar type of coordination found for Ni{[H$_4$](H)L^1}. In both complexes

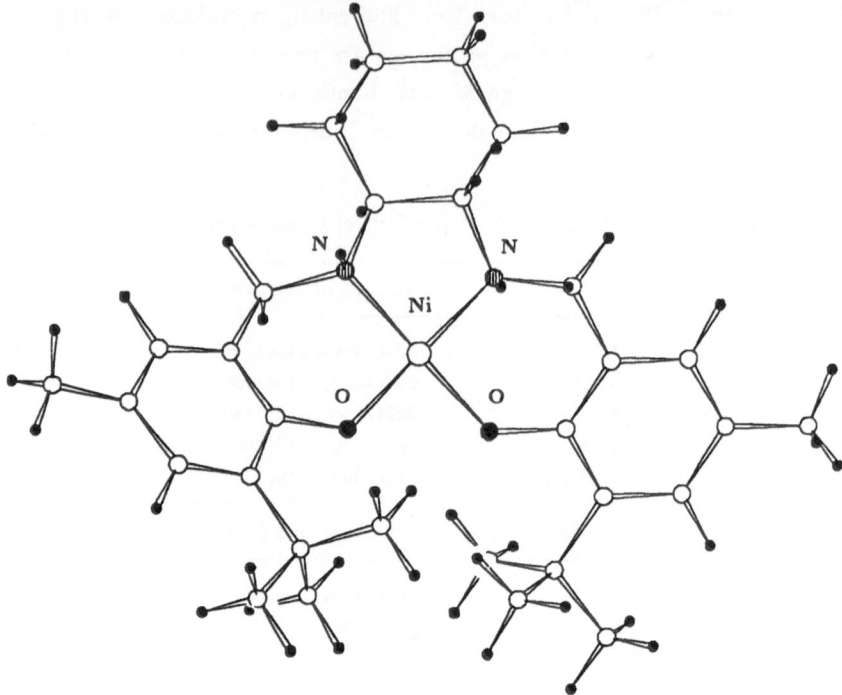

Figure 1. View of the coordination geometry in the complex Ni{(+)-trans-[H$_4$]L^5}.[23]

the sterically demanding t-butyl groups seem to prevent Ni···O interactions between two or more complex units, as observed for the complex Ni([H$_4$]salen).[9]

The coordination geometry of the dihydrosalen complex Ni{[H$_2$]L^3}[24] is shown in Figure 2. The arrangement of the donor atoms around the nickel is again practically square-planar, but the asymmetry within the tetradentate ligand is obvious. The Schiff base part of the ligand is flat and lies within the coordination plane, whereas the hydrogenated part is much more distorted.

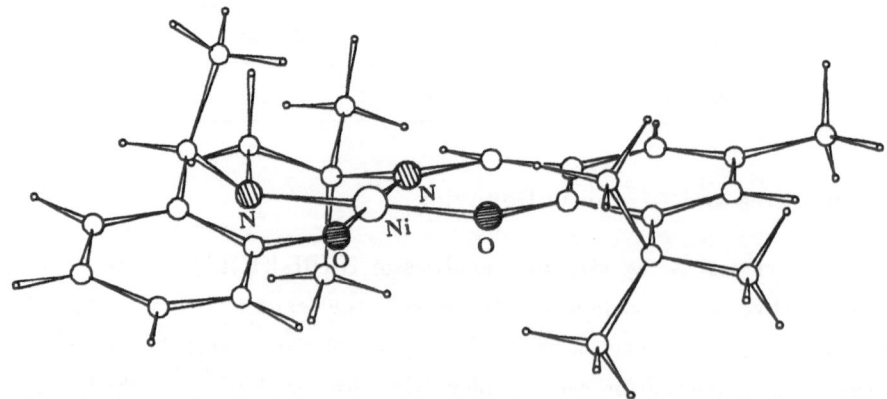

Figure 2. View of the coordination geometry in the complex Ni{[H$_2$]L^3}.[24]

Table II presents a summary of Ni-O, Ni-N, and C-N distances as found for Ni{[H$_4$](H)L^1}, Ni{(+)-trans-[H$_4$]L^5}, Ni{[H$_2$]L^3}, and Ni(salen).[25] The effect of stepwise hydrogenation of the C=N bonds (salen—>dihydrosalen—>tetrahydrosalen) is reflected in the corresponding bond lengths.

As expected for four-coordinate planar d^8 metal centers, the nickel complexes listed in Table II are practically diamagnetic. The magnetic moment at ambient temperature ranges from 0 BM (Ni(salen))[26] to 0.4 BM (Ni{[H$_4$](Me)L^1}).[6]

Table II. Bond Lengths (pm) in Tetrahydrosalen, Dihydrosalen and Salen Complexes of Nickel(II).

complex	Ni-O	Ni-N	C-N	Ref.
Ni{[H$_4$](H)L^1}	185.0	189.0	147.6	6
	186.6	193.0	139.2	
Ni{(+)-trans-[H$_4$]L^5}	184	190	143	23
	187	192	146	
Ni{[H$_2$]L^3}	184.5	184.7	132.7	24
	185.2	191.8	149.0	
Ni(salen)	185.0	184.3	129.3	25
	185.5	185.3	130.1	

Adduct Formation with Pyridine

The Lewis acidity of the nickel in complex Ni{[H$_4$](X^5)L^1} is affected by the substituents X^5. Spectrophotometric titration of the complexes with pyridine in acetone according to (2) led to the equilibrium constants listed in Table III.

$$Ni[H_4]L^1 + 2py \xrightleftharpoons{K_1} Ni[H_4]L^1 \cdot py + py \xrightleftharpoons{K_2} Ni[H_4]L^1 \cdot 2py \quad (2)$$

Adduct formation takes place stepwise and, as expected, the formation constant ß = K$_1$·K$_2$ for the octahedral adduct is greatest for the electron-withdrawing substituent X^5 = Cl. As typically observed for adduct formation of planar, four-coordinate nickel chelate complexes,[27] it is found that K$_1$ < K$_2$.

It has to be pointed out that the salen complexes Ni{(X^5)L^1} and the dihydrosalen complex Ni{[H$_2$](Me)L^1} do not add pyridine. The Lewis acidity of the nickel center is thus considerably increased by hydrogenation of the two C=N bonds in salen ligands.

Table III.

Table III. Equilibrium Constants for Adduct Formation of Complexes Ni{[H$_4$](X^5)L^1} According to (2) in Acetone at 298 K.

X^5	K$_1$, M^{-1}	K$_2$, M^{-1}	ß = K$_1$·K$_2$, M^{-2}
Me	0.4	15.7	6.0
H	0.9	6.8	6.25
Cl	0.85	36	30.5

Reaction of the Nickel(II) Tetrahydrosalen Complexes with Dioxygen in Organic Solution

As reported recently,[6] acetone solutions of complexes Ni{[H$_4$](X^5)L^1} slowly change their color from cherry-red to orange upon standing. This color change is observed also when exposure to light is strictly avoided. It is not observed, however, when the acetone solutions are carefully deoxygenated. Figure 3 shows the spectral changes recorded for Ni{[H$_4$](Me)L^1}.

There is a sharp isosbestic point but the kinetics are obviously rather complex, as indicated by the induction period at the beginning of the reaction. The final spectrum is not identical with that of Ni{(Me)L^1} (see Figure 4). Isolation and characterization

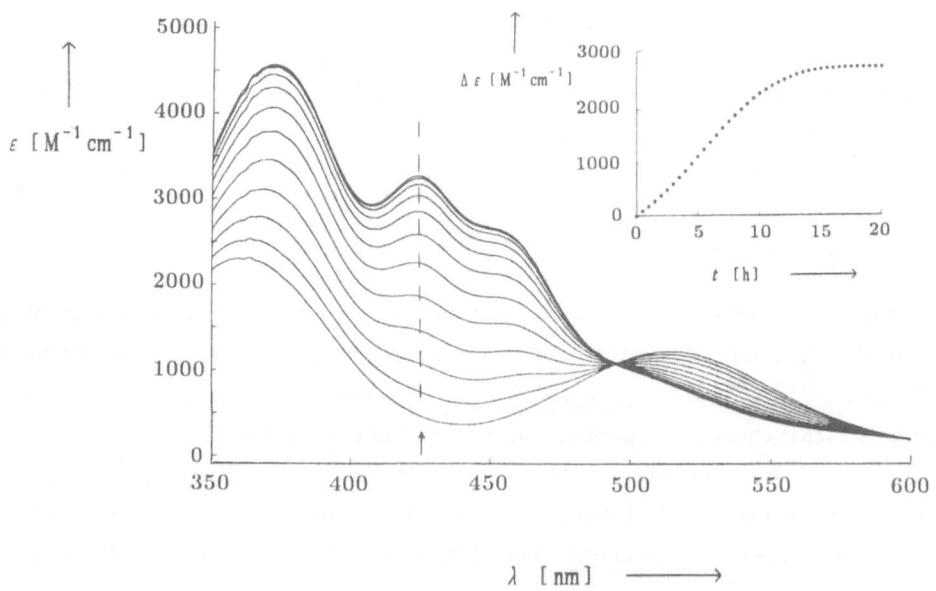

Figure 3. Spectral changes for the reaction of Ni{[H$_4$](Me)L^1} with dioxygen in acetone at 293 K and change of the absorbance at 425 nm with time.

At ambient temperature, the oxidative dehydrogenation of <u>one</u> of the two C-N bonds according to (3) is typically found for all complexes Ni{[H$_4$](X^5)L^1} and also for complexes Ni{[H$_4$](X^5)L^2}, which react however considerably slower. As shown in

$$Ni\{[H_4](Me)L^1\} \xrightarrow{+ O_2} Ni\{[H_2](Me)L^1\} \xrightarrow{//} Ni\{(Me)L^1\} \qquad (3)$$

Figure 5, substituents X^5 affect the rate of dehydrogenation and the following "order of reactivity" can be derived (based on 50 % conversion data):

$$Me : t\text{-}Bu : OMe : H : Cl = 3.3 : 3.0 : 2.3 : 1.6 : 1$$

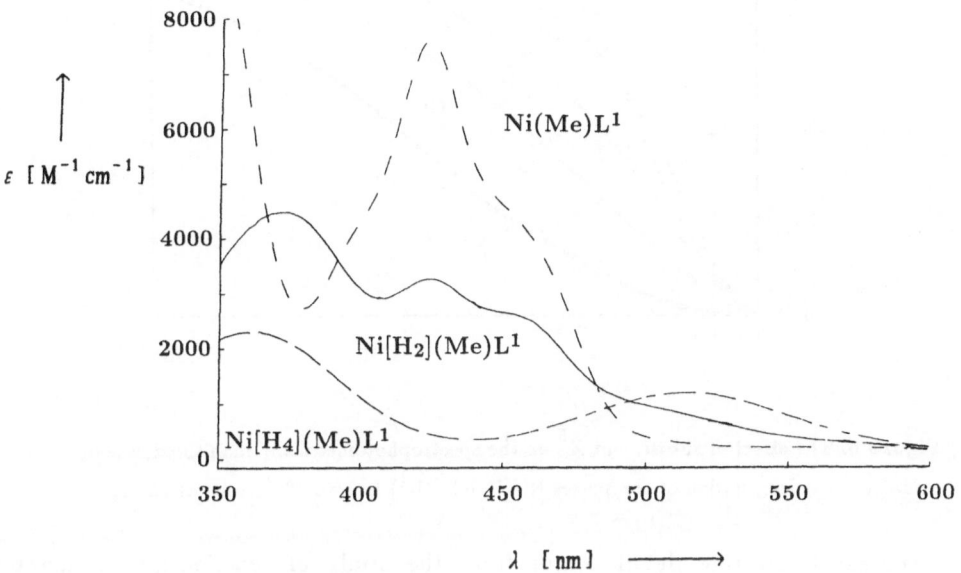

Figure 4. Vis spectra in acetone of Ni{[H$_4$](Me)L^1}, Ni{(Me)L^1} and Ni{[H$_2$](Me)L^1} (≡ product of the reaction shown in Figure 3).

of the product formed after 20 hours proved surprisingly the dihydrosalen complex Ni{[H$_2$](Me)L^1} to be the product.[6]

Complex Ni{[H$_4$](Cl)L^1} is the slowest one, although it is the one with the highest Lewis acidity (see Table III).

For a given substituent X^5, permethylation of the ethylene bridge (Ni{[H$_4$](X^5)L^1} —> Ni{[H$_4$](X^5)L^2}) reduces the rate of O$_2$-induced C=N bond formation by a factor of 20 - 30.

It is important to note that the rate of reaction (3) is very specifically solvent dependent:

$$\text{acetone : MeOH : toluene (CH}_2\text{Cl}_2\text{) : DMF} \approx 100 : 15 : 3 : 1$$

The fact that acetone and also butanone and cyclohexanone are "fast" solvents suggests that ketones possibly participate in the reaction. One might think of intermediate formation of dioxiranes (which are known to be strong oxidants) or of hydroxylation of ketones. In pyridine reaction (3) is blocked.

ΔA_{420} / %

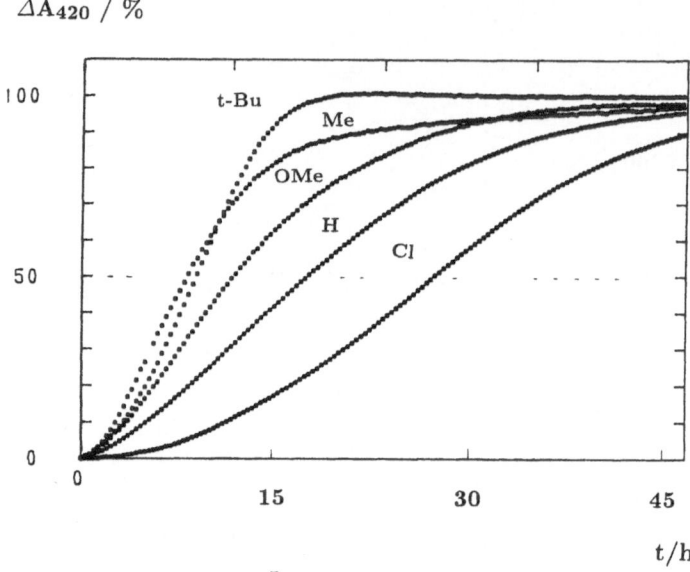

Figure 5. The effect of substituent X^5 on the spectrophotometrically monitored rate of oxidative dehydrogenation of complexes $\text{Ni}\{[H_4](X^5)L^1\}$ in aerated acetone at 298 K.

Another interesting detail comes from the study of reaction (3) in aerated acetone in the presence of PPh$_3$. As shown in Figure 6, the rate of formation of $\text{Ni}\{[H_2](\text{Me})L^1\}$ is reduced when one equivalent of PPh$_3$ is present (Ni : PPh$_3$ = 1 : 1) and further reduced by an excess of PPh$_3$ (Ni : PPh$_3$ = 1 : 10). There are no spectral changes to be observed upon addition of PPh$_3$, which means that the phosphine is not being coordinated. Interestingly enough, however, the HPLC analysis of the reaction mixture after 44 hours clearly proves the formation of O=PPh$_3$. The ratio $[\text{Ni}\{[H_2](\text{Me})L^1\}]$: [O=PPh$_3$] is found to be approximately 2 : 1 for the 1 : 1 experiment and approximately 1 : 2 for the 1 : 10 experiment. These results are preliminary and further studies are obviously necessary to clarify in detail as to which stoichiometric correlations exist. The fact of O=PPh$_3$ being formed is nevertheless an interesting aspect concerning the catalytic properties of the complex $\text{Ni}\{[H_4](\text{Me})L^1\}$.

It was not unexpected to find that the vis spectrum of the N-methylated complex NiL[4] in aerated acetone did not change upon standing. In the presence of PPh₃ solutions of this complex did not produce O=PPh₃.

The tetradentate ligand in the complex Ni{[H₄]L³} is asymmetrical, which raises therefore the question of (i) is there C=N bond formation at all and, if so, (ii) which of the two C-N bonds is being dehydrogenated in aerated acetone? The reaction of Ni{[H₄]L³} with dioxygen was monitored spectrophotometrically and the observed spectral changes clearly proved the formation of a half-salen nickel complex. The final spectrum obtained was practically identical with that of authentic Ni{[H₂]L³}. This means that, compared to the -HN-CH(Me) bond, the -HN-CH₂ bond is the favored one for oxidative dehydrogenation.

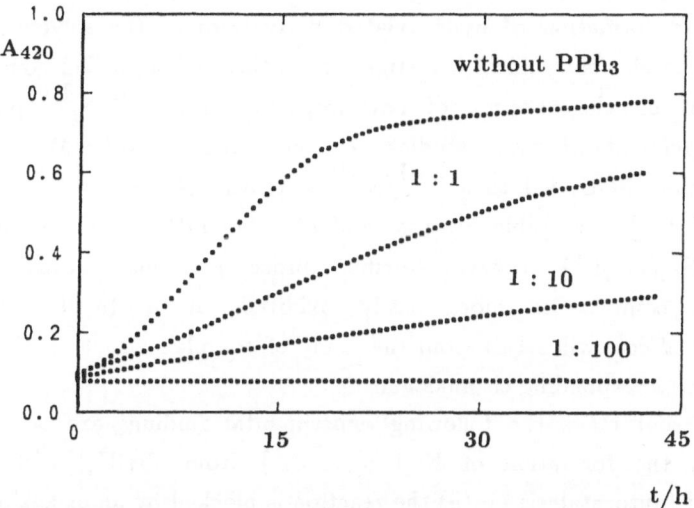

Figure 6. Spectrophotometrically monitored rate of oxidative dehydrogenation of Ni{[H₄](Me)L¹} in aerated acetone at 298 h in the presence of PPh₃ at different ratio [Ni] : [PPh₃].

Comparison of Tetrahydrosalen Complexes of Different Transition Metals

In addition to the nickel complexes discussed above the interaction of complexes Cu^{II}{[H₄](X⁵)L¹}, Cu^{II}{[H₄](X⁵)L²}, Fe^{III}{[H₄](Me)L¹}Y and Cr^{III}{[H₄](Me)L¹}Y (Y = monodentate anion) with dioxygen in aerated acetone was also studied. It was found that these complexes are obviously not dehydrogenated, since the vis spectra were stable for long periods of time.

The system $H_2[H_4](Me)L^2/Co^{2+}/DMF$ turned out to be an interesting one. When O_2 was strictly excluded, there was practically no formation of $Co\{[H_4](Me)L^2\}$ to be observed. When O_2 was admitted, however, complex formation took place and led to the salen complex $Co^{II}\{(Me)L^2\}$ with O_2 being consumed at the ratio $[Co] : [O_2] = 1 : 1$. The formation of two C=N bonds contrasts the corresponding nickel system. The resulting complex $Co^{II}\{(Me)L^2\}$ is O_2-stable even in solution but adds O_2 at low temperatures to form an adduct with a $Co^{III}\cdots O_2^-$ mode of bonding.[28]

Mechanistic Aspects

The results obtained so far are still too incomplete to allow a solid mechanistic interpretation of reaction (3). Valuable information is expected to come from experiments in which the time dependencies of several parameters are monitored simultaneously, namely, spectral changes, O_2-uptake, formation of $O=PPh_3$ from added PPh_3 and formation of hydroxylated derivatives of the solvent. It is also expected that the electrochemical investigation of the various nickel complexes will shed more light on the question of how important the Ni^{II}/Ni^{III} potential is. Preliminary cyclovoltammetric studies in acetonitrile (0.1 M tetrapropyl-ammoniumchloride; SCE) led to a Ni^{II}/Ni^{III} potential of +814 mV for the salen complex $Ni\{(Me)L^1\}$ (reversible process) and of +440 mV for the tetrahydrosalen complex $Ni\{[H_4](Me)L^1\}$ (quasi-reversible process). This means that the tetrahydrosalen complex is more easily oxidized. It is finally hoped that stereospecificity effects will result from the study of complexes such as $Ni\{(+)\text{-trans-}[H_4]L^5\}$ and the corresponding cis analogue.

At the present stage the following experimental findings are of mechanistic importance: (i) the formation of $Ni\{[H_2](X^5)L^1\}$ from $Ni\{[H_4](X^5)L^1\}$ has an induction period (autocatalysis ?), (ii) the reaction is blocked by an excess of pyridine, (iii) in the presence of PPh_3 the oxide $O=PPh_3$ is formed parallel to the dihydrosalen species $Ni\{[H_2](Me)L^1\}$, (iv) the rate of oxidative dehydrogenation does not correlate with the Lewis acidity of the complexes, and (v) when ketones are used as solvents the rate is highest. On the basis of these findings two sets of reactions appear to be plausible (S = substrate, such as PPh_3). The oxygen adduct formed in (a) is obviously

$$Ni\{[H_4]salen\} + O_2 \;\rightleftharpoons\; Ni\{[H_4]salen\}\cdot O_2 \qquad (a)$$

$$Ni\{[H_4]salen\}\cdot O_2 \;\longrightarrow\; Ni\{[H_2]salen\} + [H_2O_2] \qquad (b)$$

$$[H_2O_2] + Ni\{[H_4]salen\} \;\longrightarrow\; H_2O + Ni\{[H_2]salen\} \qquad (c)$$

$$[H_2O_2] + Ni\{[H_2]salen\} \;\longrightarrow\; H_2O + Ni(salen) \qquad (d)$$

$$[H_2O_2] + S \;\longrightarrow\; H_2O + S \text{ (oxidized)} \qquad (e)$$

of very low stability. It could be of the $Ni^{III}\cdots O_2^-$ type but experimental evidence for its formation is still missing. H_2O_2 could be an intermediate the formation of which would explain the induction period (see Figure 3). Reaction (d) could become important after most of the complex $Ni\{[H_4]salen\}$ has been converted to $Ni\{[H_2]salen\}$. Reaction (e) finally explains the oxidation of PPh_3 parallel to the formation of $Ni\{[H_2]salen\}$.

The possible participation of the solvent acetone (or other ketones) is indicated by reactions (f) - (i).

$$Ni\{[H_4]salen\}\cdot O_2 + R_2CO \longrightarrow Ni\{[H_2]salen\} + H_2O + [R_2CO_2] \quad\quad (f)$$

$$[R_2CO_2] + Ni\{[H_4]salen\} \longrightarrow R_2CO + Ni\{[H_2]salen\} + H_2O \quad\quad (g)$$

$$[R_2CO_2] + S \longrightarrow R_2CO + S \text{ (oxidized)} \quad\quad (h)$$

$$Ni\{[H_4]salen\}\cdot O_2 + R_2CO \longrightarrow Ni\{[H_2]salen\} + H_2O + R_2CO \text{ (hydroxylated)} (i)$$

The species R_2CO_2 formed in (f) could be a dioxirane or, alternatively, a hydroxylated ketone according to (i). Further experimental work is necessary to gain more insight.

CONCLUSIONS

Nickel(II) complexes of tetrahydrosalen ligands carrying t-butyl groups in 3-position interact with dioxygen under surprisingly mild conditions to become oxidatively dehydrogenated. At ambient temperature the corresponding dihydrosalen species are formed, the dehydrogenation of the second C-N bond obviously being much slower. In this process reactive oxygen species are formed (H_2O_2 ?) that can attack the tetrahydrosalen complex (more effectively than O_2), added substrates (such as PPh_3) or ketones (such as acetone).

The nickel(II) tetrahydrosalen complexes under study can thus be classified as mild O_2-activators. Their catalytic potential is probably small, as far as industrial applications are concerned. As shown very recently by C.J. Burrows et al.,[29] mildly O_2-activating nickel complexes are able to cleave DNA specifically. From the biochemical point of view it is also noteworthy that, in aerated methanol, the nickel containing coenzyme F_{430} is subject to slow oxidative dehydrogenation of a C-C bond to form 12,13-didehydro-F_{430}.[30]

ACKNOWLEDGMENTS

Sponsorship of this work by the Deutsche Forschungsgemeinschaft, Verband der Chemischen Industrie e.V., and Otto-Röhm-Stiftung is gratefully acknowledged. A. B. thanks the Studienstiftung des Deutschen Volkes and the Verband der Chemischen

Industrie e.V. for a scholarship. The authors appreciate the cooperation with Prof. Kniep, Dr. Eisenmann, and Dr. Röhr, Darmstadt (X-ray structures), Prof. Haase and Miss Hüber, Darmstadt (magnetic moments), Prof. Jäger and Dr. Rudolph, Jena (cyclovoltammetry), Prof. Pelikan and Prof. Langfelderova, Bratislava (oxygen uptake studies), Prof. Springborg, Copenhagen (stereospecific 1,2-diaminocyclohexane), and the help of Miss Hilms, Darmstadt, in the synthesis of the complexes.

REFERENCES

1. A.E. Martell, D.T. Sawyer, *Oxygen Complexes and Oxygen Activation by Transition Metals*, Plenum Press, New York (1988).

2. a) E. Kimura, A. Sakonaka, R. Machida, *J. Am. Chem. Soc.* 104:4255 (1982).
   b) E. Kimura, R. Machida, M. Kodama, *ibid.* 106:5497 (1984).
   c) E. Kimura, R. Machida, *J. Chem. Soc. Chem. Commun.* 499 (1984).
   d) Y. Kushi, R. Machida, E. Kimura, *ibid.* 216 (1985). e) R. Machida, E. Kimura, Y. Kushi, *Inorg. Chem.* 25:3461 (1986). f) D. Chen., J. Motekaitis, A.E. Martell, *ibid.* 30:1396 (1991).

3. H. Kanatomi, *Bull. Chem. Soc. Jpn.* 56:99 (1983).

4. Abbreviations: salen = H_2salen = 1,2-diamino-N,N'-disalicylideneethane; half-salen = $H_2[H_2]$salen = 1,2-diamino-N-(2-hydroxybenzyl)-N'-salicylideneethane; tetrahydrosalen = $H_2[H_4]$salen = N,N'-bis(2-hydroxybenzyl)-1,2-diaminoethane; $H_2[H_4](X^5]L^1$ = N,N'-bis(3-t-butyl-2-hydroxy-X^5-benzyl)-1,2-diaminoethane; $H_2[H_4](X^5]L^2$ = N,N'-bis(3-t-butyl-2-hydroxy-X^5-benzyl)-2,3-diamino-2,3-dimethylbutane; $H_2[H_4]L^3$ = N-(3-t-butyl-2-hydroxy-5-methylbenzyl)-N'-(α-methyl-2-hydroxybenzyl)-1,2-diamino-1,1-dimethylethane; $H_2[H_2]L^3$ = N-(3-t-butyl-2-hydroxy-5-methylbenzyliden)-N'-(α-methyl-2-hydroxybenzyl)-1,2-diamino-1,1-dimethylethane; H_2L^4 = N,N'-bis(3-t-butyl-2-hydroxybenzyl)-N,N'-dimethyl-1,2-diaminoethane; $H_2(+)$-trans-$[H_4]L^5$ = N,N'-bis(3-t-butyl-2-hydroxy-5-methylbenzyl)-(+)-trans-1,2-diaminocyclohexane.

5. A. Berkessel, J.W. Bats, Ch. Schwarz, *Angew. Chem.* 102:81 (1990); *Angew. Chem. Int. Ed. Engl.* 29:106 (1990).

6. A. Böttcher, H. Elias, L. Müller, H. Paulus, *Angew. Chem.* 104:635 (1992); *Angew. Chem. Int. Ed. Engl.* 31:623 (1992).

7. P.J. Farmer, T. Solouki, D.K. Mills, T. Soma, D.H. Russell, J.H. Reibenspies, M.Y. Darensbourg, *J. Am. Chem. Soc.* 114:4601 (1992).

8. (a) D. Chen, A.E. Martell, Inorg. Chem. 26:1026 (1987). (b) D. Chen, A.E.Martell, Y. Sun, *Inorg. Chem.* 28:2647 (1989).

9. M.J. O'Connor, B.O. West, *Aust. J. Chem.* 20:2077 (1967).

10. D.W. Gruenwedel, *Inorg. Chem.* 7:495 (1968).

11. H. Mäcke, S. Fallab, *Chimia* 26:422 (1972).

12. P.C.H. Mitchell, D.A. Parker, *J. Chem. Soc. Dalton Trans.* 1828 (1972).

13. (a) P. Subramanian, J.T. Spence, R. Ortega, J.H. Enemark, *Inorg. Chem.* 23:2546 (1984). (b) C.J. Hinshaw, G. Peng, R. Singh, J.T. Spence, J.H. Enemark, J.M. Bruck, J. Kristofzski, S.L. Merbs, R.B. Ortega, P.A. Wexler, *Inorg. Chem.* 28:4483 (1989).

14. A.R. Amundsen, J. Whelan, B. Bosnich, *J. Am. Chem. Soc.* 99:6730 (1977).

15. (a) L. Borer, L. Thalken, C. Ceccarelli, M. Glick, J. Hua Zhang, W.M. Reiff, *Inorg. Chem.* 22:1719 (1983). (b) L. Borer, L. Thalken, J. Hua Zhang, W.M. Reiff, *Inorg. Chem.* 22:3174 (1983).

16. K.V. Patel, P.K. Bhattacharya, *Ind. J. Chem.* 23A:527 (1984).

17. E. Solari, C. Floriani, D. Cunningham, T. Higgins, P. Mc Ardle, *J. Chem. Soc. Dalton Trans.* 3139 (1991).

18. (a) J. Csaszar, *Acta Phys. Chem.* 30:61 (1984). (b) J. Csaszar, *Acta Chim. Hungarica* 128:255 (1991).

19. V.T. Kasumov, A.A. Medzhidov, I.A. Golubeva, T.I. Vishnyakova, D.V. Shubina, R.Z. Rzaev, *Coord. Khim.* 17:1698 (1991).

20. M.G. Djamali, K.H. Lieser, *Angew. Makromol. Chem.* 113:129 (1983).

21. P. Baran, A. Böttcher, H. Elias, W. Haase, M. Hüber, H. Fuess, H. Paulus, *Z. Naturforsch.* 47b:1681 (1992).

22. The enantiomer (+)-trans-1,2-diaminocyclohexane was kindly provided by Prof. J. Springborg, Royal Veterinary and Agricultural University, Copenhagen.

23. The details of the X-ray structure analysis will be communicated elsewhere: A. Böttcher, H. Elias, R. Kniep, C. Röhr, J. Springborg; manuscript in preparation.

24. The details of the X-ray structure analysis will be communicated elsewhere: B. Eisenmann, H. Elias, E. Hilms, A. Huber; manuscript in preparation.

25. A.G. Manfredotti, C. Guastini, *Acta Cryst.* C39:863 (1983).

26. J.B. Willis, D.P. Mellor, *J. Am. Chem. Soc.* 69:1237 (1947).

27. M. Schumann, H. Elias, *Inorg. Chem.* 24:3187 (1985).

28. A. Böttcher, H. Elias, E.-G. Jäger, H. Langfelderova, M. Mazur, L. Müller, H. Paulus, P. Pelikan, M. Rudolph, M. Valko, "A Comparative Study on the Coordination Chemistry of Cobalt(II), Nickel(II) and Copper(II) with Derivatives of "Salen" and "Tetrahydrosalen": Metal-Catalyzed Oxidative Dehydrogenation of the C-N Bond in Coordinated "Tetrahydrosalen""; submitted for publication in Inorg. Chem.

29. C.C. Cheng, S.E. Rokita, C.J. Burrows, *Angew. Chem.* 105:290 (1993); *Angew. Chem. Int. Ed. Engl.* 32:277 (1993).

30. A. Pfaltz, D.A. Livingston, B. Jaun, G. Diekert, R.K. Thauer, A. Eschenmoser, *Helv. Chim. Acta* 68:1338 (1985).

SINGLET OXYGEN DIMOL-SENSITIZED LUMINESCENCE AND REACTIONS OF SINGLET OXYGEN WITH ORGANOMETALLICS

Christopher S. Foote*, Alexander A. Krasnovsky Jr.[†] Yulan Fu, Matthias Selke, and William L. Karney

Department of Chemistry and Biochemistry
University of California, Los Angeles, California 90024-1569

INTRODUCTION

Singlet molecular dioxygen (hereafter referred to simply as singlet oxygen, $^1O_2$) can be produced by photosensitization (Scheme 1) or by thermal processes (see below). Its luminescence is widely used for detection and investigation of $^1O_2$ in systems of photobiological and photochemical importance. In scheme 1, S, ^{1}S and ^{3}S are photosensitizer in the ground and excited singlet and triplet states, respectively, and hv is light absorbed by the sensitizer.

Scheme 1

$$S \xrightarrow[hv]{} {}^1S \longrightarrow {}^3S$$

$$^3S + {}^3O_2 \longrightarrow S + {}^1O_2$$

This paper summarizes some striking new observations in the chemistry and physics of singlet oxygen. Work in two areas is reported: energy transfer from singlet oxygen dimols to organic molecules, and reactions of singlet oxygen with iridium and rhodium organometallics.

DIMOL ENERGY TRANSFER

Weak, highly-forbidden emission from the $^1\Delta_g$ state of oxygen ("monomols") appears at 1270 nm and can be detected in photosensitized and thermal systems by modern

---

[†]On leave from Department of Biology, Moscow State University, Moscow, Russia

The Activation of Dioxygen and Homogeneous Catalytic Oxidation,
Edited by D.H.R. Barton *et al.*, Plenum Press, New York, 1993

detectors.[1,2] In addition, collisions between two $^1O_2$ molecules in the gas phase generate luminescence from $^1O_2$ "dimols", $(^1O_2)_2$, formed by "energy pooling" with principal maxima at 634 and 703 nm, twice the energy of the fundamental.[3-5]

Luminescence of organic chromophores sensitized by energy transfer from singlet oxygen has also been reported for many years.[2,4-15] Several mechanisms have been suggested for this luminescence.

Khan and Kasha suggested a purely physical mechanism involving energy transfer from singlet oxygen dimols to the fluorescer (Fl), to give the singlet (^1Fl), which fluoresces (Scheme 2).[5] This mechanism should permit excitation of fluorescers with energies ≤ 44 kcal/mol, twice the energy of singlet oxygen. Wilson[7] and Stauff and Fuhr[16,17] also supported this idea. The problem with this scheme is that no evidence for a bound state of the dimol has been presented, and it is questionable whether it could have a lifetime long enough to transfer energy. In this scheme, k_{dq} is the rate constant for quenching of dimols by the fluorescer.

Scheme 2

$$2\ ^1O_2 \rightleftharpoons (^1O_2)_2$$

$$(^1O_2)_2 + Fl \xrightarrow{k_{dq}} 2\ ^3O_2 + {}^1Fl$$

$$^1Fl \longrightarrow Fl + h\nu_f$$

Ogryzlo and Pearson suggested successive energy transfer from singlet oxygen to violanthrone, producing the triplet (^3Fl), which then reacts with a second molecule of $^1O_2$ to give ^1Fl.[6] This reaction is shown in Scheme 3, where $h\nu_f$ is acceptor fluorescence and k_q is the rate constant for quenching of singlet oxygen monomols by Fl. This reaction corresponds to a well-known delayed fluorescence mechanism, but since few organic molecules have triplet energies below 22 kcal/mol, required for exothermic energy transfer, this mechanism cannot be general. However, the second step (using directly excited ^3Fl) has been demonstrated in polymer matrices[10] and in solution.[18,19]

Scheme 3

$$^1O_2 + Fl \xrightarrow{k_q} {}^3O_2 + {}^3Fl$$

$$^1O_2 + {}^3Fl \longrightarrow {}^3O_2 + {}^1Fl$$

$$^1Fl \longrightarrow Fl + h\nu_f$$

Excitation of Fl could also occur via an unstable peroxide such as a dioxetane, formed by reaction of singlet oxygen with various complounds, including Fl. A systematic study of such luminescence has been reported.[20,21]

Krasnovsky et al. reported the first visible luminescence derived from singlet oxygen produced by protoporphyrin photosensitization in solution.[2,12-15,22] Along with the "monomol" luminescence at 1270 nm, visible luminescence with intensity proportional to the square of the excitation power was observed. Certain compounds such as naphthalocyanines and phthalocyanines give luminescence with extremely high efficiency; one of the most effective is tetra-*tert*-butyl phthalocyanine (Pc).[13,22] The spectrum of the luminescence from this compound is Pc fluorescence rather than dimol luminescence.

We now report time-resolved and steady-state measurements of phthalocyanine luminescence sensitized by singlet oxygen dimols, produced both by photosensitization and by thermal processes; the detailed kinetics strongly support Scheme 2.

Photosensitized Dimol Formation[15]

We used the fullerenes C_{60} and C_{70} excited at 355 and 532 nm as sensitizers for singlet oxygen production, since they have negligible fluorescence ($\leq 10^{-4}$), generate $^1O_2$ efficiently, are stable on strong laser excitation, and do not react with singlet oxygen at an appreciable rate.[23,24] In agreement with previous reports, singlet oxygen production was efficient, as shown by emission at 1270 nm (Fig. 1). The red light emission caused by energy transfer to the phthalocyanine was filtered by 700 or 780 nm interference filters and detected with a cryogenic germanium photodetector. As reported previously, the spectrum of the delayed luminescence in Pc-containing C_6F_6 coincides with the fluorescence spectrum of Pc, with the main maximum at 703 nm and two additional weaker bands at 740 and 780 nm.[13,22]

Solutions without Pc gave very weak luminescence at 700 nm, detected only when high fullerene concentrations and laser intensities were used. Addition of as little as 10^{-7} M Pc increased the intensity by an order of magnitude. Increasing the Pc concentration caused a further increase in intensity. The lifetime of the 700 nm luminescence was shorter than that at 1270 nm by a factor of 2; Fig. 1 shows the relationship in C_6F_6; similar lifetime relationships occur in solvents with a wide range of singlet oxygen lifetimes and in the presence and absence of quenchers (Fig. 2). In all cases, the intensity at 700 nm was

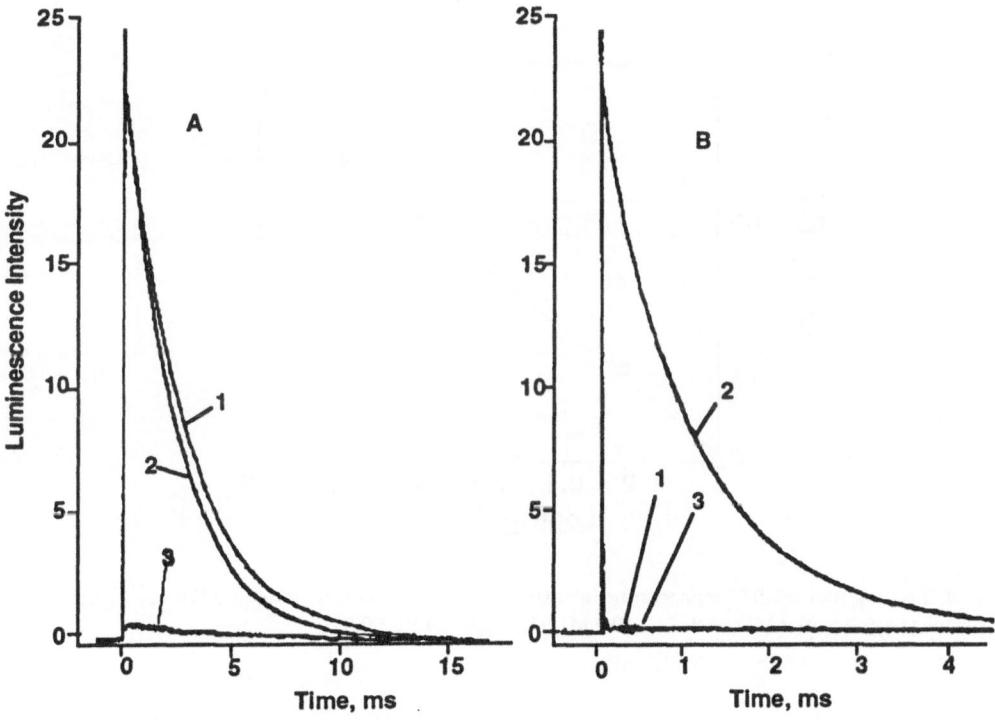

Fig. 1. Photosensitized singlet oxygen luminescence at 1270 nm (A) and 700 nm(B) after 532 nm laser pulses in C_6F_6 . Photosensitizer is C_{70}. (1) C_{70} alone; (2) C_{70} + Pc (5 x 10^{-7} M), (3) Pc alone, no C_{70}.

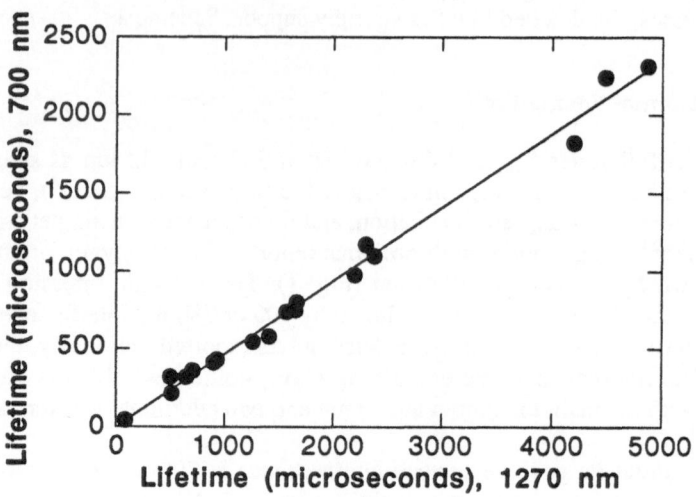

Fig. 2. Lifetime of the emission at 700 nm vs that at 1270 in a range of solvents and in the presence and absence of quenchers. The slope of the plot is 0.47 ± 0.025.

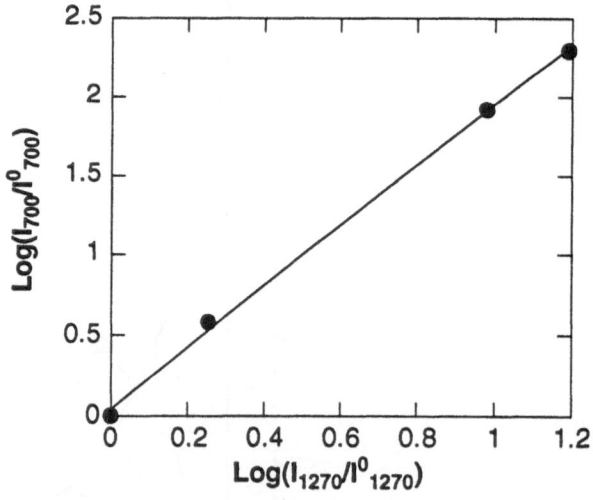

Fig. 3. Log-log plot of the luminescence intensities at 700 nm versus those at 1270 nm in solutions containing C_{60} (4×10^{-5} M) and Pc (3×10^{-6} M). The slope is 1.91 ± 0.08.

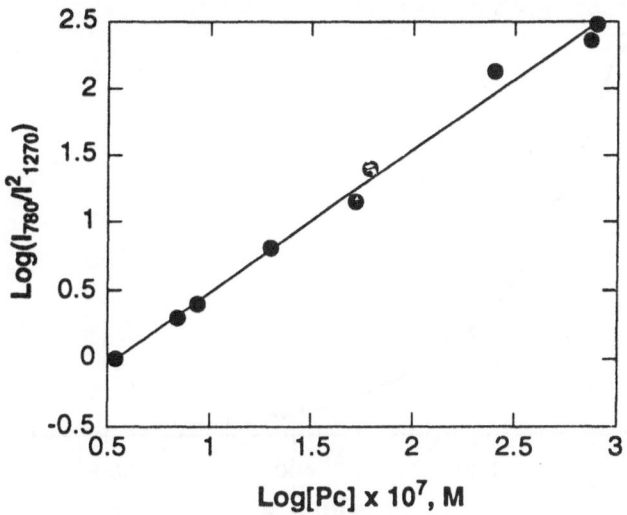

Fig. 4. Log-log plot of the ratio of the luminescence intensities at 780 nm to the square of the luminescence intensities at 1270 nm (L_{780}/L^2_{1270}) versus Pc concentrations (ratio used to normalize different laser intensities).

proportional to the square of that at 1270 nm (Fig. 3). The luminescence at 700 nm is linearly dependent on [Pc] over a wide range of concentrations (Fig. 4).

Thermally Produced Singlet Oxygen

There are many energy transfer processes possible under the above conditions which could complicate the proposed mechanism. We therefore tested the scheme in a simpler system, in which $^1O_2$ is produced by thermolysis of 1,4-dimethylnaphthalene endo-peroxide, as shown below.[25]

Thermolysis of the endoperoxide in C_6D_6 gave $^1O_2(^1\Delta_g)$ luminescence at 1270 nm which decayed exponentially as the peroxide reacted by a first-order reaction. In agreement with the photochemical work above and that reported earlier[14,15], phthalocyanine quenched $^1O_2$ monomol emission. Weak luminescence near 700 nm was strongly enhanced by the addition of phthalocyanine. The 700 nm luminescence decay is also exponential, but the rate constant for the apparent decay is twice that observed at 1270 nm, in agreement with the photochemical work. At the same temperature and endoperoxide concentration, the decay curve of the luminescence at 700 nm is almost exactly the square of that at 1270 nm (Fig. 5).

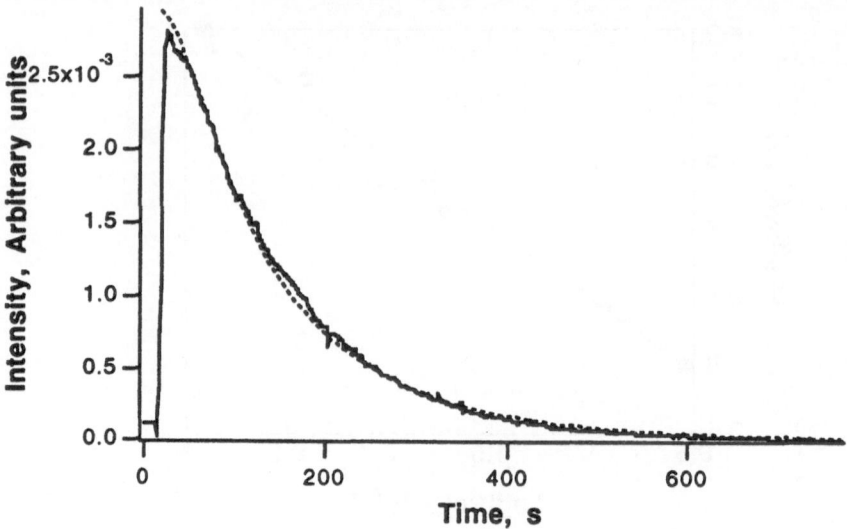

Fig. 5. Luminescence at 700 nm (dotted curve) plotted with the square of that at 1270 nm (solid). The 700 nm luminescence decay curve was normalized to the same scale as the 1270 nm.

Fig. 6 shows that at the same temperature and endoperoxide concentration, the rate constant of the 700 nm luminescence decay is twice that of the $^1O_2$ monomol emission, in agreement with the photochemical results. The initial intensity of the 700 nm luminescence was proportional to the square of the initial concentration of the endoperoxide, while that of the 1270 nm luminescence was directly proportional (data not shown).

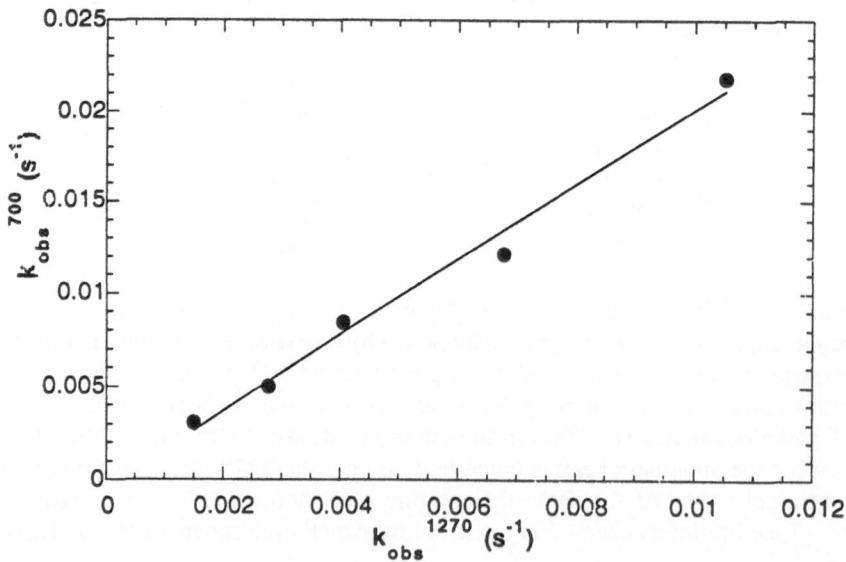

Figure 6. Rate constants of luminescence decay at 700 nm vs 1270 nm (55.0 ± 0.2 to $75.0 \pm 0.2°C$); the slope is 2.05 ± 0.42.

The activation energies for the reaction of the endoperoxide calculated from Arrhenius plots of the first-order rate constants at 1270 nm and 700 nm gave $\Delta H^{\ddagger} = 21.1 \pm 1.1$ kcal/mol, $\Delta S^{\ddagger} = -7 \pm 3$ e.u. and 21.2 ± 1.1 kcal/mol, $\Delta S^{\ddagger} = -6 \pm 3$ e.u. in C_6D_6, respectively, in reasonable agreement with the literature (24.2 ± 0.2 kcal/mol, $\Delta S^{\ddagger} = 2 \pm 1$ e.u. in dioxane).[25] In contrast, Arrhenius plots of the initial luminescence **intensities** gave $\Delta H^{\ddagger} = 22.7 \pm 4.3$ and 50.3 ± 1.4 kcal/mol for 1270 nm and 700 nm emission, respectively. Here, the activation enthalpy found at 700 nm is roughly twice that at 1270 nm.

The difference in activation enthalpies from the decay rates and that from the intensities, while at first surprising, is in accord with expectations. The decay rate of the luminescence at 700 nm is just twice that at 1270[15]; thus the Arrhenius plots have the same slopes, but are offset by a constant value of ln2. In contrast, since the intensities at 700 nm are equal to the square of those at 1270, the slope of the plot at 700 nm and the resulting activation energy are twice that at 1270 nm.

The data are consistent with the mechanism in scheme **2**, in which the 700 nm luminescence is a result of energy transfer to Pc from dimol. This mechanism predicts that if dimol formation does not lead to $^1O_2$ quenching and the lifetime of dimols is much less than of monomols, the Pc-sensitized dimol luminescence should depend quadratically on the singlet oxygen concentration, and the lifetime of the 700 nm luminescence should be half that at 1270 nm, as observed.

Note that Pc quenches singlet oxygen monomols strongly at high concentrations; however, although this decreases the $^1O_2$ lifetime, it should not affect the initial amount formed in the time-resolved experiments. This is precisely what is observed; even where lifetimes are strongly shortened by quenching, the dimol lifetime is always exactly half that of the monomol but the initial intensity at both wavelengths is independent of quencher.

$$^1O_2 + Fl \xrightarrow{\quad k_q \quad} {}^3O_2$$

In C_6F_6 at Pc concentrations $>5 \times 10^{-6}$ M, the $^1O_2$ lifetime is mostly determined by $^1O_2$ quenching by Pc. If mechanism 2 were valid, the intensity of the 780 nm band relative to the 1270 nm band should not depend on the Pc concentration above 5×10^{-6} M, since most of the $^1O_2$ is already quenched at this concentration. In fact, the dependence is linear within the experimental error over a wide range of concentrations, as shown in Fig. 4. This result argues strongly against the mechanism in Scheme 3 (intermediate fluorescer triplet).

The dissociation of $(^1O_2)_2$ to $^1O_2$ must be very fast, since the singlet oxygen lifetime is not affected by laser power (at least at the relatively low powers used in these experiments). If a significant amount of singlet oxygen were tied up as dimol, this should lead to saturation kinetics of the Pc luminescence as power increases at high Pc concentrations, which is not observed.

The data suggest that activated dimol luminescence is a universal phenomenon which can be observed in solvents of widely varying nature and can be surprisingly efficient. This type of sensitized luminescence may account for the wide distribution of red luminescence often attributed to singlet oxygen dimol emission in biological and other systems.

2. ORGANOMETALLICS

Many coordinatively unsaturated organometallic complexes are known to react with triplet dioxygen to form metal-dioxygen complexes.[26,27] The oxidative addition of triplet oxygen to Vaska's Complex ($Ir(CO)Cl(PPh_3)_2$, **1**) to give the stable oxygen complex **2** was one of the first of these reactions to be discovered.[28]

Singlet oxygen can both react with (k_R) and be physically quenched (k_Q) by substrates. Many organometallic complexes physically quench singlet oxygen in solution.[29]

However, to the best of our knowledge, there are no reports of chemical reaction of singlet oxygen with metal centers in organometallic complexes.

$$A + {}^3O_2 \xrightleftharpoons[A\ (k_Q)]{} {}^1O_2 \xrightarrow{A\ (k_R)} AO_2$$

A. Iridium[30]

The oxidative addition of triplet oxygen to **1** and related derivatives has been well studied[26-28,31,32], in part because this system is simple and the formation of the peroxide $(Ir(CO)Cl(PPh_3)_2O_2)$ (**2**) is reversible. However, we found that deoxygenation of **2** is principally photochemical. Bubbling Argon through a solution of **2** in the dark for 70 hours at room temperature gave only 0.7 % conversion to **1**, whereas complete conversion occurred within a few minutes upon irradiating with a Cermax 300-W Xenon lamp under the same conditions. That loss of oxygen was photochemical was apparently not taken into account when the kinetic and equilibrium parameters for loss of O_2 from the peroxide **2** and its derivatives were determined.[32]

1   **2**

We measured the rate constants for both physical and chemical interaction of singlet oxygen (generated by photosensitization by Methylene Blue in $CHCl_3$) with Vaska's Complex $(Ir(CO)Cl(PPh_3)_2)$ (**1**). The photooxidation of **1** gave only **2**, identical to that obtained from the reaction of **1** with triplet oxygen (characteristic absorption bands for the metal-peroxo group at 860 cm^{-1} and for the CO-group at 2000 cm^{-1} and identical UV-Vis spectra).[28] No oxidation of ligands (e.g. triphenylphosphine oxide) occurred. When a filter solution was not used so that **2** was also excited, a photostationary state was obtained.

The rate constant for total deactivation of singlet oxygen by Vaska's Complex ($k_Q + k_Q$) was determined by $^1O_2$ luminescence quenching[33] and is listed in Table 1. In order to obtain the rate constant of chemical quenching (k_R), competition experiments were carried out with 9,10-dimethylanthracene (DMA, same conditions as above). DMA is known to quench $^1O_2$ only chemically.[29,34] Loss of **1** and DMA were monitored spectrophotometrically and the results fitted to the equation of Higgins et al.[35] A second independent determination of k_R was obtained by measurements of direct disappearance of **1** compared with a tetramethylethylene (TME) reference, which quenches singlet oxygen only by a chemical mechanism.[36] Values of k_R, and by difference, of k_Q, are listed in Table 1,

The ratio of the rate constant obtained for the reaction of **1** with singlet oxygen and that with triplet oxygen obtained by Halpern et al. (2.1×10^{-2} M^{-1}sec^{-1} in benzene)[6] is approximately 10^9. In addition, the large amount of *physical* quenching of singlet oxygen (90 %) must be noted. These results show that the initial interaction between the electrophilic singlet oxygen molecule and the nucleophilic metal center can either lead to formation of the peroxide or to deactivation of singlet oxygen, and that deactivation predominates by about an order of magnitude.

Table 1. Rate constants ($\times 10^8$ M^{-1}s^{-1}) for reaction (k_R) and Quenching (k_Q) of Vaska's Complex and Rhodium Analog with Singlet Oxygen

	$(k_R + k_Q)$	k_R	k_Q
Vaska's Complex (1)	2.6 ± 0.2	0.2 ± 0.08	2.4 ± 0.3
Rhodium Analog (3)	3.3 ± 0.2	1.0 ± 0.4	2.3 ± 0.6

b. Rhodium

The rhodium analog of Vaska's complex, (Rh(CO)Cl(PPh$_3$)$_2$, **3**) does not undergo any reaction with triplet oxygen[26,37], and the corresponding rhodium-dioxygen complex is unknown. However **3** reacts with singlet oxygen at temperatures ≤ 0 °C, (Methylene Blue or C$_{60}$ sensitizer in CHCl$_3$, Cermax 300-W xenon lamp, cut-off at 550 nm) to give the novel unstable rhodium-peroxo complex Rh(CO)Cl(PPh$_3$)$_2$O$_2$, **4**, easily identified by the appearance of a new carbonyl peak at 2044 cm^{-1} (v_{CO} for **3** = 1980 cm^{-1}) as well as a weak peroxo stretch at 901 cm^{-1}. The oxygen adduct of Vaska's complex (**2**) has a peroxide band at 860 cm^{-1}. The shift in v_{CO} between **3** and **4** is 64 cm^{-1}; that between **1** and **2** is 50 cm^{-1}.[26,28] The shift suggests that peroxo-complex **4** is relatively electron poor, since the difference between v_{CO} of the starting complex and the peroxo complex *decreases* with increasing basicity of the metal for Vaska's Complex and its derivatives.[32]

Compound **4** is unstable at room temperature, with a half-life of less than a minute. It rapidly reforms the starting material **3**, losing an oxygen molecule. However, **4** is moderately stable at 0 °C with a half-life of approximately 7 min. Below - 40 °C, there was no decomposition.

The sum of the rate constants of physical (k_Q) and chemical (k_R) quenching of singlet oxygen by **3** was determined by ^{1}O$_2$ luminescence quenching as before (Fig. 7), and is approximately 30 % larger than that for Vaska's Complex in CDCl$_3$ (Table 1).

In order to obtain the rate constant of chemical quenching (k_R), the direct disappearance of **3** was measured at different concentrations. Since the resulting peroxide **4** rapidly reforms **3** at room temperature, the measurements were done by very rapidly firing a fixed number of laser shots (6 shots within two or three seconds) at the solution of **3** and determining the change in decay rate of singlet oxygen. The value of k_R thus obtained is listed in Table 1. The value for physical quenching alone is $2.3 \pm 0.6 \times 10^8$ M^{-1}s^{-1}, identical within error limits to that for Vaska's complex in CDCl$_3$. Iridium has a much higher atomic number than rhodium, yet the physical quenching rate of the rhodium analog is about the same.

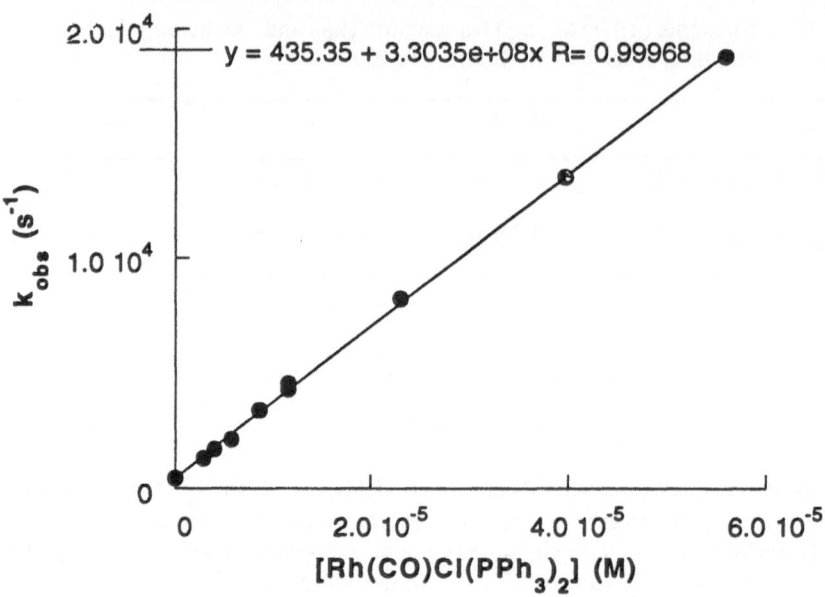

$$y = 435.35 + 3.3035e+08x \quad R = 0.99968$$

Figure 7. Luminescence quenching of singlet oxygen by **3** in CDCl$_3$. k_{obs} is the apparent rate constant of singlet oxygen luminescence decay; the slope of the plot is ($k_R + k_Q$).

Decomposition of Rh(CO)Cl(PPh$_3$)$_2$O$_2$

The decomposition of **4** was followed at various temperatures by monitoring the disappearance of the carbonyl peak of the product at 2044 cm^{-1} with the corresponding appearance of that of **3** at 1980 cm^{-1} (Fig. 8). In all cases, good first-order kinetics were observed.

The natural logs of the decay rates of the complex determined by IR at various temperatures were plotted against the reciprocal of the temperature; the resulting Arrhenius plot gave $\Delta H^{\ddagger} = 23 \pm 2$ kcal/mol, $\Delta S^{\ddagger} = 13$ e. u. atr 0 °C for the decomposition of **4**. The oxygen released by the decomposition is apparently in the triplet state, since when a sample of diphenylisobenzofuran (DPBF) was added to a 3-fold excess of **4** and warmed to room temperature, no change in the DPBF absorption spectrum was observed. Less than 1 % of the oxygen released is in the singlet state.

The enormous increase in the rate constant for reaction of singlet oxygen with both the Iridium and Rhodium complexes compared to that with triplet oxygen could be due to removal of the spin barrier. However, since both oxygen adducts release oxygen only in the triplet state, spin conservation does not seem to be a problem in these high-Z complexes. It is more likely that the high reactivity comes from the 22.4 kcal increase in energy of singlet oxygen, which is sufficient to decrease or even exceed the activation energy of the reaction with ground state oxygen. This is the case with Vaska's Complex, where $\Delta H^{\ddagger}$ is only 13.1 kcal/mol for reaction with triplet oxygen.[32] Most compounds react with singlet oxygen with very low enthalpies of activation. The reaction rate constant is usually decreased about two powers of ten by an unfavorable entropy of activation. If this is also the case with Vaska's complex, the reaction rate constant, which is two orders of magnitude below the diffusion rate constant, is limited only by the entropy of activation. This is probably also the case with the Rhodium complex.

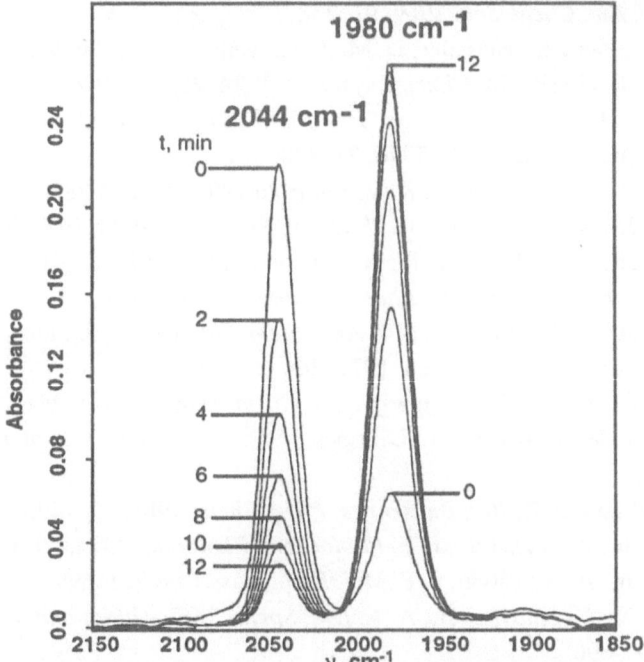

Fig. 8. Disappearance of the carbonyl band of **4** (2044 cm^{-1}) and appearance of that of **3** at 1980 cm^{-1}. Measurements taken at + 6 °C at 2-minute intervals.

These results suggest that novel energetic and unstable metal-oxygen complexes which may have interesting reactivities may be readily prepared by reaction of singlet oxygen with organometallic compounds. Because of the low activation enthalpies, reactions with singlet oxygen often go very well at temperatures as low as -120 °C, low enough to preserve even fairly unstable adducts. Further experiments with related compounds are in progress to gain a better understanding of transition states for formation of metal-peroxo complexes and the bonding of dioxygen to metal centers.

ACKNOWLEDGEMENTS

This work was supported by NSF Grant No. CHE89-14366 and NIH grant No. GM20080. Helpful discussions with Prof. J.S. Valentine on the nature of metal-peroxo complexes are gratefully acknowledged. We thank Prof. O. L. Chapman for use of the low-temperature IR.

REFERENCES

(1)   Ogilby, P. R.; Foote, C. S. *J. Am. Chem. Soc.* **1982**, *104*, 2069-2070.

(2)   Krasnovsky Jr., A. A. *Photochem. Photobiol.* **1979**, *29*, 29.

(3)   Ogryzlo, E. A. In *Singlet Oxygen*; Wasserman, H. H. Murray, R. W. Ed.; Academic Press, New York, 1979; pp 35-58.

(4)   Khan, A. U.; Kasha, M. *J. Am. Chem. Soc.* **1970**, *92*, 3293-3300.

(5)   Khan, A. U.; Kasha, M. *J. Am. Chem. Soc.* **1966**, *88*, 1574.

(6)   Ogryzlo, E. A.; Pearson, A. E. *J. Chem. Phys.* **1968**, *72*, 2913.

(7)   Wilson, T. *J. Am. Chem. Soc.* **1969**, *91*, 2387.

(8)   Abbott, S. R.; Ness, S.; Hercules, D. M. *J. Am. Chem. Soc.* **1970**, *92*, 1128.

(9)   Brabham, D. E.; Kasha, M. *Chem. Phys. Lett.* **1974**, *29*, 159-162.

(10)   Kenner, R. D.; Khan, A. U. *J. Chem. Phys.* **1977**, *67*, 1605-1613.

(11)   Krasnovsky Jr., A. A. *Biofizika* **1976**, *21*, 748.

(12)   Krasnovsky Jr., A. A.; Neverov, K. V. *Biofizika* **1988**, *26*, 884-886.

(13)   Krasnovsky Jr., A. A.; Neverov, K. V. *Chem. Phys. Lett.* **1990**, *167*, 591-596.

(14)   Krasnovsky Jr., A. A.; Rodgers, M. A. J.; Galpern, M. G.; Rihter, B.; Kenney, M. E.; Lukjanets, E. A. *Photochem. Photobiol.* **1992**, *55*, 691-696.

(15)   Krasnovsky Jr., A. A.; Foote, C. S. *J. Am. Chem. Soc.* **1993**, submitted.

(16)   Stauff, J.; Fuhr, H. *Z. Naturforsch.* **1971**, *266*, 260-263.

(17)   Stauff, J.; Fuhr, H. *Ber. Bunsenges. physik. Chem.* **1969**, *73*, 245-251.

(18)   Brukhanov, V. V.; Ketzle, G. A.; Laurinas, V. C.; Levshin, L. V. *Opt. i Spekt.* **1986**, *60*, 205-207.

(19)   Nickel, B.; Prieto, F. R. *Ber. Bunsenges. Phys. Chem.* **1988**, *92*, 1493-1503.

(20)   Krasnovskii Jr., A. A.; Litvin, F. F. *Photochem. Photobiol.* **1974**, *20*, 133-149.

(21)   Krasnovsky Jr., A. A.; Litvin, F. F. *Mol. Biol. (Russ.)* **1967**, *1*, 699.

(22)   Neverov, K. V.; Krasnovsky Jr., A. A. *Opt. Spektr. (Sov. Optics and Spectrosc.)* **1991**, *71*, 691-696.

(23)   Arbogast, J. W.; Darmanyan, A. O.; Foote, C. S.; Rubin, Y.; Diederich, F. N.; Alvarez, M. M.; Anz, S. J.; Whetten, R. L. *J. Phys. Chem.* **1991**, *95*, 11-12.

(24)   Arbogast, J. W.; Foote, C. S. *J. Am. Chem. Soc.* **1991**, *113*, 8886-8889.

(25)   Turro, N. J.; Chow, M. F.; Rigaudy, J. *J. Am. Chem. Soc.* **1981**, *103*, 7218-7224.

(26)   Valentine, J. S. *Chem. Rev.* **1973**, *73*, 235-245.

(27)   Vaska, L. *Accts. Chem. Res.* **1976**, *9*, 175-183.

(28)   Vaska, L. *Science* **1963**, *140*, 840-841.

(29)   Wilkinson, F.; Brummer, J. G. *J. Phys. Chem. Ref. Dat.* **1981**, *10*, 809-1000.

(30)   Selke, M.; Foote, C. S. *J. Am.Chem. Soc.* **1993**, *115*, 1166-1167.

(31)   Vaska, L.; Chen, L. S. *Chem. Comm.* **1971**, 1080-1081.

(32)   Chock, P. B.; Halpern, J. *J. Am. Chem. Soc.* **1966**, *1966*, 3511-3514.

(33)   Ogilby, P. R.; Foote, C. S. *J. Am. Chem. Soc.* **1983**, *105*, 3423-3430.

(34)   Stevens, B.; Perez, S. R. *Mol. Photochem.* **1974**, *6*, 1-7.

(35)   Higgins, R.; Foote, C. S.; Cheng, H. *Advan. Chem. Ser.* **1968**, *77*, 102.

(36)   Foote, C. S.; Ching, T.-Y. *J. Am. Chem. Soc.* **1975**, *97*, 6209-6214.

(37)   Vaska, L. *Inorg. Chim. Acta.* **1971**, *5*, 295.

EVIDENCES FOR SUBSTRATE ACTIVATION OF COPPER CATALYZED INTRADIOL CLEAVAGE IN CATECHOLS

Gábor Speier[1], Zoltán Tyeklár[1], Lajos Szabó, II[2], Péter Tóth[1], Cortland G. Pierpont[3] and David N. Hendrickson[4]

[1]Department of Organic Chemistry, University of Veszprém, 8201 Veszprém,
  Hungary
[2]Research Group for Petrochemistry of the Hungarian Academy of Sciences,
  8201 Veszprém, Hungary
[3]Department of Chemistry and Biochemistry, University of Colorado,
  Boulder, Colorado 80309-0215
[4]Department of Chemistry, University of California at San Diego, La Jolla,
  California 92093-0506

INTRODUCTION

The coordination chemistry of transition metal complexes containing catechol and semiquinone ligands has been developed over the past twenty years.[1,2] Much of the chemical interest in these complexes has concerned the reactivity of the coordinated quinone ligand in oxidation reactions that involve dioxygen.[3-5] Synthetic and mechanistic studies have been conducted on compounds containing a wide variety metal ions, but the two metals that have been the focus of great interest are copper and iron. Biological catechol oxidation reactions are carried out catalytically by the catechol dioxygenase enzymes. The active sites of protocatechuate-4,5-dioxygenase,[6] an extradiol dioxygenase enzyme, and protocatechuate-3,4-dioxygenase,[7] an intradiol dioxygenase enzyme, contain iron centers where dioxygen and catechol react in a process that results in ring cleavage (1).

Model studies and much of the developmental research on synthetic catechol oxidation systems has been concentrated on iron and copper catecholate complexes.

The Activation of Dioxygen and Homogeneous Catalytic Oxidation,
Edited by D.H.R. Barton *et al.*, Plenum Press, New York, 1993

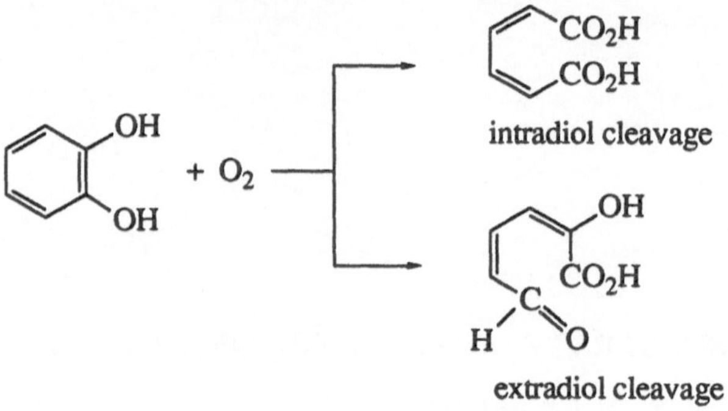

$$\text{(1)}$$

intradiol cleavage

extradiol cleavage

Brown and coworkers[8] observed reactions that may parallel the biological ring oxidation process in early studies on copper catecholate systems (2).

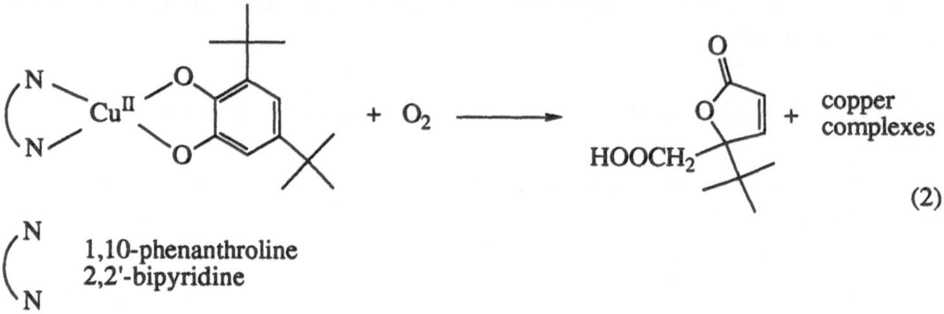

$$\text{(2)}$$

1,10-phenanthroline
2,2'-bipyridine

By far the most common reactions observed involve simple metal-mediated electron transfer, with reduction of dioxygen to peroxide accompanied by catechol oxidation to benzoquinone, without ring cleavage. Research in our laboratory has concerned the mechanistic distinction between these two fundamental catechol oxidation processes in studies on catecholate and semiquinonate complexes of copper. Tautomeric equilibria that involve electron transfer between localized metal and quinone electronic levels are known for quinone complexes of cobalt[9] and manganese,[10,11] and have been proposed to contribute to metal activation toward dioxygen coordination in synthetic and biological iron-containing catechol oxidation reactions.[7,12,13] Similar electron transfer equilibria may occur for copper-catecholate species (3).

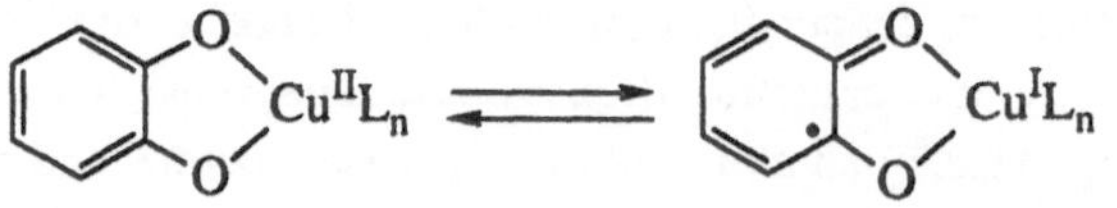

$$\text{(3)}$$

In the present paper we would like to present the results of synthetic experiments on semiquinone and catechol complexes of copper, with characterization that may provide insights on the extent to which redox activity at both the metal and the quinone ligand contributes to the synthetic course of oxidation reactions with dioxygen.

SEMIQUINONE AND CATECHOLATE COMPLEXES OF COPPER

Metallic copper may be oxidized with 9,10-phenanthrenequinone in pyridine solution. By washing the product with ether the bis(9,10-phenanthrenesemiquinone)copper(II) complex (**1**) is obtained (4). The complex has low solubility in noncoordinating solvents probably due

$$2 \quad + \quad Cu^o \quad \xrightarrow[\text{py}]{\text{Ar}} \quad \quad (4)$$

1

to intermolecular interactions in the solid state. Structural characterization on [Cu(3,5-DBSQ)2]2[14] has shown that semiquinone oxygen atoms bridge metals in the dimeric molecule, and 9,10-phenanthrenesemiquinone ligands readily form stacked structures in the solid state. Magnetic characterization on Cu(PhenSQ)2 has shown that it has a magnetic moment of 1.78 μ_B per Cu, and the X-band EPR spectrum recorded on a solid sample of the complex is shown in Figure 1. Resonances appear that may be associated with both Cu(II) and PhenSQ radical.

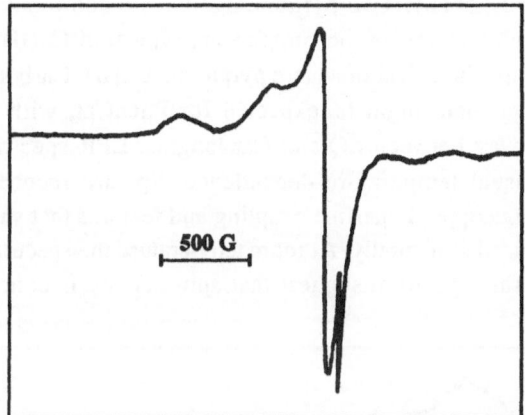

500 G

Figure 1. The solid state EPR spectrum of Cu(PhenSQ)2 (**1**)

$$\quad + \quad Cu^o \quad \xrightarrow[\text{py}]{\text{Ar}} \quad Cu^{II}(py)_2 \cdot \quad (5)$$

2

The complex is a complicated spin system;[15] in the absence of structural characterization there is not an obvious explanation for these observations. If crystals of the complex are obtained by evaporation of the pyridine solution (5) the complex **2** shown in Figure 2 is obtained. Structural

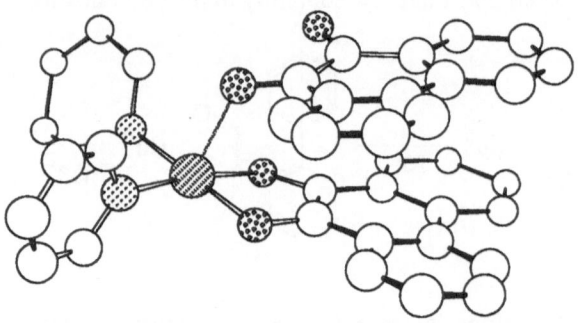

Figure 2. The molecular structure of Cu(PhenCat)(PhenBQ)(py)2 (**2**)

characterization on the molecule has revealed that the metal is five-coordinate, with a square pyramidal geometry. The oxygen of a PhenBQ ligand occupies the apical coordination site, bound with the metal relatively weakly, with Cu-O3 length of 2.297(12)Å, and with C-O lengths of 1.23(1) and 1.25(1)Å. Two pyridine ligands are coordinated at basal sites with a chelated PhenCat ligand. Features of the complex are typical of Cu(II) and the EPR spectrum recorded on a solid sample is typical of square pyramidal Cu(II). Carbon-oxygen lengths of the basal ligand are shorter than might be expected for PhenCat, with values of 1.21(1) and 1.33(1)Å that are midway between SQ and Cat lengths. EPR spectra recorded in pyridine solution show an unusual temperature dependence. Spectra recorded in solution at low temperature show strong copper hyperfine coupling and features that suggest that the molecule is in the form characterized structurally. At room temperature the spectrum shown in Figure 3 is obtained. Features of this spectrum suggest that spin density is concentrated on a PhenSQ

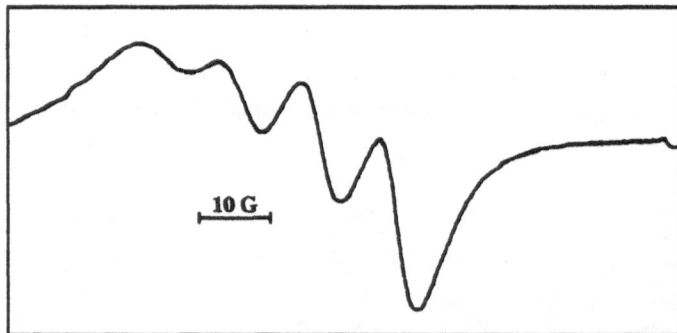

10 G

Figure 3. The EPR spectrum of Cu(PhenCat)(PhenBQ)(py)2 (**2**) at room temperature

ligand. There are two possible interpretations for this observation. First, that structural change occurs for the complex in solution so that interactions between the paramagnetic metal and radical ligands in a $Cu^{II}(py)_2(PhenSQ)_2$ species result in a S = 1/2 radical ground state. A second explanation is that the complex adopts a $Cu(py)_2(PhenSQ)$ (3) ground state that results in dissociation of the PhenBQ ligand (6).

This view is supported by the structural features of the basal quinone ligand of $Cu(py)_2(PhenCat)(PhenBQ)$. If $Cu(PhenSQ)_2$ obtained by the procedure described in (4) is redissolved in pyridine $Cu(py)_2(PhenCat)(PhenBQ)$ is obtained as shown by the characteristic EPR spectra. By using 2,2'-bipyridine (bpy) in place of pyridine $Cu^{II}(bpy)(PhenCat)$ (4) is obtained (7).

This complex displays a metal-based Cu(II) EPR spectrum at all temperatures in solution and

in the solid state. Similarly, if the reaction between metallic copper and PhenBQ is carried out in a pyridine solution containing N,N,N',N'-tetramethylethylenediamine (tmeda) the product is $Cu^{II}(tmeda)(PhenCat)$ (5) (8).

This complex may add an additional quinone ligand to give Cu(tmeda)(PhenCat)(PhenBQ) (**6**), as shown in Figure 4. Structural features of this molecule are very similar to those of

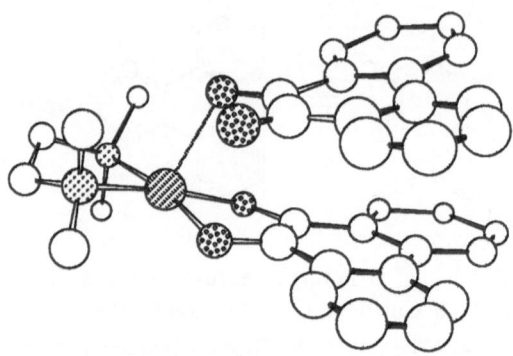

Figure 4. The molecular structure of Cu(tmeda)(PhenCat)(PhenBQ) (**6**)

Cu(py)$_2$(PhenCat)(PhenBQ) (**2**), including the unusually short C-O bond lengths (1.315(5), 1.316(5)Å) for the chelated PhenCat ligand. The EPR spectrum of Cu(tmeda)(PhenCat) in solution is solvent dependent (Figure 5). In acetonitrile spectra show evidence for the

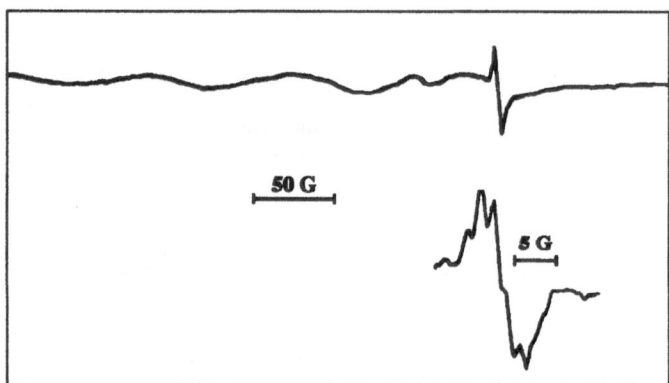

Figure 5. The EPR spectra of Cu(tmeda)(PhenCat) (**5**) in acetonitrile

Cu^{II}(PhenCat)/Cu^{I}(PhenSQ) equilibrium indicated in (9).Pyridine appears to displace the tmeda ligand reversibly.

Studies carried out with nitrogen-donor coligands have provided evidence for the existence of a redox equilibrium of the type described in equation (3) for complexes in solution.

Further preparative work has concerned complexes containing phosphine coligands as examples of Cu^{I}(PhenSQ) coordination. Reactions carried out with copper metal and PhenBQ

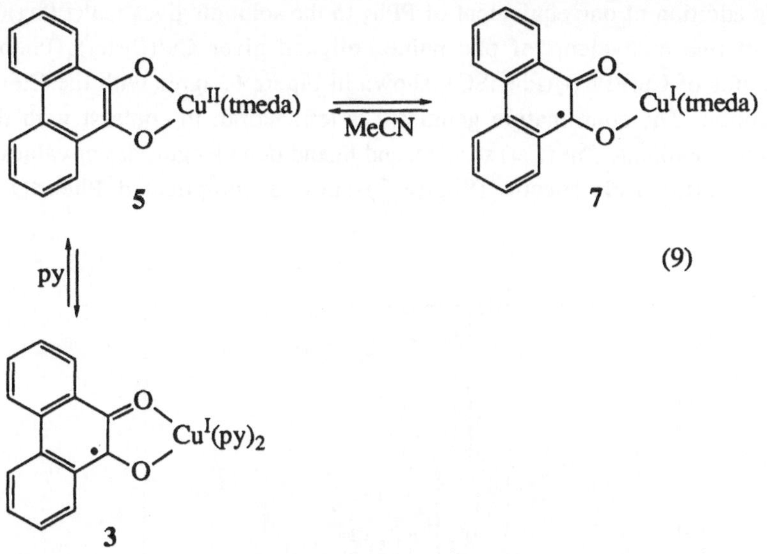

(9)

in acetonitrile solution may be used to form triphenylphosphine addition products (10).

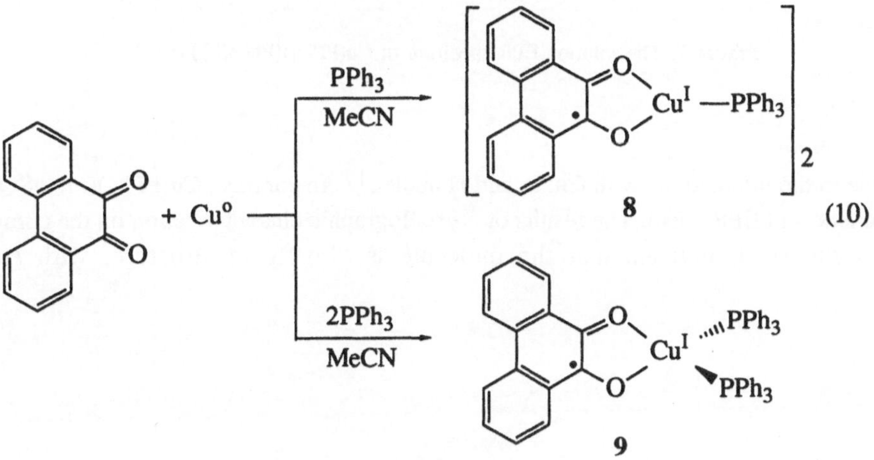

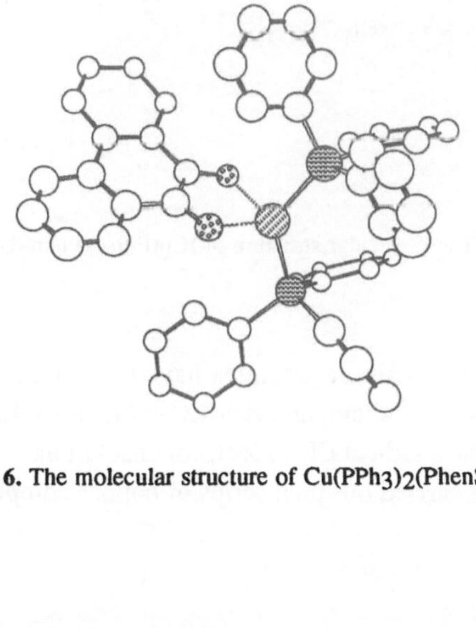

(10)

Figure 6. The molecular structure of Cu(PPh3)2(PhenSQ) (9)

Stoichiometric addition of one equivalent of PPh$_3$ to the solution gives Cu^I(PPh$_3$)(PhenSQ) (**8**), addition of two equivalents of phosphine coligand gives Cu^I(PPh$_3$)$_2$(PhenSQ) (**9**). Structural features of Cu(PPh$_3$)$_2$(PhenSQ), shown in Figure 6, agree with the Cu^I(PhenSQ) charge distribution. The coordination geometry is tetrahedral, in contrast with the planar geometry of four-coordinate Cu^{II}(Cat) species, and ligand bond lengths are of values expected for PhenSQ. Solution EPR spectra (Figure 7) show a complicated PhenSQ coupling

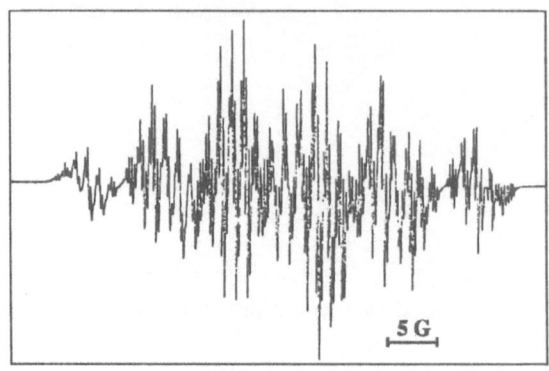

Figure 7. The solution EPR spectrum of Cu(PPh$_3$)(PhenSQ) (**9**)

pattern due to the interactions with Cu, P, and H nuclei.[17] In contrast, Cu(PPh$_3$)(PhenSQ) (**9**) is diamagnetic and EPR silent. The results of crystallographic characterization on the complex, shown in Figure 8, indicate that the molecule is dimeric in structure, with strong

Figure 8. The molecular structure of [Cu(PPh$_3$)(PhenSQ)]$_2$ (**8**)

pairing between PhenSQ ligands. Similar features have been found for Mo$_2$O$_5$(PhenSQ),[18] which is also diamagnetic and EPR silent, and Re$_2$(CO)$_7$(PhenoxSQ),[19] which is diamagnetic in the solid state but does show a radical EPR spectrum in solution.

The results of studies carried out on a series of copper complexes prepared with 9,10-

phenanthrenequinone and nitrogen- and phosphorus-donor coligands indicate that charge distribution may be directed by coligand bonding properties.[20] Examples were the charge distribution is rigidly Cu^{II}(PhenCat) and Cu^{I}(PhenSQ) have been obtained with bpy and PPh3 coligands, respectively. With pyridine and tmeda temperature-dependent EPR spectra have been interpreted to indicate the presence of a redox equilibrium between Cu(I) and Cu(II) charge distributions.

REACTIONS WITH MOLECULAR OXYGEN

It has been of interest to compare the reactivity of the three classes of complexes described in the previous section with dioxygen. Reaction between $Cu(PPh_3)_2$(PhenSQ) and O_2 occurs readily in dichloromethane to give OPPh3 and an oxidized complex product **10** (11). Accurate molecular weight measurements are consistent with dimeric species **10**. The complex

is formulated as having a peroxo bridge between PhenSQ ligands that links two complex units. Support for this formulation comes from acidification experiments that have been observed to produce hydrogen peroxide and PhenBQ.

At room temperature in pyridine solution both $Cu^{I}(py)_2$(PhenSQ) and $Cu(py)_2$(PhenCat)(PhenBQ) react rapidly with dioxygen to give the diphenate product **11** (12 and 13).

$$(13)$$

Pyridine solutions of Cu(PhenSQ)$_2$ (**1**) react similarly, through Cu(py)$_2$(PhenSQ) formed initially in the reaction (14).

$$(14)$$

Oxidation of Cu^{II}(tmeda)(PhenCat) with dioxygen, followed by hydrolysis with aqueous acid is observed to be solvent dependent. When carried out in acetonitrile (15) the products are hydrogen peroxide and PhenBQ.

$$(15)$$

In pyridine solution diphenic acid is obtained as the ring cleavage product. The reaction between Cu(tmeda)(PhenCat) and dioxygen has been studied at 0 °C in acetonitrile solution and a product has been isolated that we believe contains a peroxophenanthrenesemiquinone ligand (**12**) (16).

In the absence of structural characterization on the complex it is assumed to have features that are similar to the chelated peroxophenanthrenesemiquinone ligand of [Ir(triphos)(O$_2$PhenSQ)]$^+$ **12**.[21] The infrared spectra recorded for the complexes prepared in acetonitrile at 0° C (a) and at reflux temperature (b) are shown in Figure 9. EPR spectra recorded for **12** in acetonitrile solution and for a solid sample are shown in Figure 10. Additional evidence for the existence of a peroxosemiquinone ligand has been obtained from its chemical reactivity. Hydrolysis with protic acid is observed to produce PhenBQ and hydrogen peroxide, following the reactivity of

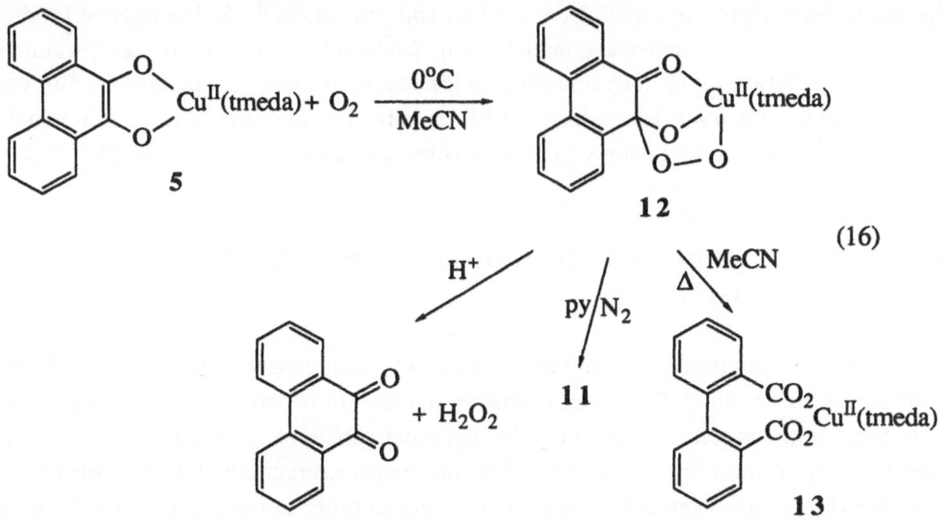

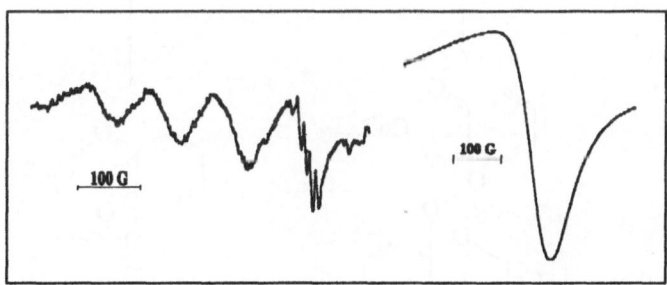

(16)

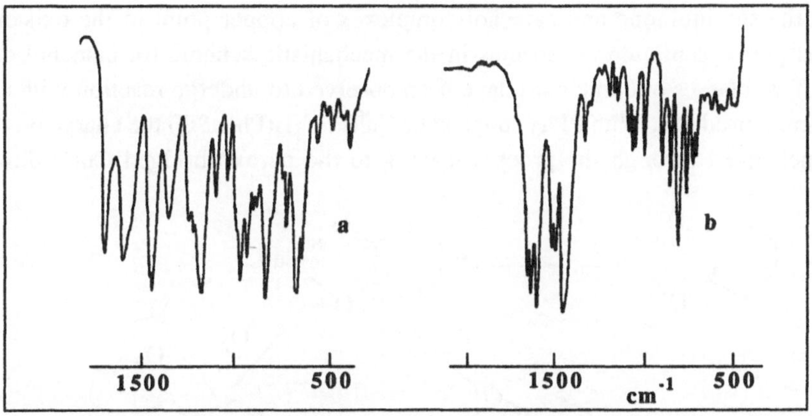

Figure 9. The IR spectra of 12 (a) and 13 (b)

Figure 10. The solution and solid state EPR spectra of $Cu^{II}(O_2PhenSQ)(tmeda)$ (12)

[Ir(triphos)(O$_2$PhenSQ)]$^+$ observed by Bianchini and coworkers.[4] A ring opened product, obtained in the form of the diphenate complex **13**, is obtained by increasing the temperature of the acetonitrile solution of **12**. The coordinated diphenate product **11** obtained in dioxygen reaction of Cu(py)$_2$(PhenSQ) is obtained if **12** is dissolved in pyridine at room temperature. These observations point convincingly to the peroxosemiquinone formulation for **12**.

MECHANISTIC ASPECTS OF DIOXYGEN ADDITION TO Cu(I) SEMIQUINONE COMPLEXES

There have been nearly as many mechanistic schemes presented for catechol oxidation reactions as systems studied. Oxidation reactions that lead to benzoquinone and ring-opened products occur in the absence of metal ion,[22] and outer sphere electron transfer[23] occurs as an important step in initiating processes that do not involve direct attack at the substrate. In cases where direct dioxygen attack does occur in reactions involving metal catecholate complexes the site of attack remains as a mechanistic consideration. Results obtained in our studies on the semiquinone and catechol complexes of copper point to the existence of a coordinated peroxosemiquinone species in the mechanistic scheme for catechol oxidation (Scheme). Two classes of complexes have been observed to undergo reaction with dioxygen giving different products. With PPh$_3$ coligand of Cu^I(PPh$_3$)$_2$(PhenSQ) the charge on the metal remains unchanged through the process leading to the peroxo-bridged Cu(I) dimer. This

Scheme. The possible mechanistic pathways for the oxygenation of catecholate and semiquinone complexes implies that dioxygen addition occurs at the semiquinone ligand as shown in the Scheme. Complexes containing pyridine and tmeda ligands show evidence for tautomeric equilibria (3) and contain metal centers that are redox active. Dioxygen attack may occur at either the metal or the semiquinone ligand of the $Cu^I(SQ)$ species, completing the intramolecular bridge to give the chelated, tridentate peroxosemiquinone species shown in the Scheme. This product may react further with acid to give hydrogen peroxide and benzoquinone or may undergo intramolecular rearrangement under conditions of higher temperature to give the coordinated ring-opened product.

ACKNOWLEDGMENT

The Hungarian Research Fund (OTKA #2326) and the U.S.-Hungarian Joint Fund (# 098/91) are gratefully acknowledged.

REFERENCES

1. C.G. Pierpont and R.M. Buchanan, Transition metal complexes of o-benzoquinone, o-semiquinone, and catecholate ligands, *Coord. Chem. Rev.* 36:45 (1981).
2. C.G. Pierpont and C.W. Lange, The chemistry of transition metal complexes containing catechol and semiquinone ligands. *Prog. Inorg. Chem.* in press.
3. M.M. Rogic, M.D. Swerdloff, and T.R. Demmin, Copper-mediated oxidations of aromatic substrates: Synthetic models for tyrosinase activities and the role of molecular oxygen, *in*::"Copper Coordination Chemistry; Biochemical and Inorganic Perspectives, " K.D. Karlin and J. Zubieta, eds, Adenine, Guilderland, New York, p.258 (1959).
4. P. Barbaro, C. Bianchini, K. Linn, C. Mealli, A. Meli, F. Vizza, F. Laschi, and P. Zanello, Dioxygen uptake and transfer by Co(III), Rh(III), and Ir(III) catecholate complexes, *Inorg. Chim. Acta*, 198-200:31 (1992).
5. L. Que, Jr., The catechol dioxygenases, *in*:"Iron Carriers and Iron Proteins," T.M. Loehr, ed., VCH Publishers, New York. p. 467 (1989).
6. D.M. Arciero and J.D. Lipscomb, Binding of [17]O-labeled substrate and inhibitors to protocatechuate 4,5-dioxygenase-nitrosyl complex, *J. Biol. Chem.* 267:2170 (1986).
7. D.D. Cox and L. Que, Jr., Functional models for catechol 1,2-dioxygenase. The role of the iron(III) center, *J. Am. Chem. Soc.* 110:8085 (1988).
8. D.G. Brown, L. Beckmann, C.H. Ashby, G.C. Vogel, and J.T. Reinprecht, Oxygen dependent ring cleavage in a copper coordinated catechol, *Tetrahedron Lett.* 1363 (1977).
9. R.M. Buchanan and C.G. Pierpont, Tautomeric catecholate-semiquinone interconversion via metal-ligand electron transfer. Structural, spectral, and magnetic properties of (3,5-di-*tert*-butylcatecholato)-(3,5-di-*tert*-butylsemiquinone)(bipyridyl)cobalt(III), a complex containing mixed-valence organic ligands, *J. Am. Chem. Soc.* 102:4951 (1980).
10. M.W. Lynch, D.N. Hendrickson, B.J. Fitzgerald, and C.G. Pierpont, Ligand-induced valence tautomerism in manganese-quinone complexes, *J. Am. Chem. Soc.* 103:3961 (1981).
11. M.W. Lynch, D.N. Hendrickson, B.J. Fitzgerald, and C.G. Pierpont, Intramolecular two-electron transfer between manganese(III) and semiquinone ligands. Synthesis and characterization of manganese 3,5-di-*tert*-butylquinone complexes and their relationship to the photosynthetic water oxidation system, *J. Am. Chem. Soc.* 106:2041 (1984).

12. T. Funabiki, T. Toyoda, H. Ishida, M. Tsujimoto, S. Ozawa, and S. Yoshida, Oxygenase model reactions. Part II. Hydroxylation of aromatic compounds by catecholatoiron complex / hydroquinones / O_2 in acetonitrile, *J. Mol. Catal.* 61:235 (1990).

13. T. Funabiki, H. Ishida, and Y. Yoshida, Catalytic monooxygenation of saturated hydrocarbons with O_2 by a nonheme mono-iron complex using hydroquinones as reductants, *Chem. Lett.* 1819 (1991).

14. J.S. Thomson and J.C. Calabrese, Copper-catechol chemistry. Synthesis, spectroscopy, and structure of bis(3,5-di-*tert*-butyl-*o*-semiquinonato)copper(II), *J. Am. Chem. Soc.* 108:1903 (1986).

15. O. Kahn, R. Prins, J. Reedijk, and J.S. Thomson, Orbital symmetries and magnetic interaction betweencopper(II) ions and the *o*-semiquinone radical. Magnetic studies of (di-2-pyridylamine)(3,5-di-*tert*-butyl-*o*-semiquinonato)copper(II) perchlorate and bis(bis(3,5-di-*tert*-butyl-*o*-semiquinonato)-copper(II)), *Inorg. Chem.* 26:3557 (1987).

16. G. Speier, S. Tisza, A. Rockenbauer, S.R. Boone, and C.G. Pierpont, Synthesis and characterization of (tetramethylethylenediamine)(9,10-phenanthrenequinone) (phenanthrenediolato)copper(II), a copper complex containing mixed charge 9,10-phenanthrenequinone ligands, *Inorg. Chem.* 31:1017 (1992).

17. A. Rockenbauer, M. Győr, G. Speier, and Z. Tyeklár, Electron spin resonance investigation of copper(I)-9,10-phenanthrenesemiquinonate complexes, *Inorg. Chem.* 26:3293 (1987).

18. C.G. Pierpont and R.M. Buchanan, Radical anion coordination of 9,10-phenantrenequinone in $Mo_2O_5(PQ)_2$, *J. Am. Chem. Soc.* 97:6450 (1975).

19. L.A. deLearie and C.G Pierpont, Rhenium carbonyl semiquinone complexes. Photochemical addition of 9,10-phenanthrenequinone to $Re_2(CO)_{10}$, *J. Am. Chem. Soc.* 109:7031 (1987).

20. R.M. Buchanan, C. Wilson-Blumenberg, C. Trapp, S.K. Larsen, D.L. Greene and C.G. Pierpont, Counterligand dependence of charge distribution in copper quinone complexes. Structural and magnetic properties of (3,5-di-*tert*-butylcatecholato)(bipyridine)copper(II), *Inorg. Chem.* 25:3070 (1986).

21. P. Barbaro, C. Bianchini, C. Mealli, and A. Meli, Synthetic models for catechol 1,2-dioxygenases. Interception of a metal catecholate-dioxygen adduct, *J. Am. Chem. Soc.* 113:3181 (1991).

22. G. Speier and Z. Tyeklár, The reaction of alkali metal catecholates with dioxygen in aprotic solvent, *J. Mol. Catal.* 57:L17 (1990).

23. C.A. Tyson and A.E. Martell, Kinetics and mechanism of the metal chelate catalyzed oxidation of pyrocatechols, *J. Am. Chem. Soc.* 94:939 (1972).

CATALYTIC, AEROBIC OXYGENATIONS WITH

METALLOPORPHYRIN CATALYSTS. NEW STRUCTURAL INSIGHTS

INTO STEREOSELECTIVITY

John T. Groves

Department of Chemistry, Princeton University

Princeton, New Jersey 08544

We have described the preparation and characterization of a *trans*-dioxo ruthenium(VI) porphyrin complexes which have been shown to be an efficient catalyst for the <u>aerobic</u> epoxidation of olefins. We have also shown that this complex will catalyze the hydroxylation of alkanes under appropriate conditions. It is now apparent that the dioxoruthenium(VI) complex is the species responsible for oxygen transfer and that an intermediate oxoruthenium(IV) complex is then reoxidized in air. We have shown that the oxidation of ruthenium(IV) to ruthenium(VI) proceeds *via* an oxygen atom transfer <u>disproportionation</u> to produce both dioxoruthenium(VI) and ruthenium(II) porphyrin complexes. The ruthenium(II) complex reacts rapidly with oxygen to regenerate the oxoruthenium(IV) intermediate which then continues the catalytic cycle. In this lecture recent results will be described which delineate the mechanism of this catalytic oxygenation and establish a connection between the observed selectivity in catalytic olefin epoxidation reactions to selectivities for the coordination of the product epoxide with ruthenium porphyrins. Specifically, X-ray structural evidence will be presented that the transition state geometry for olefin epoxidation is similar to the conformation of ruthenium(II) complexes of epoxides, thioepoxides and aziridines; the primary non-bonded interaction between the substrate and the catalyst occurs on the porphyrin plane in the immediate vicinity of the first coordination sphere of the metal; and the pronounced cis-selectivity for olefin epoxidation can be predicted quantitatively by comparing the binding constants of stereoisomeric epoxides or aziridines. These relationships establish the first detailed structural picture of metalloporphyrin catalyst interactions with substrates.

The Activation of Dioxygen and Homogeneous Catalytic Oxidation,
Edited by D.H.R. Barton *et al.*, Plenum Press, New York, 1993

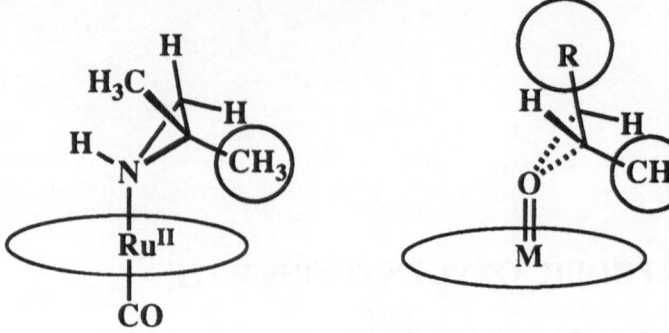

Transition State Geometry

POSTERS

COBALT BROMIDE CATALYSIS OF THE OXIDATION OF ORGANIC COMPOUNDS

Victor A. Adamian,[*] Yurii V. Geletii, Igor V. Zakharov

Moscow Physical and Technical Institute, Dolgoprudnyi, Moscow Region, 141700, Russia;
Institute of Chemical Physics in Chernogolovka, Chernogolovka, Moscow Region, 142432,
Russia.

Metal bromide catalysis is widely used in industry for oxidation of alkylaromatic compounds to their corresponding acids. The mechanism for this catalytic process has been investigated for over 20 years but it is still not established despite numerous papers on this subject. This work presents results of our investigations on the kinetics and mechanisms of cobalt bromide catalysis.

We have found that cobalt bromide catalysis is based on a reaction involving peroxy radical and Co(II) leading to hydroperoxide formation and Co(III) in a rate determining step (eq. 1) which is then followed by a rapid catalytic reduction of the generated Co(III) by bromide ions in the presence of hydrocarbon (eq. 2-4). The overall scheme involves two catalytic cycles and is given by equations (1)-(4) below.

$$RO_2{\cdot} + Co(II) \xrightarrow{k_4} ROOH + Co(III) \tag{1}$$
$$H^+ \quad \xrightarrow[Co(II),\ Br^-,\ RH,\ O_2]{} 2RO_2{\cdot} \tag{1a}$$

$$Co(III) + Br^- \longrightarrow Co(II) + Br{\cdot} \tag{2}$$
$$Br{\cdot} + RH \longrightarrow H^+ + Br^- + R{\cdot} \tag{3}$$

$$Co(III) + RH \xrightarrow{Br^-} Co(II) + H^+ + R{\cdot} \xrightarrow{O_2} RO_2{\cdot} \tag{4}$$

The hydroperoxide formed in reaction (1) is catalytically decomposed by Co(II) with formation of two peroxy radicals (eq. 1a). Thus, the reaction of $RO_2{\cdot}$ with Co(II) in protic media containing a hydrocarbon and bromide ions leads to formation of three more peroxy radicals and a branching radical chain oxidation of the hydrocarbon which in turn leads to a very high rate of oxidation with the cobalt bromide catalyst. The oxidation rate, W, is determined by both k_4 and the Co(II) concentration and is independent of hydrocarbon concentration:

$$W = 2k_4^2[Co(II)]^2/k_6.$$

Reactions between $RO_2{\cdot}$ and Co(II) or Mn(II) are discussed in terms of the methods of rate constant determination, the values of the rate constants and the effect of media on rate constants. A synergistic effect of Mn(II) addition is also discussed. The effect of the media is discussed in terms of the coordination ability of the solvent with Co(II) ion as well as the proton donor property of the solvent. It is concluded that the presence of a proton donor in the coordination sphere of Co(II) is required for hydroperoxide formation but at the same time this prevents the peroxy radical from reacting with the cobalt(II) ion. As a result, a bell-shaped dependence of the rate constant on the acetic acid and water concentration is observed. Our data explain some of the controversial literature data on solvent effects for the kinetics of metal bromide catalysis.

----
* Present address: Department of Chemistry, University of Houston, Houston, TX, 77204-5641, U.S.A.

HOMOGENEOUS CATALYTIC OXIDATION OF ORGANIC SUBSTRATES BY COPPER-PYRAZOLATO COMPLEXES

G.A.Ardizzoia[a], G.La Monica[a], N.Masciocchi[b], M.Moret[b]

(a)   *Dipartimento di Chimica Inorganica e Metallorganica, Centro C.N.R., Universita' di Milano, Via Venezian 21, Milano, Italy.*

(b)   *Istituto di Chimica Strutturistica Inorganica, Universita' di Milano, Via Venezian 21, Milano, Italy.*

The synthesis of di- and polynuclear complexes having ligands which maintain the metal centers in close proximity is an important objective in transition metal chemistry, owing to their potential role in multi-metal centered catalysis. Metal assisted reactions of dioxygen are very important processes and there is increasing interest in their products and mechanisms.

As a part of a systematic study of the chemical and structural properties of new and already known copper complexes containing pyrazolato groups as ligands, we have extensively explored the reactivity of such systems with respect to dioxygen, with the aim to ascertain their potential catalytic properties.

Indeed, some copper-pyrazolato complexes allowed the catalytic oxidation of organic substrates such as primary and secondary amines, isocyanides and phosphines in mild conditions.[1]

In an attempt to gain more insight into the reactions of copper(I) pyrazolato complexes with dioxygen, we decided to use different pyrazoles, whose donor capabilities were modified by introducing substituents in the 3,5-positions.

The synthesis of the tetranuclear copper(I) complex, [Cu(dppz)]$_4$, (Hdppz = 3,5-diphenylpyrazole), its crystal structure and catalytic properties in the oxidation of settled organic molecules will be presented and discussed.

An useful comparison with the catalytic behavior evidenced by other copper-pyrazolato systems will be drawn.

---

(1) G.A.Ardizzoia, M.A.Angaroni, G.La Monica, F.Cariati, S.Cenini, M.Moret, N.Masciocchi, *Inorg.Chem.*,**39**, (1991), 4347.

DIOXYGEN LIGAND TRANSFER FROM PLATINUM TO MOLYBDENUM. ISOLATION OF A HIGHLY REACTIVE MOLYBDENUM(VI) OXOPEROXO DIMER

H. Arzoumanian,[a] J. Sanchez,[a] G. Strukul,[b] and R. Zennaro[b]

a Ecole Nationale Superieure de Syntheses, de Procedes et d'Ingenierie Chimiques, d'Aix-Marseille, URA du CNRS 1410 Universite d'Aix-Marseille III, 13397 Marseille, France.
b Department of Chemistry, The University of Venice, Dorsoduro 2137, 30123 Venezia, Italy

The reaction of $Pt(O_2)(PPh_3)_2$ with $Mo(O)_2(mesityl)_2$ (mesityl = $2,4,6\text{-}Me_3C_6H_2$) in pyridine results in the transfer of the dioxygen ligand from platinum to molybdenum giving, in the presence of PPh_4Cl, a molybdenum peroxo compound (1) in which all organic moieties have been lost.

Elemental analysis, IR and ^{17}O NMR indicates that 1 is the tetraphenylphosphonium salt of an oxoperoxomolybdenum(VI) dimer. This is confirmed by an X-Ray analysis although a high thermal agitation causing an important disorder on the peroxo moiety of the molecule is present. The presence of two bonding patterns for the peroxydic ligand is proposed: one classical "side-on" peroxo and one open-end superoxide type structure. It epoxidizes tetracyanoethylene at a rate estimated 10^4 greater than the known oxodiperoxomolybdenum ("MIMOUN's butterfly complex"). It exhibits, furthermore, a high selectivity for electron poor olefins (TCNE/cyclooctene = 10^3). This is in part rationalized by the novel bonding mode of the peroxo moiety but also by the presence of the closeby molybdenum trioxo portion in the dimer. Extended Huckel theory calculations provide some additional support for this.

COPPER IMIDAZOLE COMPLEXES CATALYZE THE OXIDATIVE POLYMERIZATION OF 2,6-DIMETHYLPHENOL WITH DIOXYGEN

Patrick J. Baesjou[a], Ger Challa[b], Willem L. Driessen[a] and Jan Reedijk[a]

a) *Department of Chemistry, Leiden University, P.O. Box 9502, 2300 RA Leiden, The Netherlands*
b) *Laboratory of Polymer Chemistry, University of Groningen, Nijenborgh 16, 9747 Groningen, the Netherlands*

The oxidative dehydrogenation polymerization of 2,6-dialkylphenoles has been known for many years to be catalyzed by copper amine complexes [1]; see figure for 2,6-dimethylphenol oxidation. The mechanism of action of this industrially very important reaction has been studied for some time by several groups.

The mechanistic details are of interest for several reasons:

1. The amount of DPQ, the undesired side product, needs to be reduced as far as possible.
2. The resemblance with certain bioinorganic dehydrogenases is of great interest and knowledge in one area is likely to benefit also the other area.
3. The reoxidation step appears to involve a very reactive copper-dioxygen intermediate, which recently has been studied spectroscopically and kinetically for certain biomimetic systems.

G.E. Proces (Hay)

PPE

DPQ (side product)

Our efforts in this area are focusing on:
1. Study of the very first stages of the reaction (electron transfer).
2. Finding out whether the catalysts are dinuclear or mononuclear.
3. Trapping dioxygen adducts.

We have selected Cu imidazole complexes, such as $Cu(Meim)_n(NO_3)_2$ (n = 1-8), because of our earlier experiences with this type of ligands [2]. Some important conclusions already obtained are:

* a base is needed as co-catalyst to dehydronate the phenol;
* an optimal kinetics (fastest dioxygen uptake) is observed for n = 3;
* the reaction follows Michaelis-Menten kinetics with respect to the phenol concentration;
* the selectivity of the reaction (expressed as DPQ vs. PPE formation) is strongly influenced by the size of the ortho substituent of the phenol, the amine ligand, the temperature, the solvent polarity, and the concentration of the starting phenol;
* the catalytically active species seems to be a dinuclear Cu(II) species to which phenolato ligands bind in a bridging mode;
* a two-electron transfer to form a coordinated phenoxonium ion seems to be rate determining.

Acknowledgements

The research is sponsored by the Netherlands Organization for Chemical Research (SON), with financial aid of the Netherlands Organization for the Advancement of Research (NWO).

References:

[1] A.S. Hay et al. J.Am.Chem. Soc., **81**, 6335 (1959).
[2] G. Challa, W. Chen and J. Reedijk, Makromol. Chem. Macromol. Symp. **59**, 59 (1992).

Andreja Bakac, Susannah L. Scott, Adam Marchaj and James H. Espenson
Department of Chemistry and the Ames Laboratory, Iowa State University,
Ames, IA 50011

Complexes of chromium(II) and cobalt(II) react rapidly with dioxygen to yield metal superoxo ions as initial products,

$$(L)M + O_2 \rightleftharpoons (L)MO_2 \qquad\qquad K \qquad\qquad (1)$$

The kinetic and thermodynamic stabilities of $(L)MO_2$ vary greatly with the metal, the ligand system and the nature of anions present. For example, in the cobalt series the binding constant K decreases by three orders of magnitude as the macrocyclic ligand [14]aneN$_4$ is replaced by the bulkier Me$_6$[14]aneN$_4$. The anion stabilization of the superoxo complexes through inductive and charge effects is responsible for a 10^5-fold increase in the binding constant for $Co(Me_6[14]aneN_4)O_2^{2+}$ as the pH is raised from 6 to 13.

The $(H_2O)_5CrO_2^{2+}$ (hereafter CrO_2^{2+}) reacts in both stoichiometric and catalytic processes. It catalyzes the autoxidation of $CrCH_2OH^{2+}$ in a sequence of reactions that yield CH_2O, Cr^{2+} and a hydroperoxochromium(III) ion, $CrOOH^{2+}$. The capture of Cr^{2+} by O_2 regenerates the catalyst.

$$CrO_2^{2+} + CrCH_2OH^{2+} \rightarrow CrO_2H^{2+} + Cr^{2+} + CH_2O \qquad (2)$$
$$Cr^{2+} + O_2 \rightarrow CrO_2^{2+} \qquad\qquad\qquad\qquad\qquad (3)$$

The reduction of CrO_2^{2+} by Cr^{2+} yields the chromyl ion, CrO^{2+}. Unexpectedly, CrO^{2+} has a lifetime of >1 min and can be prepared in solution in submillimolar concentrations. It reacts with alcohols and aldehydes by hydride abstraction, and with PPh$_3$ by oxygen transfer, eq 4-5. Both reactions produce Cr^{2+}, which is rapidly converted to CrO_2^{2+} in the reaction with O_2, eq 3. The intense UV spectrum of CrO_2^{2+} provides the means both to monitor the kinetics and identify Cr^{2+} as the initial chromium product.

$$CrO^{2+} + R_2CHOH \rightarrow Cr^{2+} + CR_2O + H_2O \qquad (4)$$
$$CrO^{2+} + PPh_3 \rightarrow Cr^{2+} + O=PPh_3 \qquad\qquad\quad (5)$$

The combination of reactions 4 and 3 has proved to be a sensitive and unerring test for the chromyl. We have used it to confirm the formation of CrO^{2+} in the controversial reaction of chromate with alcohols.

FORMATION OF ISOXAZOLINES AND ISOXAZOLES *VIA* DIPOLAR CYCLOADDITION BY CATALYTIC OXIDATION OF ALDOXIMES WITH HYDROGEN PEROXIDE AND W(VI) PEROXO COMPLEXES

Francesco Paolo Ballistreri, Ugo Chiacchio, Antonio Rescifina, Gaetano Andrea Tomaselli and Rosa Maria Toscano

Dipartimento Scienze Chimiche, University of Catania, Viale A. Doria, 6-95125 Catania, Italy

The chemistry of oxidation of Mo(VI) and W(VI) peroxopolyoxo complexes is becoming of increasing interest in the field of oxygen transfer processes. Alkenes, alkynes, alcohols, diols, sulfides, solfoxides undergo easy oxidation by these reagents both under stoichiometric and catalytic conditions.[1]

Recently we have focused our attention to the oxidation of nitrogen containing compounds by these oxidants.[2] We have observed that secondary amines can be nicely converted into nitrones in one pot procedure whereas imines yield the corresponding aldehydes and nitroso derivatives.

Hereinafter we wish to report the results relative to the oxidation of aldoximes by hydrogen peroxide catalyzed by W(VI) peroxopolyoxo complexes. In the presence of alkynes or alkenes it is possible to obtain isoxazolines and isoxazoles respectively in very good yields. Implication of such findings will be discussed at the light of nitrile oxides cycloaddition chemistry and on the basis of W(VI) peroxo complex reactive behavior.

References

(1) Ballistreri, F.P.; Failla, S.; Tomaselli, G.A.*J.Org.Chem.* **1988**, 53, 830; Ballistreri, F.P.; Failla, S.; Spina, E.; Tomaselli, G.A.*J.Org.Chem.* **1989**, 54, 947.

(2) Ballistreri, F.P.; Chiacchio, U.; Rescifina, A.; Tomaselli G.A.; Toscano, R.M. *Tetrahedron* **1992**, 48, 8677.

OXIDATION OF AMINES AND IMINES BY Mo(VI) AND W(VI) PEROXOPOLYOXO COMPLEXES.

Francesco Paolo Ballistreri, Gaetano Andrea Tomaselli and Rosa Maria Toscano

Dipartimento Scienze Chimiche, University of Catania, Viale A. Doria, 6-95125 Catania, Italy

Mo(VI) and W(VI) peroxopolyoxo complexes have been shown to be versatile oxidant agents toward a large variety of organic compounds. They can be employed as stoichiometric reagents or catalitically in the presence of dilute hydrogen peroxide as oxygen source. The application of suitable phase transfer techniques has played a relevant role in the development of such oxidation catalytic processes.

Various substrates, e.g. alkenes, alkynes, alcohols, diols, sulfides, sulfoxides, amines were shown to be good candidates for such oxidation reactions.[1]

Recent mechanistic studies from our laboratory point to peroxopolyoxo complexes behave as electrophilic oxidants toward nucleophilic substrates such as thioethers and alkenes, whereas with sulfoxides incursion of SET processes might be a likely event.[2]

In this communication we present the results of a kinetic study concerning the oxidation of secondary amines and of imines by Mo(VI) and W(VI) peroxo complexes.

References

(1) Ballistreri, F.P.; Failla, S.; Tomaselli, G.A. *J.Org.Chem.* **1988**, 53, 830; Ballistreri, F.P.; Failla, S.; Spina, E.; Tomaselli, G.A. *J.Org.Chem.* **1989**, 54, 947.

(2) Ballistreri, F.P.; Tomaselli, G.A.; Toscano, R.M. *J.Mol.Cat.* **1991**, 68, 269; Ballistreri, F.P.; Tomaselli, G.A.; Toscano, R.M.; Conte, V.; Di Furia, F. *J.Am.Chem.Soc.* **1991**, 113, 6209; Ballistreri, F.P.; Bazzo, A.; Tomaselli, G.A.; Toscano, R.M. *J.Org.Chem.* in press.

D.H.R. Barton, S. Bévière, W. Chavasiri, D.R. Hill, B. Hu, T.L. Wang and C. Bardin.

Department of Chemistry, Texas A&M University, College Station, TX 77843, USA

The selective functionalization of saturated hydrocarbons is an important and challenging academic and industrial objective. Our own contributions to this area are based on Gif-type systems. Recent studies have shown that *tert*-butyl hydroperoxide (TBHP) is an efficient oxidizing agent in these systems.

While TBHP induces the ketonization of saturated hydrocarbons in the presence of an iron(III) catalyst, the use of copper(II) salts as catalysts gives olefination. Differences between the iron-catalyzed and copper-catalyzed reactions support a metal dependent reaction pathway.

In addition, during the course of recent mechanistic studies, new methodology for the functionalization of saturated hydrocarbons based on Gif-type protocol has been invented. These reactions are of interest as mechanistic probes and also from the point of view of synthetic utility. Thus, cycloalkanes are transformed with varying efficiency into mono-substituted cycloalkyl derivatives (chlorides, bromides, azides, cyanides, thiocyanates, dicycloalkyl sulfides, or nitroalkanes) by conducting iron or copper catalyzed reactions in the presence of alkali metal salts (scheme).

Scheme

Paraffins have historically been regarded as an inert class of compounds. These new results coupled with our prior work prove that belief to be a myth. Indeed, when confronted by the appropriate chemical species, paraffins display a rich variety of functionalization reactions.

Sponsors: The National Science Foundation, BP America and Quest International.

OXIDATION OF CYCLOHEXANE CATALYZED BY POLYHALOGENATED AND PERHALOGENATED IRON PORPHYRINS

P.Battioni[*], J.Haber[**], R.Iwanejko[**] D. Mansuy[*] and T. Mlodnicka [**]

[*]Laboratoire de Chimie et Biochimie Pharmacologiques et Toxicologiques, Universite Rene Descartes , Paris, France.

[**]Institute of Catalysis and Surface Chemistry, Polish Academy of Sciences, Krakow, Poland.

Model systems composed of metalloporphyrins and oxidizing agents have often been used for the liquid phase oxidation of hydrocarbons. The yields of the products and selectivity of the system appeared to be dependent on the character of the metal center, the oxidant used as well as on the structure of the porphyrin ligand. Simple metalloporphyrins are not very stable under the oxidizing conditions and undergo decomposition during the reaction course. In order to improve their stability efforts have been done to introduce different substituents into the porphyrin macrocycle. It appeared that the metalloporphyrins with halogen substituents in the phenyl and/or pyrrole rings are more resistant against the oxidative degradation.

The systems composed of iron tetraphenylporphyrins with halogenated phenyl and pyrrole rings in dichloromethane solutions were used for oxidation of cyclohexane with monoperoxyphthalic acid magnesium salt at room temperature. Bu_4NCl was used as the phase transfer agent. Cyclohexanol and cyclohexanone were the main reaction products. The activity of the system appeared to be dependent on the degree of halogenation of the porphyrin macrocycle as well as on the presence of N-base such as 4- tertbutyl-pyridine. The highest activity and selectivity was found for iron tetrapentafluoroporphyrin and the order of activities for other iron porphyrins is following:

$$TPFPP > TPFP\beta Br_8P > TDCP\beta Br_8P > TDCPP > TMP$$
$$(43,5) \quad (36) \quad (4) \quad (3,2) \quad (0)$$

where : $TPFP\beta Br_8P$ - tetrapentafluorophenyl-β-octabromoporphyrin, TDCPP - tetra-ortho-dichlorophenylporphyrin, $TDCP\beta Br_8P$ - tetra-ortho-dichlorophenyl-β-octabromoporphyrin. The numbers in the parentheses correspond to turnovers after 20 minutes.

The porphyrins halogenated on phenyl rings are more stable and active than tetra tolyl and mesityl porphyrins. Introduction of halogen substituents to the pyrrole rings enhances considerably the stability of these porphyrins under oxidizing conditions and in some cases the catalytic activity and selectivity to alcohol.

Addition of the N-base increases slightly the activity and selectivity of the investigated systems. However, a negative effect is observed for tetrapentafluoroporphyrin.

When compared with the corresponding manganese porphyrins, iron porphyrins show lower activity but higher selectivity to alcohol, which sometimes is the sole reaction product.

NEW OXIDATIONS OF ORGANIC SUBSTRATES USING HYDROGEN PEROXIDE AND SUPPORTED HETEROPOLYACID CATALYSTS

S W Brown, A Hackett, A Johnstone and W R Sanderson, Solvay Interox Research and Development, Widnes, England

Oxidation processes employing aqueous hydrogen peroxide in the presence of heterogeneous catalysts offer significant advantages over traditional liquid-phase processes in terms of cost, safety and environmental impact. We have previously shown[1] that catalysts prepared by depositing phosphotungstic acid on gamma-alumina, followed by calcination, are active for the selective epoxidation of olefins using 35% H_2O_2. These catalysts have been demonstrated conclusively to operate heterogeneously, with essentially no observable leaching of tungsten into the reaction solvent (aqueous t-butanol). They may also be recycled many times without significant loss of activity. Best results are obtained with cyclic alkenes, there being no detectable steric limitations on the substrates which can react.

We have now investigated other uses of these catalysts for hydrogen peroxide oxidation reactions, and present results with a wide range of organic substrates. It is often possible to obtain high selectivity to specific products in both carbon and sulphur oxidations. The importance of choosing the right reaction conditions, and in particular the appropriate solvent, will be highlighted. In all cases, the use of chlorinated solvents, which are now widely regarded as environmentally unacceptable, is avoided.

---

1. International Patent Application PCT/GB 92/01154

POLYMER BOUND MANGANESE PORPHYRIN CATALYSTS FOR OLEFIN EPOXIDATIONS WITH HYDROGEN PEROXIDE[1])

J. W. Buchler[a)], G. Goor[b)], A. von Kienle[a)], U. Mayer[a)], G. F. Thiele[b)]

a)   Institut für Anorganische Chemie, Technische Hochschule Darmstadt,
      D6100 Darmstadt, Germany
b)   Degussa AG, Abt. IC-FE-A, Rodenbacher Chaussee 4, D6450 Hanau 1

The sterically demanding manganese porphyrin Mn(TDClPP)Cl (**1**) is one of the most effective homogeneous catalysts for the direct epoxidation of olefins with hydrogen peroxide.[2)]

We have developed heterogeneous catalysts by anchoring this manganese porphyrin to a polymeric support in order to facilitate the recovery and improve the stability of the catalyst.

Simple and high yield procedures for bonding the porphyrin to a polymeric support are presented, which give materials with minimum catalyst leaching. The heterogeneous catalysts are compared with the homogeneous catalyst **1** in terms of catalytic activity, selectivity with respect to olefin and hydrogen peroxide and catalyst lifetime. The influence of the catalyst support on hydrogen peroxide utilization and the mechanisms of catalyst deactivation will be discussed.

---

1)   This work was supported by BMFT grants 03C264A and 03C264B.
2)   S. Banfi, A. Maiocchi, F. Montanari, S. Quici, *Chim. Ind. (Milan)* **72** (1990) 304 and references cited therein. G. Legemaat, dissertation, Univ. Utrecht 1990.

OLEFIN EPOXIDATION BY Ph$_4$PHSO$_5$ CATALYZED BY Mn(III)-PORPHYRINS UNDER HOMOGENEOUS CONDITIONS

S. Campestrini
Centro CNR di Studio sui Meccanismi delle Reazioni Organiche, Dipartimento di Chimica Organica, Universita' di Padova, via Marzolo 1, 35131 Padova, Italy.
F. Novello
Department of Chemistry, Brown University, Providence, R.I., 02912 USA

The oxidation of organic substrates catalyzed by Mn(III)-porphyrins with a large variety of oxygen donors has been extensively studied, mainly in biphasic systems. The complexity of such systems is, presumably, the reason why a widely accepted detailed mechanistic description of the oxidation processes is not yet available. Therefore, we decided to investigate a model system as simple as possible, *i.e.* the epoxidation of olefins by Ph$_4$PHSO$_5$, a peroxide soluble in organic solvents, which can be obtained with a high degree of purity, in the presence of a series of Mn(III)-porphyrins in the solvent dichloroethane.

We found that, in the absence of the catalyst, no epoxidation reaction takes place; moreover, also in the absence of a species that can act as an axial ligand for the porphyrin, *e.g.* 4-tert-butylpyridine or imidazole, the oxidation does not occur. Stopped-flow experiments indicate that the presence of the axial ligand is a necessary requisite for the formation of the "oxo" species of the porphyrin derived from an oxygen transfer from Ph$_4$PHSO$_5$ to the catalyst.

On the basis of a detailed investigation on the effect of various additives on this process, a mechanistic scheme for the "oxo" forming reaction may be suggested. Also the subsequent oxygen transfer from the "oxo" species to the olefin has been studied by kinetic experiments. This allows to establish the role played by the various species in the two consecutive oxygen transfer reactions. A comparison of the results thus obtained with those provided by the study of the overall epoxidation reaction, carried out by measuring the rates of epoxide formation, will be made.

SYNTHESIS AND ELECTROCHEMISTRY OF DIAQUARUTHENIUM(II) AND DIAQUAIRON(II) COMPLEXES OF QUATERPYRIDINES AND X-RAY CRYSTAL STRUCTURES OF [Ru(pQP)PPh₃)₂](ClO₄)₂, [Fe(pQP)(H₂O)](ClO₄)₂, and [Fe(QP)(H₂O)](ClO₄)₂ (pQP = 3′,5,5′ 5‴-TETRAMETHYL-2-2′:6′ 2″:6″,2‴-QUATERPYRINE; QP = 2,2′:6′,2′:6″,2′″-QUATERPYRIDINE)

Chin-Wing Chan, Ting-Fong Lai and Chi-Ming Che

Department of Chemistry, The University of Hong Kong, Hong Kong

Perchlorate salts of [Ru](QP)(H₂O)₂]²⁺, **1**, and [Ru(*p*QP(H₂O)₂]²⁺, **2**, (*p*QP = 3′,5,5′ 5′″-tetra-methyl-2-2′:6′ 2″:6″,2′″-quaterpyridine; QP = 2,2′:6′,2′:6″,2′″-quaterpyridine), have been isolated. Dioxo derivatives of **1** and **2** can be generated by electrochemical oxidation in degassed pH 1 buffer at 1.12 and 1.05 B (*vs* SCE) respectively. The $E_{1/2}$-pH relation has been determined for both diaqua complexes. They are quite stable in acidic media and exhibit reversible Ru(VI/IV), Ru(III/II) couples, though the Ru(IV/III) couples are only quais-reversible.

The X-ray crystal structure of [Ru(pQP)(PPh₃)₂](ClO₄)₂ has been determined: The quaterpyridine unit was puckered around ruthenium atom. [Ru(QP)(H₂O)₂]²⁺ is an active catalyst for the electrochemical oxidation of alcohol. Electrode surface adsorption of[Ru(QP)(H₂O)₂]²⁺ has been observed which proves it to be a promising candidate for fabrication of chemically-modified electrodes.

Analogous complexes of iron(II), namely [Fe(QP)(H₂O)₂]²⁺, **6**, and [Fe(pQP)(H₂O)₂]²⁺, **7**, have been isolated as perchlorate salts and characterized with X-ray crystallography. The quaterpyridine unit in compound **7** is twisted about the two central pyridyl rings while those in **6** are almost coplanar. This feature may partly explain the smaller stability of **7** in aqueous solutions.

Cyclic voltametric studies have been performed on **6**. *Only one* quasi-reversible couple was observed in pH 1 to pH 11 aqueous buffers and it was assigned Fe(III)/II one electron one proton couple. The absence of high-valent iron(IV)/(III) couple in **6** provides evidence against the generation of ferryl (Fe=O) species under pyridine environment. The activity of olefin epoxidations effected with 6 in H₂O/CH₃CN mixture indicates a radical rather than high valent iron species as working oxidation intermediate.

6   7

Acknowledgement: We acknowledge support from the Hong Kong Research Grants Council and the University of Hong Kong

Oxidation Of H_2S To S By Air With Fe(III)-NTA As A Catalyst: Catalyst Degradation

DIAN CHEN, RAMUNAS J. MOTEKAITIS AND ARTHUR E. MARTELL*, *Department of Chemistry, Texas A&M University, Texas 77843-3255 USA.*

DEREK MCMANUS, *ARI Technologies, Palatine, Illinois 60067 USA*

The oxidation of H_2S to S by air is represented by the overall reaction:

$$2H_2S + O_2 \longrightarrow 1/4S_8\downarrow + 2H_2O$$

The process is used with a catalyst for the removal of H_2S from natural gas. A typical catalyst would be an iron(II, III) chelate system such as iron-NTA. The catalyst reactions are:

$$2FeNTA + H_2S \longrightarrow 2FeNTA^- + 1/8S_8\downarrow + 2H^+$$

$$2FeNTA^- + 1/2O_2 + H_2O \longrightarrow 2FeNTA + 2OH^-$$

The main difficulty with the process is the degradation of the metal chelate catalyst. Thus the Fe(II,III)-NTA system degrades through the formation of a number of weaker chelating agents and eventually to carbon dioxide, water, and ammonia. A possible scheme for ligand degradation would be:

$$NTA \longrightarrow IDA \longrightarrow glycine \longrightarrow NH_3 + CO_2 + H_2O$$

$$^-OCH-COO^- \longrightarrow {}^-OOC-COO^- \longrightarrow 2CO_2$$

All of these degradation products except glyoxylic acid were detected quantitatively by HPLC, and the rate of degradation scheme is presented for the catalytic process. The fact that thiosulfate greatly decreases the degradation rate, and the fact that the degradation products were observed only in the stoichiometric reactions in which the ferrous NTA was oxidized by air, led to the interpretation that the hydroxyl radical or the superoxide radical ion is the reactive species that attacks the NTA and the intermediate degradation products and is responsible for the degradation of the organic ligands present. A comparison of the rates of degradation of the three ligands NTA, EDTA, and HEDTA shows the sequence HEDTA > EDTA > NTA, indicating that of these three ligands, NTA is least subject to degradation, and that its Fe(III) chelate is the best catalyst for the oxidation of hydrogen sulfide to sulfur by air.

CrAPO-5-CATALYZED LIQUID PHASE OXIDATION
OF SECONDARY ALCOHOLS WITH O_2

J. D. Chen, J. Dakka and R. A. Sheldon

Laboratory of Organic Chemistry and Catalysis
Delft University of Technology, Julianalaan 136
2628 BL Delft, The Netherlands

Liquid phase oxidations generally employ soluble catalysts, usually a metal salt. The use of heterogeneous catalysts in the liquid phase offers several advantages compared to their homogeneous counterparts, e.g. ease of recovery and recycling and stability.

One approach to creating heterogeneous oxidation catalysts with novel activities and selectivities is to incorporate redox metals, by isomorphous substitution, into the lattice framework of zeolites and related molecular sieves. Site-isolation of redox metals in inorganic lattices prevents the dimerization or oligomerization of active oxometal species which is characteristic of many homogeneous oxometal complexes and leads to their deactivation in solution. We coined the term 'redox molecular sieves' to describe such catalysts[1-3]. The first and most well-known example is titanium silicalite (TS-1) which has been shown to catalyze a variety of systhetically useful oxidations with H_2O_2[4].

As part of an ongoing programme on redox molecular sieves we are investigating the use of metal substituted alumino-phosphates (MeAPOs) in liquid phase oxidations. We have found that CrAPO-5 is an active and selective catalyst for the liquid phase oxidation of secondary alcohols with TBHP or O_2.

Oxidations were carried out at 85 °C in chlorobenzene as solvent and substrate:catalyst molar ratios of ca. 100:1. Cyclohexanol, and α-methylbenzyl alcohol were converted to the corresponding ketones in high selectivities using TBHP under a N_2 atmosphere. The selectivity on TBHP was also > 90%. Good results were also obtained for the selective oxidations of Cyclohexanol, α-methylbenzyl alcohol, tetralol and indanol using O_2 as the oxidant and 10 mol% TBHP as initiator. The catalyst could be recycled several times without any loss of activity. A mechanism is proposed for the CrAPO-5 catalyzed oxidation of alcohols.

References

1. R. A. Sheldon, in "Heterogeneous Catalysis and Fine Chemicals II", (M. Guisnet et al., Eds.), Elsevier, Amsterdam, 1991, pp. 33-54.
2. R. A. Sheldon, CHEMTECH, 566-576 (1991).
3. R. A. Sheldon, in Topics in Current Chemistry, in press.
4. U. Romano, A. Esposito, F. Maspero, C. Neri and M. Clerici, Chim. Ind. (Milan), **72**, 610-616 (1990).

SOLVENT EFFECTS ON THE OXYGENATION OF NI-BOUND THIOLATES, Patrick J. Farmer, Marcetta Darensbourg, Takako Soma, Department of Chemistry, Texas A&M University, College Station, TX 77843

The air-oxidation of N,N'-bis(mercaptoethyl)-1,5-diazocyclooctanenickel(II), ($\underline{1}$), yielding sulfur-oxygenated products, displays a pronounced solvent effect with the rate decreasing in the order DMF > CH_3CN > H_2O >> ROH . Square wave voltammetry has been used to monitor the reaction in various solvents. The reactions in nonprotic organic solvents result in mixtures of mono- ($\underline{2}$) and bis-sulfinato ($\underline{3}$) complexes and a trinuclear species resulting from free Ni^{2+}. Mechanistic considerations are complicated by the fact that isolated monosulfinate $\underline{2}$ does not

react with O_2 to form the bis-sulfinate, $\underline{3}$. In water the reaction of $\underline{1}$ with O_2 yields a different oxygenate, I, as the sole product which, when isolated and redissolved in nonprotic solvents, is prone to further oxygenation to form the $\underline{3}$ or decomposition to form $\underline{2}$ and $\underline{1}$. The nature of I and a comparison of the solvent effect to those seen in organic systems will be discussed.

MONOOXYGENATATION OF ALKANES AND AROMATICS CATALYZED BY NONHEME-IRON COMPLEX/HYDROQUINONE/O$_2$ SYSTEMS

Takuzo Funabiki, Takehiro Toyoda, Koji Kashiba,
Jun Yorita, and Satohiro Yoshida

Division of Molecular Engineering, Graduated School of Engineering
Kyoto University, Kyoto 606-01, Japan

Monooxygenation of aromatics, alicyclic and linear alkanes with molecular oxygen is catalyzed by nonheme iron complexes in the anhydrous organic solvents in the presence of hydroquinones as electron and proton donors. Iron complexes are prepared in situ by stirring FeCl$_3$, pyrocatechol, and pyridine (mole ratio is 1:1:2) in acetonitrile or in pyridine. Isolated catecholatoiron complex is also used. Catalytic activity is greatly dependent on the kinds of hydroquinone and increases in the order of 2,5-di-t-butyl- ≈ t-butyl->methyl->H-hydroquinone. Non-substituted hydroquinone hardly exhibits activity, and the activity is controlled by the oxidation potential and steric effect of hydroquinones.

The reactivity is also very much dependent on the pyridine concentration. At the low concentration of pyridine in acetonitrile, hydroxylation to form alcohol or phenol proceeds selectively, but at the high pyridine concentration, e.g. in pyridine, carbonylation to form ketones and aldehydes becomes dominant. This suggests the formation of the different species depending on the pyridine concentration.

The effect of hydroquinones on the selectivity is also characteristic. The C^2/C^3 ratio (the relative reactivity of the secondary C-H bond to the tertiary) is dependent on hydroquinones in the reactions of adamantane and isopentane. The reactivity is also affected by the substituent on pyrocatechol.

The reaction mechanism is studied by estimation of the NIH shift of toluene-4-D and Me-NIH shifts of xylenes, and also by the kinetic isotope effects. These results clearly indicate that the results can be explained by assuming the presence of the iron-oxygen active species. The C-H bond may be cloven homolytically in the solvent cage, but the reaction may proceed in the non-free radical process.

T.Funabiki et al., *Chem.Lett.*, 1267 (1989); *J.Mol.Catal,*, **61**, 235 (1990); *Chem.Lett.*, 1819 (1991); *ibid.*, 2143 (1991); *ibid.*, 1279 (1992); *ibid.*, 2303.

EFFECTS OF SENSITIZER AND SOLVENT ON SINGLET OXYGENATION OF β-METHYLSTYRENES

Shigeru Futamura

Research Center for Advanced Science and Technology, The University of Tokyo

It has been reported that benzaldehydes (6) and allylic hydroperoxides (13) are obtained in photosensitized oxygenation of β-methylstyrenes (1) and that dioxetanes (12) and 1,2-dioxenes (2) are the possible intermediates of (6). Thorough reinvestigation on singlet oxygenation of (1), however, has clearly shown that 1,4-cycloaddition predominates over 1,2-cycloaddition and the ene-reaction.

Tetraphenylporphine-sensitized photooxygenation of *trans*-anethole (1b) in CCl_4 afforded a 1,2-dioxene (2b) and an α-diketo epoxide (4b) along with a dihydropyranol (3b), 2-hydroxy-4-methoxybenzaldehyde (5b), 4-methoxybenzaldehyde (6b), 4-methoxyphenyl vinyl ketone (7b) and *trans*-anethole ᴐ. Different product distributions were obtained for β-methylstyrene and isosafroi.

Quenching of this reaction by 1,4-diazabicyclooctane supports that singlet oxygen is the active oxygen species. The product distribution for (1b) indicates the intermediacy of a 4-hydroperoxy-1,2-dioxene derivative (10b).

This paper will present the effects of sensitizer and solvent on singlet oxygenation of (1) and the substituent effect on the chemical behavior of (2).

HETEROPOLYPEROXO COMPLEXES AS CATALYSTS FOR ORGANIC OXIDATIONS

W. P. Griffith, B. C. Parkin, M. Spiro and K. M. Thompson

Imperial College of Science, Technology and Medicine, London SW7 2AY, United Kingdom.

A recent area of developing interest is that of low nuclearity polyperoxometallates, especially the "heteropolyperoxo" species, $[PO_4\{WO(O_2)_2\}_4]^{3-}$, which was isolated and found to have a highly symmetrical structure by Venturello and co-workers[1]. This is an active catalyst for the epoxidation of unactivated alkenes, by H_2O_2 under phase transfer conditions, notably as salts containing large lipophilic quaternary ammonium cations[2,3].

A variety of compounds of the type $[Q^{n+}]_{4/n}[XO_4\{MO(O_2)_2\}_4]$, where $Q = (^nC_6H_{13})_4N^+$, PPN^+ and $[Co(en)_3]^{3+}$; $X = P$ or As and $M = Mo$ or W, has been synthesised and characterized by IR, Raman and ^{31}P NMR spectroscopy where applicable. Structural characterization of these species and their analogues is underway. A comparison of the oxidizing abilities of $[(^nC_6H_{13})_4N]_3[XO_4\{MO(O_2)_2\}_4]$; $X = P$ or As; $M = Mo$ or W towards the epoxidation of both terminal and cyclic alkenes has been investigated, as well as the ability of $[(^nC_6H_{13})_4N]_3[PO_4\{WO(O_2)_2\}_4]$ to transform primary and secondary alcohols to aldehydes and ketones respectively.

Epoxidation of cycloalkenes by $[(^nC_6H_{13})_4N]_3[XO_4\{MO(O_2)_2\}_4]$ was markedly more successful than for their linear analogues. Furthermore, the larger the cycloalkene ring, the greater the degree of conversion to the epoxide. In all cases, the order of increasing activity was $[PO_4\{MoO(O_2)_2\}_4]^3 < [AsO_4\{MoO(O_2)_2\}_4]^{3-} << [PO_4\{WO(O_2)_2\}_4]^{3-} < [AsO_4\{WO(O_2)_2\}_4]^{3-}$. Oxidation of primary alcohols to aldehydes is in general faster than the conversion of secondary alcohols to ketones. Oxidation of benzylamine by $[(^nC_6H_{13})_4N]_3[PO_4\{WO(O_2)_2\}_4]$ leads to a mixture of benzaldehyde and benzyloxime The latter presumably is the primary oxidation product, which is further transformed to the aldehyde

Peroxide oxidations of natural colouring materials by such species have also been investigated Studies have been carried out on the oxidation of malvidin chloride by water soluble salts of $[XO_4\{MO(O_2)_2\}_4]^{3-}$ with H_2O_2 at pH 5.

1 C. Venturello, R. D'Aloisio, J. C. J. Bart and M. Ricci, *J. Mol. Catal.*, 1985, **32**, 107.
2 C Venturello and R. D'Aloisio, *J. Org. Chem.*, 1988, **53**, 1553 and later references.
3. C. Aubry, G Chottard, N. Platzer, J-M. Brégeault, R. Thouvenot, F. Chauveau, C. Huet and H. Ledon, *Inorg. Chem.*, 1991, **30**, 4409.

IRON COMPLEXES OF CYCLAM AND CYCLAM-LIKE LIGANDS AS MODELS FOR NON-HEME IRON ENZYMES

Raymond Ho and Joan Selverstone Valentine, Department of Chemistry and Biochemistry, University of California, Los Angeles, Los Angeles, California, USA

Metal-containing monooxygenase enzymes catalyze the oxygenation of a variety of substrates. In our models for non-heme iron enzymes we have concentrated on mimicking the reactivity of these metalloenzymes. We began our study by looking for non-heme iron complexes that catalyze the epoxidation of olefins, a reaction that is rarely seen with non-heme iron complexes. Our initial studies using the iron complex of cyclam and the oxygen donors: hydrogen peroxide, iodosylbenzene (PhIO), and *meta*-chloroperoxybenzoic acid (MCPBA) demonstrated that this iron complex is an efficient catalyst for the epoxidation of olefins. Although the reactions of all three of these oxygen donors with iron cyclam leads to epoxidation, further studies using a variety of substrates including *cis*-stilbene, styrene and *para*-substituted styrene, and norborene suggest that different intermediates are formed from the different oxygen donors. The nature of these intermediates is currently being explored.

Because of the success with iron cyclam, our search for other non-heme iron complexes involved a survey of cyclam like ligands including HMCD (5,7,7,12,14,14-hexamethyl-1,4,8,11-tetraazocyclodec-4,11-diene), TIM (2,4,9,10-tetramethyl-1,4,8,11-tetraazocyclodec-1,3,8,11-tetraene), TDO (1,4,8,11-tetraazocyclodecane-5,7-dione), etc. Under the same reaction conditions as for iron cyclam, the iron complexes of these other ligands also proved to be efficient catalysts for olefin epoxidation with PhIO and MCPBA, but not for H_2O_2. These results further support the notion that the three oxidants behave differently in their reactions with these non-heme iron complexes.

Other studies of the iron complexes of cyclam and cyclam like ligands with H_2O_2 have included changing the reaction conditions; addition of radical traps, axial ligands, acids, and bases to the reaction mixture; and studying the hydroxylation of aromatic compounds and alkanes. Preliminary results indicated that addition of axial ligands such as imidazole, acids, or bases inhibits the iron cyclam-catalyzed epoxidation of olefins by H_2O_2 and that the addition of the radical trap 2,4,6-tri-*t*-butylphenol does not inhibit the reaction. Also, preliminary experiments using napthalene and toluene as substrates gave napthoquinone and cresol.

Further study of the iron complexes of cyclam and cyclam like ligands may provide new insight into the mechanism of non-heme iron enzymes.

This research was supported by the NSF

OXYGEN ACTIVATION BY 3d METAL COMPLEXES OF OPEN CHAIN AND MACROCYCLIC SCHIFF BASE LIGANDS

E.-G. Jäger, M. Rost, A. Schneider

University of Jena, Germany

A big number of metal complexes of the types **1 - 4**[1] (M = Cu, Ni, Co, Fe, Mn; R^1 = Me, Ph, OEt; R^2 = H, COMe, COOEt, CN) have been compared with respect to i) the uptake of dioxygen (measured by precision gas volumetry), and ii) their catalytic activity for the oxidation of hydroquinones and olefines in different solvents (toluene, pyridine). The time dependent oxygen uptake can be described by the following general characteristics ($x = n_{dioxygen} / n_{metal}$):

i) no significant oxygen uptake ($x < 0.05$ after 10 hours); observed with the most complexes of Cu(II) and Ni(II) but also with the Co(II)-, Fe(II)-, and Mn(II)-complexes of **1**, R^1 = OEt, R^2 = CN, in pyridine;

ii) irreversible oxygen uptake with a definite and time stable limiting value of x = 0.25, indicating the formation of a μ-oxo derivative; observed with the Fe(II)-complexes **2**, R^1 = Me, R^2 = COOEt, COMe, and **3**, R^2 = COOEt, as well as with some Cu(I)-complexes;

iii) irreversible oxygenation with a definite and time stable limit of x = 0.5, indicating the formation of a oxo- or μ-peroxo derivative respectively; observed with the Fe(II)- and Mn(II)-complexes **1**, R^1 = Me, R^2 = COOEt, COMe in pyridine, and **1**, R^1 = OEt, R^2 = CN in toluene;

iv) irreversible oxygenation with a definite limit of x = 1, indicating the formation of a superoxo derivative (detectable by EPR spectroscopy); observed with the most electrochemically formed Ni(I) complexes **1** and **2**;

v) reversible oxygenation with a definite and time stable limit between $0 \leq x \leq 1$, depending on the special complex and the strength and concentration of additional bases, often characterized by a sharp maximum in the time dependent oxygenation curve; observed with the most Co(II) complexes **1**, **2** ($R^2 \neq H$), and **3** ($R^2 \neq H$);

vi) slow continuous oxygen uptake without a definite and time stable limiting value ($x > 1$), indicating an irreversible oxidation of the ligand or the solvent; observed with the Co(II)-complexes **2** ($R^2 = H$) and **4** and the Fe(II)-complexes **4** with special axial bases (e.g. imidazole).

The best catalytic activity is observed with complexes showing oxygenations of type ii) (for hydroquinones) or type iii) (for olefines).

[1] The formulas will be given at the poster. The authors thank the *Deutsche Forschungsgemeinschaft* and the *Fonds der Chemischen Industrie* for financial support.

SELECTIVE MONOEPOXIDATION OF 1,3-DIENES CATALYZED BY TRANSITION-METAL COMPLEXES

Karl Anker Jørgensen
Department of Chemistry
Aarhus University, DK-8000 Aarhus C
Denmark.

The chemistry of the selective monoepoxidation of 1,3-dienes is presented. By using transition-metal complexes and a terminal oxidant as *e. g.* hypochlorite it is possible to perform both regioselective and enantioselective expoxidation of a selected double-bond of the 1,3-diene. This procedure allows *e. g.* one to perform regioselective epoxidation of the less-substituted double bond of the 1,3-diene, and, furthemore, to avoid the polymerization of the 1,3-diene which is in contrast to conventional oxidation reagents. The scope of this reaction will be discussed and attempts to understand the oxygen-transfer from an oxo-transition-metal complex intermediate to only one of the double bonds of the 1,3-diene will also be discussed.

The transition-metal catalyzed oxidation of 1,3-dienes giving the corresponding vinyl epoxide, in principle an 1,2-addition of the oxygen atom to the 1,3-diene, will also be compared to the oxidation of 1,3-dienes taking place on early, as well as, late transition-metal surfaces, as the vanadyl pyrophosphate- and Ag(110) surfaces, where an 1,4-addition of the oxygen atom to the 1,3-butadiene takes place. The difference of oxygen-transfer taking place in the sphere of a discrete metal complex and the oxygen-transfer taking place on the metal surface will also be discussed.

The oxidation of 1,3-dienes catalyzed by transition-metal complexes will also in more general terms be discussed in realtion to the above-menthioned work.

HYDROCARBON OXIDATION BY A POLYNUCLEAR IRON SANDWICH POLYOXOTUNGSTATE - HYDROGEN PEROXIDE SYSTEM

Alexander M. Khenkin and Craig L. Hill

Department of Chemistry, Emory University, Atlanta, GA 30322 U.S.A.

The synthesis and characterization of isolated, isomerically pure tetranuclear iron sandwich polyoxotungstates $[(Fe^{II})_4(B-PW_9O_{34})_2]^{10-}$ and $[(Fe^{II})_4(B-P_2W_{15}O_{56})_2]^{16-}$ as organic-soluble tetra-*n*-butylammonium salts are reported. These complexes have been characterized by ^{31}P NMR, elemental analysis, magnetic susceptibility, UV-visible, infrared and EPR.

These complexes catalyze the epoxidation of alkenes by hydrogen peroxide in acetonitrile. They afford much better yields and selectivities than related mononuclear iron polyoxotungstates $[FePW_{11}O_{39}]^{5-}$ and $[FeP_2W_{17}O_{61}]^{7-}$. *cis*-Stilbene was epoxidized nonstereoselectively. Under high turnover conditions the tetranuclear iron sandwich complexes, unlike $PW_{12}O_{40}^{3-}$ and most polytungstophosphates are quite stable with respect of solvolytic degradation by H_2O_2.

$[(Fe^{II})_4(B-PW_9O_{34})_2]^{10-}$ also catalyzes the oxidation of cyclohexane, methylcyclohexane and hexane with formation of the corresponding alcohols and ketones. The mechanism of hydrocarbon oxidation in the tetranuclear iron sandwich polyoxotungstate - hydrogen peroxide system will be discussed.

THIANTHRENE OXIDE AS A MECHANISTIC PROBE FOR BLEACH INGREDIENTS

J.H. Koek, E.W.J.M. Kohlen, L. vd Wolf, K. Hissink
Unilever Research Laboratorium, P.O. Box 114, 3130 AC Vlaardingen, The Netherlands

Introduction

Thianthrene oxide, described in lit. [W.Adam, J.Am.Chem.Soc 113, 6202-6208 (1991)] as a mechanistic probe for oxidation processes, was applied under aqueous conditions to investigate the behaviour of several bleach active materials of detergent formulations.

Results

An HPLC method was developed for characterisation of the oxidation products. Peracids were found to be more reactive at 40 °C than hydrogen peroxide. The peracids were found to give nucleophilic oxidation at high pH and electrophilic oxidation at low pH. Under wash conditions (pH = 10) mainly nucleophilic oxidation was observed. For full implications to the bleach process at wash conditions the heterogenic nature of the last process should also be taken into account

Conclusions

Thianthrene oxide can be applied as oxidation probe also under aqueous conditions. The higher reactivity of the peracids compared with hydrogen peroxide at 40C correlate with the better bleach these peracids give at wash conditions.
Peracid bleach under wash conditions involves a nucleophilic oxidation mechanism.

THE SELECTIVE HYDROXYLATION OF CYCLOHEXANE BY ALKYL HYDROPEROXIDES
A CHALLENGE THAT CAN BE SOLVED ?

U.F. Kragten[1], C.B. Hansen[1], J. Nener[2]

[1]DSM Research, CP-Chemicals/CP-Intermediates, PO Box 18 6160 MD Geleen, The Netherlands
[2]Present address, University of York, Dep. of Chemistry, Heslington, York YO1 5DD, England

Abstract

The development of efficient routes towards cyclohexanone and cyclohexanol is one of the challenges for industrial researchers. Both chemicals are important feedstocks for the production of caprolactam and adipic acid. The latter two are precursors for Nylon-6 and Nylon-6,6

In our lab the hydroxylation of cyclohexane by cyclohexyl hydroperoxide (CHHP) and cumyl hydroperoxide (CumOOH) was studied. As catalysts sterically hindered metalloporphyrines (Metals = Co, Mn, Cr, Porphyrines = TPP, TDCIPP, TPyP[1]) were used. A comparison is made with data from literature and the possible impact on industrial scale production of cyclohexanone is discussed briefly. To distinguish between formation of cyclohexanone *via* the decomposition of the CHHP and the oxidation of the formed alcohol experiments were performed with both 3,3,5,5,-tetramethylcyclohexanol and cyclohexanol-d_{12}. The results indicated that the added alcohols were not oxidized to the corresponding ketone. Based on both epoxidation experiments with styrene and *cis*-stilbene and the hydroxylation experiments it was concluded that for both MnTDCIPP and CrTDCIPP the reaction mainly goes *via* a heterolytic mechanism. A third indication for a non radical pathway with Mn porphyrines was found by using a radical trapping agent, galvinoxyl (2,6-di-*tert*-butyl-α-(3,5-di-*tert*-butyl-4-oxo-2,5-cyclohexadien-1-ylidene)-p-tolyloxy, free radical). The addition of galvinoxyl did not influence the rate of reaction which is a strong indication of the absence of free radicals in the solution for the tested Mn porphyrin. With Co porphyrines the reaction is believed to proceed via a homolytic mechanism in which non sterically hindered porphyrines are readily converted to non-porphyrin species.

A comparison of the results for CumOOH and CHHP clearly indicated that CumOOH is a better hydroxylating agent than cyclohexylhydroperoxide. It is believed that this is due to the absence of an α-hydrogen in cumyl alcohol which makes it a less reactive molecule than cyclohexane. With a sterically hindered Mn porphyrin it was possible to hydroxylate with a high selectively cyclohexane (yield > 40 %). The formed cyclohexanol was almost not oxidized which resulted in very high alcohol : ketone ratios (>10). Despite these high selectivities there is still a strong competition between hydroxylation and normal decomposition of the peroxide.

Acknowledgements

We would like to thank DSM for their financial support and DSM Chemicals North America for providing us with the CHHP. We would also like to thank Prof. W. Drenth (University of Utrecht, NL) and Dr. J.R. Lindsay Smith (University of York, UK) for the fruitful discussions.

---

1. TDCIPP = <u>meso</u>-tetra(2,6-dichloro phenyl) porphyrin
   TPP   = <u>meso</u>-tetraphenylporphyrin
   TPyP   = <u>meso</u>-tetrapyridylporphyrin

DIOXYGEN OXIDATION BY THE METAL-IMMOBILIZED CATALYST DERIVED FROM SILICA AND MONTMORILLONITE MODIFIED BY SILANE COUPLING REAGENT

Yasuhiko Kurusu (Department of Chemistry, Sophia University, 7-1 Kioicho,Chiyoda-ku, Tokyo Japan 102)

The silica and montmorillonite modified with silane coupling reagents(SCR) were treated by metal ions(Fe,Cu,Co) in order to obtain effective dioxygen oxidation catalyst. The immobilized catalyst derived from silica was useful for the dioxygen oxidation of cyclohexanes. Also, the immobilized catalyst from montmorillonite was effective for oxidation of ethylbenzene.

1. Functionalization of silica and montmorillonite: These inorganic polymers are not suitable for organic reaction in solution because of its hydrophilicity. To remove this defect, it has been tried to construct inorganic-organic hybrid type compound by the treatment with SCR. By this process, the functional group in SCR becomes an effective ligand for metal complex formation. The functionalized silica or montmorillonite was treated with metal ion in order to obtain immobilized catalyst. N,N-dimethyl-3-aminopropyltrimethoxysilane and N-2-aminoethyl-3-aminopropyltrimethoxysilane were used as SCR. From the silica with higher surface area immobilized catalyst FS-1(Fe) and FS-1(Cu) were obtained. From montmorillonite cobalt immobilized catalyst([I] and [II]) was derived.

2. Oxidation of hydrocarbon: Cyclohexane and methyl cyclohexane were oxidized in the presence of FS-1(Fe) and FS-1(Cu).[1] In this case zinc and acetic acid were indispensable for this oxidation. Ethylbenzene was oxidized with cobalt immobilized catalyst[I] and [II]. This oxidation started after the addition of a little content of hydogen peroxide.

Table 1. Oxidation of ethylbenzene in the presence of various immobilized catalysts

No	Catalyst	Reaction time / h	Conversion / %	Selectivity / %		
				PhCOMe	PhCH(OH)Me	PhCH$_2$CHO
1[a)	I	24	15 0	65	35	0
2[a)	II	24	0	0	0	0
3[b)	c)	20	0	0	0	0
4[b)	CoBr$_2$	1	0 5	100	0	0
		3	3 8	50	50	0
		44	41 3	57	24	19
5[b)	I	1	2 8	61	39	0
		3	8 3	60	40	0
		20	38 5	63	19	18
	I(NaBr)[d)	1	7 0	55	45	0
		3	20 2	55	33	19
		20	52 1	54	18	28
	I(LiBr)[d)	1	5 4	38	52	10
		20	7 4	23	33	44
6[b)	II	20	40 0	70	12	18

Reaction condition Temp 90℃, O$_2$(1atm), Catalyst (Co, 0 17 mmol), Acetic acid (40 mmol),
Ethylbenzene (32 mmol), Hydrogen peroxide (2 mmol)
a) reaction with no addition of H$_2$O$_2$ b) reaction with addition of H$_2$O$_2$ c) without Co catalyst(blank)
d) salt content (1 mmol)

REFERENCES

1) Y.Kurusu, *Makromol.Chem. Makromol.Symp.* **59**, 313-318(1992) and the references cited therein.

BIOMIMETIC OXIDATION OF CYCLOHEXANE USING THE GIF-KRICT SYSTEM

Kyu-Wan Lee, Seong-Bo Kim and Ki-Won Jun
Catalysis Research Division, Korea Research Institute of Chemical Technology, P.O. Box 9 Daedeog-Danji, Taejon 305-606 (Korea)

Gif systems were originally designed to mimic non-heme enzymatic oxidation of alkanes. All of them involve a pyridine-acetic acid (or other carboxylic acid) solution of the hydrocarbon being oxidized, an iron-based catalyst, and an electron source. However, Gif systems have a ploblem in the separation of catalyst because the homogeneous catalytic systems are employed. Thus, we have modified the GifIV system by using heterogeneous catalysts, iron oxide supported on silica, and then the results are reported in this paper.

Fe/silica catalyst was prepared by the impregnation of $FeCl_3 \cdot 6H_2O$ onto silica gel(Kiesel gel 60). The catalyst was subsequently dried at 150℃ for 2h and calcined at 400℃ for 3h in an air stream. The prepared catalyst was analyzed by the spectroscopic methods of X-ray fluorescence, X-ray diffraction, X-ray photoelectron and Mössbauer. For the reaction, cycloheaxne, pyridine, acetic acid, zinc and the corresponding heterogeneous iron catalyst were placed in a 125 ml Erlenmeyer flask. The reaction mixture was stirred vigorously under air (1 atm) at room temperature for 16 hr. Quantitative analysis of reaction mixture was performed on a gas chromatograph.

The experimental results indicated that Fe/silica gives good yields in spite of being insoluble. The characterization data of the Fe/silica catalyst showed that the iron species of the Fe/silica catalyst exists mainly as Fe_2O_3 type. The pure iron oxides FeO and Fe_2O_3 can also catalyze GifIV oxidation and the oxidation state of iron gives the large effects on the activity and selectivity. The oxidic iron has better activity than metallic iron and favors production of cyclohexanone. We also investigated the effect of amount of supported iron catalyst on the reaction. The ratio of cyclohexanone/ cyclohexanol shows a marked dependence on the amount of catalyst. As the amount of catalyst increasing, the formation of alcohol decreases and the formation of ketone increases.

NON-SYMMETRIC DINUCLEATING LIGANDS FOR OXIDATION CATALYSTS TEMPLATES FOR REALISTIC DINUCLEAR METALLOENZYME MIMICS

M. Lubben, Auke Meetsma and Ben L. Feringa
Department of Organic and Molecular Inorganic Chemistry, University of Groningen, Nijenborgh 4, 9747 AG Groningen, The Netherlands

Abstract. One of the most striking features in many dinuclear metalloenzymes is the different coordination environments for the two metal centers. Nevertheless most of the reported enzyme mimics which are small molecule analogues for the active sites of these complicated enzymes, are based on symmetric dinucleating ligands. These ligands generally result in the formation of dinuclear coordination complexes with identical coordination environments for the metal centers. However, a limited number of non-symmetric dinuclear complexes is known to be formed by spontaneous self-assembly in solution.

We present here a new entry into the synthesis of non-symmetric dinuclear transition metal complexes by making use of non-symmetric dinucleating ligands. The use of a non-symmetric dinucleating ligand will force the two metal ions in the resulting complex into different chemical or coordination environments. This will result in the formation of a more realistic enzyme mimic.

The synthesis of the non-symmetric ligands is based on the Bimolecular Aromatic Mannich reaction. Non-symmetric ligands can be obtained via two different routes. The first route comprises a sequential Mannich reaction on a phenol using two different secondary amines. An alternative route is a Mannich reaction using a secondary amine and a substituted salicylaldehyde followed by either condensation of the aldehyde functionality with a primary amine and subsequent reduction or reductive amination with a secondary amine.

The excellent ligating ability of these non-symmetric ligand systems is illustrated by the presentation of the crystal and molecular structure of a non-symmetric dicopper(II) complex.

Preliminary results on the use of non-symmetric dinuclear complexes in catalytic oxidations will be presented.

Acknowledgement. This research is sponsored by Unilever Research.

BINUCLEAR COBALT(II) COMPLEXES OF MACROCYCLIC AND MACROBICYCLIC LIGANDS AS OXYGEN CARRIERS

Arthur E. Martell, Ramunas J. Motekaitis and Dian Chen

Department of Chemistry, Texas A&M University, College Station, Texas 77843-3255

The macrocyclic and macrobicyclic ligands, O-BISDIEN, and O-BISTREN, form thermo-dynamically stable hydroxo-bridged binuclear cobalt(II) complexes. Both of these complexes react with dioxygen to form kinetically stable peroxo-bridged complexes that do not undergo (metal-centered) degradation at appreciable rates. The loss of dioxygen from the μ-peroxo-μ-hydroxo-dicobalt(III) O-BISDIEN complex is estimated to have the first order rate of about 2×10^{-8} sec^{-1}. The ligand is not attacked in this reaction but is converted to the binuclear hydroxo-bridged cobalt(III) complex. The rate of degradation of the dioxygen complex formed from O-BISTREN is kinetically even more stable, with no measurable rate of degradation in water solution from ~0 °C to 75 °C. Even if the degradation were to take place (a reaction which is very slow for the O-BISDIEN complex and even slower for that formed from O-BISTREN) the ligands are not attacked during the degradation reactions, and the active Co(II) complexes can be regenerated by electrolytic reduction of the corresponding Co(III) complexes. In any case, the very slow rates of degradation of and the lack of ligand attack, make these binuclear cobalt(II) complexes the most effective and efficient oxygen carriers ever reported, and are highly recommended for the separation of dioxygen from air.

The dioxygen complex formed from O-BISDIEN, has a much higher thermodynamic stability than that of O-BISTREN, in spite of the fact that each metal center in the O-BISTREN complex is coordinated by four basic amino groups while the metal centers of O-BISDIEN are coordinated by only three such groups. This reversal of the usual correlation between basicity of the donor and thermodynamic stability of the dioxygen complex formed is ascribed to the flexibility of the macrocyclic O-BISDIEN ring compared to the relative inflexibility (pre-organization) of the O-BISTREN ligand, and the fact that considerable bending and distortion of the macrocyclic or macrobicyclic ligand is necessary in order to form the dioxygen complex.

Because the two ligands O-BISDIEN and O-BISTREN are difficult to synthesize, an attempt has been made to form dioxygen complexes of dinuclear cobalt(II) complexes that are readily prepared. A number of macrocyclic and macrobicyclic ligands are formed in good yield by a two-step process: (2+2 or 3+2) condensation of a dialdehyde with a bis- or tris-primary amine to give a macrocyclic or macrobicyclic tetra- or hexa Schiff Base, respectively, followed by hydrogenation to give the macrocyclic hexamine or the macrobicyclic octamine. The macrocyclic hexamines whose binuclear Co(II) complexes form dioxygen adducts were obtained by the initial condensation of DIEN (diethylenetriamine) with the four following aldehydes: pyrrole-2,5-dicarboxaldehyde, benzene-1,4-dicarboxaldehyde, pyridine-2,6-dicarboxaldehyde, and furan-2,5-dicarboxaldehyde. The macrobicyclic octamines whose binuclear Co(II) complexes form dioxygen adducts were obtained by the initial condensation of TREN (2,2′,2″-triaminotriethylamine) with pyridine-2,6-dicarboxaldehyde and pyrrole-2,5-dicarboxaldehyde. Of these six dioxygen complexes, all were found to undergo metal-centered degradation reactions (with the release of hydrogen peroxide) at appreciable rates; in fact none have a half-life longer than one day. These findings indicate that the binuclear cobalt(II) complexes of O-BISDIEN and O-BISTREN remain as the best oxygen carriers thus far reported.

Sponsors: The Robert A. Welch Foundation (Grant No. A-259) and L'Air Liquide.

REACTIONS OF KMnO$_4$ WITH VARIOUS SCHIFF BASE LIGANDS IN APROTIC SOLVENTS

Takayuki Matsushita

Department of Materials Chemistry, Faculty of Science and Technology, Ryukoku University, Seta, Otsu 520-21, Japan

Recently, high-valent manganese complexes have been investigated in connection with biological functions of manganese redox enzymes such as the manganese superoxide dismutase and the oxygen-evolving complex of photosystem II. We have studied on the reactions of a series of manganese(III) and (IV) Schiff base complexes with superoxide ion (O$_2^-$) and water, respectively, in order to clarify the functions of these manganese co-factors. In the consecutive study, we have attempted to prepare novel high-valent manganese complexes.

In this paper, we describe the reactions of KMnO$_4$ with various Schiff base ligands in organic solvents, mainly CH$_3$CN. KMnO$_4$ is soluble in it about 5x10^{-3} M in concentration and its solution is relatively stable for ca. 24 h at room temperature. A cyclic voltammogram of the solution exhibits a redox wave at a potential of -1.0 V(vs.SCE), which may be assigned to a redox couple of MnO$_4^-$/MnO$_4^{2-}$. This means that KMnO$_4$ in CH$_3$CN is not so a strong oxidant than in H$_2$O. The reactions of KMnO$_4$ with the Schiff base ligands in CH$_3$CN were monitored by measuring the visible absorption spectra. For example,with N,N'-propylenebis-(salicylideneimine) (salpnH$_2$), the absorption bands characteristic to MnO$_4^-$ in the visible region disappeared gradually and a broad absorption band appeared around 500 nm with a moderate intensity. This spectral changes suggested that a high-valent manganese complex having Mn-O bonds is formed, because the absorption band can be assigned to the charge transfer transition between manganese and oxygen atoms. By the reactions of KMnO$_4$ with Schiff base ligands, several manganese complexes were isolated and characterized by the elemental analyses and physicochemical properties such as IR, magnetic susceptabilities, and redox potentials. All data are consistent with that these complexes have a di-μ-oxo structure of manganese(IV).

These high-valent manganese complexes having Mn-O bonds are considered to be useful for the oxidation of organic substances.

CHROMIUM(VI) OXIDE CATALYZED OXIDATIONS BY *TERT*-BUTYLHYDROPEROXIDE: BENZYLIC OXIDATION *VERSUS TERT*-BUTYLHYDROPEROXIDE DECOMPOSITION

Jacques Muzart and Abdelaziz N'Ait Ajjou

Unité "Réarrangements Thermiques et Photochimiques" Associée au CNRS
Université de Reims Champagne-Ardenne, B.P. 347, 51062 Reims Cedex, France

For the last few years, we have been involved in hydrocarbon oxidation using *tert*-butylhydroperoxide and catalytic quantities of chromium(VI).[1,2] We have precedently observed that an excess of oxygen source is required to reach high or full conversion of the organic substrate owing to the concomittant unproductive decomposition of *t*-BuOOH by chromium.[3]

It would seem interesting to know if an efficient transfer of oxygen from *t*-BuOOH to the substrate could be obtained under suitable conditions. The simpler conditions are to use an excess of substrate vis à vis the oxygen source as often done in many papers where the high yields claimed for catalytic procedures were based on the amount of the oxidant. Such a procedure is generally less valuable for the bench organic chemist but can become worth-while on industrial scales if the starting material is easily separable from the oxidized compounds. In this poster, we will report the benzylic oxidations of diphenylmethane, indane, tetraline and fluorene pursued under these selected conditions.

Although the expected ketones have been usually obtained with good selectivities, we have nevertheless observed the peroxidation of the benzylic carbon to give *t*-butylperoxy and hydroperoxy derivatives in low amounts. The presence of the air atmosphere could *a priori* explain the formation of the latter compound but the hydroperoxytetraline was again produced when working under an argon atmosphere. Furthermore, the reaction carried out under an oxygen atmosphere did not give this hydroperoxyde. We have however observed that the peroxytetralines were decomposed under the reaction conditions. The formation of the hydroperoxytetraline seems to result, at least under anerobic conditions, from a reaction with the oxygen produced by decomposition of *t*-BuOOH. Otherwise, experiments using 1,3-diphenylisobenzofuran as substrate suggest the *in-situ* formation of singlet oxygen from *t*-BuOOH.

Acknowledgment: We thank the "Comité National de la Chimie" for travel facilities.

REFERENCES

[1] Muzart J. *Tetrahedron Lett.* **1986**, *27*, 3139; **1987**, *28*, 2131.
[2] Muzart J. *Chem. Rev.* **1992**, *92*, 113.
[3] Muzart J., N'ait Ajjou A. *J. Mol. Catal.* **1991**, *66*, 155.

CYTOCHROME P-450 LIKE SUBSTRATE OXIDATION CATALYZED BY HEME-UNDECAPEPTIDE, MICROPEROXIDASE-11

Shigeo Nakamura, Tadahiko Mashino, Masaaki Hirobe
Faculty of Pharmaceutical Sciences, University of Tokyo,
7-3-1 Hongo, Bunkyo-ku, Tokyo 113, Japan

Microperoxidase-11, MP-11, heme-undecapeptide, prepared by pepsin digestion of cytochrome *c*, retains the proximal His ligand and two thioether bonds between iron-protoporphyrin IX and two Cys residues. The peroxidase activities of MPs have been demonstrated in the oxidation of so-called peroxidase's substrates, phenolic compounds, in the presence of H_2O_2. MPs are unique and simple hemoprotein which are not restricted by consideration of the apoprotein structure. In this study, cytochrome P-450-like reactivities of MP-11, sulfide oxidation, N-dealkylation, and olefin epoxidation, were investigated[1,2].

Methyl phenyl sulfide was stoichiometrically converted to the corresponding sulfoxide by MP-11 based on H_2O_2 in 50% methanol-aqueous solution[1]. The relative reactivities of *p*-substituted methyl phenyl sulfide correlated well with the Hammett σ_p^+ values. The sufide oxidation with $H_2{}^{18}O_2$ gave completely ^{18}O-labeled methyl phenyl sulfoxide, which indicated that the oxygen atom had originated from iron-oxenoid active species, as it does with peroxidases and cytochrome P-450[2].

Demethylation of *N,N*-dimethylaniline was also effectively catalyzed by MP-11. Contrary to the case of sulfide oxidation, horseradish peroxidase (HRP) and cytochrome P-450 catalyze *N*-demethylation reaction in the different mechanisms. The source of the oxygen atom incorporated into the carbinolamine intermediate in the cytochrome P-450-catalyzed reaction is molecular oxygen (this is origin of iron-oxenoid active species of cytochrome P-450), however, that in the HRP-catalyzed reaction is not oxidant but solvent hydroxide or molecular oxygen. *N*-Methylcarbazole was used for studying the *N*-demethylation mechanism because its carbinolamine intermediate was stable. We found that ^{18}O incorporation into *N*-hydroxymethylcarbazole from $H_2{}^{18}O_2$ in the MP-11 reaction was 80%. This indicates that the MP-11-catalyzed *N*-demethylation mechanism is similar to the cytochrome P-450-catalyzed reaction rather than the HRP-catalyzed one[2]. These results suggest that MP-11 has a characteristically high ability to rebind the oxygen atom on the heme iron and the one-electron-oxidized intermediate.

An olefin oxidation is also of interest because olefin is not oxidized by HRP directly, whereas cytochrome P-450 catalyzes the epoxidation with retention of stereochemistry of the substrate. *cis*-Stilbene was converted stereospecifically to *cis*-epoxide and to the rearrangement product, diphenylacetaldehyde, by MP-11 and H_2O_2[1].

In conclusion, MP-11 catalyzed *N*-demethylation, sulfide oxidation and olefin epoxidation in the cytochrome P-450-like manner.

References: 1) T. Mashino, S. Nakamura, and M. Hirobe, *Tetrahedron Lett.*, **31**, 3163 (1990). 2) S. Nakamura, T. Mashino, and M. Hirobe, *Tetrahedron Lett.*, **33**, 5409 (1992).

OXIDATION OF ORGANIC COMPOUNDS WITH MOLECULAR OXYGEN CATALYZED BY HETEROPOLYOXOMETALLATES

Yutaka Nishiyama, Kouichi Nakayama, and Yasutaka Ishii
Department of Applied Chemistry, Faculty of Engineering, Kansai University
Suita, Osaka 564, Japan

It is increasing interested in the utilization of heteropoly compounds as catalysts for the oxidation of various organic compounds. Recently, we have found that the mixed addenda heteropolyoxometallates such as $(NH_4)_5H_4PV_6Mo_6O_{40} \cdot 6H_2O$ (PV_6Mo_6) was efficient catalysts for the aerobic oxidation of olefinic compounds in the presence of isobutyraldehyde. We now present here the direct oxidation of amines and hydrocarbons with molecular oxygen by PV_6Mo_6 catalyst.

When a toluene solution of benzylamine was stirred at 100°C for 20 h under an atmospheric pressure of molecular oxygen, the oxidative dehydrogenation was performed to give the corresponding Schiff base imine ($PhCH=NCH_2Ph$) (eq. 1). In this reaction system, the similar oxidative dehydrogenation of various benzylic amines was performed to give the corresponding imines in moderate yields. In the case of cyclic amine such as 1,2,3,4-tetrahydroisoquinoline, 3,4-dihydroisoquinoline was formed in good yield (eq. 2). The application of the $PV_6Mo_6-O_2$ oxidation system to tetraline provided the mixture of tetralone and 1,2,3,4-tetrahydro-1-naphthol (eq. 3).

We will discuss about the scope and limitation of the $PV_6Mo_6-O_2$ oxidation system.

"THERMODYNAMIC AND KINETIC STUDIES TO ELUCIDATE THE AMOCO Co/Mn/Br AUTOXIDATION ('MC') Catalyst"

Walt Partenheimer
Amoco Chemical Corporation
Naperville, IL 60563

An efficient, general,and high yield method to prepare aromatic acids is to pass dioxygen through an acetic acid solution of Co/Mn/Br salts containing alkylaromatic compounds. This method has been licensed world-wide to prepare terephthalic acid from p-xylene. We have used the reaction of m-chloroperbenzoic acid (MCPBA) with mixtures of Co(II) acetate/ Mn(II) acetate/bromide in acetic acid/water solutions to understand the functions of each catalyst component. The sequence of redox reactions that occurs is first the reaction of MCPBA with Co(II) to give Co(III); Co(III) then oxidizes Mn(II) to Mn(III) and finally, Mn(III) oxidizes bromide to bromine. Some of the functions of each component are 1) the cobalt rapidly reacts (very selectively) with the MCPBA (Mn and Br react slowly), 2) Mn lowers the steady state of Co(III) which significantly reduces solvent decomposition and also avoids Co(III) re-arranging into a less reactive form, and 3) bromine reacts rapidly with the methylaromatic compound to generate methylaromatic radicals (Co(III) and Mn(III) react slowly). The dimeric structure of Co(II) in acetic acid is partly responsible for the highly selective nature of the MCPBA oxidation of Co(II). The order of the redox reactions is the opposit to that expected from thermodynamics.

IRON(III) AND MANGANESE(III) TETRA-ARYLPORPHYRINS CATALYZE THE OXIDATION OF PRIMARY AROMATIC AMINES TO THE CORRESPONDING NITRO DERIVATIVES

Francesca Porta and Stefano Tollari

Dip. Chimica Inorganica, Metallorganica e Analitica, and CNR Center University of Milano, Via Venezian 21, 20133, Milano, Italy.

The catalytic reaction of the metallo (III) porphyrins, M(P)Cl (1-4), in the presence of an appropriate amount of an axial ligand L and a 3 M isooctane solution of t-BuOOH give, after 15 minutes, the total conversion of the amines 5-12, to the the corresponding nitro derivatives (5a-12a), whose yields range between 93% and 15% (molar ratios: cat./oxidant/amine = 1/300/100) (eq.1):

$$RC_6H_4NH_2 \xrightarrow[t\text{-BuOOH, L}]{M(P)Cl} RC_6H_4NO_2 \quad (1)$$

M(P) = Fe^{III}(TPP) (1); Mn^{III}(TPP) (2); Fe^{III}(T2,6Cl$_2$PP) (3); Mn^{III}(T2,6Cl$_2$PP) (4); ; R = H (5), p-Me (6), p-Br (7), p-Cl (8), p-NO$_2$ (9), p-F (10), m-Me (11), p-OMe (12), L = imidazole, 1-methylimidazole, 2-methylimidazole, 4-$t$butylpyridine.

The appropriate amount of the added ligand changes with the nature of the amine used, but the best results were obtained, as a rule, when the ligand/amine molar ratios were in the 0.3-0.5 range, which corresponds to an initial 30-50 ligand/metallo-porphyrin molar ratio. The effect of various ligands was studied. Indeed, the best results were obtained by using the most strongly coordinating ligand, 1-Meimidazole, whose effect is analogous to that reported in the literature on olefin epoxidation by Fe^{3+} or Mn^{3+} porphyrins in the presence of hydrogen peroxide or hypochlorite[1]. The difference in the coordination capability of the ligand L and amines is evidenced by the spectral changes of the Fe(TPP)Cl Soret band at 416 nm, in the presence of the imidazole ligand or the amine.

The catalytic system still proved to be efficient when 10^4 molar equivalents of amine with respect to the catalyst were used . Very high turnovers/ hour (18900) were observed.

The same reactions carried out with the manganese(III) porphyrins 2 and 4 with amines 6, 8 and 12 gave good yields of the nitroderivatives 6a and 8a , but longer reaction times were requested   (2 and 5 h, respectively). Comparison with the ferric system yields shows that both manganese(III) complexes 2 and 4 are less efficient in the oxidation of the aromatic amines to the corresponding nitroderivatives .

References

1) P.Battioni, J.P.Renard, J.F.Bartoli, M.Reina-Artiles, M.Fert, D.Mansuy, *J.Amer.Chem.Soc.*, 1988, **110**, 8462.

STOICHIOMETRIC AND CATALYTIC OXIDATION BY A BINUCLEAR Cu(I) DIOXYGEN MACROCYCLIC COMPLEX DERIVED FROM PYRIDINE-2,6-DICARBOXALDEHYDE AND BISDIEN

DAVID A. ROCKCLIFFE AND ARTHUR E. MARTELL

Department of Chemistry, Texas A&M University College Station, Texas 77843-3255

The 2+2 condensation of pyridine-2,6-dicarboxaldehyde with BISDIEN produces a macrocyclic tetra Schiff base with a 24-membered ring in good yield. The macrocycle reacted with two equivalents of $Cu(CH_3CN)_4PF_6$ to form a binuclear Cu(I) complex. The latter complex reacted with one equivalent of dioxygen to produce a complex which is believed to contain a μ-peroxo group bridging the metal centers. The half-life of this dioxygen complex was found to be about 100 minutes at 25 °C. The dioxygen complex was found to oxidize the substrates 2,6-dimethoxyphenol, 2,6-ditertiarybutylphenol, hydroquinone, tertiary-butylhydroquinone, 3,5-ditertiarybutylcatechol, 4-tertiarybutylcatechol, 4-methylcatechol, and 3,4-dimethyl-aniline. All of the redox reactions are catalytic, except for those of the last three substrates, with from three to five turnovers. The cycling of these catalytic reactions require that the binuclear Cu(II) complex formed in the initial (two-electron) substrate oxidation also be an active oxidant of the same substrate. The binuclear Cu(II) complex of the tetra Schiff base ligand was also synthesized and used as an oxidant for the substrates listed above. Only the substrates that were catalytically oxidized were also oxidized by the binuclear Cu(II) macrocyclic complexes. The substrates that were not catalytically oxidized were not oxidized at measurable rates by the Cu(II) complex. Comparison of the initial rate constants for oxidations by the binuclear Cu(I) dioxygen complex and for the binuclear Cu(II) macrocyclic tetra Schiff base complex show that for the substrates examined, the Cu(I) dioxygen complex exhibits a significantly higher rate than does the Cu(II) complex.

Acknowledgement: This work was support by the Office of Naval Research

Fenton Reagents (1:1 $Fe^{II}L_x$/HOOH) React Via [$L_xFe^{II}OOH(BH^+)$, 1] as Hydroxylases (RH → ROH); *NOT* as Generators of Free Hydroxyl Radicals (HO·)[a]

Donald T. Sawyer, Chan Kang, Antoni Llobet, and Chad Redman

Department of Chemistry, Texas A&M University,
College Station, Texas 77843, USA

Abstract: The one-to-one combination of hydrogen peroxide and $Fe^{II}(PA)_2$ [PAH = picolinic acid (2-carboxyl-pyridine)], a Fenton reagent, forms an adduct

$$\{Fe^{II}(PA)_2 + HOOH \xrightarrow[\text{(py)}_2HOAc]{k, 2 \times 10^3 M^{-1}s^{-1}} [(PA)_2\text{-}Fe^{II}OOH + pyH^+] \text{ (1)}\}$$ that is

the primary reactant with (a) excess $Fe^{II}(PA)_2$ to give two $(PA)_2Fe^{III}OH$,

(b) excess HOOH to give O_2 plus two H_2O, and (c) excess cyclohexane ($c\text{-}C_6H_{12}$)

to give ($c\text{-}C_6H_{11}$)py (or $c\text{-}C_6H_{11}OH$ in the absence of pyridine). The presence of a

carbon-radical trap (PhSeSePh) with the organic substrate yields $c\text{-}C_6H_{11}SePh$

with an efficiency of >90% (relative to HOOH). This Fenton reagent has

reactivity ratios (k_A/k_B) [in comparison with free HO·] with $c\text{-}C_6H_{12}/c\text{-}C_6D_{12}$

(KIE) of 1.7 [versus 1.0], with 1°/2°/3° carbon centers (per C–H, normalized) of

0.07/0.44/1.0 [0.41/0.50/1.0], and with $c\text{-}C_6H_{12}/PhCH_2CH_3$ of 2.0 [0.6]. In the

presence of O_2 (1 atm, 3.4 mM) **1** forms an adduct [**1**(O_2), **5**], which reacts with

$c\text{-}C_6H_{12}$ to give $c\text{-}C_6H_{10}(O)$ (KIE, 2.1), cyclohexene ($c\text{-}C_6H_{10}$) to give

$c\text{-}C_6H_8(O)$, and $PhCH_2CH_3$ to give $PhC(O)CH_3$ (reactivity ratio for

$c\text{-}C_6H_{12}/PhCH_2CH_3$, 0.6). Similar chemistry results for HOOH/O_2 with

$Fe^{II}(bpy)_2^{2+}$ [reactivity ratio for $c\text{-}C_6H_{12}/PhCH_2CH_3$ (0.4)], $Fe^{II}(OPPh_3)_4^{2+}$ (1.0),

and $Cu^I(bpy)_2^+$ (1.0)]. All of this is compelling evidence that Fenton reagents

do not produce (a) free hydroxyl radicals (HO·) or (b) free carbon radicals (R·), but

(c) can exhibit catalytic turnovers with respect to HOOH and O_2.

---

[a]Sawyer, D. T.; Kang, C.; Llobet, A.; Redman, C. *J. Am. Chem. Soc.* **1993**, *115*, 0000 (June, 1993).

IRON(II)-INDUCED GENERATION OF HYDROGEN PEROXIDE FROM DIOXYGEN; INDUCTION OF FENTON CHEMISTRY AND THE ACTIVATION OF O$_2$ FOR THE KETONIZATION OF HYDROCARBONS; Fe^{II}(PA)$_2$/(t-BuOOH)-INDUCED ACTIVATION OF O$_2$[a]

Donald T. Sawyer, Chan Kang, Andrzej Sobkowiak, and Chad Redman
Department of Chemistry
Texas A&M University, College Station, Texas 77843

Abstract: The combination of Fe^{II}(DPAH)$_2$ (DPAH$_2$ = 2,6-dicarboxyl pyridine) and O$_2$ in a 2:1 pyridine/acetic acid solution results in a rapid autoxidation to produce HOOH and Fe^{III}(DPA)(DPAH) (k$_1$, 1.8±0.5 M^{-1}s^{-1}). The resultant HOOH reacts with excess Fe^{II}(DPAH)$_2$ via a nucleophilic addition to give a Fenton reagent [(DPAH)$_2$ Fe^{II}OOH + pyH$^+$] (1) that reacts with (a) excess Fe^{II}(DPAH)$_2$ to give Fe^{III}(DPA)(DPAH) [k$_2$, (2±1) × 10^3 M^{-1}s^{-1}], (b) excess c-C$_6$H$_{12}$ and PhSeSePh (a carbon radical trap) to give c-C$_6$H$_{11}$SePh [kinetic-isotope-effect (k$_{c-C_6H_{12}}$/k$_{c-C_6D_{12}}$), 1.7], and (c) excess O$_2$ to form an adduct [(DPAH)$_2$ Fe^{III}OOH(O$_2$) + pyH$^+$](5), which reacts with excess c-C$_6$H$_{12}$ via an intermediate [(DPAH)$_2$Fe^{IV}(OH)(OOC$_6$H$_{11}$)] (6) and Fe^{II}(DPAH)$_2$ to give c-C$_6$H$_{10}$(O) [KIE, 2.6] and Fe^{III}(DPA)(DPAH) [32 mM Fe^{II}(DPAH)$_2$, O$_2$ (1 atm, 3.4 mM), and 1 M c-C$_6$H$_{12}$ yield 6 mM c-C$_6$H$_{10}$(O)]. The related Fe^{II}(PA)$_2$ complex in combination with t-BuOOH (or HOOH) catalytically activates O$_2$ to oxygenate hydrocarbons {e.g., c-C$_6$H$_{12}$ ⟶ c-C$_6$H$_{10}$(O) [9 O$_2$ turnovers per Fe^{II}(PA)$_2$]; PhCH$_2$CH$_3$ ⟶ PhC(O)CH$_3$ (25 O$_2$ turnovers); c-C$_6$H$_{10}$ ⟶ c-C$_6$H$_8$(O) (5 O$_2$ turnovers); and PhCH(Me)$_2$ ⟶ PhC(O)Me, Ph(Me)$_2$COH, and Ph(Me)C=CH$_2$ (5 O$_2$ turnovers). With large t-BuOOH (or HOOH)/Fe^{II}(PA)$_2$ ratios spontaneous decomposition occurs to give free O$_2$ that is incorporated into the substrates. The first-formed intermediate is a one-to-one t-BuOOH/FeL$_x$ adduct {[(PA)$_2$ Fe^{II}OOBu-t + pyH$^+$] (1)}, which reacts with (a) excess t-BuOOH to give O$_2$ and t-BuOH, (b) excess Fe^{II}(PA)$_2$ to give (PA)$_2$Fe^{III}OH(Bu-t), (c) excess c-C$_6$H$_{12}$ to give (c-C$_6$H$_{11}$)py [KIE, 4.6 with t-BuOOH and 1.7 with HOOH], and (d) O$_2$ to form an adduct, [(PA)$_2$ Fe^{III}(O$_2$)(OOBu-t) + pyH$^+$](5), that reacts with c-C$_6$H$_{12}$ to form c-C$_6$H$_{10}$(O) [KIE, 8.7 (2.5 with HOOH)]. When PhCH$_2$Me or c-C$_6$H$_{10}$ are the substrates (RH), 5 reacts to form [(PA)$_2$Fe^{IV}(OH)(OOR)] (6), which in turn reacts with RH and O$_2$ in a catalytic cycle to give PhC(O)Me or c-C$_6$H$_8$(O) [2–3 O$_2$ turnovers per Fe^{II}(PA)$_2$]. The Fe^{II}(bpy)$_2$$^{2+}$, Fe^{II}(OPPh$_3$)$_4$$^{2+}$, (Cl$_8$TPP)Fe^{II}(py)$_2$, Fe^{III}Cl$_3$, and Cu^I(bpy)$_2$$^+$ complexes in MeCN are similar effective catalysts.

[a]Kang, C.; Redman, C.; Cepak, V.; Sawyer, D. T. *Bioorg. Med. Chem.* **1993**, *1*, 000 (June, 1993).

THE UNIQUE ROLE OF IRON CATIONS IN THE ACTIVATION OF MOLECULAR OXYGEN IN THE GAS PHASE

D. Schröder, H. Schwarz

Institut für Organische Chemie der Technischen Universität Berlin, 1000 Berlin 12 (Germany)

The activation of molecular oxygen by transition metal complexes is of fundamental interest for chemistry, biology, and medicine. As it is well known from biochemistry, iron is crucial in the transportation, storage, and activation of oxygen, and with respect to these processes only copper is of comparable relevance.

The origin of the exceptional role of the element iron with respect to oxygen-activation is not well understood yet. It can be either due to the intrinsic properties of this metal or be dominated by other effects, eg. ligand sphere, solvation etc.

In the gas phase the intrinsic properties of an individual transition metal ion M^+ can be studied in the absence of additional ligands and similar effects. These studies may provide information about the elementary steps of oxidation reactions related to biological processes.

To this end, we have studied the gas phase reactions of ethylene complexes $M(C_2H_4)^+$ of the 1st row transition metal cations (M = Sc - Zn) with molecular oxygen by means of *Fourier Transform Ion Cyclotron Resonance* (FTICR) Mass Spectrometry. This model reaction migth serve as a test for the capability of the individual metal cations with respect to oxygen and C-H-activation as well. For comparison the reactions of the radical cation $C_2H_4^{+\circ}$ with oxygen as well as those of $M(C_2H_4)^+$ with M = H, Na, and Mg have also been included in this comprehensive gas phase study.

Most of the olefin complexes examined in this study exhibit an unspectacular reactivity towards molecular oxygen, i. e. either ligand exchange reactions, O-O-bond activation by highly oxophilic metals Sc, Ti, and V, or even complete absence of any reaction are observed (eg. even $Cu(C_2H_4)^+$ is unreactive). However, in the case of the iron complexes extensive oxidation reactions are observed. Indeed, not only olefins attached to an iron cation react effectively with molecular oxygen, even stable molecules like benzene and acetone are rapidly oxidized in the presence of Fe^+.

In comparison, Fe^+ is the only cation of the 1st row transition metal cations, which mediates both, the activation of molecular oxygen as well as that of the olefin. Thus the experimental findings indicate that the extraordinary role of iron with respect to the activation of molecular oxygen is - at least partially - due to an intrinsic property of this element.

The results of these oxidation processes are presented and discussed on the basis of mechanistical and energetical arguments.

LOW TEMPERATURE OXIDATION OF p-XYLENE TO TEREPHTHALIC ACID

Ulf Schuchardt, Carol H. Collins and Regiane L. Ambrósio

Instituto de Química, Universidade Estadual de Campinas
Caixa Postal 6154, 13081-970 Campinas, SP (Brazil)

The oxidation of p-xylene by air to give terephthalic acid is almost universally carried out with cobalt and manganese salts as catalysts. In the Amoco process, 95% acetic acid is used as a solvent and bromide ion as a promotor. The conditions are drastic (195-205°C, 15-30 bar of air) in order to oxidize the intermediates p-toluic acid and terephthalic aldehyde, giving a selectivity for terephthalic acid better than 90%. We reinvestigated this reaction at low temperature with an oxygen flow (1 bar) in an attempt to achieve a similar selectivity for terephthalic acid. The reactions were carried out in 50 ml of acetic acid using 56 mmol of p-xylene and 5 mmol of cobalt and manganese acetate (9:1) under an oxygen flow (60 ml/min) for 12 h. Preliminary experiments showed that it was necessary to initially add acetic anhydride (120 mmol), which eliminates the reaction water formed, and a promotor to maintain the radical chain process. Hydrobromic acid was found to be the best promotor. A quantity of 2.5 mmol optimized the selectivity without producing brominated products. A co-oxidant (such as paraldehyde) was not necessary under these conditions. Varying the reaction temperature, we found that the conversion of p-xylene was complete at 95°C. At 100°C we obtained terephthalic acid with a selectivity of 90% which was maintained up to 108°C. At higher temperatures the selectivity was reduced due to decarboxylation reactions. We conclude that p-xylene can be effectively oxidized to terephthalic acid under an oxygen flow at 100°C with a hydrobromic acid promotor, if the acetic acid used as a solvent is maintained water-free. A reaction mechanism which is in agreement with these results is discussed.

This work was supported by FAPESP and CNPq.

COPPER(I)/(t-BuOOH)-INDUCED ACTIVATION OF DIOXYGEN
FOR THE KETONIZATION OF METHYLENIC CARBONS[a]

Andrzej Sobkowiak, Aimin Qiu, Xiu Liu, Antoni Llobet, and Donald T. Sawyer

Department of Chemistry

Texas A&M University, College Station, Texas 77843

In acetonitrile/pyridine bis(bipyridine)copper(I) [$Cu^I(bpy)_2^+$] activates HOOH and t-BuOOH for the selective ketonization of methylenic carbons. With 5 mM $Cu^I(bpy)_2^+$/100 mM HOOH(Bu) the conversion efficiencies [product per 2 HOOH(Bu)] for c-C_6H_{12} are 31% (HOOH) and 59% (t-BuOOH, argon atmosphere), and for $PhCH_2CH_3$ are 24% (HOOH) and 64% (t-BuOOH, argon). With 5 mM $Cu^I(bpy)_2$ and 10 mM t-BuOOH under argon the conversion efficiency for c-C_6H_{12} is 10% and for $PhCH_2CH_3$ is 140%.   However, in the presence of O_2 (1 atm, 7 mM) the conversion efficiency for c-C_6H_{12} increases to 67%, and for $PhCH_2CH_3$ to 440% [all PhC(O)Me (22 mM)], respectively. The latter result represents a $Cu^I(bpy)_2^+$/ t-BuOOH-induced *auto-oxygenation* with at least 2.2 O_2/catalyst turnovers. The first-formed intermediate is a one-to-one adduct [$(bpy)_2Cu^IOOBu$-t + pyH^+, **1**], which reacts with (a) excess $Cu^I(bpy)_2^+$ to give 2 $(bpy)_2^+Cu^{II}OH$, (b) excess t-BuOOH to give O_2 and t-BuOH, (c) excess c-C_6H_{12} to give c-$C_6H_{11}OH$, (d) excess c-C_6H_{12} and excess t-BuOOH to give c-$C_6H_{11}OOBu$-t, and (e) O_2 to form an adduct, [$(bpy)_2Cu^{III}(O_2)(OOBu$-t) + pyH^+, **5**], that reacts with (i) c-C_6H_{12} to give c-$C_6H_{10}(O)$ and c-$C_6H_{11}OH$ and (ii) $PhCH_2CH_3$ to give $PhC(O)CH_3$.

---

[a]Sobkowiak, A.; Qiu, A.; Liu, X.; Llobet, A.; Sawyer, D. T. *J. Am. Chem. Soc.*. **1993**, *115*, 609-614.

ENANTIOSELECTIVE LACTONIZATION OF CYCLIC KETONES WITH H_2O_2 CATALYZED BY COMPLEXES OF PLATINUM(II)

G. Strukul, A. Gusso and F. Pinna

Department of Chemistry, University of Venice, 30123 Venice, Italy.

The Baeyer-Villiger oxidation of ketones to give esters is an oxygen transfer reaction typical of organic peroxy acids. We have recently reported that this reaction can be easily accomplished with hydrogen peroxide as the oxygen source and a series of $[(P-P)Pt(CF_3)(solv)]^+$ complexes as catalysts (P-P = several diphosphines). This allows the easy catalytic conversion of a variety of cyclic ketones (like e.g. cyclohexanone, cyclopentanone, cyclobutanone, etc.) into the corrsponding lactones via a concerted mechanism involving the activation of the substrate at the metal center.

In this communication we report a study of the stereochemical features of this catalytic process and the use of chiral diphosphine modified catalysts for the enantioselective synthesis of certain naturally ocurring lactones through kinetic resolution of a racemic mixture of the starting ketones.

ACKNOWLEDGEMENTS

CNR (Rome) and Murst (Rome) are gratefully aknowledged for financial support.

AEROBIC OXYGENATION OF OLEFINS
CATALYZED BY TRANSITION-METAL COMPLEXES
- PREPARATION OF EPOXIDES AND α-HYDROXY KETONES -

T. Takai, T. Yamada, S. Inoki, E. Hata, and T. Mukaiyama[†]

Basic Research Laboratories for Organic Synthesis, Mitsui Petrochemical Industries, Ltd., Nagaura, Sodegaura-shi, Chiba-ken 299-02 Japan.
[†]Department of Applied Chemistry, Faculty of Science, Science University of Tokyo, Kagurazaka, Shinjuku-ku, Tokyo 162 Japan.

By combined use of molecular oxygen and aldehyde, various olefins are smoothly oxygenated into the corresponding epoxides catalyzed by bis(1,3-diketonato)nickel(II) complexes under an atmospheric pressure of oxygen at room temperature.[1a, b] Transition-metal complex catalyst is quite influential over the reaction system of the present oxygenation, therefore, in order to develop the useful procedures, suitable selection of the catalyst is crucial.

When tris(1,3-diketonato)iron(III) complex was employed as a catalyst, styrene and its analogues which are highly reactive toward oxygenation, were converted into the corresponding epoxides in high yield without accompanying undesirable C=C bond cleavage.[3] α,β-Unsaturated carboxamides, electron deficient olefins, were oxygenated into the corresponding epoxides in good yields when bis(1,3-diketonato)oxovanadium(IV) complex catalyst was used.[4]

By using the catalyst system of OsO_4 and bis(1,3-diketonato)nickel(II) complex, various olefinic compounds were directly converted into the corresponding α-hydroxy ketones in good yields.[5] This procedure provides a convenient method for the preparation of α-hydroxy ketones just starting from olefinic compounds.

References

[1] a) T. Yamada, T. Takai, O. Rhode, and T. Mukaiyama, *Chem. Lett.*, **1991**, 1. b) *idem.*, *Bull. Chem Soc. Jpn.*, **64**, 2109 (1991). [2] T. Takai, E. Hata, T. Yamada, and T. Mukaiyama, *Bull. Chem. Soc. Jpn.*, **64**, 2513 (1991). [3] S. Inoki, T. Takai, T. Yamada, and T. Mukaiyama, *Chem. Lett.*, **1991**, 941. [4] T. Takai, T. Yamada, and T. Mukaiyama, *Chem. Lett.*, **1991**, 1499.

OXIDATION OF PHENOLIC ARYLALKENES IN THE PRESENCE OF MOLECULAR OXYGEN

Yasuomi Takizawa, Tadanori Sasaki, Kazuaki Nakamura and
Nobutoshi Yoshihara

Department of Chemistry, Tokyo Gakugei University,
4-1-1, Nukuikita-machi, Koganei-shi, Tokyo 184 , Japan

From the view point of finding the oxidation mechanism
and biomimetic synthesis of naturally occuring phenolic
compounds in vivo, it is important to study on the oxidation
of the phenolic compounds related to melanin, lignin, and
vitamin· E. The authors have studied on the oxidation of mono-
phenols with copper(II) compounds in various systems.
Various phenolic arylalkens and the related compounds were
oxidized with (BPA)Cu(II)Cl$_2$:bis(1,3-Propandiaminato)Copper
(II)dichloride, Cu(OAc)-Pyridine, and DDQ(2,3-dichloro-5,6-
dicyanobenzoquinone) in the presence of molecular oxygen
using with the various solvents: ethanol, acetic acid and
acetic acid-water.
Substituted phenols(3.0 mmol) were dissolved in the above
solvents and (BPA)Cu(II)Cl$_2$(0.35 mmol) was added into the
solution. Oxygen was bubbled through the solution.
Oxidation of isoeugenol with (BPA)Cu(II)Cl$_2$in ethanol gave
the ortho-ortho coupling dimer and the aldehyde derivatives.
On the other hands, oxidation of isoeugenol in acetic acid
- H$_2$O gave the adduct of two molar isoeugenol and one
molar acetic acid. When 2,6-dibutyl-4-methylphenol was oxidized in the system of copper(II) acetate-pyridine
quinone methide derivatives were produced.
However, 2-t-butyl-4-methoxyphenol was oxidized in the same
system to give dibenzofuran derivatives concomitantly with
dimer. It was useful to use copper(II)compounds in order
to get these phenols by one step synthesis.

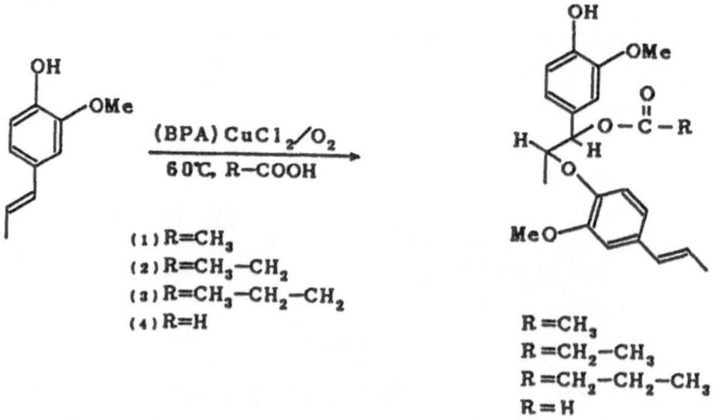

$$(1) R=CH_3$$
$$(2) R=CH_3-CH_2$$
$$(3) R=CH_3-CH_2-CH_2$$
$$(4) R=H$$

$$R=CH_3$$
$$R=CH_2-CH_3$$
$$R=CH_2-CH_2-CH_3$$
$$R=H$$

REFERENCES

1) Y. Takizawa, A. Tateishi, J. Sugiyama, H. Yoshida, and
   N. Yoshihara, J. Chem. Soc.,Chem. Commun., 104(1991).

OXIDATION OF HYDROCARBONS BY OXYGEN AND HYDRO-GEN OVER Pd-CONTAINING TITANIUM SILICALITES

T. Tatsumi, K. Yuasa and K. Asano

Department of Synthetic Chemistry, Faculty of Engineering, The University of Tokyo, Hongo, Tokyo 113,

The direct introduction of hydroxyl groups into an aromatic ring and unreactive alkanes has been attracting much interest. Titanium silicalite is a recently developed new type molecular sieve which incorporates Ti in the framework. It has been found to be effective in the hydroxylation of aromatic compounds such as benzene and phenol and also of alkanes[1] by H_2O_2 as an oxidant under mild conditions. Since using molecular oxygen in place of H_2O_2 should have significant advantages, we have sought to establish procedures through which molecular oxygen could be used. We have now designed a system containing Pd metal particles in the titanium silicalite which, in an oxygen/hydrogen atmosphere, should generate H_2O_2 at the palladium site and then use that H_2O_2 to oxidize various organic substrate present in the pore system.

Catalyst (50-100 mg), substrate (10 ml), and 3-30 mM hydrochloric acid (5 ml) was placed in the flask and stirred with a magnetic stirrer. The reactant gas mixture (O_2/H_2 = 14/3) was bubbled into the liquid at a flow rate of 17 ml min^{-1} at 25 to 40 °C.

Hydroxylation of benzene to phenol took place over the ATS(alumino-titanium silicalite)-1 and TS-1 catalysts containing Pd in the presence of hydrochloric acid. Although Pd/HZSM-5 was found to be slightly active in the phenol formation, it is clearly seen that Pd-Ti dual functionality resulted in much better catalytic performance. Pd/TS-1(B) prepared by impregnation with $[Pd(NH_3)_4]Cl_2$ exhibited improved catalytic performance in terms of turnover number, which amounted to 13.5 per Ti at 35 °C in 3 h. Sulfuric acid was able to substitute for hydrochloric acid. They are known to be required additives to form H_2O_2 from O_2 and H_2 on Pd catalysts. Physical mixture of Pd/C and TS-1 turned out to be active when a cosolvent such as *t*-BuOH was added. Without a cosolvent hydrophobic Pd/C was present in the organic phase, resulting in difficult contact with the aqueous phase, where H_2O_2 formation should take place.

These Pd-Ti systems were active in the oxidation of other substrates such as alkanes, alkenes and alcohols. Hexane was hydroxylated into 2- and 3-hexanols, which were further oxidized in part to the corresponding ketones. In this case the product turnover was sensitive to the concentration of HCl. The addition of MeOH was effective as in the case of oxidation by H_2O_2 over TS-1.[1] Finally we note that shape selectivity was found in the oxidation of alkanes and alkenes similarly to what was observed for the oxidation where H_2O_2 was used as oxidant[1,2]; the rates for oxidation of cyclic alkanes and alkenes were much lower than those of linear alkanes and alkenes.

REFERENCES

1) T. Tatsumi, M. Nakamura, S. Negishi and H. Tominaga, *J. Chem. Soc., Chem. Commun.*, 1990, 476
2) T. Tatsumi, M. Nakamura, K. Yuasa and H. Tominaga, *Chem. Lett.*, 1990, 297.

PREPARATION, CHARACTERIZATION, AND REACTION OF AN OXO-FE(V)-PORPHYRIN COMPLEX

Yoshihito Watanabe, Kazuya Yamaguchi, and Isao Morishima

Division of Molecular Engineering, Graduate School of Engineering, Kyoto University, Kyoto 606-01, Japan

Oxidation of Fe^{III}TDCPP (m-chlorobenzoate) (1-mCB, 2.2 x 10^{-5} M, TDCPP: 5,10,15,20-tetrakis-2,6-dichlorophenylporphyrin) with 1.8 equiv amount of p-nitroperoxybenzoic acid in dry methylene chloride at -90°C produced an oxo-ferryl porphyrin cation radical (2). The addition of 4 equiv of methanol to the solution afforded a red species (3) with the loss of the characteristic broad band (600-750 nm) for porphyrin cation radicals. The spectrum of 3 is rather similar to that of oxo-ferryl (O=Fe^{IV}) porphyrin species (4) but very different from those of ferric porphyrin dications and ferric porphyrin N-oxides. Whereas 4 is stable at -5°C, the absorption spectrum of 3 changed to that of ferric high-spin porphyrin (1) even at -70°C in several hours. The addition of norbornylene to a methylene chloride solution of 3 caused the absorption spectrum change of 3 to that of 1 at -90°C, while 4 does not react with the olefin under similar condition. Titration of 3 with iodide ion indicates the oxidation state of 3 to be two-electron oxidized from the ferric state. That ^{2}H-NMR spectra of 3 gave pyrrole deuterium and $meta$-phenyl deuterium signals at -35.1 and 8 ppm at -95°C, respectively, is consistent with the oxo-perferryl (O=Fe(V)) porphyrin formulation. ESR spectrum (g = 4.33, 3.69, and 1.99 at 4.2K) and solution magnetic susceptibility (μ_{eff} = 4.0 ± 0.2 μ_B) of 3 indicate that 3 is in a high-spin state (S = 3/2). While 3 catalyzes the epoxidation of olefins at -90°C, 3 is about 10 times less reactive than 2.

NON-CHLORINE BLEACHING OF WOOD PULPS MEDIATED BY HETEROPOLYOXOMETALATES

Ira A. Weinstock, James L. Minor and Rajai H. Atalla; U.S. Dept. of Agriculture, Forest Products Laboratory, One Gifford Pinchot Drive, Madison, WI 53705-2398.

Wood consists primarily of carbohydrates (cellulose and some hemicelluloses) and lignin. In the production of chemical pulps, used in the manufacture of high quality paper, most of the lignin is removed by reaction with alkaline sulfide. At the elevated temperatures used in pulping, chemical reactions of lignin give rise to highly colored conjugated aromatic structures that remain within the wood cell (fiber) walls. The purpose of bleaching is to degrade or remove these chromophores along with remaining lignin.

Chemical pulps are currently bleached with elemental chlorine. Reaction of elemental chlorine with chromophores and residual lignin generates chlorinated aromatics and dioxins, which pose a threat to the environment. Despite world-wide consumer and regulatory pressures to eliminate the use of elemental chlorine in bleaching, no effective alternatives are currently available. The reactions of obvious alternatives such as dioxygen and peroxides are difficult to control under the extreme conditions required to achieve significant bleaching. Cellulose degradation occurs, resulting in unacceptably weak fibers. One solution to this problem might be to use water soluble catalysts to enhance the selectivity and effectiveness of dioxygen or peroxides.

Since the discovery of heme-containing lignin peroxidases, excreted by wood-rotting fungi, a variety of biomimetic lignin degrading systems have been prepared using halogenated, water soluble porphyrin complexes. However, these synthetic materials are expensive and, while more robust than simple porphyrin complexes, are inherently susceptible to oxidative degradation. For large scale commercial bleaching, there is a need for catalysts that are water soluble, oxidatively robust, and easily prepared from inexpensive, non-toxic materials. As a class, the polyoxometalates include a variety of materials that meet these criteria. Vanadium-containing mixed addenda heteropolyoxometalates, such as the phosphomolybdovanadates, may be used as direct oxidants. Transition metal substituted heteropolyoxometalates, some of which may be viewed as inorganic analogues of metalloporphyrins, may be used as catalysts in the presence of peroxide compounds or other oxygen donors.

The vanadium-substituted polyoxometalates are applied to chemical pulps as stoichiometric oxidants, much as elemental chlorine is currently used. The reduced polyoxometalate bleaching solutions are regenerated in a second step by treatment with chlorine-free terminal oxidants, preferably air or dioxygen. By exclusion of dioxygen during the bleaching step, exposure of the pulp to non-selective species such as hydroxyl radicals is avoided entirely.

Near Infrared (NIR) Fourier Transform (FT) Raman spectroscopy was used to observe changes in residual lignin in kraft pulp upon exposure to $Na_{5-x}H_x[PV_2Mo_{10}O_{40}]$. The extent of reduction of this material during bleaching was determined titrametrically, and its reoxidation by air was monitored by UV-vis spectroscopy. The integrity of the reoxidized polyoxometalate was then checked by ^{31}P NMR spectroscopy. FT Raman and UV-vis spectra of bleached pulps, combined with lignin-model studies and industry standards such as kappa numbers and brightness measurements, were used to develop hypotheses regarding the structural transformations that occur in residual lignin during stages of a complete bleaching sequence. These studies suggest that during the polyoxometalate treatment benzylic alcohols are oxidized to α-carbonyls, and phenols to quinones. These are easily degraded by subsequent treatment with alkali or alkaline hydrogen peroxide.

INDEX